SIGHT, TOUCH, AND IMAGINATION IN BYZANTIUM

Considering the interrelations between sight, touch, and imagination, this book surveys classical, late antique, and medieval theories of vision to elaborate on how various spheres of the Byzantine world categorized and comprehended sensation and perception. Revisiting scholarly assumptions about the tactility of sight in the Byzantine world, it demonstrates how the haptic language associated with vision referred to the cognitive actions of the viewer as they grasped sensory data in the mind in order to comprehend and produce working imaginations of objects for thought and memory. At stake is how the affordances and limitations of the senses came to delineate and cultivate the manner in which art and rhetoric were understood as mediating the realities they wished to convey. This would similarly come to contour how Byzantine religious culture could also go about accessing the sacred, the image serving as a site of desire for the mediated representation of the divine.

ROLAND BETANCOURT is an Assistant Professor of Art History at the University of California, Irvine. In the 2016–2017 academic year, he was the Elizabeth and J. Richardson Dilworth Fellow at the Institute for Advanced Study in Princeton. He has co-edited *Byzantium/Modernism: The Byzantine as Method in Modernity* (2015), and is the author of a forthcoming book on the intersectionality of race, sexuality, and gender identity in the medieval world.

SIGHT, TOUCH, AND IMAGINATION IN BYZANTIUM

ROLAND BETANCOURT
University of California, Irvine

CAMBRIDGE
UNIVERSITY PRESS

CAMBRIDGE
UNIVERSITY PRESS

University Printing House, Cambridge CB2 8BS, United Kingdom

One Liberty Plaza, 20th Floor, New York, NY 10006, USA

477 Williamstown Road, Port Melbourne, VIC 3207, Australia

314–321, 3rd Floor, Plot 3, Splendor Forum, Jasola District Centre, New Delhi – 110025, India

79 Anson Road, #06–04/06, Singapore 079906

Cambridge University Press is part of the University of Cambridge.

It furthers the University's mission by disseminating knowledge in the pursuit of education, learning, and research at the highest international levels of excellence.

www.cambridge.org
Information on this title: www.cambridge.org/9781108424745
DOI: 10.1017/9781108344067

First published 2018

Printed in the United Kingdom by Clays, St Ives plc

A catalogue record for this publication is available from the British Library.

ISBN 978-1-108-42474-5 Hardback

Publication of this book has been supported by a grant from the International Center of Medieval Art and the Samuel H. Kress Foundation.

To unrequited love

Let me touch you with my gaze,
optic nerves and visual rays.
– Flula Borg (feat. Mamrie Hart), "Eye Love You" (2015)

Contents

Tables

Acknowledgements

Drawn to all those things ostentatiously displayed before our eyes, it is sight itself that often reminds us of its own inability to touch. These are lessons intimately familiar to any millennial, our pop culture being replete with language and tropes that stress the limitations of sight. The icon draws us to them, yet this figure lies always beyond our grasp and sensation, except in the cognitive perceptions that we construe to enliven them, so that they might become tangible and tactile for us in our imagination. Often, the distinction between touch and sight is a prohibition of access, as in M. C. Hammer's reminder that while we can look into his eyes, we "can't touch this," which is reiterated by various artists, including Fergie, who states that while one can see her, one can still not squeeze her. Yet, similarly, the distinction between touch and sight confronts us with the reality of one's thirst for contact, as in Rupee's contouring of wanting and yearning at the sight of one's lover, which leaves one "feeling for a taste" and "so tempted to touch." Enlivened by the actions of the mind, this gazing, staring, seeing even seems to give way to the very tactility it craves, as when a lover through their stare is able to "touch on you more and more every time," in Beyoncé's words. This triangulation of seeing, touching, and desiring has opened spaces for me to think through Byzantine art, religion, and culture that merit acknowledgement.

Methodologically, this book owes a great deal to the contemporary pop-music to which it was written, namely: Beyoncé, Nicki Minaj, Troye Sivan, Lana Del Rey, Ellie Goulding, Years & Years, Walk The Moon, Carly Rae Jepsen, Miley Cyrus, Daft Punk, Betty Who, The Weeknd, Sia, Grimes, Katy Perry, Taylor Swift, Sam Smith, Zarah Mahler, Ben Cosgrove; and, especially, Clean Bandit and Lady Gaga, who have taught me more about Byzantium than I could have ever hoped. The intellectual complexity, richness, and brilliance of their songs have always kept my analytical skills sharp and agile. Even when Byzantium seemed too dull or too impenetrable, their words have helped me see the light and solve puzzles in ways that I alone could never have imagined.

This project has benefited from conversations, assistance, and support from many colleagues, including: Matt Abrams, Joe Ackley, Elizabeth Allen, Amy Anderson, Ben Anderson, Sonja Anderson, Traci Arden, Heather Badamo, William Earl Betsch, Deb Brown, Magdalene Breidenthal, Rebecca Brienen, Gregory Bryda, Börje Bydén, Annemarie Weyl Carr, Francesco Cassetti, Paroma Chatterjee, Kristen Collins, Alexander Coyle, Clayton Critcher, Tony Cutler, Beatrice Daskas, Marianna Davison, Ivan Drpić, Helen Evans, Chelsea Foxwell, Rico Franses, Milette Gaifman, Fani Gargova, Izabel Gass, Sharon Gerstel, Nina Glibetić, James Goebel, Jorge Gómez-Tejada, Cecily Hilsdale, Brad Hostetler, Aaron Hyman, Sharrissa Iqbal, Liz James, Kaelin Jewell, David Joselit, Jacqueline Jung, Machi Karanika, Bryan Keene, Joy Kim, Katherine King, Beatrice Kitzinger, Kostis Kourelis, Derek Krueger, Anna Kryczka, Annie Labatt, Christopher Lakey, Henrike Lange, Sean Leatherbury, Amanda Luyster, Alexandra Marraccini, Lyle Massey, Elaine McLemore, Alice Lynn McMichael, Margaret Mitchell, Margaret Mullett, Jennifer Nelson, Bob Ousterhout, Stratis Papaioannou, Alka Patel, Glenn Peers, Julia Perratore, Bissera Pentcheva, Lucy Pick, Jordan Pickett, Jacob Potash, Amy Knight Powell, Tara Prescott, Gabriel Radle, Kishwar Rizvi, Laurence Ross, Mike Ryan, Nancy Patterson Ševčenko, Shannon Steiner, Nicole Paxton Sullo, Maria Taroutina, Nancy Thebaut, Rebecca Uchill, Tolga Uyar, Francisco Prado Vilar, Alicia Walker, Shannon Wearing, Moira Weigel, Cécile Whiting, and Ann Marie Yasin. My co-conspirators in the BABEL Working Group, the Material Collective, and their orbits merit special mention for making the task of being a medievalist a little more ethical, a little less myopic, especially: Lauren Berlant, Jennifer Borland, Brantley Bryant, Sakina Bryant, Joseph Derosier, Irina Dumitrescu, Elizabeth Freeman, Anne Harris, Eileen A. Joy, Dan Kline, Anna Klosowska, Asa Mittman, Julie Orlemanski, Karen Overby, Christopher Michael Roman, Arthur Russell, Myra Seaman, Ben Tilghman, and Maggie Williams. And, to all those whom I have surely missed, please remind me so that I can apologize.

The following deserve special mention: Aden Kumler and Suzanne Conklin Akbari for being two exceptional interlocutors, the likes of which I have never quite encountered; Dorothy Kim for our Skype sessions, which kept me on track and motivated as I worked on my edits; and Charles Barber for his support and scholarship. Finally, this would not have been possible without Robert S. Nelson. His scholarship is the sole reason why I became a Byzantinist, his foresight made it possible for me to pursue this as my life's work, and his character is why I remained a Byzantinist. He has always sought to show me all the ways in which I could pursue my ideas, no matter how experimental or how much he might have disagreed. The ability to support those whose desires you cannot fully understand is perhaps the most ethical commitment that he instilled in me through example.

My colleagues at the University of California, Irvine, need to be thanked for their enthusiastic support and unwavering collegiality. And, of course, my colleagues – and dear friends – at the Institute for Advanced Study who have had an immense impact on my scholarship and quality of life, particularly Karen Engle, Emine Fetvaci, Patrick Geary, Rebecca Maloy, Sabine Schmidtke, Despina Stratigakos, María Mercedes Tuya, and Marian Zelazny. Special mention must be given to Jennifer Davis, Dan Sherman, and Columba Stewart for being the best friends and colleagues I could have ever asked for: Jennifer and Dan for keeping me balanced, Columba for everything else.

As an Elizabeth and J. Richardson Dilworth Fellow at the Institute for Advanced Study in Princeton, I was able to complete this volume and prepare it for publication. There, Brett Savage offered helpful assistance with proofreading, and Alexis May assisted in the compilation of the bibliography. Joel Kalvesmaki has served as an intrepid and studious editor for the volume. He made the tedious ordeal of revising a first book into an exciting process of growth and exchange. My proofreader, Aleah Hernandez, offered a sharp eye with an exceptional mastery of Greek. Subsidies for production costs were made possible by grants from the Art History Department, the Humanities Commons at the University of California, Irvine, and the International Center of Medieval Art and the Samuel H. Kress Foundation.

Jaś Elsner, Michael Squire, Simon Goldhill, and an anonymous reader provided extensive, thorough, and thought-provoking feedback on this manuscript. Working between antiquity, late antiquity, and the medieval world can be a perilous journey, but their comments, critiques, and nudges made this a far stronger and more exciting volume. I am endlessly grateful to them, in particular to Jaś whose enthusiasm and support for this at times whimsical, at times iconoclastic project I cherish beyond words.

At Cambridge University Press, I must thank Michael Sharp for his editorial guidance, as well as Emma Collison and Kay McKechnie for their exceptional work in producing and copy-editing the manuscript, and Raymond Howard for proofreading.

In part, this volume emerges from two earlier articles on Byzantine visuality. As such, it behooves me to acknowledge the insightful feedback of the anonymous readers for those pieces, which helped shape this project's scope, breadth, and balance. In this vein, I wish to thank Dumbarton Oaks for permission to reproduce an expanded version of my article, which serves as the footprint for Part II, "Why Sight Is Not Touch: Reconsidering the Tactility of Vision in Byzantium," *Dumbarton Oaks Papers* 70 (2016): 1–23; and the Medieval Academy for permitting me to similarly expand my article in Part III, "Tempted to Touch: Tactility, Ritual, and Mediation in Byzantine Visuality," *Speculum* 91:3 (2016): 660–689. I also must thank Flula Borg and Steven Gunther for allowing me to reproduce Flula's song lyrics as an epigraph to the volume.

My friends continuously unfold all the then and there's of queer utopias for me, and are a luminous horizon of endless possibility. They are my sun, and make me shine, like diamonds: especially Alex Brey, Eddie McCaffray, Devin O'Neill, Dan Zolli, Luke Fidler, Meghan Vicks, Kate Durbin, and Tatsiana Zhurauliova.

Naturally, I end here with all those who have done nothing but smother me with their love. A mi familia le quiero dar gracias por haberme dado de todo de lo que ellos mismos no tenían, y por haber compartido en la responsabilidad de mi crianza: Aida Echevarria Valdés, Mayda Josefina Valdés-Betancourt, Haymee Josefina Prieto, Rene Prieto, y Rolando Esteban Betancourt. Aunque malcriado, bien criado fui. And, to Chloe LeGendre, for putting up with each other like the siblings that neither one of us ever had. And, finally, I wish to thank Sam* Jacobson for giving me the chance to love them in more than one gender.

To me, this book has become a personal tribute to unrequited love; and, all those imaginative and obsessive fantasies that allow us to cope with their absence from our embrace. In a sense, this book is begrudgingly dedicated to them.

Note to the Reader

Since this volume is ultimately of a technical nature, featuring diligent close analyses of primary texts in their original language, it is understood that challenges could emerge for novice readers or those coming from cognate fields and disciplines. Keeping these challenges in mind, every effort has been made to aid in the accessibility of this book. The names of late antique and Byzantine authors have been standardized according to their appearance in the *Oxford Dictionary of Byzantium*, both for consistency and for the sake of looking up any less well-known figures discussed herein. All primary sources are cited by their Latin titles as they appear in the *Thesaurus Linguae Graecae* (*TLG*). This will enable the reader to readily look up the full text in that repository through a title search without the complex guesswork of having to extrapolate a Byzantine author's Latinized name. For both names and source titles, the reader is particularly encouraged to consult the *TLG* bibliography.

The goal has been to provide as full citations as possible so that students of Classics, Medieval, and Byzantine Studies can readily find texts without the hindrance of limited citations, of field-specific abbreviations and idiosyncratic customs, or of cryptic numerations of volumes, passages, and text. Therefore, every effort has been made to provide all the necessary information for finding cited texts in the *TLG*, as well as in the published editions and translations. Given the lack of standardization as to how texts are numbered, the reader is provided in all instances with both page numbers and book, section, etc. numbers that agree with both digital and hardcopy versions. Redundancies, however, have been avoided. Thus, for example, in cases where page numbers alone are used to numerate the text, pages alone are listed; and other possible numerations that are either superfluous or unnecessary for either cross-checking the original text or its translations have been judiciously left out. All this information has been compiled and presented with particular sensitivity as to how the texts can be searched and found in their online iterations for ease of use; and also, so that both original texts and translations can be compared with other editions and/or alternative translations.

In the hopes of providing greater ease of accessibility for independent scholars and those at institutions with limited resources, diligent care has been paid to citing editions of texts and translations that are available as open-source materials or on large-scale textual databases, such as the Hathi Trust Digital Library, Google Books, the Perseus Digital Library, the digital Loeb Classical Library, and the *TLG*. My hope is that scholars from all parts within and outside the academy who are able to access this tome will be able to critically engage with my arguments by readily cross-examining its texts and sources. This book is intended to be as much a contribution to our scholarship and an intervention in our methods, as a roadmap of constellations for further research, study, and thought.

Abbreviations

LSJ	H. G. Liddell, R. Scott, and H. S. Jones, *A Greek–English Lexicon*, 9th edition (Oxford University Press, 1996).
PG	Ed. Jacques-Paul Migne, *Patrologia Cursus Completus: Series Graeca* (Paris: Migne, 1857-1866).
TLG	Ed. Maria C. Pantelia, *Thesaurus Linguae Graecae Digital Library* (University of California, Irvine), www.tlg.uci.edu

Introduction: Can't Touch This

> This embracing or kissing the image visually, I submit, was meant literally as well as metaphorically. Like all successful religious symbolism, this metaphor was grounded in perception and perceptual theory. Because the optical rays that issue forth from the eyes were thought to touch the object seen, vision was haptic, as well as optic, tactile as well as visual. Vision thus connected one with the object seen, and, according to extramission, that action was initiated by the viewer.[1]
>
> – Robert S. Nelson

In 2000, Robert S. Nelson's groundbreaking essay "To Say and to See: Ekphrasis and Vision in Byzantium" raised the theory of haptic sight to a privileged position in the history of Byzantine art and culture. Positioning extramission (where the eye emits rays that contact the object) against intromission (where rays enter the eye passively from an outside source), Nelson argued for a theory of vision wherein sight was a species of touch.[2] Despite acknowledging that competing antique and late antique theories

[1] Robert S. Nelson, "To Say and to See: Ekphrasis and Vision in Byzantium" in Robert S. Nelson (ed.), *Visuality Before and Beyond the Renaissance: Seeing as Others Saw* (Cambridge University Press, 2000), 143–68, at 153.

[2] In a later book review, Nelson provides a concise summary of these two theories of vision, while detailing the normalization of intromission in the early modern world with the birth of modern optics, writing: "This is why intromission [*sic*], the principal Byzantine theory of vision, was ideally suited for icon devotion and vice versa. Shortly after the end of the iconoclastic controversy, Patriarch Photios evoked such vision in a celebrated sermon at the inauguration of the new apse mosaic of Hagia Sophia. In intromission [*sic*], the viewer sends out a ray that touches the object seen and returns to the eye. Intromission [*sic*] is thus tactile and active and motivated by the viewer who does something to something else in order to see. Extramission [*sic*], a more passive sense of seeing, is the proper term for the generally accepted notion of vision in our world. Light rays reflect off the object seen, enter the eye, and pass to the brain. In the West, extramission [*sic*] began to gain adherents during the late Middle Ages, due to the translation of Arabic texts. Perspectival, optical, and physiological studies of vision during the early modern period gradually made extramission [*sic*] the norm in the West" (N.B. Here Nelson has inverted the terms intromission and extramission in his summary). See Robert S. Nelson, "Review of Charles Barber, *Figure and Likeness: On the Limits of Representation in Byzantine Iconoclasm*," *Art Bulletin* 85:4 (2003): 797–800, at 799.

of vision circulated in Byzantium, Nelson's analysis of classical and medieval sources, particularly their metaphors and literary tropes, concluded that the Byzantine world preferred a theory of vision whereby the eyes emitted rays that grazed the body of the object and returned to the viewer, thus enabling perception. Therefore, sight was understood to be a species of touch and an active rather than passive engagement with visual culture. The Byzantine viewer was able to visually touch the icon – the most important part of the visual culture of veneration – as if kissing and embracing its surface. Weaving metaphor, science, theology, and philosophy, Nelson claimed to have captured an all-encompassing theory of visual perception for the Byzantines, whereby viewer and object were not distinct entities but physically connected through the tactility of sight.

This haptic theory of vision, however, is by no means exclusive to Byzantine Studies or to Nelson's influential piece. It had already been prevalent in the study of ancient theories of vision for decades, as well as in (western) Medieval Studies. In 1983, for example, Margaret Miles described Augustine's theory of vision in a manner that presages Nelson's sentiments two decades later, almost verbatim, writing:

> For the classical people. . ., sight was an accurate and fruitful metaphor for knowledge because they relied on the physics of vision, subscribed to by Plato and many others, that a ray of light, energized and projected by the mind toward an object, actually touches its object, thereby connecting viewer and object.[3]

Miles goes on to stress, much like Nelson above, that a haptic model of sight produced a "connection" between viewer and object, wherein vision was "initiated by the viewer," and that in the act of seeing viewer and object became "united."[4] In other words, Miles emphasized the same keywords as Nelson, demonstrating the importance that the viewer's initiation of the process and connection with the object had in the Christian episteme. And, just like Nelson, Miles's summary also erroneously associates the intersection of sight and tactility with "Plato and many others," even though Plato's writings offer little to presume as much. In other words, Nelson's perspectives on Byzantine sight certainly did not develop in a vacuum, nor were they by any means unique or idiosyncratic. This is attested by the context of his essay's publication, emerging from an

[3] Margaret R. Miles, "Vision: The Eye of the Body and the Eye of the Mind in Saint Augustine's *De trinitate* and *Confessions*," *Journal of Religion* 63:2 (1983): 125–142, at 127. See also Margaret R. Miles, *Augustine on the Body* (Missoula, MT: Scholars Press, 1979), 9–39.

[4] Miles, "The Eye of the Body," 127–128.

interdisciplinary conference on vision and visuality at the University of California in Los Angeles in the spring of 1995.[5] As Nelson's edited volume demonstrates, his fellow medievalist contributors, including Georgia Frank, Cynthia Hahn, and Michael Camille, all followed in the haptic model of extramissive sight, as did their Classicist counterparts, Jaś Elsner and Shadi Bartsch.[6]

Research into ancient and late antique theories of vision has long featured the same steadfast adherence to tactile sight (whether it be defined as extramissive, intromissive, or interactionist in some capacity), even though the primary sources demonstrate that there was no single visual theory that was by any means restrictive or canonical for the ancient world – and, even less so any theory that could have been seen as tactile. As the works of Jaś Elsner, Shadi Bartsch, Helen Morales, Gérard Simon, David Lindberg, Olivier Darrigol, A. Mark Smith, Suzanne Conklin Akbari, and others have observed, theories of vision from the ancient through medieval worlds were immensely eclectic and diverse, depending on the philosophical, religious, and cultural matters that a particular author or thinker wished to stress.[7] And as Sue Blundell, Douglas Cairns, Elizabeth Craik, and Nancy Sorkin Rabinowitz caution, the operation of sight often appears to overstep its prescribed physiological or theoretical function, "such as the role of the eyes in the expression of anger, [where] the relevant interpretive frame is not the 'haptic' model of vision, but specific Greek understandings of the expressive, emotional, and social role of the eyes in ordinary social interactions."[8] In the Byzantine world, we witness the same matters at play. Byzantium comprised a diverse, multi-lingual, and fluid empire that lasted well over a millennium.

[5] Robert S. Nelson, "Introduction: Descartes's Cow and Other Domestications of the Visual" in Robert S. Nelson (ed.), *Visuality Before and Beyond the Renaissance: Seeing as Others Saw* (Cambridge University Press, 2000), 1–21, at 14.

[6] See Robert S. Nelson (ed.), *Visuality Before and Beyond the Renaissance: Seeing as Others Saw* (Cambridge University Press, 2000).

[7] See Jaś Elsner, *Roman Eyes: Visuality and Subjectivity in Art and Text* (Cambridge University Press, 2007); Shadi Bartsch, *The Mirror of the Self: Sexuality, Self-Knowledge, and the Gaze in the Early Roman Empire* (University of Chicago Press, 2006), esp. 57–114; Helen Morales, *Vision and Narrative in Achilles Tatius' Leucippe and Cleitophon* (Cambridge University Press, 2004), 8–35; Gérard Simon, *Le regard, l'être et l'apparence dans l'Optique de l'Antiquité* (Paris: Éditions du Seuil, 1988); David C. Lindberg, *Theories of Vision from Al-Kindi to Kepler*, revised edition (University of Chicago Press, 1996); Olivier Darrigol, *A History of Optics: From Greek Antiquity to the Nineteenth Century* (Oxford University Press, 2012); A. Mark Smith, *From Sight to Light: The Passage from Ancient to Modern Optics* (University of Chicago Press, 2015); Suzanne Conklin Akbari, *Seeing through the Veil: Optical Theory and Medieval Allegory* (University of Toronto Press, 2004).

[8] Sue Blundell, Douglas Cairns, Elizabeth Craik, and Nancy Sorkin Rabinowitz, "Introduction: Vision and Viewing in Ancient Greece," *Helios* 40:1–2 (2013): 3–37, at 22.

It is impossible to ascribe a monolithic theory of vision to the Byzantine worlds, which, as Michael Squire points out in the case of antiquity, "lacked any singular, culturally dominant model for explaining the mechanics of sight."[9] Byzantine writers demonstrate a great deal of coherence and unity in their syntheses and permutations of visual theories, but they are each uniquely varied in crucially nuanced ways.

In the context of Medieval Studies, it is perhaps possible to attribute the rise of haptic extramission to the Latin-speaking West, wherein the transmission of ancient theories of vision did at times explicitly deploy tactile language in a manner that was unparalleled in the Greek-speaking Byzantine world. Perhaps the most prominent exemplar of this tradition is Augustine, whose statements have served as a rallying cry for a tactile visuality, as suggested already in Margaret Miles's work. Namely, we witness this in Augustine's often-cited line describing how the "rays" (*radios*) of the eyes "shine through them and touch whatever we see" (*qui per eos emicant et quidquid cernimus tangunt*).[10] While we might propose reading this grasping cognitively, in the case of Augustine *tangunt* here appears to be resolutely haptic. Therefore, it is necessary to concede that in some circles sight might have been understood to be haptic – beyond the limits of metaphor alone – given Augustine's popularity and, to a lesser extent, the relative inaccessibility of the classic Greek texts and their commentaries to his Latin-speaking audience. Likewise, earlier Latin sources similarly attest to this haptic depiction of vision, as in the case of Apuleius's *Apologia*, where, in a summary on vision, he attributed to the Stoics the idea that the effluxes from the eye "touch and contact the object" (*quod extra tangant ac uisant*) through attenuated air.[11] In these two examples, we might appreciate the inklings of a haptic extramission in the medieval Latin-speaking West, but such statements have no true parallel in the Greek-speaking East.

Thus, it is permissible to propose that the haptic theory of extramission as a visual *lingua franca* for Byzantium originates in the West, but particularly in its historiography. In other words, haptic sight in

[9] Michael Squire, "Introductory Reflections: Making Sense of Ancient Sight" in Michael Squire (ed.), *Sight and the Ancient Senses* (London: Routledge, 2016), 1–35, at 15.

[10] Augustine, *De trinitate*, 9.3.3, ed. Johann Kreuzer, *De trinitate (Bücher VIII–XI, XIV–XV, Anhang: Buch V)* (Hamburg: Felix Meiner Verlag, 2001), 54; trans. Stephen McKenna, *On the Trinity: Books 8–15*, ed. Gareth B. Matthews (Cambridge University Press, 2002), 27.

[11] Apuleius, *Apologia*, 15, ed. and trans. Christopher P. Jones, *Apologia*, Loeb Classical Library 534 (Cambridge, MA: Harvard University Press, 2017). I wish to thank Christopher P. Jones for recommending this source.

Byzantium appears to be a projection from a certain body of comments made (in passing) in the Latin world and used as adages for vision in the secondary literature. By the late twentieth century, western medieval theories of vision had been far more studied and elucidated in the historiographic record than any of their Byzantine counterparts. Thus, it is understandable that Byzantinists would have grafted such presumptions about vision onto the study of Byzantine art and culture.

Yet even the articulations of haptic sight in the West seem to betray a certain confusion emerging as by-product of a crude cropping and reduction of Greek theories. I would say that in the examples of Augustine and Apuleius we encounter the fragments of more complex arguments and metaphors that have simply gone awry in their excerption and translation, rather than constituting any properly fleshed-out theory of haptic sight. This matter, however, goes well beyond the goals of the present book. Nevertheless, we can even witness the unease with which Michael Camille approached haptic vision in his contribution to Nelson's volume, for example. There, Camille struggled with a more nuanced perspective on the differences in theories of perception at play in the western world, giving due focus to the work of Avicenna.[12] His essay rightly placed attention upon an interactionist theory of vision, wherein both the eye and the object performed mutual actions in the process of sight, while still oriented around the misguided presumption that extramission always denotes a tactile theory of sight. However, in shifting his attention from sight alone to the internal senses, such as the "common sense" (*sensus communis vel sensatio*) and the "cognitive imagination" (*imaginatio vel formalis*) in Avicenna's *De anima* commentary, Camille was able to better grasp and grapple with the problem of tactility from the perspective of the mind's perceptions, rather than the senses' sensations alone.[13] In a sense, Camille's essay indirectly responded to Nelson's contribution by demonstrating that the Aristotelian (and Platonic) heritage was never quite founded upon haptic sight, understood as the touch of the optical rays, but rooted in an interactionist theory wherein the cognitive gaze grasped the object in the mind. Camille, however, did not explicitly demonstrate just how tactility emerged from the operations of the common sense and the imagination, nor did he endeavor to take down haptic sight fully, as I aim to do in

[12] Michael Camille, "Before the Gaze: The Internal Senses and Late Medieval Practices of Seeing" in Nelson (ed.), *Visuality Before and Beyond the Renaissance*, 197–223.

[13] See Camille, "Before the Gaze," 198–202, 204–205, 211–214.

this volume. Instead, Camille only politely side-steps haptic vision's problematic implications for medieval art and culture.

In projecting stereotypes of western visuality onto Byzantium, Nelson and others designated tactility to be the defining aspect of Byzantine visual theories of sight, but, as I argue in this book, the one thing that most theories of vision in the Byzantine period seemed to generally agree upon was that touch proper did not occur through sight, despite the frequent haptic language and metaphors that characterized the process of perception. By looking closely at classical, late antique, and Byzantine sources, I aim to show that the notion of haptic vision in both the classical and Byzantine worlds was a product of metaphors that in reality were describing cognition and the mental processes of perception. In other words, the idea of sight being haptic does not emerge from a Byzantine theory that sight is a species of touch, but due to the shorthand describing the processes of perception. It must be clarified at the outset that this book's argument is not that one should favor one ancient theory of vision over another, such as intromission over extramission. Instead, the goal is to show overwhelmingly that the Byzantine rhetoric of haptic sight pointed to the mind's grasp and apprehension, not the literal grasp of the eyes. Many of the usual suspects in discussions of sight will be brought into the dialogue, particularly those that relate sight to touch.

This book's launching point is Nelson's essay and its primary sources. I actively read his argument and the sources closely, against the grain, and alongside other crucial texts, to demonstrate how Byzantine authors writing on medicine, the natural world, and theology produced nuanced articulations of sight. By distancing sight from touch we can then attend to what is actually being described by this haptic language, namely, the cognitive unfolding of perception, a process in which the imagination is the centerpiece. This shift in focus allows us to recover an image theory that did not fetishize the illusion of presence in the icon, but rather expressed the desire and "directed absence"[14] upon which the icon relied theologically. Turning to the importance that the imaginative faculty played in perception and the viewer's investment in the image, we are able to read anew key texts about Byzantine art, both before and after the Iconoclastic period, to appreciate how the imagination was responsible for animating the icon in the viewer's mind. It is the imagination that allowed the viewer to feel as if they were in the presence of a holy person depicted

[14] Charles Barber, *Figure and Likeness: On the Limits of Representation in Byzantine Iconoclasm* (Princeton University Press, 2002), 121.

in the icon, speaking with and touching them, but never under the illusion that they actually were in said person's presence.

A Brief History of Byzantine Theories of Vision

Modern scholars have often divided classical, medieval, and early modern theories of vision into two major camps: extramission and intromission. Thinkers like Euclid, Empedocles, Hipparchus, and the Stoics generally fall into the category of extramission, whereby the eyes emit flames, rays, or some form of efflux that flows to the object and contacts it somehow. It is from this theory that we get the notion of haptic extramission, because many of the metaphors explaining the operation of the effluxes are haptic. For example, the effluxes are rays that extend "just as by the touch of hands" (καθάπερ χειρῶν ἐπαφαῖς) to the object, in the case of Hipparchus, or they are a fiery mixture of *pneuma* extending to the object, seeing "as if through the walking stick" (ὡς διὰ βακτηρίας) of the blind touching the object, in the case of the Stoics. Intromission, on the other hand, details a series of divergent processes whereby rays, colors, images, or light emitted from the object stream into the eyes. For Aristotle, the diaphanous medium betwixt the viewer and the object conducts colors from the object's surface to the eyes when activated by light, or in the case of the atomists, "replicas" (εἴδωλα) are shed by the objects and impinge themselves into the eyes of the viewer, according to the theories of Leucippus and Democritus, popular with the Epicureans.

However, as Squire and others have noted, there are other theories that are far more "interactionist" than the binary terms extramission and intromission would suggest.[15] For example, Plato, who is often taken as a paradigmatic extramissionist, actually believed that vision occurred through the union of a flame (φλόγα) coming from the object and the visual stream of the eyes, a union activated by external light. Even the Stoic theory of *pneuma* held that the fiery mixture flowing through the eyes could not reach the object of sight by itself, but did so because the mixture put the surrounding air in a state of tension that allowed the two together to conduct an object's colors back to the eyes, a process that Galen expounds and repeatedly clarifies. Furthermore, throughout the antique and post-antique worlds, there was an immense variety of language describing theories of vision. Even when clearly describing the theory of a particular author, sources often varied in their characterization of the

[15] Squire, "Introductory Reflections," 16.

effluxes, diversely described as rays, light, fire, flames, or, more generally, as an efflux or emanation. This great variation in an eclectic body of sources from the ancient to early modern period shows that authors freely reworked the tradition.

Byzantinists and Classicists alike have stressed the importance of tactility in premodern theories of vision, often stressing the operation of sight in relation to these haptic metaphors. Nelson's summary of Byzantine theories of vision relied on this general consensus, and his words, quoted above, echo the very model that Hipparchus describes in the extant fragments of his theory, where "rays from the eyes, stretched out to the ends just as by the touch of the hands themselves, seizing the bodies outside to deliver an apprehension (ἀντίληψιν) of them to the faculty of sight."[16] Preserved in the work of Aetios of Antioch, whose doxography was used by a long line of writers, from Nemesios of Emesa in the late fourth century to Michael Psellos in the eleventh, this single-sentence description was merely one of many. And it is the only theory that explicitly relates a theory of vision in explicitly haptic terms. Yet even here touch is used as a metaphor. The thing returned is the "apprehension" or "comprehension" (ἀντίληψιν) of the object, not the object itself, which suggests that Hipparchus meant to describe a cognitive process, not a physical imprint on the eyes. Late antique and Byzantine sources use Hipparchus's line only in passing, to lay the groundwork for the later theories of Plato, Aristotle, and Galen. Thus, while sight certainly had haptic valences in the pre-Socratic world, later thinkers from Plato onward actively fought against this tendency, or at least tried to restrict it to metaphor.

Despite the great diversity of theories, it is nevertheless possible to loosely define two major groups of theories that held sway in the Byzantine world. First, in circles educated in philosophy (a matter to which I shall return in the following section), up through the twelfth century, we witness a preference for the Platonic or Aristotelian theories of vision. Oftentimes, Plato's theory of a union of rays allowed authors a middle ground to articulate the agency of both the object and the viewer, permitting them to mediate the divide between intromission and extramission. Aristotle's theory of the transparent, while downplaying the agency of the viewer, still allowed for the notion that an intervening medium served as

[16] "ἀκτῖνας ἀπὸ τῶν ὀφθαλμῶν ἀποτεινομένας τοῖς πέρασιν ἑαυτῶν καθάπερ χειρῶν ἐπαφαῖς καθαπτούσας τοῖς ἐκτὸς σώμασι τὴν ἀντίληψιν αὐτῶν πρὸς τὸ ὁρατικὸν ἀναδιδόναι." Aetios of Antioch, *De placitis reliquiae (Theodori et Nemesii excerpta)*, IV:13.9, ed. H. Diels, *Doxographi Graeci* (Berlin: Reimer, 1879), 404. Unless otherwise indicated, all translations are my own.

the point of communion between viewer and object. The two theories of Plato and Aristotle could be mixed together or unified. One such theory, for example, claimed that the intermediary of air or water is the medium in which sight occurs, but rays or fires also emerge from the eyes or objects. Second, it is quite clear that the Galenic approach to *pneuma* was the theory par excellence for medical authors. Many of the physiological treatises that come down to us are quite silent regarding the exact physics of this visual efflux in the world, but they devote their attention to ameliorating diseases of the eye by enhancing or stanching the flow of the pneumatic emission. Much of this work, and its assumptions, depends directly upon Galen's writings. These two camps – those who focused on the emission or reception of rays and fires and others who focused on the outpouring and operation of the optic *pneuma* – comprise the two major approaches to theories of vision in the Byzantine world.

Two other minor groups, however, still make their appearances throughout the literature. The first of these is represented mainly by Euclid and, to a lesser extent, Ptolemy. Euclid's theories often featured in the teaching of the *quadrivium*, the four key subjects of Byzantine education, composed of arithmetic, geometry, music, and astronomy. Although Euclid was an extramissionist, whose linear rays are the bedrock of his mathematical teachings, his work says little about the physiology of sensation. His work was used primarily in service to mathematical optics, that is, to understand the geometry and mathematics of the rays of vision, calculating the effects of reflections, refraction, and so on. A curious aspect of Euclid's work in the medieval world is that even authors who disagreed with the theory of extramissive rays nevertheless retained his models because of their mathematical accuracy in studying visual phenomena. This happened in the Arabic world, which developed new intromissive theories, and also with Ptolemy, who built upon Euclidian teachings for his optical calculations. Whereas Ptolemy's astronomical corpus and optics were known in the Byzantine world but rarely used, Euclid's presence was far more prominent in Byzantine education, but it was relegated to the mathematical geometry of optics alone. In other words, it carried little (if any) weight on notions regarding the physiology of vision, and certainly had nothing to say on the cognitive aspects of perception. Lastly, the final popular camp in Byzantine theories of vision is comprised of the atomists, those who believed that objects emit replicas or images (εἴδωλα) of themselves that are passively received by the eyes. While we see more popular references to the atomists than we do to Euclid or Ptolemy, these often occur in passing and are often meant to exploit the marginal status of that theory. The atomists are treated as a curiosity in the history of theories of vision, and there is not much evidence to suggest a sincere belief in their theory.

Classical Education and the Transmission of Knowledge

The great challenge in constructing a survey of Byzantine theories of vision and their relation to religious and secular practices is that most of the sources that come down to us are from elite circles tied to centers of knowledge and study. To this day, our understanding of education in Constantinople, and Byzantium more broadly, is lacking. However, we can see how widespread theories of vision were throughout texts like homilies, aimed at broad audiences, for which reason a third of this volume is dedicated to the close study of Photios's Homily 17 on the icon of the Theotokos in the apse of Hagia Sophia. It is also worth contextualizing the work, education, and teaching of some important figures that appear prominently in this book, such as Michael Psellos, who attests – both in his sources and his influence – to a tradition of visual knowledge unbroken from late antiquity. Let us briefly sketch out some important points regarding Byzantine education and the reception and transmission of the classical heritage.

Over the course of the empire, there was never any form of compulsory education, yet the resources were there for students who had the ability to pay for it. Higher education itself was relegated to the major cities and from the ninth century onwards associated with imperial efforts and patronage, and the works of prominent private tutors.[17] Basic literacy was relatively widespread, however, with particularly high levels amongst the elites.[18] Byzantine schooling overall was divided loosely into what we might call "primary" and "secondary" education.[19] Primary education or *propaideia* began around the ages of six to eight, lasted for three to four

[17] See Athanasios Markopoulos, "Education" in Robin Cormack, John F. Haldon, and Elizabeth Jeffreys (eds.), *The Oxford Handbook of Byzantine Studies* (Oxford University Press, 2008), 785–795, esp. 786–788.

[18] On literacy in Byzantium, see Michael Jeffreys, "Literacy" in Robin Cormack, John F. Haldon, and Elizabeth Jeffreys (eds.), *The Oxford Handbook of Byzantine Studies* (Oxford University Press, 2008), 796–802. See also Robert Browning, "Literacy in the Byzantine World," *Byzantine and Modern Greek Studies* 4 (1978): 39–54; Robert Browning, "Further Reflections on Literacy in Byzantium" in S. Reinert et al. (eds.), *To Hellenikon: Studies in Honor of Speros Vryonis, Jr.* (New Rochelle, NY: Aristide D. Caratzas, 1993), 69–84.

[19] On the use of these terms, see R. A. Kaster, "Notes on 'Primary' and 'Secondary' Schools in Late Antiquity," *Transactions of the American Philological Association* 113 (1983): 323–346.

years, and consisted of basic skills, including arithmetic, reading, and writing.[20] Here the Psalms served as the key educational text for students to practice writing and memorization.[21] Secondary education or the *enkyklios paideia* began around the ages of twelve to fourteen, lasted three to four years, and consisted of the *trivium* of grammar, rhetoric, and philosophy, and the *quadrivium* of arithmetic, music, geometry, and astronomy.[22] While it is uncertain to what extent of completion these curricula were taught, especially in the case of the *quadrivium*, it is generally accepted from the existing evidence that there was a great deal of stability and continuity in the training and literary canons of the Byzantine educational tradition.[23]

Following the late antique model, during the course of the *trivium*, students would have begun their rhetorical education with Homer's *Iliad*, along with nine ancient tragedies drawn from the three playwrights Aeschylus, Sophocles, and Euripides, and three comedies from Aristophanes.[24] This also was augmented by works from Hesiod, Pindar,

[20] For a general overview of the timing and curriculum of primary and secondary education, see Markopoulos, "Education," 788–789. See also Ann Moffatt, "Schooling in the Iconoclast Centuries" in Anthony Bryer and Judith Herrin (eds.), *Iconoclasm: Papers Given at the Ninth Spring Symposium of Byzantine Studies, University of Birmingham, March 1975* (Birmingham: Centre for Byzantine Studies, University of Birmingham, 1977), 85–92; Ann Moffatt, "Early Byzantine School Curricula and a Liberal Education" in *Byzance et les Slaves* (Paris: Association des Amis des Etudes Archéologiques des Mondes Byzantino-Slaves et du Christianisme Oriental, 1979), 275–288; M. J. Kyriakis, "Student Life in Eleventh Century Constantinople," *Byzantina* 7 (1975): 375–388.

[21] Herbert Hunger, *Schreiben und Lesen in Byzanz: Die byzantinische Buchkultur* (Munich: C. H. Beck, 1989), esp. 76–77.

[22] For an overview of the *trivium* and *quadrivium*, see Markopoulos, "Education," 788–789. See also Michel Cacouros, "La philosophie et les sciences du *trivium* et du *quadrivium* à Byzance de 1204 à 1453 entre tradition et innovation: les textes et l'enseignement, le cas de l'école du Prodrome (Pétra)" in Michel Cacouros and M.-H. Congourdeau (eds.), *Philosophie et sciences à Byzance de 1204 à 1453: les textes, les doctrines et leur transmission* (Leuven: Peeters, 2006), 1–51.

On the structure and content of the *enkyklios paideia* since the Graeco-Roman world, see Teresa Morgan, *Literate Education in the Hellenistic and Roman Worlds* (Cambridge University Press, 1998), esp. 50–89.

[23] See Athanasios Markopoulos, "Teachers and Textbooks in Byzantium: Ninth to Eleventh Centuries" in Sita Steckel, Niels Gaul, and Michael Grünbart (eds.), *Networks of Learning: Perspectives on Scholars in Byzantine East and Latin West, c. 1000–1200* (Berlin: Lit Verlag, 2014), 3–15, esp. 3–4. See also Václav Ježek, "Education as a Unifying and 'Uplifting' Force in Byzantium," *Byzantinoslavica* 65 (2007): 167–200; Nikolaos M. Kalogeras, "Byzantine Childhood Education and Its Social Role from the Sixth Century until the End of Iconoclasm," unpublished PhD dissertation, University of Chicago (2000).

[24] See Rafaella Cribiore, "The Grammarian's Choice: The Popularity of Euripides' *Phoenissae* in Hellenistic and Roman Education" in Yun Lee Too and Niall Livingston (eds.), *Pedagogy and Power: Rhetorics of Classical Learning* (Cambridge University Press, 1998), 241–259. See also Raffaella Cribiore, *Gymnastics of the Mind: Greek Education in Hellenistic and Roman Egypt* (Princeton University Press, 2001), esp. 185–219.

Theocritus, Lucian, Demosthenes, Aeschines, Xenophon, Philostratus, Aelian, Plato, the poems of Gregory of Nazianzos, and others.[25] In other words, students would have had a strong and concrete survey of classical texts, including the dialogues of Plato, upon which to build further higher education that took a more concerted approach to classical philosophy, particularly the works of Plato, Aristotle, and the latter's commentators as evidenced by the educational narratives of learned figures like Plotinos and Michael Psellos, as shall be discussed later. The educational grounding of the *quadrivium* would have likewise exposed students to Euclidian and perhaps even Ptolemaic optics, given that Euclid served as the basis for geometry, and Ptolemy's *Mathematical Composition* also appeared frequently in the context of astronomical learning along with Theon of Alexandria's commentaries.[26] Thus, those boys and girls completing the curricula of the *trivium* and *quadrivium* would surely have had a nuanced grasp not only on theories on human sensation and perception through philosophical and literary Graeco-Roman texts, but they would also have had a foundation, at least, of Euclidian geometry around optics, as evidenced in the *quadrivium* of George Pachymeres.

In the context of imperial circles, education also included the further education or *paideia* of the learned elite. Particularly during the Palaeologan period, this could last as long as a decade and involved the acquisition of an Atticizing social dialect, different even from that of the time, and at times the careful study of philosophy and, more rarely, astronomy.[27] Yet, as Niels Gaul has noted, this did not simply mean that these learned circles were closed off from others, but rather that the ability to produce writing and knowledge at these higher levels did not preclude the social literacy of those in other strata of society, most notably below them.[28] Many of those

[25] See Markopoulos, "Education," 788–789. See Paul Lemerle, *Le premier humanisme byzantin* (Paris: Presses Universitaires de France, 1971), esp. 100–104; Athanasios Markopoulos, "De la structure de l'école byzantine: le maître, les livres, et le processus éducatif" in Brigitte Mondrain (ed.), *Lire et écrire à Byzance* (Paris: Association des amis du Centre d'histoire et civilisation de Byzance, 2006), 85–96.

[26] On mathematical and scientific learning in the *quadrivium*, see Anne Tihon, "Numeracy and Science" in Robin Cormack, John F. Haldon, and Elizabeth Jeffreys (eds.), *The Oxford Handbook of Byzantine Studies* (Oxford University Press, 2008), 803–819, esp. 805–810.

[27] See Niels Gaul, "All the Emperor's Men (and His Nephews): Paideia and Networking Strategies at the Court of Andronikos II Palaiologos, 1290–1320," *Dumbarton Oaks Papers* 70 (2016): 245–270, esp. 262 n 77. See also P. Agapitos, "Teachers, Pupils and Imperial Power in Eleventh-Century Byzantium" in Yun Lee Too and Niall Livingston (eds.), *Pedagogy and Power: Rhetorics of Classical Learning* (Cambridge University Press, 1998), 170–191; Niels Gaul, *Thomas Magistros und die spätbyzantinische Sophistik: Studien zum Humanismus urbaner Eliten der frühen Palaiologenzeit* (Wiesbaden: Harrassowitz, 2011), esp. 229–310.

[28] Gaul, *Thomas Magistros*, 163–168.

writers who underwent higher education would have relied upon and grounded their understanding of the natural world through the gaze of the commentaries on Aristotle and the Aristotelian corpus, building as well on any earlier exposure to the Platonic dialogues throughout the course of the *trivium*. This may surprise anyone who regards the Byzantine and medieval world at large as being largely Neoplatonic, but keep in mind, Aristotle often served in late antique and Byzantine education as a pathway into Platonism. Because he carefully surveyed the opinions of his predecessors, Aristotle was seen less as a polemicist and more as an explicator, thus often serving as a prerequisite for a proper and disciplined understanding of Plato. This trajectory, seen as early as Plotinos, coaxed Byzantine thinkers into synthesizing Aristotelian and Platonic theories of vision.

Commentators and other writers such as John Philoponos or Nemesios of Emesa served as some of the most crucial sources for theories of sight and perception, not only because they presented ancient learning within a Christian context, but, more importantly, because through their extensive erudition they produced important surveys that set their own teachings alongside summaries of a very broad wealth of ancient learning. Plotinos, Alexander of Aphrodisias, Iamblichus, Themistios, Simplikios, and John Philoponos are examples of both readers and producers of such treatises, giving us theories of vision that regarded sight as a union of effluxes from eye and object. The specifics varied from source to source, but nevertheless evidenced a great deal of consistency and similarity in their approaches. For example, these commentators all stress the division and uniqueness of the individual senses, stressing often the fallacy of atomic or some earlier extramissive theories conflating sight with touch. The idea being that an atomic emission would bear the material essence of the thing seen and is thus similar to the experience of touch or taste, which is often understood as a close variant of touch. Rather than stressing the tactility of sight, they often accuse earlier thinkers of this logical fallacy, since touch cannot sense colors and since each sense is particular to itself.

In the ninth century, in the aftermath of the Iconoclastic controversy, many figures actively cultivated, preserved, and transmitted many classical and late antique texts related to philosophy and natural science.[29] Photios, for example, actively taught Aristotelian logic, known from his extant comments on Aristotle's *Categories* and related scholia. His younger contemporary Arethas of Caesarea was also a teacher of Aristotelian philosophy

[29] See Alphonse Dain, "La transmission des textes littéraires classiques de Photius à Constantin Porphyrogénète," *Dumbarton Oaks Papers* 8 (1954): 31–47.

and is a crucial figure for the preservation of the Platonic corpus through his patronage of complete copies of Plato's works. And neither was alone. Other contemporaries such as the ninth-century iconoclast Patriarch John the Grammarian, Leo the Mathematician, and Constantine the Philosopher were active educators of rhetoric and philosophy in Constantinople.[30]

Furthermore, interest in the codification of knowledge in the tenth century, during the so-called Macedonian Renaissance, blossomed. Ecclesiastical and imperial scribes, the vanguard of a society that valued research and study, were charged with the task of compiling and preserving knowledge on a variety of topics. For example, in one treatise on imperial military campaigns, compiled for Leo at the behest of his father Constantine VII Porphyrogennetos, the unnamed author writes: "Having completed a great deal of research, yet finding no memorandum deposited in the palace, we were at last just able to discover one which dealt with these matters in the monastery called Sigrianē..."[31] From there, the author further recounts how the found text was written in poor grammar and syntax, lacking in "Hellenic learning" (παιδείας Ἑλληνικῆς), and thus he had rewritten and revised it.[32] This military treatise, like comparable contemporary works in other fields, depended on earlier sources derived from antiquity for conveying knowledge about military weaponry, siege warfare, tactics, and so on.[33] This passing remark suggests an image of tenth-century intellectual life in Constantinople that encompassed not merely elite scholars but also a wide swathe of the literate class, including educated scribes tasked with the duties of researching and compiling knowledge for the imperial elite.

30 See John Duffy, "Hellenic Philosophy in Byzantium and the Lonely Mission of Michael Psellos" in Katerina Ierodiakonou (ed.), *Byzantine Philosophy and Its Ancient Sources* (Oxford University Press, 2002), 139–156, esp. 144–145.

31 "ὅθεν πολλὰ περὶ τούτων ἀνερευνήσαντες καὶ μηδεμίαν ὑπόμνησιν ἐναποκειμένην τῷ παλατίῳ εὑρίσκοντες, ὀψὲ καὶ μόλις ποτὲ περὶ τούτων ὑπόμνημα ἐν τῇ μονῇ τῇ καλουμένῃ Σιγριανῆς εὑρεῖν ἠδυνήθημεν." Constantine VII Porphyrogennetos, *Treatise on Imperial Military Expeditions*, C:456.24–26, ed. and trans. John Haldon, *Three Treatises on Imperial Military Expeditions* (Vienna: Verlag der Österreichischen Akademie der Wissenschaften, 1990), 94–95.

32 Constantine VII Porphyrogennetos, *Treatise on Imperial Military Expeditions*, C:457.33, ed. and trans. Haldon, 96–97.

33 See, for example, Denis F. Sullivan, *Siegecraft: Two Tenth-Century Instructional Manuals by "Heron of Byzantium"* (Washington, DC: Dumbarton Oaks, 2000), 1–23. See P. H. Blyth, "Apollodorus of Damascus and the *Poliorcetica*," *Greek, Roman, and Byzantine Studies* 33 (1992): 127–158; E. Lacoste, "Les poliorcétiques d'Apollodore de Damas," *Revue des Études grecques* 3 (1890): 230–281; F. Lammert, "Zu den Poliorketikern Apollodoros und Athenaios und zur Poliorketik des Vitruvius," *Rheinisches Museum für Philologie* 87 (1938): 304–332.

Many of the key materials that we shall consider in this book were to be found not only in broad compendia of ancient knowledge and the educational notes and orations of educators, such as Michael Psellos's summary on vision, but also within popular religious texts, for example, John of Damascus's *Expositio fidei,* one of the most prominent texts found in the manuscript tradition.[34] In this text, to which we shall return throughout this volume, John gives a cogent rundown of the five senses, placing them in dialogue with one another and also within the broader context of cognitive perception through imagination, discursive thought, and memory. Furthermore, the *Progymnasmata,* schoolbooks used for rhetorical exercises, reflected and refracted complex Byzantine understandings about representation and the operations of the imagination in the processing and handling of sense perception.

In the eleventh century, we arrive at the elusively erudite figure of Michael Psellos, whose intimacy with Hellenic philosophy might make one question how representative he was of the broader spheres of Byzantium. Are we to dismiss his work as a "lonely mission,"[35] as John Duffy has called it? Psellos himself poignantly stated: "I am a lone philosopher in a time without philosophy" (φιλοσοφῶ μόνος ἐν ἀφιλοσόφοις καιροῖς).[36] While Psellos is well known for such hubristic statements exclaiming his grandeur and contributions, as Duffy and others have noted, to some extent he was perhaps not being so extreme in his self-appraisal.[37] Psellos was a voracious reader, whose education he himself recounts in the *Chronographia* with equal boasting, writing:

> ὅτι ἐκπνεύσασαν τὴν σοφίαν καταλαβών, ὅσον ἐπὶ τοῖς μετέχουσιν, αὐτὸς ἀνεζωπύρησα οἴκοθεν, οὔτε διδασκάλοις ἀξιολόγοις περιτυχών· οὔτε σπέρμα σοφίας ἐν τῇ Ἑλλάδι ἢ τῇ βαρβάρῳ τὸ ξύμπαν διερευνησάμενος εὑρηκώς. ἀλλ᾽ ἐπειδὴ μέγά τι περὶ φιλοσοφίας ἐν τῇ Ἑλλάδι ἀκούων, ἐν φωναῖς τισιν ἁπλαῖς καὶ προτάσεσι κατεμάνθανον (καὶ ἦν ταῦτα ὡσανεὶ στῆλαι καὶ ὅρια), καταγνοὺς τῶν περὶ ταῦτα σμικρολογουμένων, ἐζήτησά τι πλέον εὑρεῖν. ὡς δέ τισι τῶν ἐξηγησαμένων τὴν ἐπιστήμην ἐνέτυχον, τὴν ὁδὸν παρ᾽ αὐτῶν ἐδιδασκόμην τῆς γνώσεως· καί με ἄλλος εἰς ἄλλον

[34] See Stratis Papaioannou, *Michael Psellos: Rhetoric and Authorship in Byzantium* (Cambridge University Press, 2013), 15–17.

[35] Duffy, "Hellenic Philosophy in Byzantium," 152.

[36] Michael Psellos, *Oratoria minora,* 6.52–53, ed. A. R. Littlewood, *Oratoria minora* (Leipzig: Teubner, 1985), 20.

[37] Duffy, "Hellenic Philosophy in Byzantium," esp. 154–155. See also Dominic J. O'Meara, "Aspects du travail philosophique de Michel Psellus" in C. F. Collatz et al. (eds.), *Dissertatiunculae criticae: Festschrift für Günther Christian Hansen* (Würzburg: Königshausen und Neumann, 1998), 431–439.

> παρέπεμπεν, ὁ χείρων πρὸς τὸν κρείττονα· κἀκεῖνος αὖθις εἰς ἕτερον· καὶ οὗτος εἰς Ἀριστοτέλην καὶ Πλάτωνα. ὧν δὴ καὶ οἱ πρὸ ἐκείνων ἠγάπησαν, εἰ εὐθὺς τὰ δευτερεῖα μετ᾽ ἐκείνους εἰλήχασιν. ἐντεῦθεν οὖν ὁρμηθεὶς αὖθις ὥσπερ περίοδον ἐκπληρῶν, ἐς Πλωτίνους καὶ Πορφυρίους καὶ Ἰαμβλίχους κατῄειν. μεθ᾽ οὓς ὁδῷ προβαίνων εἰς τὸν θαυμασιώτατον Πρόκλον ὡς ἐπὶ λιμένα μέγιστον κατασχών, πᾶσαν ἐκεῖθεν ἐπιστήμην τε καὶ νοήσεων ἀκρίβειαν ἔσπασα· μέλλων δὲ μετὰ ταῦτα ἐπὶ τὴν πρώτην ἀναβαίνειν φιλοσοφίαν· καὶ τὴν καθαρὰν ἐπιστήμην μυεῖσθαι, τὴν περὶ τῶν ἀσωμάτων θεωρίαν προὔλαβον ἐν τοῖς λεγομένοις μαθήμασι (ἃ δὴ μέσην τινὰ τάξιν τετάχαται, τῆς τε περὶ τὰ σώματα φύσεως· καὶ τῆς ἀσχέτου πρὸς ταῦτα νοήσεως· καὶ αὐτῶν δὴ τῶν οὐσιῶν, αἷς ἡ καθαρὰ συμβαίνει νόησις), ἵν᾽ ἐντεῦθεν εἴ τι καὶ ὑπὲρ ταῦτα ὑπέρνουν ἢ ὑπερούσιον καταλήψομαι.
>
> Philosophy, when I first studied it, was moribund as far as its professors were concerned, and I alone revived it, untutored by any masters worthy of mention, and despite my thorough research, finding no germ of philosophy either in Greece or in the barbarian world. I had heard that Greece had a great reputation for philosophy, expressed in simple words and simple propositions, and their work in this field set a standard and criterion for the future. There were some who belittled the simplicity of the Greeks, but I sought to learn more, and as I met some of the experts in the art, I was instructed by them how to pursue my studies in a methodical way. One passed me on to another for tuition, the lesser light to the greater, and he again recommended me to a third, and he to Aristotle and Plato. Doubtless my former teachers were well satisfied to take second place to these two. Starting from these authors I completed a cycle, so to speak, by coming down to Plotinus, Porphyry, and Iamblichus. Then, continuing my voyage, I put in at the mighty harbour of the admirable Proclus, eagerly picking up there his doctrine of perception, both in its broad principles and in its exact interpretation. From Proclus I intended to proceed to more advanced studies – metaphysics, with an introduction to pure science, – so I began with an examination of abstract conceptions in the so-called mathematics, which hold a position midway between the science of corporeal nature, with the external apprehension of these bodies, and the ideas themselves, the object of pure thought. I hoped from this study to apprehend something that was beyond the reach of mind, something that was not subject to the limitations of substance[38]

Psellos here depicts a dire picture of the state of philosophy in eleventh-century Byzantium, claiming he had to search far and wide for proper tutelage and guidance. His path provides us an intricate image of his

[38] Michael Psellos, *Chronographia*, VI.37–38, ed. D. R. Reinsch, *Chronographia*, vol. 1 (Berlin: De Gruyter, 2014), 122; trans. E. R. A. Sewter, *Fourteen Byzantine Rulers* (New York: Penguin Books, 1979), 173–174.

philosophical training, particularly its process and unfolding over time. In his arrogance, he frames this as a unique process, but the stages of his education are identical to those taken by many before him and they suggest as well that he had at hand an abundant amount of resources – tutors and primary sources – to undertake such study.

While Psellos may have been exemplary in erudition, he was also not unrivaled in the post-iconoclastic Byzantine world before the eleventh century. Like his ninth- and tenth-century predecessors, Psellos progressed from Aristotle and Plato (more than likely, from Aristotle *to* Plato), and on to Plotinos, Porphyry, Iamblichus, and eventually Proklos.[39] He explains his philosophical interests, and, beyond the extract quoted above, goes on to explain his further studies in philosophy, rhetoric, and so on. As far as his schooling is concerned, Psellos's education was fairly standard for a member of the Byzantine intelligentsia of the time, such as John Mauropous. Theirs was an education based on basic grammar, orthography, Homeric poetry, and composition, before moving on to more specialized topics.[40]

Psellos's corpus itself certainly attests to his careful study and engagement with the works of far more thinkers than the very few mentioned above. He had a keen and fairly comprehensive grasp of commentators and doxographers such as Alexander of Aphrodisias, Aetios of Antioch, and John Philoponos, which he used and cited for his own works and philosophical compendia. And if Psellos's extant literary corpus is significant, surely even more so is that which is lost to us. That is, the entire oeuvre, both extant and lost, shows us that Psellos had deeply ingested classical, late antique, and Byzantine sources, and can be relied upon as a critical reader of the breadth of the tradition. And even though his erudition was exceptional, it nevertheless is representative of Byzantine culture, both in his own time and in centuries to come. The bulk of his writings on philosophy were produced as educational materials. Psellos taught everything from basic grammar to rhetoric and philosophy in what were private schools funded by the emperor around the city of Constantinople, which often were associated with monasteries and churches.[41]

39 On Psellos's study of Plotinos, see Frederick Lauritzen, "Psellos and Plotinos," *Byzantinische Zeitschrift* 107:2 (2014): 711–724.

40 On Psellos's education, see Papaioannou, *Michael Psellos*, 4–5.

41 On Psellos and teaching, see Papaioannou, *Michael Psellos*, 7. On schools in Constantinople, see Floris Bernard, *Writing and Reading Byzantine Secular Poetry, 1025–1081* (Oxford University Press, 2014), 210–251, esp. 210–213. See also V. Katsaros, "Προδρομικοί «θεσμοί» για την οργάνωση της ανώτερης εκπαίδευσης της εποχής των Κομνηνών από την προκομνήνεια περίοδο" in

His influence extended beyond elite Constantinopolitan circles, evident in a letter he wrote to Patriarch Michael Keroularios in the 1050s. Psellos boasts that he had attracted students from both west and east – from the Celtic, Arabian, Egyptian, Babylonian, Persian, and Ethiopian worlds:

> ἀλλὰ Κελτοὺς μὲν καὶ Ἄρραβας ἁλωσίμους ἡμῖν πεποιήκαμεν καὶ καταπεφοιτήκασι κατὰ κλέος ἡμέτερον κἀκ τῆς ἑτέρας ἠπείρου, καὶ ὁ μὲν Νεῖλος τὴν γῆν ἐπάρδει τοῖς Αἰγυπτίοις, ἡ δὲ ἐμὴ γλῶττα τὴν ἐκείνων ψυχήν· κἂν πύθοιο τῶν Περσῶν κἂν τῶν Αἰθιόπων, ἐροῦσιν ὡς ἴσασί με καὶ τεθαυμάκασι καὶ τεθήρανται. Καὶ νῦν δέ τις ἐκ τῶν Βαβυλῶνος ὁρίων ὥστε πιεῖν τῶν ἐμῶν ναμάτων ἀσχέτοις ταῖς προθυμίαις ἐλήλυθε.
>
> But I have made Celts and Arabs yield to me, and on account of my fame they regularly come down here to study even from the other continent; and while the Nile irrigates the land of the Egyptians, my speech irrigates their souls. And if you ask a Persian or an Ethiopian, they will say that they have known me and admired me and sought me out. And now someone from across the boundary of Babylon has come to drink from my springs through an insatiable desire of his.[42]

While peppered with his unmistakable pomp, this claim is certainly feasible, especially when considering the international cachet of Constantinople in the eleventh century. Possibly two international students of his were George and John of Petritzos, from Georgia.[43] In another letter to a friend, Psellos wrote that he was the one responsible for "adorning the City [i.e. Constantinople] with words," referring both to his activities as a

V. Vlyssidou (ed.), *The Empire in Crisis? Byzantium in the 11th Century, 1025–1081* (Athens: Hellenic National Research Foundation, 2003), 452–455; Paul Lemerle, *Cinq études sur le XIe siècle byzantin* (Paris: Presses Universitaires de France, 1977), 227–235; Markopoulos, "De la structure de l'école byzantine," 85–86; P. Agapitos, "Teachers, Pupils, and Imperial Power in Eleventh Century Byzantium" in Yun Lee Too and Niall Livingstone (eds.), *Pedagogy and Power: Rhetorics of Classical Learning* (Cambridge University Press, 1998), 170–191.

[42] Michael Psellos, *Epistula ad Michaelem Cerularium*, 96–103, ed. U. Criscuolo, *Epistola a Michele Cerulario*, Hellenica et byzantina neapolitana 15, reprint (Naples: Bibliopolis, 1990), 21–31; trans. Anthony Kaldellis and Ioannis Polemis, "Letter to the Patriarch Kyr Michael Keroullarios" in Anthony Kaldellis and Ioannis Polemis (eds.), *Psellos and the Patriarchs: Letters and Funeral Orations for Keroullarios, Leichoudes, and Xiphilinos* (South Bend, IN: University of Notre Dame, 2015), 37–48, at 41–42.

[43] On Psellos and his students, see Robert Volk, *Der medizinische Inhalt der Schriften des Michael Psellos* (Munich: Institut für Byzantinistik und neugriechische Philologie der Universität, 1990), 15–20. On his possible Georgian students, see B. Martin-Hisard, "Georgian Hagiography" in *The Ashgate Research Companion to Byzantine Hagiography*, vol. 1 (Burlington, VT: Ashgate, 2011), 285–298, esp. 288–289; L. Gigineishvili and G. Van Riel, "Ioane Petritsi: A Witness to Proclus' Works in the School of Psellus" in A.-Ph. Segonds and C. Steel (eds.), *Proclus et la théologie platonicienne* (Leuven University Press, 2000), 571–587.

teacher and to those of his pupils.[44] In his writings, Psellos described his pedagogical task as being to fully educate his students in everything vital, to make them into philosophers in his own mold.[45]

Furthermore, Psellos's students spanned the imperial and intellectual elite of the city, from the controversial John Italos to the future Emperor Michael VII, whom Psellos tutored at the behest of Emperor Constantine X Doukas. The *De omnifaria doctrina*, Psellos's compendium of natural and philosophical knowledge, including an important entry on the senses, was composed for the education of Michael VII. Psellos's reputation as a thinker, commentator, and educator came to be well known, and his name is cited on a list of Aristotle's works and his commentators in one thirteenth-century manuscript (Jerusalem, Taphou 106) as part of a sequence of illustrious antique and late antique commentators.[46] Thus, if Psellos began as a lonely philosopher (doubtful) that surely is not how he finished. Even if Psellos's erudition and talent were truly unique and not matched by any contemporary, he nevertheless inculcated those skills within his students and pupils, as well as future admirers such as Anna Komnene.[47] His considerable impact speaks to a wide promotion of philosophical thought in the Byzantine world, and many emulators and followers that continued in that tradition.

In the early twelfth century, following Psellos's influence and model, Anna Komnene assembled an elite circle of scholars, including Eustratios of Nicaea and Michael of Ephesus. In fourteenth-century Constantinople, particularly around the imperial court, intellectual life thrived, with figures as diverse as Nikephoros Blemmydes and his imperial pupil Theodore II Doukas Laskaris, Nikephoros Gregoras, Manuel Bryennios, Theodore Metochites, and Sophonias.[48] Throughout this volume, it will be seen

[44] "οἱ τὴν Πόλιν τοῖς λόγοις κοσμήσαντες." Michael Psellos, *Epistula*, 198, ed. Konstantinos Sathas, *Μεσαιωνικὴ Βιβλιοθήκη*, vol. v (Venice: Typois tou Chronou, 1876), 491–492.

[45] See Duffy, "Hellenic Philosophy in Byzantium," 152. See Michael Psellos, *Theologica*, 20.2–3, 89.85–86, 91.3, 76.11–12, ed. P. Gautier, *Theologica*, vol. I (Leipzig: Teubner, 1989), 76, 353, 356, 302.

[46] Duffy, "Hellenic Philosophy in Byzantium," 154. See P. Wendland, *Commentaria in Aristotelem Graeca*, vol. III.1 (Berlin: Akademie der Wissenschaften, 1901), xv-xix, esp. xviii.

[47] On Anna Komnene and Michael Psellos, see S. Linnér, "Psellus' *Chronographia* and the *Alexias*: Some Textual Parallels," *Byzantinische Zeitschrift* 76 (1983): 1–9; Stratis Papaioannou, "Anna Komnene's Will" in D. Sullivan, E. Fisher, and S. Papaioannou (eds.), *Byzantine Religious Culture: Studies in Honor of Alice-Mary Talbot* (Leiden: Brill, 2011), 99–121.

[48] On higher education in the thirteenth through fifteenth centuries, see C. N. Constantinides, *Higher Education in Byzantium in the Thirteenth and Early Fourteenth Centuries* (Nicosia: Cyprus Research Center, 1982); Sophia Mergiali, *L'enseignement et les lettrés pendant l'époque des Paléologues (1261–1453)* (Athens: Société des Amis du Peuple, Centre d'Études Byzantines, 1996). See also

how the theories of vision Psellos discussed and expounded were part of a larger transmission from late antiquity onward, across different kinds of audiences.

John of Damascus in the eighth century, Photios in the ninth century, and (the allegedly unique) Psellos in the eleventh century, despite the different genres and types of readers they were addressing, shared many clear and direct parallels. We will repeatedly encounter a shared body of sources, citations, and readings that suggest that the tradition Psellos represented was far less discontinuous than he may have wanted to admit. We see that tradition exhibited by the defenders of icons in the Iconoclastic period such as John of Damascus, Theodore of Stoudios, and Patriarch Nikephoros of Constantinople. They discussed and expounded a common pool of theories of vision and perception that are fundamental for us in comprehending how they thought images operated. Oftentimes, these writers make this explicitly clear. By explaining how human vision and perception worked, they went on to mold and contour the theology of the icon based on what they thought perception could do, and used this understanding of vision and perception to expand and amplify the icon's ability to create an illusion of presence through representation.

Methodology: Mining Down, Close Reading, and Translation

Given its aim, to consider the impact of theories of vision on the treatment of the icon, this book is centered on the Iconoclastic controversies and the period immediately after their resolution. Hence, I shall often begin a topic by turning to seventh-to-ninth-century sources, analyze them, and then explore how they engaged with their sources specifically and more generally with the classical and late antique heritage. From there, I shall turn to consider their influence upon later periods. Yet, my focus will always be on what we would term the early and middle Byzantine periods. The book's temporal scope, however, falls short of the fourteenth century, providing often only brief glimpses into how the trajectory continued. By doing so I acknowledge the different way vision operated in the context of Hesychasm. Later Byzantium merits its own separate study, attuned to the complexities of Hesychastic theology and nuanced theories of perception and representation.

Fevronia Nousia, *Byzantine Textbooks of the Palaeologan Period*, Studi e Testi, 505 (Vatican City: Biblioteca Apostolica Vaticana, 2016).

This volume explicitly resists a more traditional order – a chronologically linear survey of the classical theories of vision, Byzantium's manipulations of them, and then their later afterlives. Such a linear approach would pit the late antique and Byzantine worlds in a false dichotomy, suggesting a body of classical thought on one side and the mere rearrangement, excerption, and corruption of these earlier theories on the other. These theories of sight were in constant flux not just in Byzantium but in the classical period. My effort to ascribe a particular classical heritage to any one Byzantine theory is more for structure and clarification, not to judge or define it against its classical Ur-sources. Thus, as I have just said, throughout this volume, we shall begin with Byzantine authors from the period of iconoclasm, or immediately thereafter, a kind of triptych centered upon iconoclasm through which to gaze across both directions of the tradition.

By moving back in time to mine the Byzantine archives and to demonstrate the nestled complexities within their thinking, we will see a world of thinkers who read voraciously and widely, citing their sources in various ways, while developing their own arguments and research projects. The Byzantine texts discussed here shall reveal layers and complexities of Byzantine theories of sense perception. Some expositions are founded upon ancient theories of sight, and consequently seek to summarize the classical doxography. Some classical teachings were revised and refracted through the writings of the late antique commentators, early Christian thinkers, Church Fathers, contemporaries, and so on. Only after this long process did select theories of vision come to be staged and curated by our Byzantine authors, who, in attempting to support and promote their own intellectual projects and achieve specific rhetorical goals, produced a creative and intricate worldview of sensory experience, perception, and cognition.

Throughout the course of the book, we shall study a host of different texts, from late antique pagan novels to Christian homilies, ancient scholia to philosophical commentaries. These texts feature immense variety in content, style, audience, and delivery. They also encompassed a great variety of audiences in background and education. The summaries of ancient philosophy produced for an elite pupil in the highest inner circles of Constantinopolitan nobility are not interchangeable with a homily delivered before the congregation at Hagia Sophia on a major feast day. Nevertheless, these different texts participate in a shared intellectual tradition and have common themes of sight, tactility, and imagination. Rather than presuming that all the members of this eclectic assortment of sources are the same or comparable, the study seeks to reveal the congruities that emerge across texts written in very different contexts and intended for radically different groups. My goal is to show that even if we cannot delineate a single and specific intellectual tradition for the transmissions of these various texts, Byzantine writers were clearly tuned to broad, longstanding debates about the interrelations of vision, touch, and perception. This method privileges a varied breadth of sources, where a deep and clear lineage of ideas is not always, if ever, possible. In sketching out these networks, I wish, not to posit positivist lines of transmission, but to stress that the fact that at any given point in time two writers producing very different texts for diametrically opposed audiences, writing with great nuance and complexity, could still agree on the deeper matters of Byzantine visuality.

In its methodology, this book is primarily a form of close reading within a massive and expansive archive. I take close reading, eloquently described by Elizabeth Freeman, to be "the decision to unfold, slowly, a small number of imaginative texts rather than amass a weighty archive of or around texts, and to treat these texts and their formal work as theories of their own, interventions upon both critical theory and historiography."[49] My goal has been to actively resist the desire to make generalizing claims or to produce a singular survey, which would obscure the exciting variety of thought in the Byzantine world. This, however, has not always been possible; as a caveat, any such claims (either of a unitary narrative, or of the classical categorization of a Byzantine articulation of sight) that make their way into the book do so for narrative convenience, because of scholarly convention, or to provide some rhetorical scaffolding for the reader.

Oftentimes, I structure my arguments as responses to extant translations, particularly popular and highly accessible ones, such as those found in the Loeb Classical Library for the ancient world or those of Cyril Mango for the Byzantine world. As a rhetorical device, I clearly signpost the moments in which my response to a translation plays an important role. As I shall explain, using often-cited translations allows me to push back against the assumptions that have become commonplace in the field, and beyond it, and permits me to clearly explain how such texts have colored our views of Byzantine vision and how my interventions begin to open new spaces. I see influential translations, and my slow and careful critiques, as products of a crucial and critical subversive act. At times, I use the close

[49] Elizabeth Freeman, *Time Binds: Queer Temporalities, Queer Histories* (Durham, NC: Duke University Press, 2010), xvii.

study and response to translations as a narrative structure in this book. This is primarily the case in my close analysis of Photios's Homily 17 (in Part II), dedicated to carefully unfolding a single sentence of the homily a few words at a time. This will emphatically demonstrate the richness, complexity, and nestled knowledge of our Byzantine counterparts.

Structure

This book has been structured as a triptych. The first side-panel, Part I, provides readers with a representative survey of Byzantine writings on vision and its relationship to touch, culled from a variety of areas and sources spanning from the fourth to the fourteenth century, as well as a careful discussion of the classical and late antique texts on vision that grounded these discussions for the Byzantines. This provides a variety of discussion about sight's relationship to tactility in late antiquity and Byzantium. This section, by no means exhaustive, lays the conceptual and thematic foundation upon which the rest of this volume rests. This problem of sight and tactility, which is the proper focus of this book, is divided into three main concerns: (1) the centrality of medium and mediation in perception; (2) the tendency to compare, equate, and conflate tactility with sight, often to explain the operation of mental perception and thought, not bodily sensation; and (3) the cognitive commonalities and rhetorical transgressions of the senses, where each sense, while retaining its distinctive power, produces sensory experience that exceeds its own bounds.

The central panel is my close reading of Photios's Homily 17, where I flesh out the nuances of his vision theory and excavate his theory of cognitive perception, drawing our attention to the role of the imagination.[50] By closely unfolding Photios's litanies regarding visual and aural perception, I sketch out a five-stage process of perception. This begins with sensation, then moves on to an initial apprehension or grasp by the mind that names and categorizes the sensory data being passed along by the senses. This information is then offered up to the imagination, which is tasked with imagining or visualizing sense perception. At this stage the imagination synthesizes the various imprints of each of the senses. Here, the notion of the common sense appears, whereby the imagination

[50] For an earlier and abridged version of this section, see Roland Betancourt, "Why Sight Is Not Touch: Reconsidering the Tactility of Vision in Byzantium," *Dumbarton Oaks Papers* 70 (2016): 1–23.

recognizes that honey, for example, is not only amber, but also sticky, sweet, scented, and produces sound when stirred or poured. As the imagination processes sensory data that comes to it through, for example, hearing, it imagines those words as visual narratives as if those events or persons described are being reenacted.

Another stage is memory, which preserves knowledge and works with the imagination to enhance the common sense. When one merely sees fire, for example, the memory "tells" the imagination that it is hot to the touch. Thus, memory and imagination allow sight to conjure the illusion of touch, cued by memory from previous encounters. However, Photios also indicates another stage of perception sandwiched between the imagination and memory, namely the process of judgment and assent. Here, the mind debates what has come to it through the senses and the imagination, considering whether these impulses and imprints are to be considered valid and truthful. If they are deemed to be truthful and not some figment of the imagination or some deception of the senses, then one consents and assents to the truths which they impart. For Photios, this process is missing from vision. The actuality and vividness of images, which come readymade into the imagination, are simply passed on "effortlessly" to the memory. This is unlike hearing, in which the imagination must first convert words into sights and then the mind must judge these things "with sober attention" so as to ascertain that they have been properly visualized before they can be deposited in memory for future use and the preservation of knowledge.

This five-stage process (sensation, apprehension, imagination, judgment, and memory) has its precedents in the works of late antique thinkers such as Porphyry of Tyre and Nemesios of Emesa, and is fundamental to John of Damascus's and Theodore of Stoudios's articulations of the icon's effectiveness. By closely reading and mining Photios's sentence, as well as its sources and his fellow thinkers, Part II allows us to redefine the icon's effectiveness. We can stop putting undue emphasis on tactile sight, and distribute our attention to *all* the mental faculties that partake in perception. The icon has power because it serves as a starting point for the imagination to generate the illusion of the presence of the holy in the mind. This approach radically differs from one that relies upon haptic extramission and treats the icon as a sacred site in itself, which tactile sight can actually touch and grasp, as if the person represented were actually and palpably present before the physical eyes. Rather, the language of grasping or apprehending refers to the process of imagining the sacred, rather than actually touching it; since the ancient and medieval worlds describe cognitive comprehension through haptic language, just as we do when we say that we have "grasped" or "apprehended" something. The heuristic I advocate helps us to distinguish between sensation alone, that is, the senses' individual contact with the object, and perception proper, a process that encompasses sensation and the ensuing cognitive stages as the mind comprehends the sensible object.

The second side-panel, Part III, contains three chapters, each of which concludes a corresponding chapter of Part I. The aim of this last section is to demonstrate what is at stake in the distancing of sight from tactility. It shows the ramifications of my analysis of the senses in terms of (1) theories of representation and mediation in both literature and art; (2) literal forms of touch deployed within the context of icon veneration in ritual and the imagination; and (3) the broader context of cross-sensory metaphors and language found throughout the Byzantine world.[51] While these three chapters respond to the three initial chapters that began our investigation, they also enjoy their own internal cohesion, building upon one another to develop how we think of presence and the limits of representation in the icon when we elevate the faculties of the imagination and the so-called "common sense" to a privileged position in the history of Byzantine art and culture.

In the conclusion, I step back to consider the intellectual context in which Robert Nelson's proposition of haptic and extramissive sight emerged in Byzantine Studies. There I compare historiographic trends around the turn of the millennium with the Byzantines' own uneasy desires and expectations regarding images and their operations. This methodological and historiographic reflection enables me to close the investigation of this introduction by returning to Nelson's important words, which are the foundation, inspiration, and impetus of this work. My contribution, like his, is not a new, decisive theory of vision for Byzantine Studies, but rather a notable intervention in the current conversation. We should take our attention away from mere sensation and notions of presence in the icon and instead focus on the imaginative processes by which we perceive and through which we invest ourselves in the images that surround us. Like us, when we encounter a work of art denied our touch by a museum's case, so the Byzantines, when they saw, they felt that inalienable temptation to touch. Sight could not supplant touch; quite the contrary. My central proposition, in fact, is that Byzantine

[51] For an earlier and abridged rendition of some of these matters, see Roland Betancourt, "Tempted to Touch: Tactility, Ritual, and Mediation in Byzantine Visuality," *Speculum* 91:3 (2016): 660–689.

image theory to some extent had to actively deny the possibility of haptic sight, or else risk losing the effectiveness of the icon itself. Gazing upon the icon generated and sustained an impressive desire to touch and commune with the sacred, to imagine oneself to be in the presence of the holy, to touch and dialogue with their flesh. These are notions that are still very present to us. In popular culture we repeatedly deploy tropes of sexual desire. The gaze alone of the desired one does not satiate that pull but rather makes one ever more so intimately "tempted to touch," and so viscerally aware that one "can't touch this."

PART I

How Sight Is Not Touch

CHAPTER 1

The Medium of Sight

In seeking to qualify the degree of contact inherent in ancient and medieval theories of sight, it is necessary to understand how vision's medium was understood and how said medium mediated sensory perception. In this chapter, my goal is to consider the intervening medium or interval in the construction of classical theories of vision. I wish to show that whether extramissive or intromissive, most theories of vision emphasized the separation between the object of sight and the seer. Sight never allowed direct contact with the object of sight and was always mediated. That medium took on interestingly different features in the theories of Aristotle and Plato. What for Plato is an intervallic space where the fusion of rays from the seer and the object unite is for Aristotle reified as spatial, material. After exploring this issue of mediation in Aristotle and Plato, I will turn to consider the later iterations of these ideas through the work of Philo of Alexandria and a work dubiously attributed to Basil of Caesarea. We will see that these classical teachings endured with nuance, albeit often in passing and without clear articulation, in an unbroken tradition leading up to the later Byzantine periods. The chapter closes with a fourteenth-century thinker, Nikephoros Choumnos, who claimed to reconcile differences between Plato and Aristotle on explaining of the interval of sight.

Aristotle

In the mid eleventh century, Michael Psellos, presumably in his role as educator, provided his student reader with a well-curated composition of excerpts from Alexander of Aphrodisias's popular commentary on Aristotle's *On Sense and Sensible Objects*.[1] After summarizing the theories

[1] Alexander of Aphrodisias, *In librum de sensu commentarium*, ed. P. Wendland, *In librum de sensu commentarium*, Commentaria in Aristotelem Graeca 3.1 (Berlin: Reimer, 1901); trans. Alan Towey, *On Aristotle On Sense Perception* (London: Duckworth, 2000).

of Plato, Empedocles, and Democritus, and explaining Aristotle's theories of sensual perception at large, Psellos summarized Aristotle's theory of vision:

> Δόξα δὲ Ἀριστοτέλους περὶ τῆς αἰσθήσεως παρὰ τὰς τῶν ἄλλων αὕτη ἐστί: πάσχουσα γάρ, φησίν, ἡ ὄψις ὑπὸ τῶν ὁρατῶν αἰσθάνεται, ὥσπερ καὶ ἑκάστη τῶν ἄλλων αἰσθήσεων, ἀλλ' οὐχὶ ποιοῦσα καὶ ἐκπέμπουσα, πάσχουσα δὲ οὐ τῷ δέχεσθαι ἀπορρέοντά τινα ἀπὸ τῶν αἰσθητῶν, ἀλλὰ τῷ τὸ μεταξὺ τῆς τε ὄψεως καὶ τοῦ ὁρωμένου διαφανές, ὅταν κατ' ἐνέργειαν ᾖ τοιοῦτον (τοιοῦτον δέ ἐστι τὸ πεφωτισμένον), κινεῖσθαι ὑπὸ τῶν ὁρατῶν, τουτέστι τῶν χρωμάτων (κινητικὸν γὰρ τὸ χρῶμα τοῦ κατ' ἐνέργειαν διαφανοῦς). κινούμενον γὰρ καὶ διατιθέμενον τὸ κατ' ἐνέργειαν διαφανὲς ὑπὸ τῶν ὁρατῶν τὸ εἶδος αὐτοῦ διαδίδωσι τῇ κόρῃ, οὔσῃ καὶ αὐτῇ διαφανεῖ. καὶ οὕτως διὰ ταύτης δεχομένης διὰ τοῦ διαφανοῦς τὸ μεταξὺ εἶδος τοῦ ὁρωμένου καὶ μέχρι τοῦ πρώτου αἰσθητικοῦ διαδιδούσης αὐτὸ τῷ τὸν μεταξὺ πόρον τοιούτου σώματος εἶναι πλήρη τὸ ὁρᾶν γίνεσθαι: οὐ γὰρ ταῖς ἀπορροίαις, [ὡς ᾤοντο οἱ] πρὸ αὐτοῦ (ἁφὴ γὰρ οὕτω καὶ ἡ ὄψις ἔσται), ἀλλὰ τῇ τοῦ μεταξὺ τῆς ὄψεως καὶ {τῆς} τοῦ ὁρωμένου διαφανοῦς κινήσει ὑπὸ τῶν ὁρατῶν.
>
> The Aristotelian teaching on perception is this among other things. For, he says, sight perceives by being affected by the things seen, just as each of the other [senses] perceives, but not by making and sending out, and not by being affected by admitting what flows from the things perceived. Rather, the transparent (διαφανές), which is a medium between sight and the thing seen, when it is this [transparent] in actuality (and it is this in actuality when it has become light), is moved by the things seen, these being the colors (for color moves that which is transparent in actuality). For the transparent in actuality, being moved and arranged by the things seen, transmits its form to the pupil, which is also transparent. And hence vision comes about in this way as the form of the thing seen is received through the transparent medium and is transmitted as far as the primary sense by means of the intermediate passage which is to be full of this body. This is not caused by emissions, as those before him suppose (for on this view sight would also be touch), but because the transparent medium between that which sees and that which is seen is moved by the things seen.[2]

Here, Psellos clarifies how the views of Aristotle differed from those of his predecessors, who believed rays were either shot out from the eyes or passively received. Earlier in the text, Psellos suggested that Aristotle newly unified these positions. His theory of vision lacks emissions from either viewer or viewed, but posits that colors of an object are transmitted

[2] Michael Psellos, *Opuscula psychologica, theologica, daemonologica*, 8, ed. D. J. O'Meara, *Philosophica Minora*, vol. II (Leipzig: Teubner, 1989), 16–17; trans. Charles Barber, *Contesting the Logic of Painting: Art and Understanding in Eleventh-Century Byzantium* (Leiden: Brill, 2007), 94.

through an intervening medium. Color, being the proper sensible element of sight, affects this medium between the viewer and object, by causing alterations in said medium that ripple through it and thus passes the quality of colors on to the eye.

Aristotle terms this transparent medium the "diaphanous" (τὸ διαφανές), which is activated by light and by virtue of this activation is able to conduct colors to the eyes, allowing for the transmission of color through itself by virtue of the transparent's affective movement, not by the transmission of effluxes or emanations from viewers or things. As a result of this, Aristotle distances sight from touch since tactility cannot transmit color, and also because the transparent medium produces a necessary interval of separation between the object and the viewer as the primary actor in the act of visual perception.

Michael Psellos's text challenges quite concisely the haptic theory of extramission that has dominated the work of art historians since around the turn of the millennium.[3] However, this challenge comes not because of Aristotle's form of intromission, but rather because he insists on sight as a form of perception mediated by a third party between object and viewer, in this case the transparent. Uniquely, Charles Barber aptly views this passage as a pivotal source in Psellos's image theory, given that Psellos uses Aristotle to focus attention on this physical space between the viewer and the icon; in Barber's own words, the "medium that both separates and links the one seeing and the things seen."[4] Thus, the literal medium of the Byzantine image in this model is the intervallic space between the perception of the viewer and the image in the icon, a space that denies the instant gratification that haptic sight would allow.

If Psellos's passage is to be considered anything more than just a curiosity in the work of a uniquely and idiosyncratically educated figure, then it is necessary to understand the role of Aristotle's thinking and its transmission, critically important because it drew the attention of many late antique commentators, in particular John Philoponos and Alexander of Aphrodisias, whose works enjoyed immense popularity in intellectual

[3] See Suzannah Biernoff, *Sight and Embodiment in the Middle Ages* (New York: Palgrave Macmillan, 2002); Georgia Frank, *The Memory of the Eyes: Pilgrims to Living Saints in Christian Late Antiquity* (Berkeley: University of California Press, 2000); Jacqueline Jung, "The Tactile and the Visionary: Notes on the Place of Sculpture in the Medieval Religious Imagination" in C. Hourihane (ed.), *Looking Beyond: Visions, Dreams, and Insights in Medieval Art and History* (Princeton: Index of Christian Art, 2010), 203–40; Herbert L. Kessler, "Seeing" in *Seeing Medieval Art* (Orchard Park, NY: Broadview Press, 2004), 165–179; Christopher Lakey, "The Materiality of Light in Medieval Italian Painting," *English Language Notes* 53:2 (2015): 119–136.

[4] Barber, *Contesting the Logic of Painting*, 95.

circles up through the fourteenth century. As A. Mark Smith has argued, Aristotle was seen more as an explicator than a polemicist in his own right, thus his work was treated as a gateway to the classics.[5] When reading Plato and earlier writers, students and scholars alike would start with Aristotle.

Aristotle himself states repeatedly throughout his writing on vision that Empedocles, Democritus, and Plato were in error in believing in variations of both extramission and intromission, precisely because their views evidenced an absence or muddling of a medium for the senses. Aristotle writes:

> πάσχοντος γάρ τι τοῦ αἰσθητικοῦ γίνεται τὸ ὁρᾶν: ὑπ᾽ αὐτοῦ μὲν οὖν τοῦ ὁρωμένου χρώματος ἀδύνατον: λείπεται δὴ ὑπὸ τοῦ μεταξύ, ὥστ᾽ ἀναγκαῖόν τι εἶναι μεταξύ: κενοῦ δὲ γενομένου οὐχ ὅτι ἀκριβῶς, ἀλλ᾽ ὅλως οὐθὲν ὀφθήσεται.
>
> For vision occurs when the sensitive faculty is acted upon; as it cannot be acted upon by the actual colour which is seen, there only remains the medium to act on it, so that some medium must exist; in fact, if the intervening space were void, not merely would accurate vision be impossible, but nothing would be seen at all.[6]

That is, the medium was essential for vision. In the same passage, Aristotle clarifies that if one were to press a colored object to one's eye, one would be wholly unable to see it, given that an interval of difference is necessary between the object and the organ of sight. Hence, sight is not a species of touch. The same applies to sound and smell, which are both perceived through a medium. Only touch – and taste, which is a form of touch – does not require a medium because of the immediacy of those kinds of perception. However, Aristotle goes on to say that even this is misleading since touch is indeed mediated, but not because a medium acts upon us, but rather because we perceive tangible things "at the same time as the medium" (ἅμα τῷ μεταξύ).[7] Touch occurs because both the sense organ and the sensible thing are fused together, not because one was moved toward the other through a medium. For Aristotle, all sensation is mediated, but touch, due to its instantaneity, produces the illusion of unmediated contact. As he states in *De anima*:

[5] A. Mark Smith, *From Sight to Light: The Passage from Ancient to Modern Optics* (University of Chicago Press, 2014), 130–34.

[6] Aristotle, *De anima*, 419a18–22, ed. and trans. W. S. Hett, *On the Soul*, Loeb Classical Library 288 (Cambridge, MA: Harvard University Press, 1957), 106–107.

[7] Aristotle, *De anima*, 423a23–423b27, at 423b15, ed. and trans. Hett, 130–135, at 132–133.

ἀλλὰ διαφέρει τὸ ἁπτὸν τῶν ὁρατῶν καὶ τῶν ψοφητικῶν, ὅτι ἐκείνων μὲν αἰσθανόμεθα τῷ τὸ μεταξὺ ποιεῖν τι ἡμᾶς, τῶν δὲ ἁπτῶν οὐχ ὑπὸ τοῦ μεταξὺ ἀλλ' ἅμα τῷ μεταξύ, ὥσπερ ὁ δι' ἀσπίδος πληγείς· οὐ γὰρ ἡ ἀσπὶς πληγεῖσα ἐπάταξεν, ἀλλ' ἅμ' ἄμφω συνέβη πληγῆναι.

But a tangible thing differs from visible or audible things. We perceive the latter because some medium acts on us, but we perceive tangible things not by a medium, but at the same time as the medium, like a man wounded through his shield; for it is not the stricken shield that struck him, but both he and the shield were struck simultaneously.[8]

Aristotle refutes the notion that all senses are based on touch, precisely by saying that touch itself is a mediated form of contact, one that occurs simultaneously. He inverts his logic later in the work:

πάντα δὲ τῷ δι' ἑτέρου αἰσθάνεσθαι ποιεῖ τὴν αἴσθησιν καὶ διὰ τῶν μεταξύ· ἡ δ' ἁφὴ τῷ αὐτῶν ἅπτεσθαί ἐστιν, διὸ καὶ τοὔνομα τοῦτο ἔχει. καίτοι καὶ τὰ ἄλλα αἰσθητήρια ἁφῇ αἰσθάνεται, ἀλλὰ δι' ἑτέρου· αὕτη δὲ δοκεῖ μόνη δι' αὑτῆς.

[The senses] all produce sensation by means of something else, that is through a medium. But touch occurs by direct contact with its objects, and that is why it has its name. The other sense organs perceive by contact too, but through something else; touch alone seems to perceive through itself.[9]

In this passage, we see some intricate aerobatics as he grapples with the degree of mediation offered by touch. Here, he is attempting to clarify precisely the muddled language of touch. First, he states that all sensation must occur through a medium, but that the so-called "direct contact (ἅπτεσθαί)" with objects is what makes touch unique. This is not to say that this direct contact is unmediated, but rather in Aristotle's words, that it *seems* to perceive through itself (αὕτη δὲ δοκεῖ μόνη δι' αὑτῆς). Then, he reasons that the other senses also apprehend things through a form of "contact" (ἁφῇ). Yet this so-called "touch" is more accurately understood as a form of mere perceptual contact that is mediated through something else (δι' ἑτέρου). While certainly touch can occur through a medium, such as the touch felt through a walking stick, touch perceives through itself (δι' αὑτῆς). The other senses, however, perceive through something else (δι' ἑτέρου), which is to say that they always rely on an external medium that is fundamentally different to the organ of sensation itself. The

8 Aristotle, *De anima*, 423b11–18, ed. and trans. Hett, 132–133.
9 Aristotle, *De anima*, 435a16–20, ed. and modified trans. Hett, 198–201.

different senses are particular to themselves, and to conflate them would be unreasonable given the efforts Aristotle makes to avoid such conflations.

Plato

Despite being portrayed by modern literature as a form of haptic sight, extramission was also grounded by Plato on a similar notion of the medium. Plato's theory of extramission postulated vision as occurring when rays projected from both the eye and the object coalesced in the intervallic space between them. As Plato writes in the *Timaeus*:

> τῶν δὲ ὀργάνων πρῶτον μὲν φωσφόρα ξυνετεκτήναντο ὄμματα, τοιᾷδε ἐνδήσαντες αἰτίᾳ. τοῦ πυρὸς ὅσον τὸ μὲν κάειν οὐκ ἔσχε, τὸ δὲ παρέχειν φῶς ἥμερον, οἰκεῖον ἑκάστης ἡμέρας, σῶμα ἐμηχανήσαντο γίγνεσθαι. τὸ γὰρ ἐντὸς ἡμῶν ἀδελφὸν ὂν τούτου πῦρ εἰλικρινὲς ἐποίησαν διὰ τῶν ὀμμάτων ῥεῖν λεῖον καὶ πυκνὸν ὅλον μέν, μάλιστα δὲ τὸ μέσον ξυμπιλήσαντες τῶν ὀμμάτων, ὥστε τὸ μὲν ἄλλο ὅσον παχύτερον στέγειν πᾶν, τὸ τοιοῦτον δὲ μόνον αὐτὸ καθαρὸν διηθεῖν. ὅταν οὖν μεθημερινὸν ᾖ φῶς περὶ τὸ τῆς ὄψεως ῥεῦμα, τότ᾽ ἐκπῖπτον ὅμοιον πρὸς ὅμοιον, ξυμπαγὲς γενόμενον, ἓν σῶμα οἰκειωθὲν ξυνέστη κατὰ τὴν τῶν ὀμμάτων εὐθυωρίαν, ὅπῃπερ ἂν ἀντερείδῃ τὸ προσπῖπτον ἔνδοθεν πρὸς ὃ τῶν ἔξω ξυνέπεσεν. ὁμοιοπαθὲς δὴ δι᾽ ὁμοιότητα πᾶν γενόμενον, ὅτου τε ἂν αὐτό ποτε ἐφάπτηται καὶ ὃ ἂν ἄλλο ἐκείνου, τούτων τὰς κινήσεις διαδιδὸν εἰς ἅπαν τὸ σῶμα μέχρι τῆς ψυχῆς αἴσθησιν παρέσχετο ταύτην, ᾗ δὴ ὁρᾶν φαμέν.
>
> And of the organs they constructed first light-bearing eyes, and these they fixed in the face for the reason following. They contrived that all such fire as had the property not of burning but of giving a mild light should form a body akin to the light of every day. For they caused the pure fire within us, which is akin to that of day, to flow through the eyes in a smooth and dense stream; and they compressed the whole substance, and especially the centre, of the eyes, so that they occluded all other fire that was coarser and allowed only this pure kind of fire to filter through. So whenever the stream of vision is surrounded by mid-day light, it flows out like unto like, and coalescing therewith it forms one kindred substance along the path of the eyes' vision, wheresoever the fire which streams from within meets with an obstructing object without. And this substance, having all become similar in its properties because of its similar nature, distributes the motions of every object it contacts, or whereby it is contacted, throughout all the body even unto the Soul, and brings about that sensation which we now term "seeing."[10]

[10] Plato, *Timaeus*, 45B–D, ed. and trans. R. G. Bury, *Timaeus*, Loeb Classical Library 234 (Cambridge, MA: Harvard University Press, 1929), 100–103.

This passage presents sight as consisting of a commingled flow of rays, compounded in the space betwixt seer and seen to enable sight. The excerpt makes Platonic sight akin to touch itself, for as Plato explicitly states, the visual stream passes the motions of all the objects it "touches" (ἐφάπτηται) or that touch it, and that the fire streaming from the eye meets (ξυνέπεσεν) the objects outside. The visual stream here is not merely the fiery effluxes of the eyes that "flow out" or "stream out" (ἐκπῖπτον or προσπῖπτον), but rather the substance coalesced from the fiery emissions and the outside light, which, when united, produce an activated medium through which the "motions" (κινήσεις) of the objects can be sent back to the eyes. These motions indicate that rather than allowing for a proximate, unmediated contact with the sensible object, sight is moved by those objects through the coalescing of external light and the fire of the eyes.

Plato's broader account in the *Timaeus*, however, offers further problems, namely, that sight occurs not only through the fiery effluxes of the eyes fusing with the outside light, but also through that coalesced medium's interaction with the object's own emissions. Consider Plato's discussion of the mirror, which comes soon after the passage cited above. The mirror encapsulates the intervallic space between viewer and object as a plane of reflection whereby the two rays (from viewer and object) meet. There, Plato tells us:

> ἐκ γὰρ τῆς ἐντὸς ἐκτός τε τοῦ πυρὸς ἑκατέρου κοινωνίας ἀλλήλοις, ἑνός τε αὖ περὶ τὴν λειότητα ἑκάστοτε γενομένου καὶ πολλαχῇ μεταρρυθμισθέντος, πάντα τὰ τοιαῦτα ἐξ ἀνάγκης ἐμφαίνεται, τοῦ περὶ τὸ πρόσωπον πυρὸς τῷ περὶ τὴν ὄψιν πυρὶ περὶ τὸ λεῖον καὶ λαμπρὸν συμπαγοῦς γιγνομένου.
>
> It is from the communion with each other of the inner and outer fires, every time that they unite on the smooth surface [of a mirror] and are variously reflected, that all such reflections necessarily result, owing to the fire of the reflected face coalescing with the fire of vision.[11]

There is an important problem here. In his earlier account, Plato states that "the fire of the reflected face" (τοῦ περὶ τὸ πρόσωπον πυρός) coalesces with "the fire of the vision" (τῷ περὶ τὴν ὄψιν πυρί). While the optical fire is to be expected, the fire of the face is unaccounted for, since earlier Plato made no mention of a fiery emission coming from perceptible objects. However, this second excerpt implies that an efflux comes from visible things to become coalesced (συμπαγοῦς γιγνομένου) with the visual fire,

[11] Plato, *Timaeus*, 46A–B, ed. and modified trans. Bury, 102–105.

just as the visual fire coalesced with the external light. It is through this partnership or "communion" (κοινωνίας) of effluxes, a term that will be reiterated in Galen, that vision comes about.

Plato articulates what is missing from these earlier and cryptic passages only much later in the *Timaeus*, when he discusses the various senses and focuses on the issue of colors (χρόας), which are the sensible property responsible for vision. There, Plato finally supplies his optics the missing piece:

> φλόγα τῶν σωμάτων ἑκάστων ἀπορρέουσαν, ὄψει ξύμμετρα μόρια ἔχουσαν πρὸς αἴσθησιν· ὄψεως δὲ ἐν τοῖς πρόσθεν αὐτῶν περὶ τῶν αἰτίων τῆς γενέσεως ἐρρήθη.
>
> [Color] consists of a flame which emanates from the several bodies, and possesses particles so proportioned to the visual [stream] as to produce sensation; and as regards the visual [stream], we have already stated merely the causes which produced it.[12]

This resolves the conundrum regarding the emanations from the face that reflect in the mirror. Those emanations are shown to be the colors of the face, which are characterized as a flame (φλόγα) and thus consubstantial with the fiery effluxes of the eyes. This flame of color emanates (ἀπορρέουσαν) from the objects, a matter to which we shall return in future chapters. Because that emanation is commensurate with the visual efflux, both unite to produce sight. Following this excerpt, Plato goes on to explain various theories as to how such flames emerge and how they relate to the visual stream. Yet before doing so, he notes that the details of the visual stream have already been discussed, as if acknowledging that the earlier account on sight was partial, discussing only the effluxes from the eyes and failing to articulate how rays emanate from objects, despite the fact that the mirror example clearly implies such a phenomenon.

This model is what later came to be referred to as the "Platonic united rays" (Πλατωνικὴ συναύγεια) in the doxography of Aetios of Antioch.[13] Early on, in the fourth or third century BCE, Theophrastus neatly summarized Plato's theories in his commentary *De sensu et sensibilibus*:

[12] Plato, *Timaeus*, 67C–D, ed. and modified trans. Bury, 172–173.

[13] This is the term adopted by the *De placita philosophorum* of Aetios of Antioch (Pseudo-Plutarch) written in the first or second century. This text enjoyed great popularity in the Byzantine world, as we shall see throughout this volume, and is actively quoted across the writings of Michael Psellos. See Pseudo-Plutarch (Aetios of Antioch), *De placita philosophorum*, IV.13, ed. Jürgen Mau, *Plutarchi moralia*, vol. V.2.1 (Leipzig: Teubner, 1971), 123. Cf. Michael Psellos, *De omnifaria doctrina*, 108, ed. L. G. Westernik, *De omnifaria doctrina* (Nijmegen: Centrale Drukkerij, 1948), 60.

καὶ τὴν μὲν ὄψιν ποιεῖ πυρός (διὸ καὶ τὸ χρῶμα φλόγα τιν' ἀπὸ τῶν σωμάτων σύμμετρα μόρια τῇ ὄψει ἔχουσαν), ὡς ἀπορροῆς τε γινομένης καὶ δέον συναρμόττειν ἀλλήλοις ἐξιοῦσαν μέχρι τινὸς συμφύεσθαι τῇ ἀπορροῇ καὶ οὕτως ὁρᾶν ἡμᾶς· ὥσπερ ἂν εἰς τὸ μέσον τιθεὶς τὴν ἑαυτοῦ δόξαν τῶν τε φασκόντων προσπίπτειν τὴν ὄψιν καὶ τῶν φέρεσθαι πρὸς αὐτὴν ἀπὸ τῶν ὁρατῶν.

Vision he makes of fire (and this is why he regards color also as a flame given off from bodies, having particles commensurate with the visual [stream]) on the ground that an emanation occurs and they [the emanation and the visual stream] must unite with each other; [hence, the visual stream] passes out of the eye for a certain distance and coalesces with the effluence, and that is how we see. Therefore, his view may be said to lie midway between the theories of those who hold that vision falls upon [perceptible objects] and of those who hold that something is borne from visible objects to [the organ of sight].[14]

Theophrastus, as an early and popular commentator, set the standard for the conceptualization of Plato's visual theory, stating unequivocally that Plato's model was midway between extramission and intromission. His correct reading of Plato is that vision emerges from two effluxes, denoted by the same term (ἀπορροῆς, ἀπορροῇ), which come together, and which is resolutely different from theories that saw vision as sending out something that "falls upon" (προσπίπτειν) visual objects or wherein something is brought from those objects to the eye alone (φέρεσθαι πρὸς αὐτὴν ἀπὸ τῶν ὁρατῶν). Popular commentators like Alexander of Aphrodisias and John Philoponos embraced this synthesis, using Aristotle to develop and enhance Plato's fragmented explanation. This syncretic understanding of Plato's hybrid theory of intromission and extramission may well have accounted for the popularity of his theory of vision above that of others in Byzantium, but it had an amorphous character. Because of Plato's disconnected inclusion of both extramissive and intromissive features, writers could stress one side or another to articulate their own ideas regarding the agency of the object or viewer in the process of sight, or to make a point regarding the operation of the soul and its relationship with material bodies.

[14] Theophrastus, *De sensu et sensibilibus*, 5–6, ed. and modified trans. G. M. Stratton, *Theophrastus and the Greek Physiological Psychology before Aristotle* (New York: Macmillan, 1927), 68–71. See also David Sedley, "Empedocles' Theory of Vision and Theophrastus' *De sensibus*" in William W. Fortenbaugh and Dimitri Gutas (eds.), *Theophrastus: His Psychological, Doxographical, and Scientific Writings* (New Brunswick, NJ: Transaction Publishers, 1992), 20–31, esp. 30.

Platonic Afterlives, 1: Philo of Alexandria

Plato's theory lived on through late antiquity. Early on in the first century, the Jewish philosopher Philo of Alexandria, whose work had a significant impact on the Church Fathers, used it several times.[15] In one such instance, Philo writes about the power of the eyes, contrasting it to the timidity of the ears, which are seen as hesitant in apprehending their perceptibles since they wait to be acted upon rather than actively reaching out to them. Philo writes:

> βραδύτερα δέ πως καὶ θηλύτερα ὦτα ὀφθαλμῶν ἐπὶ τὰ ὁρατὰ φθανόντων ὑπὸ εὐτολμίας καὶ οὐκ ἀναμενόντων, ἄχρις ἂν ἐκεῖνα κινήσῃ, προϋπαντιαζόντων δὲ καὶ ἀντικινῆσαι γλιχομένων.
>
> The ears are more sluggish and timid than the eyes since the eyes anticipate the visible thing with boldness and do not delay, moving up to those things, encountering them in advance, and striving to act upon them instead.[16]

Gerhard Delling, Maren Niehoff, and most recently Rachel Neis have aptly noted that Philo's theory of perception here is resolutely Platonic, but they all read it as implying haptic sight, pointing to the use of "reaching out" (φθανόντων).[17] But if we do take for granted that Philo is indebted to Plato, then this passage becomes more nuanced. It certainly depicts the active agency of sight, and its swift movements in attempting to comprehend the sensible object, but it does not make sight tactile. Philo's language meticulously indicates a mirrored movement of perceiver and perceptible since the external movement of the sensibles (κινήσῃ) is precisely echoed in the case of sight through a countermovement (ἀντικινῆσαι), which not only anticipates (φθανόντων) but also advances to meet (προϋπαντιαζόντων) the object. While some aspects of Philo's understanding are unclear, it seems that he structures optical agency upon the model of united rays, where the rays are always rooted on an action and counteraction that are mutual and immediate. Thus, we can appreciate how the eye echoes the agency of the object. It sends out its rays, and stages its own efflux as a countermovement to the motions of the object.

[15] See David T. Runia, *Philo and the Church Fathers: A Collection of Papers* (Leiden: Brill, 1995).

[16] Philo of Alexandria, *De Abrahamo*, 149–151, ed. and modified trans. F. H. Colson, *Philo*, vol. VI, Loeb Classical Library 289 (Cambridge, MA: Harvard University Press, 1935), 76–77.

[17] See Maren Niehoff, *Philo on Jewish Identity and Culture* (Tübingen: Mohr Siebeck, 2001), 197–199; Gerhard Delling, "The 'One Who Sees God' in Philo" in Frederick E. Greenspahn, Earle Hilgert, and Burton L. Mack (eds.), *Nourished with Peace: Studies in Hellenistic Judaism in Memory of Samuel Sandmel* (Chico, CA: Scholars Press, 1984), 27–41. See also Rachel Neis, *The Sense of Sight in Rabbinic Culture* (Cambridge University Press, 2013), esp. 174–181.

Elsewhere, however, Philo qualifies this "bold" agency of the eyes, choosing to stress instead the mediation of the cognitive faculties in perception in order to demonstrate how sensation can lead one astray. Philo explains:

> τυφλὸν γὰρ φύσει ἡ αἴσθησις ἅτε ἄλογος οὖσα, ἐπεὶ τὸ λογικὸν ἐξομματοῦται· παρὸ καὶ μόνῳ τούτῳ τὰ πράγματα καταλαμβάνομεν, αἰσθήσει δ' οὐκέτι, μόνα γὰρ τὰ σώματα φαντασιούμεθα δι' αἰσθήσεως.
>
> For sense by itself is a blind thing, inasmuch as it is irrational, for it is the reasoning faculty that confers sight. Accordingly it is with the reason only that we apprehend matters; sense does not carry us so far; for by means of sense we gain impressions only of the material forms of things.[18]

This alteration in Philo's tone demonstrates how malleable sensory perception was for ancient writers, who by focusing on different aspects or facets of sight could either exalt the venerable powers of vision, or condemn their puerile or earthly nature. Here Philo attacks the excesses of pleasure and its sway through sensual experience, demonstrating how the senses, being inherently irrational, must be governed through reason. Reason apprehends (καταλαμβάνομεν) matters, but senses can only gain the impressions or images (φαντασιούμεθα) of material forms. So also in the same treatise, Philo emphasizes the importance of an individual's will in the face of temptation, deploying the relationship between the senses and the mind to explain moral responsibility. Philo writes:

> ἅπαντα αἴσθησις μετὰ τοῦ ἡμετέρου νοῦ γνωρίζει καὶ ἅμα αὐτῷ· οἷον ἡ ὅρασις ἅμα τῷ νῷ ἐπιβάλλει τῷ ὁρατῷ· εἶδέ τε γὰρ ὁ ὀφθαλμὸς τὸ σῶμα καὶ εὐθὺς ὁ νοῦς κατέλαβε τὸ ὁραθέν, ὅτι μέλαν ἢ λευκὸν ἢ ὠχρὸν ἢ φοινικοῦν ἢ τρίγωνον ἢ τετράγωνον ἢ στρογγύλον ἢ τὰ ἄλλα χρώματά τε καὶ σχήματα· καὶ πάλιν ἡ ἀκοὴ ἐπλήχθη ὑπὸ τῆς φωνῆς καὶ μετ' αὐτῆς ὁ νοῦς· τεκμήριον δέ, ἔκρινεν εὐθὺς τὴν φωνήν, ὅτι λεπτὴ ἢ μέγεθος ἔχουσα ἢ ἐμμελὴς καὶ εὔρυθμος καὶ πάλιν εἰ ἐκμελής τε καὶ οὐκ ἐναρμόνιος· καὶ ταὐτὸν ἐπὶ τῶν ἄλλων αἰσθήσεων εὑρίσκεται.
>
> Sense becomes aware of all things with our mind and simultaneously with it. For instance, vision falls upon both [the object] of the sight and of the mind; for the eye saw the [visible] body, and at once the mind took in the thing that had been seen, that it was black or white or yellow or red or triangular or square or round, or some other color or shape. Again the hearing and the mind with it was struck by the sound: in proof that it did, it

[18] Philo of Alexandria, *Legum allegoriarum*, III.108, ed. and trans. F. H. Colson and G. H. Whitaker, *Philo*, vol. 1, Loeb Classical Library 226 (Cambridge, MA: Harvard University Press, 1929), 372–375.

immediately judged that the sound was weak or loud, tuneful and rhythmical, and on the other hand whether it is out of tune and a discord. We find the same thing in the case of the other senses.[19]

Clearly, Philo retains certain elements of our first example, such as the immediacy of perception and the active language of seizing (καταλαμβάνομεν, in the former, and κατέλαβε, in the latter), which in both these latter examples denotes the activity of the mind in comprehending the sensory data of perception being passed along to it by the senses. The haptic language here describes the cognitive apprehension of the mind, and matches the fact that vision "falls upon" (ἐπιβάλλει) both object and mind, just as the mind is vividly "struck" (ἐπλήχθη) by sound as well. The focus on cognition removes the eye from the center of perception, relocating haptic language onto the mental processes of perceptual apprehension. This demonstrates how quickly vision transmits sensory imprints from the eye on to the intellectual faculty. None of this diverges from the Platonic model of united rays; rather it underscores different sequences of vision for varying rhetorical effects.

In the first of these three examples, Philo stresses the power of hearing and sight in the perception of loftier matters, such as philosophical thought and the realm of the divine; in the second and third examples, Philo writes against godless pleasures, which maim and pervert the operation of the senses and the soul. In the second example, sight is praised as an active agent and countermovement, in contrast to the passivity of the other senses, and treated as the highest of the senses for praiseworthy deeds. In the third case, Philo qualifies sensory experience and so constructs vision differently. This third example does not disagree with or divert from the Platonic model in any noticeable way; it simply focuses on the perceptual processes that follow sensation. This reminds the reader that the senses will always be mediated by the cognitive functions, no matter how active or dynamic a given sensory organ might be. This pertains not to sight or sensation but to the cognitive side of perception.

Platonic Afterlives, II: 'Basil'

In the late fourth century, Christian thinkers used Plato's theory of vision to consider the nature of man and the perception of the divine. Take for instance the second of two homilies on the origin of humanity, which

[19] Philo of Alexandria, *Legum allegoriarum*, III.57–58, ed. and modified trans. Colson and Whitaker, 338–339.

complement Basil's *Hexameron* and are thus referred to as the tenth and eleventh homilies of that work, although they are most likely by an unknown author.[20] In the second homily, the writer provides an eloquent thesis on vision that adopts the Platonic theory of united rays, where effluxes pour out from each of the eyes, converge over the nose, and then unite with that coming from the object seen. The author, reflecting on the senses and their organs, writes:

> Οὐκ ἤρκει ἐπ᾽ ἐμοὶ εἷς; Ἀλλὰ δύο ὀφθαλμοὶ ἀλλήλων διάδοχοι, ἵνα ἡ τοῦ ἑνὸς διάλυσις τὴν τοῦ ἑτέρου παραμυθίαν ἔχῃ. Ἔπειτα καὶ ἀπὸ ἑνὸς προερχόμεν τὸ ὁρατικὸν ἀσθενέστερον· ἐκ δὲ δύο πηγῶν συμβαλλόμενος ὁ ὁλκὸς εὐτονώτερος γίνεται. Προσϊοῦσα γὰρ ἡ ὄψις, ὁμοῦ μὲν προσαναπαύεται τῷ διατειχίσματι τῆς ῥινός, ὁμοῦ δὲ προκύπτουσα ἑνοῦται. Οἷον γάρ τις ὀχετός, ἀφ᾽ ἑκατέρας τῆς ὄψεως πρόεισιν ἐντεῦθεν τοῦτο τὸ ὁρατικόν, κἀκεῖθεν ἐκεῖνο· πόρρωθεν δὲ τῇ συμβολῇ ἑνοῦται· ἑνούμενον δὲ ἰσχυρότερον γίνεται. Τίς ἡ ἀπόδειξις ὅτι ἑνοῦται; Οὐχ ὁρᾷς τοὺς γέροντας ὅτι πρὸς τὰ μὲν ἐγγὺς οὐ βλέπουσιν; Ἐπειδὴ γὰρ ἀσθενές ἐστι τὸ ὁρατικὸν σχιζόμενον, τὸ ἐγγὺς οὐχ ὁρᾷ· ὅπου δὲ ἡ συμβολὴ τῆς ὄψεως, οἷον ἐπὶ πλημμύρας τινὸς κατὰ τὴν ὄψιν γινομένης, εὐτονωτέρα γίνεται τῶν αἰσθητῶν ἡ ἀνάληψις.
>
> Why was one eye not enough for me? Two eyes take each other's place, that the failure of one may have relief from the other. Then indeed the vision coming forth from one is weaker; but the stream converging from two sources becomes more vigorous. For the vision is added, it rests together with the support of the nose, and peering out together, unites itself. For like some current, what is seen comes forth from each visible object, this from here, that from there; and coming forward it unites from what converges.

[20] The homilies are in some manuscripts unattributed, and in others attributed to Basil of Caesarea or Gregory of Nyssa. The attribution to the latter is generally disregarded, given that there are notable discrepancies between the views espoused in these homilies and those expressed by Gregory. The general consensus is that the homilies are in keeping with Basil's theological anthropology, but that they were most likely not directly composed or edited by him for circulation. In other words, they are either the writings of a close follower of Basil based on his homilies and/or notes or by Basil himself but written late in life and left unfinished, given that they do not match Basil's other works stylistically. On this matter, see Alexis Smets and Michel van Esbroeck, *Sur l'origine de l'homme (Hom. X et XI de l'Hexaéméron)*, Sources chrétiennes 160 (Paris: Éditions du Cerf, 1970), 13–26. See also Emmanuel Amand de Mendieta, "Les deux homélies sur la création de l'homme que les manuscrits attribuent à Basil Césarée ou à Grégoire de Nysse: Le problème de leur rédaction," *Zetesis: Album amicorum door vrienden en collega's aangeboden aan Prof. Dr. E. de Strycker ter gelegenheid van zijn 65e verjaardag* (Antwerp: De Nederlandsche Boekhandel, 1973), 695–716; Emmanuel Amand de Mendieta, "Les neuf homélies de Basile de Césarée sur l'Hexaéméron," *Byzantion* 48 (1978): 337–368, esp. 339; Maximos Aghiorgoussis, "Applications of the Theme 'Eikon Theou' (Image of God) according to Saint Basil the Great," *Greek Orthodox Theological Review* 21:3 (1976): 265–288, esp. 272 n. 28; Philip Rousseau, "Human Nature and Its Material Setting in Basil of Caesarea's Sermons on Creation," *Heythrop Journal* 49 (2008): 222–239, esp. 222–223.

> And what is united becomes stronger. What is the proof that it unites? Do you not see that the elderly do not see things nearby? For the vision, being divided, is weak, it does not see what is near; but where there is a convergence of vision, as if some flood of vision occurs, the grasp of things perceived becomes more vigorous.[21]

In this model, the author accounts for the union of the rays coming not just from object and seer but also from both of the seer's eyes. He places himself within the extramissionist side of the Platonic model, where the eye is an active participant in sight, but so is the object that issues its own rays. Here, sight occurs because both the viewer and the object have actively contributed to the act of perception.

Yet beyond homologies of united rays, the author's Platonic inclination is also evidenced by the similar role played by the thought experiment of the mirror, just as it does in the *Timaeus*. In the opening passage of the first homily on the human condition, the author reflects on the function of sight as an allegory of the mental faculties, writing:

> Ὥσπερ γὰρ οἱ ὀφθαλμοὶ ἡμῶν τὰ ἔξω βλέποντες, ἑαυτοὺς οὐκ ὁρῶσιν, ἐὰν μήπου λείου τινὸς ἅψωνται στερεοῦ, κἀκεῖθεν ἀνακλασθεῖσα ἡ ὄψις ὥσπερ ἀπὸ παλιρροίας ὁρᾶν αὑτοὺς ποιήσῃ τὰ ἑαυτῶν κατόπιν· οὕτω καὶ ὁ νοῦς ὁ ἡμέτερος ἄλλως ἑαυτὸν οὐ βλέπει, ἐὰν μὴ ταῖς Γραφαῖς ἐγκύψῃ. Τὸ γὰρ ἐνταῦθα φῶς ἀνακλώμενον τοῦ καθορᾶσθαι ἕκαστον ἡμῶν αἴτιον γίνεται.
>
> For just as our eyes see external things but do not see themselves except when they encounter (ἅψωνται) something smooth and hard, then the image reflected as if by reflux makes them see things that are behind them; so also our mind does not see itself otherwise than by examining the Scriptures. For the light reflected there becomes the cause of vision for each of us.[22]

Just as the Scriptures allow for their readers to contemplate themselves through the stories of human action and deeds, the mirror allows sight to contemplate the seer through the reflection that occurs upon the mirror's shiny surface. However, the author presents us with a difficult passage in its passing statement on the occurrence of mirrored sight. The haptic term "encounter" (ἅψωνται), from which we get the term "haptic" itself, details the encounter between sight and the mirror, thus suggesting that the

[21] Basil of Caesarea, *Hexameron*, XI.16, ed. Smets and van Esbroeck, *Sur l'origine de l'homme*, 270–273; trans. Nonna Verna Harrison, *On the Human Condition* (Yonkers, NY: St. Vladimir's Seminary Press, 2005), 62.

[22] Basil of Caesarea, *Hexameron*, X.1, ed. Smets and van Esbroeck, 166–168; modified trans. Harrison, 31.

optical effluxes literally fasten or join the object seen. Yet, the neat and eloquently argued theory of vision in the second homily qualifies that judgment. The verb cannot be read simply as denoting tactile contact between seer and object; rather it describes the act of contact between sensation and the sensible object. In Plato's theory, certainly there is contact, as in all theories of perception, otherwise it would be impossible to perceive external things. But this contact is not reliant on touch because it is mediated by the fusion of optical rays.

Here, the line "the image reflected as if by reflux" (ἀνακλασθεῖσα ἡ ὄψις ὥσπερ ἀπὸ παλιρροίας) demonstrates the author's indebtedness to Plato. Were we to read ἡ ὄψις as vision or sight itself, we might understand that the rays that have shot out from the eyes are here bending back to grasp not the object but the image of the viewer's countenance. Yet, this reading would be fallacious, since in the popularized notion of extramission the rays always reach out and bound back to the viewer, carrying with them the tactile traces of the object's touch. The mirror is crucial because it subverts the notion that any particle or ray comes from the eyes and bounces back from the object. If this were true, then the vision of one's face in a mirror might be impossible, those tactile rays having accurately touched and transmitted the essence of the mirror itself rather than a reflection. What is being reflected is not vision or its rays but rather the perceptible rays coming off the face, which are now bouncing off the mirror through a form of reflux. As these rays bounce off the reflective surface, they encounter the viewer's optical rays and fuse to produce sight.

The nuanced language of this "reflux" (παλιρροίας) is similar to other precise terms, such as *aporroe*, to be discussed later, that describe effluxes coming from the objects seen. The root, *rhoia*, "flow," with its prefix *pali-*, "again," precisely describes an efflux that turns back or backwards on itself. That is, while literally denoting this act of reflection, *palirroia* suggests a flowing back and forth. It is often used in Greek literature to describe the ebbs and flows of the ocean and the tides, as in Ptolemy's *Apotelesmatica*, Cassius Dio's *Historiae Romanae*, or John Stobaios's *Anthologium*.[23] It is also used to describe the formation of eddies in

[23] Claudius Ptolemy, *Apotelesmatica* (or, *Tetrabiblos*), II.9.7, ed. E. Boer and F. Boll, *Opera quae exstant omnia*, vol. III.1 (Leipzig: Teubner, 1940), 137; Cassius Dio, *Historiae Romanae*, XXXIX.41.1, ed. U. P. Boissevain, *Historiarum Romanarum quae supersunt*, vol. I (Berlin: Weidmann, 1895), 488; John Stobaios, *Anthologium*, I.38.8, ed. O. Hense and C. Wachsmuth, *Anthologium*, vol. I (Berlin: Weidmann, 1884), 253.

narrow straits, as in the *Souda*'s entry for "Cataracts" (Καταρράκται), Gregory of Nazianzos's first oration on peace, and Michael Psellos's *De omnifaria doctrina.*[24]

These uses, therefore, speak not only of a simple reflux or back-flow as the literal term itself suggests, but rather an ebbing back and forth at sites where different tides crash together and unite. This metaphorical language aptly depicts not only the efflux emanating from sensible objects being reflected back, but also the flow of rays from the eyes crashing upon that flow in the stream of vision. In keeping with the Platonic theory he espouses in the second homily, the author deftly describes the confluence of visual streams as a turbulence of tidal flows and waves, capturing the tumult of optical rays like a mirror. The author's theory is clearly Platonic. But, even more, the rhetoric emphasizes a mediating interval where visual rays not only fuse, but also tumultuously collide. This plays with the language of flowing, rushing water and subverts the idea that the rays of sight follow a direct, linear path. For example, in the first quotation (from the second homily), the author uses metaphors like the gushing springs (πηγῶν), the waterpipe (ὀχετός), or the flood tide (πλημμύρας) to describe the flow of the optical rays. And in the second quotation, through an aquatic metaphor, the word παλιρροίας fittingly and crucially establishes this mediating space of contact, whose usual conditions the mirror alters, as a kind of tumultuous frenzy of waves in the space between mirror, seers, and the visages seen.

Plato's theory relies heavily upon the notion of an interval for sight, and so is quite close to Aristotle's eventual theorization of sight, except that it does not reify the intervallic space as does Aristotle's notion of the transparent (διαφανής).[25] Instead, Plato's medium is merely a point of encounter and fusion of the optical rays that emerge both from the eyes and the object. Despite the variation in the physical mechanics, both theories, as well as those of Empedocles and Democritus, rely on the primacy of a medium for its underlying structure. Thus, no matter to which theory a Byzantine may have subscribed, a degree of separation was central to the Byzantine theory of vision. Sight was *never* a vehicle for unmediated, haptic contact with an image.

[24] Anonymous, *Souda*, "Καταρράκται," kappa 745, ed. and trans. Catharine Roth, *Souda-On-Line*, http://www.stoa.org/sol-entries/kappa/745; Gregory of Nazianzos, *De pace* 1 (Orat. 6.19), ed. PG 35:748A; Michael Psellos, *De omnifaria doctrina*, 171, ed. Westerink, 86.

[25] On the theory of the transparent, see Anca Vasiliu, *Du diaphane: image, milieu, lumière dans la pensée antique et médiévale* (Paris: J. Vrin, 1997).

The Platonic-Aristotelian Synthesis: Nikephoros Choumnos

The question regarding whether sight was a product of the efflux from the eyes and the emanation of rays from objects endured through the teachings of the ancient and late antique commentators. Oftentimes, as we have seen already in the case of Philo and "Basil," this operates less as an active debate than as a preference for one side or another of the theory, for rhetorical, theological, or psychological effect. While a difference was certainly understood between Plato and Aristotle, most Byzantines framed sight somewhere between them. Sight was mediated by an intervening medium, objects contributed something to this process, either through the emanation of rays or colors, and the eyes played an active part in the process of human cognition and perception. This produced a sort of middle ground between Aristotle and Plato, where differences were downplayed. This perspective, for example, is conveyed in Nemesios's *De natura hominis*, which treated sight as being the cause of emanations from both the object and the eye, and stated that Galen agreed with this model.[26] Nemesios bases this synthesis on Aetios of Antioch's summary on sight in *De placita philosophorum*. That synthesis also found its way into works such as John Stobaios's *Anthologium* and Michael Psellos's *De omnifaria doctrina*.[27]

In later intellectual and philosophical circles, however, the differences between Plato and Aristotle were acknowledged and stirred lively debate, particularly in the fourteenth century. For example, in his *ekphrasis* on the Church of Holy Apostles in Constantinople, written sometime between 1198 and 1203, Nicholas Mesarites notes that student physicians from the adjacent school were spending time in the courtyard of the church debating "whether we see according to a sending out or a coming in" (εἰ κατ' ἐκπομπὴν ὁρῶμεν ἢ εἰσπομπήν).[28] This suggests a general distinction between some form of extramission and intromission, which is not to indicate that sight was believed to be the product of solely one or the other

[26] Nemesios, *De natura hominis*, 7, ed. M. Morani, *De natura hominis* (Leipzig: Teubner, 1987), 58; trans. R. W. Sharples and P. J. van der Eijk, *On the Nature of Man* (Liverpool University Press, 2008), 105. On this author, see the next chapter.

[27] Michael Psellos, *De omnifaria doctrina*, 108, ed. Westerik, 60. See John Stobaios, *Anthologium*, I.52.7, ed. Hense and Wachsmuth, 484; Pseudo-Plutarch (Aetios of Antioch), *De placita philosophorum*, IV.13, ed. Mau, 123; trans. William W. Goodwin, *Plutarch's Morals* (Boston: Colonial Press, 1905), 168.

[28] Nicholas Mesarites, *Descriptio ecclesiae SS. Apostolorum*, 42.5, ed. and modified trans. Glanville Downey, "Nikolaos Mesarites: Description of the Church of Holy Apostles," *Transactions of the American Philosophical Society* n.s. 47:6 (1957): 855–924, at 895, 917.

since that strict division is a modern one. Instead, the discussion here seemed to be contoured according to what the critical principle or agent of sight was considered to be, the eye's effluxes or those coming from the object. This could have been as much of a debate between the atomists and Plato, as between Aristotle and Plato, as between Galen and some hybrid of Aristotle and Plato, or some permutation of these categories. And, notably, this topic was but one of the many debates regarding human cognition and psychology that Mesarites describes occurring in the courtyard of the church, as students dealt with the differences and nuances of the classical teachings.

In the early fourteenth century, the Constantinopolitan scholar and statesman Nikephoros Choumnos turned his attention to the discrepancies on sight between Aristotle and Plato, in his treatise *On the Nutritive and Sensitive Soul.*[29] There, Choumnos proudly boasted that he has successfully unified Plato's theory of vision with that of Aristotle through a dialectical synthesis. In the *protheoria* to his commentary, summarizing the arguments that are to follow in the treatise, Choumnos writes:

> Αἰσθήσεων δὲ πέντε οὐσῶν, ὁράσεως, ἀκοῆς, ὀσφρήσεως, γεύσεως καὶ ἁφῆς, δείκνυσιν ἑκάστην ὅθεν καὶ ὅπως ἐστί, καὶ ὅπως αὖθις φθειρομένη ἐστὶ καὶ τὰς ἐνεργείας δείκνυσιν ἑκάστης. καὶ γοῦν περὶ ὄψεως Πλάτωνι καὶ Ἀριστοτέλει μὴ τῶν αὐτῶν καὶ ἴσων δεδογμένων, ἀλλὰ δόξης μὲν τοῦ Πλάτωνος οὔσης, κατ' ἐκπομπὴν ταύτην ἐνεργεῖν καὶ τὴν αἰτίαν καὶ ἀρχὴν ἔχειν ἐκ πυρός, τοῦ δ' Ἀριστοτέλους πᾶν τοὐναντίον ἐκ διαφανοῦς ἀέρος καὶ ὕδατος ὁμοίως τῷ ὀφθαλμῷ ἐνόντος, διαφανοῦς καὶ αὐτοῦ ὄντος ταύτην εἶναι, καὶ τῶν χρωμάτων ἐμπιπτόντων αὐτῇ ἐνεργεῖν· προσόντος δ' ἀμφοῖν οὐκ ἀπόρου δή τινος μόνου, ἀλλά τί γε καὶ ἀδυνάτου, ὁ λόγος καταλλάττει καὶ εἰς ἓν ἄμφω τὼ δόγματε ἄγει, λύων ἅμα καὶ τὰς ἀπορίας, καὶ τὴν ὄψιν δεικνὺς σαφῶς, ὡς ἔστιν ὀρθῶς ἐνεργοῦσα τὸ ἑαυτῆς.
>
> The senses being five – sight, hearing, smell, taste, and touch – [Aristotle] shows whence and how each is, and how in turn each is destroyed and shows the operation of each one. At any rate, the same and equal things are not taught by Plato and Aristotle concerning sight. On the one hand, the opinion of Plato is that the responsibility for the operation of [sight] is due to a sending-out and that it has its source from fire. Aristotle, on the other hand, [believes] the complete opposite, that [sight] operates from transparent air and water becoming like to the eye, and that [sight] exists when it [i.e., the air or water] becomes transparent and that it gets its operation

[29] See J. Verpeaux, *Nicéphore Choumnos: homme d'état et humaniste byzantin* (Paris: Editions A. et J. Picard, 1959), esp. 143–145. See also Nikephoros Choumnos, *Opera philosophica*, 6, ed. Konstantinos P. Chrestou, *Τὸ φιλοσοφικὸ ἔργο τοῦ Νικηφόρου Χούμνου* (Thessaloniki: Ekdoseis Kyromanos, 2002), 88–142.

> when colors fall upon it. Since presenting only one of the two is not just difficult but impossible, this discourse reconciles and leads both opinions into one, at the same time dissolving the differences, and clearly explicating sight, that it is correctly operating in its own right.[30]

In other words, with his treatise, Choumnos seeks to resolve the differences between Plato and Aristotle to show that while some fiery essence is being "sent out" (ἐκπομπήν) from the eyes in Plato, there are also colors that "fall upon" (ἐμπιπτόντων) the eyes through the transparent air or water. By considering matters like reflections in a mirror, the sight of objects at far distances, and the visual perception of scale and magnitude, Choumnos goes on to show that vision occurs through a hybrid model in which the perspectives of both Plato and Aristotle are at play; Galen expounded similar ideas, as discussed in the following chapter.

Choumnos goes about carefully proving his theory in the treatise itself, while following closely the logic of Aristotle and the theory of Plato. He begins by considering light, rejecting its corporeality, and then elucidating how the transparent, when activated by light, is affected by the colors of visible objects, as in Aristotle.[31] The challenge that soon emerges, and one reiterated in the work of the commentators, is how sight could possibly see objects at a far distance since the optical lights or rays could not possibly be said to reach as far as the stars. And, furthermore to this point, even if sight could reach such distances: how then could one see things immediately when sight had to cross such immense distances? And, how could one perceive scale if the rays equally reached all visible things? Choumnos zeroes in on these two matters given that they are both as valid in a Platonic theory, wherein rays travelled to the object, as in an Aristotelian theory, where the object's colors flowed through the transparent to the pupil.

Clarifying that Plato never said that the rays actually reach all the way to the object itself from the eyes, Choumnos smooths over the differences between the two theories writing:

> Οὐ γὰρ δὴ τοῦτο Πλάτων οὔτ᾽ ἐνόμισεν οὔτ᾽ εἴρηκεν, ἀχρόνως εἶναί τινα κίνησιν ἀλλάττουσαν τόπον, οὐδέ γε τὸ ἐκ τῆς ὄψεως ὡρμημένον πῦρ ἢ

[30] Nikephoros Choumnos, *Opera philosophica*, 6, ed. Chrestou, 88.

[31] Börje Bydén has carefully elucidated Choumnos's view by placing it within the history of Greek optics, whereby the illuminated air accepts and transmits colors while at the same time light flows from the eyes, uniting with the rays in the transparent. For a summary of Choumnos's theory of sight, see Börje Bydén, *Theodore Metochites'* Stoicheiosis Astronomike *and the Study of Natural Philosophy and Mathematics in Early Palaiologan Byzantium* (Gothenburg: Acta Universitatis Gothoburgensis, 2003), 206–207.

φῶς ὄν, αὐτὸ δὴ τοῦτο διαβαῖνον ἥκειν μέχρι τῶν ὁρωμένων, ἀλλ' ἐπειδὴ χρῶμα πᾶν ἐν φωτί, σκότους γὰρ οὐδέν ἐστι φαινόμενον τούτων, πληροῖ μὲν ταῦτα τὸν ἀέρα, διαφανῆ ἤδη τῷ φωτὶ γεγενημένον, καὶ αὐτοῖς πεπονθότα τοῖς χρώμασι καὶ κατ' αὐτὰ μορφωθέντα. οὕτω δ' ἀέρος, χρωμάτων καὶ φωτὸς ἐσχηκότος καὶ μέχρις ὀφθαλμῶν ἀφικνουμένου, φῶς αὖθις ἐκεῖθεν ὥσπερ ἐκ λαμπτῆρος προσαυγάσαν, ἐς αὐτὸν ἀέρα διαχεῖται. καὶ τῷ γ' ἐν τούτῳ φωτί, ὡς ἔφη Πλάτων, συμφυέν, ὅλον ὅσον ἐν αὐτῷ καὶ σχῆμα καὶ μόρφωμα καὶ μέγεθος βλέπει καὶ διαβιβάζει τοῦτ' αὖθις πρὸς αἴσθησιν τοῦ ἐντὸς ὄντος καὶ ἡγεμονοῦντος· καὶ γίνεται οὕτως τὸ ὁρᾶν, μηδενὸς ἥκιστα ἐπιδεόμενον χρόνου.

Now then, Plato neither believed nor said that any movement changes place absent time, nor that something from sight, being fire or light, having shot out, itself crosses over to reach the things being seen. But since all color [appears] in light, for in darkness nothing visible is evident, these [i.e. the colors of visible things] fill the air, which actually becomes transparent with light, having been affected by colors themselves and having taken on their form. In this way, air, partaking of colors and light reaches the eyes. In turn, light, gleaming from there as if from a lantern, diffuses into the same air. And so one sees there united in that light, just as Plato said, everything – as much figure, form, and magnitude – and, in turn, one transports it to the sensory-perception within, the *hegemonikon* [i.e. 'the ruling part of the soul']). And thus seeing occurs, needing the least amount of time.

Here, Choumnos presents a model wherein the eyes shoot out rays, but these do not simply reach out to visible things, but rather unite with the illuminated air that has itself been affected by the colors of the visible objects. Thus, when the effluxes from the eyes come out, they unite with the light-borne colors being passed along by the air. Thus, Choumnos rightly associates it with Plato's theory regarding the effluxes' union (ὡς ἔφη Πλάτων, συμφυέν). In other words, Choumnos adopted the model of Platonic united rays while retaining the transparent as the medium of intersection between the rays of the viewer and the object.

Through this combined theory, then, it is possible to perceive the entire whole of large objects, along with figure, form, and magnitude, since they are adequately scaled for the viewer by virtue of being united in the medium, presumably at some middle point.[32] Like in the case of the mirror's reflection, form is preserved yet size is altered, which as Choumnos argues is akin to an alteration of the external form being conducted through the air, rather than necessitating that the visual ray alone and itself be altered to grasp large objects. Likewise, since air has transmitted the

[32] See Nikephoros Choumnos, *Opera philosophica*, 6, ed. Chrestou, 111–117.

colors through the transparent into the reaches of the eye's effluxes, one is able to perceive immediately, rather than necessitating that the visual rays cross large distances before being able to make contact and communion with the object. In other words, Choumnos's theory is resolutely non-haptic, given that it is by virtue of the medium that sight is capable of operating instantaneously and efficiently at perceiving objects of a large scale. Were it the onerous task of the eye's fire to reach the stars, not only would this take a great deal of time, but if the images were being sent from such massive things, they would likewise not fit into the small opening of the eye's pupil.

Informed by what appear to be a range of sources, including Alexander of Aphrodisias, Plotinos, John Philoponos, Priscian, Galen, and others, Nikephoros Choumnos's felicitous resolution of this age-long debate deserves to be qualified in its uniqueness and thinking. In many ways, this unified theory was presaged in the late antique commentators' Aristotelian understanding of Plato, evidenced by the popularity of the term "united rays" (συναύγεια) to properly describe Plato's theory wherein both the emanations from viewer and object unite. Choumnos's sly implication that Plato's theories have been misrepresented, by Aristotle but certainly by others, is evidently nothing more than a strawman for articulating his own theory. This trope had been used by the Aristotelian commentators, but the nuanced approach to Plato's theory was already evident in the doxography of Aetios of Antioch, passed on through the works of John Stobaios and Nemesios. Likewise, as we shall see in the following chapter, similar ideas were expounded by Galen's own thoughts on the matter, passed down through figures like Nemesios and the medical writers. As Börje Bydén rightly points out, a potential Galenic influence upon Choumnos may have come from one of Choumnos's contemporaries, like the erudite Nikephoros Blemmydes.[33] A medical doctor by training, Blemmydes was a proponent of Galen's theories on sight, which is not surprising given that Galen served as the backbone of medical approaches to ophthalmology and diseases of the eye, as we shall see in the next chapter.

Thus, Choumnos is certainly the perfect capstone to this narrative, both reconciling the Aristotelian and Platonic theories of vision so as to stress the prime importance of the visual medium and a mediated visuality, and also evidencing ties to the thinking of Galen, his medical predecessors, and the late antique writers that set the foundation for the history of Byzantine theories of vision. Nevertheless, Choumnos is more than anything a

[33] Bydén, *Theodore Metochites'* Stoicheiosis Astronomike, 207.

fortuitous narrative resolution here, given that the importance of his thinking is precisely that his breakthrough is by no means unique, but articulates for us a certain visual *koine*, a *lingua franca* for the hybrid theory of vision that popularly circulated across the centuries of Byzantine thought. It is by no means representative of all the nuances, changes, and polymorphous articulations that this fusion enjoyed over different writers, intellectual circles, and centuries. But it does speak to the homogenizing and harmonizing drive of writers since the late antique period, who sought to resolve the *aporia* of their teachings and sources. And, as with everyone else discussed in this chapter, Choumnos provides us with some literal distance between sight and touch.

CHAPTER 2

The Problem of Tactility

As we have seen, sight became confused with touch. The phenomenon is not exclusive to our contemporary historiography, but had precedents in the Byzantine world. In this chapter I argue that this confusion was the exception, not the rule, and was the result of a handful of non-specialized readers interpreting Euclid, the Stoics, Ptolemy, and Galen. More influential Byzantine thinkers such as Paul of Aegina, Meletios the Monk, Nemesios of Emesa, Symeon Seth, George Pachymeres, and Nikephoros Blemmydes read Galen and the Stoics alongside Aristotle and Plato in order to continuously rearticulate the role and function of sight in relation to the other senses and their operations. Some explicitly warned against conflating sight with touch, and none ever collapsed sight and touch.

One could argue that those ancient authors who explicitly distanced sight from touch were fighting against a culture that conflated them. I shall argue that those texts, rather than confirming a cultural practice through its negation, were rhetorical tools intended to explain the unique characteristics and affordances of each sense. Antique and post-antique sources were not arguing against a zeitgeist; rather, they were wrestling with specific touch metaphors meant to describe the senses and their cognitive functions. Following Foucault, Robert Nelson attempted to extrapolate the Byzantines' "loquacious gaze" through such metaphors, for he was committed to the idea that, in his own words, "to consider the act of seeing in a certain society or period leads inextricably to general issues of cultural construction and maintenance."[1] In other words, Nelson attempted to see how far such metaphors enabled a practiced theory of vision. But to let such metaphor dominate our perspective is misguided. In this chapter I shall argue that late antique and Byzantine scholars went to great lengths to clarify the language of touch, to keep metaphors metaphorical, and to situate sight within the other senses that were also described

[1] Nelson, "To Say and to See," 145.

metaphorically. Taking these sources seriously leads to the true Byzantine zeitgeist: sight may be likened to touch, but it isn't touch.

The Optical *Pneuma*: Galen and the Stoics

David Lindberg, in his classic survey of theories of vision, rightly observes that the ancient Greeks did not suggest that sight occurred by grasping the essences of visible objects, but, in Lindberg's words, "by the qualitative changes produced by the object in a medium suitably prepared to receive them," which he notes is common to the teachings of Plato, Aristotle, and the Stoics alike.[2] As demonstrated in the last chapter, authors always stressed that sight was a mediated form of sensation occurring across an interval of space. Across history, the metaphor of vision as touch appears most prominently in the Stoics' theories of vision, which likened sight perception to the touch of a walking stick. While little evidence suggests that Platonic theories of vision led to the understanding of haptic sight, there is evidence that such confusion emerged out of Stoicism. According to the Stoic model, attributed to Chrysippus by Aetios and others, the air between the viewer and the object is placed in a state of extension or tension (συνέντασις), having been altered by the visual *pneuma* flowing from the brain to the eyes.[3] This pneumatic flux then touches the air and produces an activated conically shaped mixture that projects to the object of sight.

The crucial factor is the *pneuma*, an all-pervasive medium that is physiologically composed of a mixture of air and fire.[4] The *pneuma* was seen to be a spatial extension of the body, comparable to a walking stick. As Diogenes Laertius says:

> ὁρᾶν δὲ τοῦ μεταξὺ τῆς ὁράσεως καὶ τοῦ ὑποκειμένου φωτὸς ἐντεινομένου κωνοειδῶς, καθά φησι Χρύσιππος ἐν δευτέρῳ τῶν Φυσικῶν καὶ Ἀπολλόδωρος. γίνεσθαι μέντοι τὸ κωνοειδὲς τοῦ ἀέρος πρὸς τῇ ὄψει, τὴν δὲ βάσιν πρὸς τῷ ὁρωμένῳ· ὡς διὰ βακτηρίας οὖν τοῦ ταθέντος ἀέρος τὸ βλεπόμενον ἀναγγέλλεσθαι.
>
> [They hold that we] see when the light between the visual organ and the object stretches in the form of a cone: so Chrysippus in the second book of

[2] Lindberg, *Theories of Vision* , 9–11, at 10.

[3] See Robert B. Todd, "ΣΥΝΕΝΤΑΣΙΣ and the Stoic Theory of Perception," *Grazer Beiträge* 2 (1974): 251–261. See also Aetios of Antioch, *De placitis reliquiae*, IV.15.3, ed. Diels, 405–406.

[4] See Samuel Sambursky, *Physics of the Stoics* (Princeton University Press, 1959), 1–11; Gérard Verbeke, *L'évolution de la doctrine du pneuma: du stoicisme à s. Augustin* (Paris: Desclée de Brouwer, 1945), esp. 1–11.

> his *Physics* and Apollodorus. The apex of the cone of air is at the eye, the base at the object seen. Thus the thing seen is reported to us as if through a stick of air stretching out towards it.[5]

The optical *pneuma* flows from the seat of ruling faculty of the mind (τὸ ἡγεμονικόν) to the eye and activates the air adjacent to the eye by placing it in a state of tension or stress. When that is illuminated, it makes contact with the visible object. This process is manifested through the operation of light (φωτός). In Diogenes Laertius's summary, light is perplexingly used as a stand-in for the *pneuma* 's union with the surrounding air (ἀέρος), being said to extend in a conical shape toward the objects of sight. The analogy to touch comes via the phrase "as if through a stick" (ὡς διὰ βακτηρίας), which was intended to help readers grapple with and understand the *pneuma*'s deployment of air as an intervening medium.[6]

In the Stoics' view, sight may be likened to touch, but only to indicate a type of mediated contact between perceiver and object. Furthermore, sight was not the only sense understood to depend upon percussions and movements in the air. For example, immediately after the discussion on sight, Diogenes Laertius goes on to describe hearing:

> Ἀκούειν δὲ τοῦ μεταξὺ τοῦ τε φωνοῦντος καὶ τοῦ ἀκούοντος ἀέρος πληττομένου σφαιροειδῶς, εἶτα κυματουμένου καὶ ταῖς ἀκοαῖς προσπίπτοντος, ὡς κυματοῦται τὸ ἐν τῇ δεξαμενῇ ὕδωρ κατὰ κύκλους ὑπὸ τοῦ ἐμβληθέντος λίθου.
>
> We hear when the air between the sonant body and the hearer is struck in the form of a sphere and then forms waves and strikes upon the hearing, just as the water in a reservoir forms wavy circles when a stone is thrown into it.[7]

This statement is surgically precise in its parallelism with the previous description of sight. In each passage, the relationship between the sense and its medium is described within an elemental theory, wherein each sense is associated with a particular natural element: sight with light and

[5] Diogenes Laertius, *Vitae philosophorum*, VII.1.157, ed. and modified trans. R. D. Hicks, *Lives of Eminent Philosophers*, vol. II, Loeb Classical Library 185 (Cambridge, MA: Harvard University Press, 1925), 260–261.

[6] See David E. Hahm, "Early Hellenistic Theories of Vison and the Perception of Color" in P. K. Machamer and R. G. Turnbull (eds.), *Studies in Perception: Interrelations in the History of Philosophy and Science* (Columbus: Ohio State University Press, 1978), 60–95; Bernard Saint-Pierre, "La physique de la vision dans l'Antiquité: contribution à l'établissement des sources anciennes de l'optique médiévale," unpublished PhD dissertation, University of Montreal (1972), esp. 107–114; Sambursky, *Physics of the Stoics*, esp. 23–29.

[7] Diogenes Laertius, *Vitae philosophorum*, VII.1.158, ed. and modified trans. Hicks, 260–261.

hearing with air.[8] Such parallelism between sight and other senses, particularly hearing, points to one of the challenges in the study of ancient and medieval visual theories of perception, namely, that sight often has been surveyed and considered on its own, at times without substantial concern for how writers described and characterized the other senses. Here, we see that just as in hearing it is air (ἀέρος) that serves as a crucial medium, what truly distinguishes sight from hearing is light. Thus, light is the proper sensible of vision. The author stresses in both cases even the unique geometry of each sense, wherein sound spreads as a spherical (σφαιροειδῶς) movement, but sight in a conical (κωνοειδῶς) projection. And, the passage likewise ends with an easily comprehensible metaphor for the reader to visualize how the process works, this time "just as water is swept in waves in a reservoir" (ὡς κυματοῦται τὸ ἐν τῇ δεξαμενῇ ὕδωρ). The latter is the equivalent to the former's comparison to the walking stick (ὡς διὰ βακτηρίας), stressing the pedagogical and rhetorical function of each comparison, both of which model the operations of the senses without ever suggesting that the metaphors describe literal physics or physiology.

Nevertheless, the walking stick metaphor within the Stoic corpus did not denote any form of tactility. Although modern translators render the line "by the medium of air as if through a walking stick" (Hicks's original translation), the reality is that the walking stick refers precisely to that "cone of air" (τὸ κωνοειδὲς τοῦ ἀέρος) that Apollodorus tells us we perceive through. As Diogenes Laertius clearly states, the thing seen is not touched or grasped but "reported" to us (τὸ βλεπόμενον ἀναγγέλλεσθαι), suggesting that the walking stick does not allow for any form of unmediated contact but is rather a messenger of what lies outside. More important, however, is the fact that the walking stick is not a metaphor for the act of seeing, but rather for that stretched or tensed air that mediates sight. In other words, just as touch can be transmitted through a medium, so vision is also a form of perception that occurs through the transmission of sensible qualities through a medium, in this case thought of as a long, tapered cone.

Galen built upon these teachings, but took issue with the metaphor's crude allusion to tactility.[9] He dismissed the walking stick metaphor,

[8] On the association of the particular senses with the elements, see John I. Beare, *Greek Theories of Elementary Cognition: From Alcmaeon to Aristotle* (Oxford: Clarendon Press, 1906), 236–239.

[9] For a succinct introduction to Galen's theory, see Katerina Ierodiakonou, "On Galen's Theory of Vision," *Bulletin of the Institute of Classical Studies. Supplement* 114 (2014): 197–211.

where he criticizes its foolish association of sight with touch, but he did not think that the term could aptly serve as a metaphor for vision as a whole and the complex ways in which *pneuma*, air, and light unite. Rather, he understood the metaphor to apply specifically to that attenuated air, writing that the Stoics were wrong to say that we see "as if by means of a walking stick of nearby air" (ὡς διὰ βακτηρίας τοῦ πέριξ ἀέρος).[10] In the spurious *De anima libri mantissa* associated with Alexander of Aphrodisias, the author devotes an entire section to arguing against the notion that vision comes about through any form of tensed air and seems to be responding at times as much to the Stoics as to Galen.[11] In this text, however, we find one of the earliest interpretations of the walking stick metaphor as applying to tactility. The metaphor here is used, as with Diogenes Laertius, in connection with the conical air, but in a new construction: "as by touch through a walking stick" (καθάπερ καὶ τῇ ἁφῇ διὰ βακτηρίας).[12] The metaphor's association with tactility even continues into the work of Plotinos, who again like all others dismisses the crude and violent action of this metaphor in his tract on sight:

> Εἰ δὲ μένει μὲν ἡ ψυχὴ ἐφ' ἑαυτῆς, φωτὸς δὲ δεῖται ὥσπερ βακτηρίας πρὸς τὸ φθάσαι, ἔδει τὴν ἀντίληψιν βίαιον καὶ ἀντερείδοντος εἶναι καὶ τεταμένου τοῦ φωτός, καὶ τὸ αἰσθητόν, τὸ χρῶμα, ᾗ χρῶμα, ἀντιτυποῦν καὶ αὐτὸ εἶναι· οὕτω γὰρ διὰ μέσου αἱ ἁφαί.
>
> But if the soul stays in its own place, but needs light like a stick to reach the object with, then the apprehension would be a violent business, with the light stretched out and pushing against the object and the object of perception, the colour as colour, itself pressing back: for this is how sensations of touch occur through a medium.[13]

Note, the walking stick is not associated with the sight generally, but specifically with light. Plotinos nevertheless attacks the metaphor with an argument *ad absurdum*, asserting that the metaphor would imply the colors work like tensile properties that push back against vision, just as touch itself unfolds. It is possible then to surmise that over the course of

[10] Galen, *De placitis Hippocratis et Platonis*, VII.7.19, ed. and trans. Phillip De Lacy, *On the Doctrines of Hippocrates and Plato*, Corpus Medicorum Graecorum 4.1.2 (Berlin: Akademie Verlag, 1980), 474–475.

[11] Alexander of Aphrodisias, *De anima libri mantissa*, 130.13–134.26b, ed. Robert W. Sharples, *De anima libri mantissa*, Peripatoi 21 (Berlin: De Gruyter, 2008), 72–77; partial trans. Sambursky, *Physics of the Stoics*, 124–125.

[12] Alexander of Aphrodisias, *De anima libri mantissa*, 130.17, ed. Sharples, 72.

[13] Plotinos, *Enneades*, IV.5.4.40 (29), ed. and trans. H. A. Armstrong, *Enneads*, 7 vols., Loeb Classical Library 440–445, 468 (Cambridge, MA: Harvard University Press, 1966–1988), IV: 296–297.

the late antique period the metaphor took on increasingly haptic implications as it became reinterpreted. That trend occurred despite, not because of, the earliest uses by the Stoics, who saw it as a metaphor for intervallic elements, not touch.

Returning to Galen, we find that his true problem with the Stoic walking stick metaphor was the role that it assigned to air in the process of vision. Galen complained that the Stoics had not properly accounted for the role that light plays to activate that walking stick of air for the optic *pneuma*. At best, the Stoics used air and light interchangeably without any clear or extensive distinction, a matter that is evidenced by the fact that Aetios's best description of Stoic theories of vision occurs not in his summary on sight, but rather in his entry on whether darkness itself is visible.[14] Galen's preferred theory elaborated on the instrumentality of the intervening medium of air, but not without scrupulous clarification that this air was activated as a receptive medium through the influx of light.[15] What we find in Galen's thinking is loosely a theory of extramission, but one that strongly and eloquently opposes the pseudo-haptic operation of the Stoic *pneuma* and has more in common with the teaching of Aristotle's transparent, which he continually praises above all other theories. Galen sees the air as an extension of the senses through the outpouring of the *pneuma*, explaining that "the air becomes for the eye the same kind of instrument for the proper discrimination of its sense-objects, as the nerve is for the brain: therefore, as the brain is to the nerve, so the eye is to the air."[16]

Galen elaborates futher, to contextualize and delimit what he means by this comparison between the nerves and the air:

> εἴπερ οὖν ἡ ὄψις μόνη τῶν ἄλλων αἰσθήσεων αἰσθάνεται τοῦ κινοῦντος αὐτὴν αἰσθητοῦ διὰ μέσου τοῦ ἀέρος, οὐχ ὡς βακτηρίας τινός, ἀλλ᾽ ὡς ὁμοειδοῦς τε καὶ συμφυοῦς ἑαυτῇ μορίου καὶ μόνῃ τοῦτ᾽ ἐξαίρετον αὐτῇ δέδοται μετὰ τοῦ καὶ δι᾽ ἀνακλάσεως, ὁρᾶν εἰκότως ἐδεήθη πνεύματος ἄνωθεν ἐπιρρέοντος αὐγοειδοῦς, ὃ προσπῖπτον τῷ πέριξ ἀέρι καὶ οἷον

14 Aetios of Antioch, *De placitis reliquiae*, IV.15.3, ed. Diels, 405–406.

15 See Rudolph E. Siegel, *Galen on Sense Perception: His Doctrines, Observations, and Experiments on Vision, Hearing, Smell, Taste, Touch, and Pain, and Their Historical Sources* (Basel: Karger, 1970), esp. 37–40; Harold Cherniss, "Galen and Posidonius' Theory of Vision," *American Journal of Philology* 54 (1933): 154–61; Owsei Temkin, "On Galen's Pneumatology," *Gesnerus* 8 (1951): 180–189.

16 "καὶ γίγνεται δὲ τοιοῦτον ὄργανον αὐτῷ πρὸς τὴν τῶν αἰσθητῶν οἰκείαν διάγνωσιν οἷον ἐγκεφάλῳ τὸ νεῦρον, ὥσθ᾽ ὃν ἔχει λόγον ἐγκέφαλος πρὸς τὸ νεῦρον, τοιοῦτον ὀφθαλμὸς ἔχει πρὸς τὸν ἀέρα." Galen, *De placitis Hippocratis et Platonis*, VII.5.32–33, ed. and trans. De Lacy, 460–461.

> ἐπιπλῆττον αὐτὸν ἑαυτῷ συνεξομοιώσει. δεόντως οὖν ἐροῦμεν αὐγοειδὲς μὲν εἶναι τὸ τῆς ὄψεως ὄργανον, ἀεροειδὲς δὲ τὸ τῆς ἀκοῆς, ἀτμοειδὲς δὲ τὸ τῆς ὀδμῆς καὶ τὸ μὲν τῆς γεύσεως ὑγρόν, τὸ δὲ τῆς ἁφῆς γεῶδες.
>
> So since sight alone of all the senses perceives when sensation moves it by the medium of air – not like a certain walking stick [of air] but like a portion [of air] that is of form and nature identical to itself – and [since] to [sight] alone is given this remarkable [ability], along with that of refraction, it likely must see when there flows down from above [i.e. from the head] the sunlight-like *pneuma*, which, when it falls upon and "punishes" the nearby air, [the *pneuma*] will make [that air] like itself. So we must necessarily say that the organ of sight is sunlight-like, but that of hearing is air-like, and that of smell vapor-like, and that of taste wet, and that of touch, earth.[17]

While Galen allows for air to be the medium of vision, just as the Stoics did, he makes sure that this is only the case when that air has been fundamentally changed or altered into some form of luminous, "sun-light-like" (αὐγοειδές) thing consubstantial with the optic *pneuma*. The model presented here echoes at its core Plato's ideas that in vision like unites with like and that the efflux of the colored flame from the object unites with the visual fire of the eye. In Galen's view, however, that likeness emerges from a certain sympathy between the optic *pneuma* and the exterior air, caused by the former's forceful alteration of the latter. It is through this process that the air is "hit" (προσπῖπτον) and "assimilated" or, more literally, "punished" (ἐπιπλῆττον) into the nature of the *pneuma*. This violence exerted upon the air enacts a state of tension through which vision is possible.

However, unlike the Stoics before him, who understood that state of tension alone was responsible for sight, Galen regarded this alteration of the proximate air as wholly meaningless without the crucial action of light. Light is what fundamentally alters the air so that it can be molded into a likeness akin to that of the optic *pneuma*. This is critically explained only later in the text, where Galen returns to the analogy between brain and nerve, eye and air:

> αὐτό τε γὰρ τὸ νεῦρον ἐγκεφάλου μέρος ἐστὶν οἷόν περ ἀκρεμὼν ἢ βλάστημα δένδρου, τό τε μέλος εἰς ὃ τὸ μέρος ἐμφύεται τὴν δύναμιν αὐτοῦ δεχόμενον εἰς ὅλον ἑαυτὸ διαγνωστικὸν γίνεται τῶν ψαυόντων αὐτοῦ. παραπλήσιον οὖν τι κἀπὶ τοῦ περιέχοντος ἡμᾶς ἀέρος γίγνεται:

[17] Galen, *De placitis Hippocratis et Platonis*, VII.5.41, modified ed. and modified trans. De Lacy, 460–461.

> πεφωτισμένος γὰρ ὑφ' ἡλίου τοιοῦτόν ἐστιν ἤδη [τὸ] τῆς ὄψεως ὄργανον οἷον τὸ παραγιγνόμενον ἐξ ἐγκεφάλου πνεῦμα: πρὶν φωτισθῆναι δέ, κατὰ τὴν ὑπὸ {τῆς} τοῦ πνεύματος εἰς αὐτὸν ἐκβολῆς ἐναποτελουμένην ἀλλοίωσιν ὁμοιοπαθὲς ὄργανον οὐ γίγνεται.
>
> The nerve itself is a part of the brain, like a branch or offshoot of a tree, and the member to which the part is attached receives its power from the part into the whole of itself and thus becomes capable of discerning the things that touch it. Something similar happens also in the case of the air that surrounds us. When it has been illuminated by the sun, it is already an instrument of vision similar to the *pneuma* emanating from the brain; but before [the air] is illuminated it does not become an instrument of the same experience [as the organ of sight], by virtue of the alteration perfected from the outflow of the *pneuma* into the [light].[18]

Here, we see how Galen thinks about the nerves of the body, primarily in the case of touch felt by the skin. The nerve sends touch back to the brain, and he says that something of the same sort occurs with the air surrounding us. Once that air has become illuminated by light, it becomes a sympathetic instrument that may be deployed by the *pneuma* for vision. Light is a sine qua non. The air on its own being struck by the *pneuma* would be unable to perceive sight, as is evident in the case of darkness.

This passage leads the reader directly into Galen's climactic and eloquent dismissal of the Stoic's walking stick metaphor once and for all, writing:

> μὴ τοίνυν ὡς διὰ βακτηρίας τοῦ πέριξ ἀέρος ὁρᾶν ἡμᾶς οἱ Στωϊκοὶ λεγέτωσαν. ἡ γὰρ τοιαύτη διάγνωσις ἀντιβαινόντων ἐστὶ σωμάτων, κατὰ συλλογισμὸν <δ'> ἔτι μᾶλλον· οὐ πιλητοῦ δ' αἰσθητικὴ ἡ τοῦ ὄμματος αἴσθησις ἡμῖν ἐστιν οὔτε τῆς σκληρότητος ἢ μαλακότητος, ἀλλὰ χρόας καὶ μεγέθους καὶ θέσεως, ὧν οὐδὲν ἡ βακτηρία διαγνῶναι δύναται.
>
> The Stoics, then, must not say that we see as if through a walking stick of surrounding air. This latter kind of discernment is of resistant bodies, and it is besides more inferential than perceptive, whereas the perception of our eye is not perceptive of a thing as close packed, or of its hardness or softness, but of its color, size, and position, and none of these can be discerned by a walking stick.[19]

This rebuttal begins with Galen dismissing the tactile aspect of the metaphor on two accounts: first, that the walking stick operates on a model of touch, which is the nature of "resistant bodies"

[18] Galen, *De placitis Hippocratis et Platonis*, VII.7.19, ed. and modified trans. De Lacy, 474–475; see also 454–455.

[19] Galen, *De placitis Hippocratis et Platonis*, VII.7.19–21, ed. and trans. De Lacy, 474–475.

(ἀντιβαινόντων. . .σωμάτων): that is, objects that are able to push back and thus are capable of being apprehended through touch, a point reiterated by Plotinos, as discussed earlier. Second, Galen clarifies further that since touch cannot transmit color, size, and position, sight could never operate through any form of touching. The walking stick analogy makes the metaphor of touch too literal for Galen's comfort, suggesting that sight is simply a species of touch at a distance, the optical *pneuma* being reduced to a force through which pressures are sent back to the body. This is a fallacy for Galen given that touch cannot perceive color in a body. Thus, it is impossible to say that the optic *pneuma* extends to and envelops the object being viewed, Galen explains. Rather, it activates the air that is coextensive with it, causing this activation to ripple through the continuum of air, similar to the motion effected upon the transparent in Aristotle.[20]

However, Galen was dissatisfied not only with the metaphor's implications but also with the Stoic system teachings within which it is entrenched. The Stoics were able neither to properly account for the nature of the intervening medium of sight nor to explain how light and air could become commensurate with the *pneuma* nor how they could come together to allow vision. Galen immediately turns to other, more preferable theories of vision to consider how each one deals with the mechanics of how the eyes receive sensory impression. Galen has freshly on his mind the passage rebuking the use of the walking stick metaphor when he writes:

> πολὺ τοίνυν ἄμεινον ὡς ἐν ἐσφαλμένοις Ἐπίκουρος ἀπεφήνατο τῶν Στωϊκῶν. ἐκεῖνοι μὲν γὰρ τῶν ὁρατῶν οὐδὲν ἄγουσιν ἄχρι τῆς ὀπτικῆς δυνάμεως, ὁ δ᾽ Ἐπίκουρος ἀπεφήνατο ἄγειν· καὶ πολύ γε τούτου κρείττων Ἀριστοτέλης, οὐκ εἴδωλον σωματικὸν ἀλλὰ ποιότητα δι᾽ ἀλλοιώσεως τοῦ πέριξ ἀέρος ἀπὸ τῶν ὁρατῶν ἄγων ἄχρι τῆς ὄψεως. εὐλαβήθη γὰρ ὅλως αἰσθητικὸν ποιῆσαι τὸν πέριξ ἀέρα, καίτοι τήν γε σάρκα σαφῶς ὁρῶν αἰσθητικὴν γιγνομένην ἐκ τῆς ἀφικνουμένης εἰς αὐτὴν ἀπὸ τῆς ἀρχῆς δυνάμεως.
>
> Therefore Epicurus' view – considering that both views are mistaken – is much better than that of the Stoics. For the latter do not bring anything of the visual object up to the visual power, but Epicurus declared that he did so. Aristotle is much superior to (Epicurus); he does not bring a corporeal image but a quality from the visual object to the eyes through an alteration of the surrounding air. He avoided making the surrounding air sensitive at

[20] Galen, *De placitis Hippocratis et Platonis*, VII.4.20–5:12, ed. and trans. De Lacy, 452–455.

> all although he saw clearly that flesh becomes sensitive as a result of the power that comes to it from the ruling part.[21]

Here, Galen is carefully working out a theory of vision whereby its medium does not allow for any form of haptic or essential contact and where vision does not occur solely from a human efflux. Galen wishes to assert that sight is never simply extramissive or intromissive, by our modern categories, but always interactionist. While flawed, Epicurus is superior to the Stoics because at least in his theory of vision there was some sensory operant that approached the visual faculty, namely the *eidola* of intromissive sight. The Stoics, in seeing sight as wholly a projective process, did not allow for any feedback or sensations returning from the object of sight, and the walking stick metaphor perfectly embodied this problem. Yet, Aristotle supersedes both the Stoics and Epicurus because in his theory that sensory operant is not a "corporeal *eidolon*" (εἴδωλον σωματικόν), but an incorporeal "quality" (ποιότητα), namely colors, which are transmitted through the "surrounding air" (πέριξ ἀέρος), i.e. the transparent (διαφανές) in Aristotle, from the object to sight.

This passage demonstrates the great indebtedness that Galen's theory has to Aristotle's conceptualization of sight, whereby the transparent, composed of air or water, serves as a receptive medium that transmits the visible qualities of object, i.e., color, but only after it has been activated by light. Critically, however, Galen also resolves here the prickly situation regarding his own analogy between the brain and the nerve and the eye and the *pneuma*, which might suggest that the optic *pneuma* and the air altered by it are a sensory extension of the human being. He praises Aristotle's caution not to make the medium itself capable of feeling or sensation (εὐλαβήθη γὰρ ὅλως αἰσθητικὸν ποιῆσαι). Neither the surrounding air nor the sunlight that activates it is on its own capable of feeling. As Galen continues:

> τί δὲ χαλεπόν ἐστι τὴν ἡλιακὴν αὐγὴν αἰσθητικὴν ὑποθέσθαι, οἷον μάλιστα τὸ κατὰ τοὺς ὀφθαλμοὺς πνεῦμα τὸ παραγόμενον ἐξ ἐγκεφάλου προδήλως ὁρᾶται; φωτοειδὲς [καὶ] γάρ ἐστιν. εἰ δὲ καὶ περὶ ψυχῆς οὐσίας ἀποφήνασθαι χρή, δυοῖν θάτερον ἀναγκαῖον εἰπεῖν· ἢ τοῦτ᾽ εἶναι τὸ οἷον αὐγοειδές τε καὶ αἰθερῶδες σῶμα λεκτέον αὐτήν, εἰς ὃ κἂν μὴ βούλωνται κατ᾽ ἀκολουθίαν ἀφικνοῦνται Στωϊκοί τε καὶ Ἀριστοτέλης, ἢ αὐτὴν μὲν ἀσώματον ὑπάρχειν οὐσίαν, ὄχημα δὲ τὸ πρῶτον αὐτῆς εἶναι τουτὶ τὸ σῶμα δι᾽ οὗ μέσου τὴν πρὸς τἆλλα σώματα κοινωνίαν λαμβάνει. τοῦτο μὲν

[21] Galen, *De placitis Hippocratis et Platonis*, VII.7.21–23, ed. and trans. De Lacy, 474–475.

> οὖν αὐτὸ δι' ὅλου λεκτέον ἡμῖν ἐκτετάσθαι τοῦ ἐγκεφάλου, τῇ δέ γε πρὸς αὐτὸ κοινωνίᾳ τὸ κατὰ τὰς ὄψεις [αὐτῶν] πνεῦμα φωτοειδὲς γίγνεσθαι.
>
> But what is the difficulty in supposing that the light of the sun is sensitive? After all, the *pneuma* in the eyes that is passed along from the brain clearly sees, because it is light-like. Then, if we must speak about the substance of the soul, we must say one of two things: we must say either that [the substance of the soul] is that body of a type that is light-like and ether-like – a view to which the Stoics and Aristotle are carried in spite of themselves, as the logical consequence (of their teachings) – or that it [i.e. the substance of the soul] is a bodiless substance, and that [the *pneuma*] is the first vehicle of [the soul], through whose mediation it establishes communion with other bodies. Thus, we must say that this [i.e. *pneuma*] extends throughout the brain, and that in communion with itself, that is [with] the *pneuma* regarding the sight of things, becomes light-like.[22]

With this careful consideration on the nature of the *pneuma* and its relationship to sunlight and the soul, Galen ensures that the pneumatic emission and its union with light and air is not only possible, but also that the sunlight is not confused as a direct sensory extension of the body. The challenge here is that if sight occurs when like unites with like, then the *pneuma* must be light-like – and it is since it unites with sunlight. But where then does this leave the soul if the soul is sentient by virtue of said light-like *pneuma*?[23]

Galen resolves that one would have to say that the substance of the soul is a body consubstantial with both light and ether (αὐγοειδές τε καὶ αἰθερῶδες) – since the sun's rays are luminous and ethereal (i.e. coming from the realm of the ether wherein the sun lies) – or that substance of the soul is simply a bodiless substance, which takes its first vehicle in the corporeality of the *pneuma*, and through which sight can achieve a communion with objects given the optical *pneuma*'s light-like nature. In uniting with sunlight, the pneumatic substance of the eyes changes in nature, becoming in some way different to the pneuma of the brain and

[22] Galen, *De placitis Hippocratis et Platonis*, VII.7.24–26, ed. and modified trans. De Lacy, 474–475.

[23] Galen has already brought up this challenge earlier in the text where he considers the interrelation between *pneuma* and the substance of the soul, a passage to which this excerpt is heavily indebted and without which this excerpt becomes quite impenetrable. In that earlier passage, Galen writes: "From these phenomena you might suppose either of two things about the pneuma in the ventricles of the brain: if the soul is incorporeal, the pneuma is, so to speak, its first home; or if the soul is corporeal, this very thing [i.e. the *pneuma*] is the soul" (ἐκ τούτων οὖν τῶν φαινομένων ἴσως ἄν τις ὑπονοήσειε τὸ κατὰ τὰς κοιλίας τοῦ ἐγκεφάλου πνεῦμα δυοῖν θάτερον, εἰ μὲν ἀσώματός ἐστιν ἡ ψυχή, τὸ πρῶτον αὐτῆς ὑπάρχειν, ὡς ἂν εἴποι τις, οἰκητήριον, εἰ δὲ σῶμα, τοῦτ' αὐτὸ [πνεῦμα] τὴν ψυχὴν εἶναι). Galen, *De placitis Hippocratis et Platonis*, VII.3.19–20, ed. and trans. De Lacy, 442–445.

other parts of the body. Thus, Galen says that the soul sees by virtue of the optic *pneuma*'s communion (κοινωνίαν) with other bodies, and then in turn by that optic *pneuma*'s own communion (κοινωνίᾳ) with the psychic *pneuma* of the brain. Thus, sight is possible only through the unification of various properties as one: psychic *pneuma*, optic *pneuma*, air, and light. However, this does not mean that either sunlight or air are sensitive in themselves, but rather that through their union they are able to enable perception via the *pneuma* in the eyes, which is the only proper sensitive entity for sight. Putting this in the context of Galen's broader argument, as I have laid out here, the optic *pneuma* then does not reach the objects themselves, but rather unites with the surrounding air that has been activated by light in order to receive the visible qualities of the object being sent through that light–air mixture.

Galen unequivocally states the importance of the air as medium when he writes, "Now it is clear and agreed to by all that we see through the medium of air (διὰ μέσου τοῦ ἀέρος)," but this is only the first step, laying the foundation for a more complex relationship between air, light, and *pneuma* to enable sight.[24] In reading the Stoics, Galen was clearly dissatisfied with the haptic inklings inherent in the walking stick metaphor. He resents the Stoics – particularly Chrysippus – for not fully explaining the aporia of how air and light unite with the optic *pneuma*. Galen agreed with the Stoics on the fundamental need for an elemental theory of sensation, whereby each sense was tied to a certain element, e.g. sight with light, hearing with air. The Stoics also understood both the action of the *pneuma* in producing an effect upon surrounding air and the importance of light in this process, yet they made no concerted effort to reconcile these two steps. For this reason, Galen critiques the walking stick metaphor by carefully expounding the relationship between the brain and the *pneuma*, as well as the *pneuma* and the air, making sure not to make it seem as if the *pneuma* is an all-encompassing, all-reaching body that can touch the objects of sight, from a nearby tree to a distant star. The *pneuma* activates only the air immediately next to it – in the same way in which a single beam of sunlight can immediately affect the totality of air.[25] This ensures that the sensitive faculty is embedded in the viewer, not distributed throughout the intervening medium as some extension of the body into space. The latter

[24] "Τὸ μὲν οὖν διὰ μέσου τοῦ ἀέρος ὁρᾶν ἡμᾶς ἐναργές ἐστιν καὶ πᾶσιν ὁμολογούμενον." Galen, *De placitis Hippocratis et Platonis*, VII.7.17, ed. and trans. De Lacy, 472–473.

[25] Galen, *De placitis Hippocratis et Platonis*, VII.4.24–5.9, VII.7.1–4, ed. and trans. De Lacy, 452–455, 470–471.

operations are rooted in a certain consubstantiality of air, light, and *pneuma* that comes through their union, given the light-like and air-like qualities of the optic *pneuma*.

Mathematics and Geometry: Ptolemy and Euclid

Ptolemy, a younger contemporary of Galen, built his own theory of optical rays, one modeled closely on the teachings of Euclid and informed by the teaching of the Stoics on the *pneuma*. Significantly, Ptolemy's account does not privilege Euclid's model of independent visual rays, but rather considers the efflux of the eyes to be a unified force similar to the pneumatic emission. Ptolemy included Stoic pneumatic theory because, as far as we know, Euclid's theory of vision lacks a clear articulation of the physiology of vision, despite his notion that rays are emitted as a cone from the eye. Readers of Euclid were interested exclusively in the mathematics for calculating the optical geometry of sight. The only physiology that can be reasonably deduced from Euclid's theory is laid out in the first three definitions of the *Optics*: optical rays project from the eye, creating a cone with the eye as its vertex and the object seen as its base; only the things upon which the visual rays fall are able to be seen.[26]

Certainly, Euclid held to a theory of extramission, but the ontological and physiological realities of those optical rays, and their relationship to the other senses and perception as a whole, cannot be definitively inferred. Thus, Euclid in the Byzantine world appears almost exclusively within the context of mathematical treatises, for example, as in George Pachymeres's *Quadrivium* from around 1300. Pachymeres's mathematical text on vision also does not contribute to the question of sight's tactile operation.[27] It was intended as an introduction to the four mathematical fields of arithmetic, music, geometry, and astronomy as part of the Byzantine educational program.[28] The work precisely included the Euclidian treatise under the field of Geometry (Γεωμετρία – Στερεομετρία).[29] Thus, Euclid's presence in the Byzantine world offers little to the construction of a practiced theory

[26] Euclid, *Opticorum recensio Theonis*, ed. J. L. Heiberg, *Opera omnia*, vol. vii (Leipzig: Teubner, 1895), 154–156.

[27] George Pachymeres, *Quadrivium*, ed. P. Tannery, *Quadrivium de Georges Pachymère*, Studi e Testi 94 (Vatican City: Biblioteca Apostolica Vaticana, 1940), esp. 313–314.

[28] See G. Katsiampoura, "The *Quadrivium* of 1008 and Pachymeres' *Syntagma*: Comparing Two Byzantine *Quadrivia*" in *Libri di scuola e pratiche didattiche: Dall'Antichità al Rinascimento*, vol. ii, ed. Lucio Del Corso and Oronzo Pecere (Cassino: Università di Cassino, 2010), 409–24.

[29] Pachymeres, *Quadrivium*, ed. Tannery, 201–328.

of vision beyond optical calculations. Therefore, it is impossible to assume any indication of a haptic extramission from the *Optics* in Pachymeres, given that Euclid's text itself provides little evidence on the physiology of sight beyond a mere efflux of optical rays for geometric calculation.

Observing the lack of a physiological explanation of the visual rays in the Euclidean *Optics*, Ptolemy clarifies that sight cannot be said to be a species of touch, but at best merely analogous to it.[30] In his *Optics* he likens vision to touch only through a language of analogy.[31] For instance, when describing the eye's ability to perceive concave and convex surfaces, Ptolemy states that the visual rays comprehend those objects just as touch also can perceive their curvature.[32] While he compares the two senses' ability to identify these qualities, Ptolemy never suggests that sight is a species of touch. In fact, early in his text, Ptolemy says:

> Possibile est inueniri unicuique sensuum proprium sensibile conueniens, ut species repulse manus in tactu, et humorum in gustu, et uocum in auditu, et odorum in odoratu. In omnibus uero que secundum principum neruosum communia sunt sensibus, tactus et uisus participant sibi, excepto in colore. Color enim nullo sensuum dinoscitur nisi per uisum. Debet ergo color esse sensibile proprium uisui, et ideo factus est color id quod primo uidetur post lumen.
>
> A [sole] proper sensible can be found that is appropriate to each of the senses; e.g., the quality of "resting the hand" for touch, savors for taste, sounds for hearing, and odors for smell. But among the things that are common to the senses according to the origin of nervous activity, sight and touch share in all except color, for color is perceived by no sense but sight. Thus, color must be the proper sensible for sight, and that is why color is taken to be what is primarily visible after light.[33]

Hence, while Ptolemy is attempting to identify what is common to the senses, he is sure to distinguish sight – like others before him – for its unique ability to perceive color. At the same time, Ptolemy also argues for a form of sensual medium-specificity whereby each sense is unique by virtue of its distinctive sensibles. So he would deny that sight is touch or

30 See A. Mark Smith, *Ptolemy's Theory of Visual Perception: An English Translation of the Optics* (Philadelphia: American Philosophical Society, 1996), 22–23.

31 The text of Ptolemy's *Optics* comes down to us through a Latin translation produced by a Byzantine Greek. For the Latin text, see Claudius Ptolemy, *Optics*, ed. A. Lejeune, *L'Optique de Claude Ptolémée dans la version latine d'après l'arabe de l'émir Eugène de Sicile* (Leiden: Brill, 1989); trans. A. Mark Smith, *Ptolemy's Theory of Visual Perception: An English Translation of the Optics* (Philadelphia: American Philosophical Society, 1996).

32 Ptolemy, *Optics*, II.67, trans. Smith, 99.

33 Ptolemy, *Optics*, II.13, ed. Lejeune, 17; trans. Smith, 74–75.

operates as touch, for if that were true – because touch by definition does not perceive color – the eye would be wholly unable to perceive colors and the faculty of sight would not be possible, nor would there be any distinction between sight and touch. On this matter, it seems that all the authorities agree.

As we shall see throughout this and subsequent chapters, the mathematical texts of writers such as Ptolemy and Euclid figured little in the contouring of theories of vision because of how little they have to say about the operation of the rays. Many thinkers in the Byzantine and Arab worlds retained Euclidian geometry even when they did not ascribe to a theory of extramission, recognizing its ability to accurately model mathematical calculations. Hence, Ptolemy's caveat is important because it acknowledges the failures of Euclid, despite his dominance in the instruction of geometry, and because it translated for his contemporaries the Euclidian heritage into the language of the Stoic *pneuma*.

Byzantine Medicine

The Stoic *pneuma*, as interpreted by Galen, found popularity in the Byzantine world, particularly in medical circles. Galen's work was very well known in Byzantium: every medical treatise used his work as a foundation.[34] Galen's focus on physiology and anatomical detail influenced not just Byzantine society but tenth- and eleventh-century Arabic science.[35] And the influence stretched back in time to nearly Galen's era.

[34] On Byzantine medicine, see Owsei Temkin, "Byzantine Medicine: Tradition and Empiricism," *Dumbarton Oaks Papers* 16 (1962): 95–115; John Duffy, "Byzantine Medicine in the Sixth and Seventh Centuries: Aspects of Teaching and Practice," *Dumbarton Oaks Papers* 38 (1984): 21–27. On Byzantine doctors, see Barry Baldwin, "Beyond the House Call. Doctors in Early Byzantine History and Politics," *Dumbarton Oaks Papers* 38 (1984): 15–19; Alexander Kazhdan, "The Image of the Medical Doctor in Byzantine Literature of the Tenth to Twelfth Centuries," *Dumbarton Oaks Papers* 38 (1984): 43–51; Mary E. Keenan, "St. Gregory of Nyssa and the Medical Profession," *Bulletin of the History of Medicine* 15 (1944): 150–161. For a survey of late antique and early Christian medicine, see Owsei Temkin, *Hippocrates in a World of Pagans and Christians* (Baltimore: Johns Hopkins University Press, 1993). See also John Scarborough, "Roman Medicine to Galen," *Aufstieg und Niedergang der Römischen Welt* 37:1 (1993): 3–48. For an introduction to and Italian translations of some of the texts of Oribasios of Pergamon, Aetios of Amida, Alexander of Tralles, Paul of Aegina, and Leo the Physician, see Antonio Garzya et al., *Medici Bizantini* (Turin: Unione Tipografico – Editrice Torinese, 2006).

[35] See Smith, *From Sight to Light*, esp. 36–46. See also Emilie Savage-Smith, "Galen's Lost Ophthalmology and the *Summaria Alexandrinorum*," *Bulletin of the Institute of Classical Studies. Supplement* 77 (2002): 121–138; Vivian Nutton, "From Galen to Alexander, Aspects of Medicine and Medical Practice in Late Antiquity," *Dumbarton Oaks Papers* 38 (1984): 1–14. See also Barbara Zipser, "Deleted Text in a Manuscript: Galen *On the Eye* and the *Marc. gr. 276*," *Galenos: Rivista di Filologia dei Testi Medici Antichi* 3 (2009): 107–112; J. A. C. Greppin, "The Galenic Corpus in

In the mid fourth century, Oribasios, who served as the physician to Emperor Julian the Apostate, composed a series of medical works, including an epitome of the works of Galen and the *Medical Collection*.[36] His writings drew the attention of the Patriarch Photios in the ninth century, who, summarizing four of his works, extolled the learnedness and accessibility of the work even to those uninitiated in the medical arts, echoing Oribasios's own words and wishes.[37] Of note as well in the late fourth century is Nemesios of Emesa, whose very popular *De natura hominis* combined natural philosophy and medicine, using the Galenic foundation to theorize more generally on human physiology and nature.

In the sixth and seventh centuries we find a series of important medical writers, some of whom were in the orbit of the imperial court in Constantinople, for example, Aetios of Amida, who possibly served the court of Justinian in a medical capacity, though little information about his life survives.[38] He also drew the attention of the learned medical eye of Photios, who summarized the sixteen books of his *Tetrabiblion*, the seventh of which is dedicated to diseases of the eye.[39] Photios particularly praised Aetios's use of the august and reputable works of Oribasios, Galen, Dioscurides, Posidonius, and several others.[40] Alexander of Tralles, a younger contemporary of Aetios, is survived by a treatise on fevers, another on intestinal worms, and his twelve books on medicine, the *Therapeutics*, the second of which deals with eye diseases.[41] Another book on the eyes,

Classical Armenia: A Preliminary Report," *Society for Ancient Medicine* 9 (1982): 11–13; Vivian Nutton, "Galen in Byzantium" in Michael Grünbart et al. (ed.), *Material Culture and Well-Being in Byzantium* (Vienna: Österreichische Akademie der Wissenschaften, 2007), 171–176; Nigel Wilson, "Aspects of the Transmission of Galen,"in Guglielmo Cavallo (ed.), *Le strade del testo* (Bari: Adriatica, 1987), 47–64.

36 See Barry Baldwin, "The Career of Oribasius," *Acta Classica* 18 (1975): 85–97. See also Mark Grant, *Dieting for an Emperor: A Translation of Books 1 and 4 or Oribasius' Medical Compilations with an Introduction and Commentary* (Leiden: Brill, 1997), esp. 1–17.

37 Photios, *Bibliotheca*, 216, 217, 218, 219, ed. R. Henry, *Bibliothèque*, 8 vols. (Paris: Les Belles Lettres, 1959–1977), III:131–139.

38 See John Scarborough, "Theodora, Aëtius of Amida, and Procopius: Some Possible Connections," *Greek, Roman, and Byzantine Studies* 53 (2013): 742–762.

39 Aetios of Amida, *Iatricorum liber*, 7, ed. A. Olivieri, *Libri medicinales v-viii*, vol. II, Corpus medicorum Graecorum 8.2 (Berlin: Akademie Verlag, 1950), 253–399; trans. Julius Hirschberg, *The Ophthalmology of Aëtius of Amida*, trans. Richey L. Waugh, Jr. (Ostend, Belgium: J. P. Wayenborgh, 2000), 1–130. Note: The Greek text reprinted here has many errors, and the translation as well should be used with caution.

40 Photios, *Bibliotheca*, 221, ed. Henry, III:140–152; trans. Hirschberg, 133–146.

41 John Scarborough, "The Life and Times of Alexander of Tralles," *Expedition* 39:2 (1997): 51–60. See Petros Bouras-Vallianatos, "Clinical Experience in Late Antiquity: Alexander of Tralles and the Therapy of Epilepsy," *Medical History* 58:3 (2014): 337–353. For the Latin afterlife, see D. R. Langslow, *The Latin Alexander Trallianus: The Text and Transmission of a Late Latin Medical Book* (London: Society for the Promotion of Roman Studies, 2006).

De oculis, at times has been transmitted with the *Therapeutics*, yet this book has been deemed a spurious interpolation.[42] In the early seventh century, Paul of Aegina, apparently active in Alexandria, wrote yet another compendium of medical knowledge in seven books, which likewise includes similar sections regarding diseases of the eyes.[43]

Between the seventh and thirteenth centuries, most likely in the ninth, Meletios the Monk wrote his *De natura hominis*, which appears to have enjoyed significant popularity in the late Byzantine world as attested to by the surviving manuscripts, which date from the thirteenth century onward, and none before.[44] His longest chapter happens to be that on the eyes, which devotes two sections to the explanation of the optic *pneuma* and the causes of the colors of the eye. Robert Renehan has concluded that these discussions are closely derived from Galen in language and content via a lost intermediary.[45] Not only did Meletios's work have a significant impact on later scholars, as we shall see, but Leo the Physician also produced a synopsis of Meletios's text.[46] Leo's excerpts provide us with useful summaries of Meletios's thinking, as informed by Galen, which describes the operation of the optic *pneuma*:

[42] See Barbara Zipser, "Pseudo-Alexander Trallianus. *De oculis*: Einleitung, Text, Übersetzung und Kommentar," unpublished PhD dissertation, Universität Heidelberg (2003). See also Pseudo-Alexander of Tralles, *De oculis libri tres*, ed. T. Puschmann, *Nachträge zu Alexander Trallianus*, reprint (Amsterdam: Hakkert, 1963), 134–178.

[43] Paul of Aegina, *Epitomae medicae libri septem*, ed. J. L. Heiberg, *Paulus Aegineta*, 2 vols., Corpus medicorum Graecorum 9.1 and 9.2 (Leipzig: Teubner, 1921–1924); trans. Francis Adam, *The Seven Books of Paul Ægineta*, 3 vols. (London: Sydenham Society, 1844–1847). See also Peter E. Pormann, *The Oriental Tradition of Paul of Aegina's* Pragmateia (Leiden: Brill, 2004).

[44] Meletios, *De natura hominis*, ed. J. A. Cramer, *Anecdota Graeca e codd. manuscriptis bibliothecarum Oxoniensium*, vol. III, reprint (Amsterdam: Hakkert, 1963). See J. Lascaratos and M. Tsirou, "Ophthalmological Ideas of the Byzantine Author Meletius," *Documenta Ophthalmologica* 74 (1990): 31–35; R. Reneham, "Meletius' Chapter on the Eyes: An Unidentified Source," *Dumbarton Oaks Papers* 38 (1984): 159 168. See also S. R. Holman, "On Phoenix and Eunuchs: Sources for Meletius the Monk's Anatomy of Gender," *Journal of Early Christian Studies* 16:1 (2008): 79–101.

[45] Robert Renehan, "Meletius's Chapter on the Eyes: An Unidentified Source," *Dumbarton Oaks Papers* 38 (1984): 159–168, at 162, 168 n. 49. See Meletios, *De natura hominis*, ed. Cramer, 61.7–72.20, esp. 65.27–67.12.

[46] Preserved today in a single manuscript, see Leo the Physician, *De natura hominis synopsis*, ed. and trans. Robert Renehan, *Epitome on the Nature of Man*, Corpus Medicorum Graecorum, 10.4 (Berlin: Akademie Verlag, 1969). See Barbara Zipser, "Zu Aufbau und Quellen der Σύνοψις ἰατρικῆς des Leo medicus" in Thorsten Fögen (ed.), *Ancient Technical Texts* (Berlin: Walter de Gruyter, 2005), 107–115; Barbara Zipser, "Überlegungen zum Text des Σύνοψις ἰατρικῆς des Leo medicus" in *Studia Humanitatis ac Litterarum Trifolio Heidelbergensi dedicata* (Frankfurt am Main: Peter Lang 2004), 393–399; Robert Renehan, "On the Text of Leo Medicus: A Study in Textual Criticism," *Rheinisches Museum für Philologie*, n.s. 113:1 (1970): 79–88; Athanasios Kambylis, "Zu Leos Schrift: Σύνοψις εἰς τὴν φύσιν τῶν ἀνθρώπων," *Hermes* 101:4 (1973): 483–495.

> Τὸ δὲ ὀπτικὸν πνεῦμα τὸ ἐν τοῖς ὀφθαλμοῖς φέρεται ἀπ' ἐγκεφάλου καὶ διττὴν ἀποτελεῖ τὴν ἐνέργειαν ἢ τὴν χρείαν, ἵνα πληροῖ τὸν ὀφθαλμὸν καὶ ἁδρύνῃ αὐτὸν καὶ ἵνα ἐξακοντιζόμενον ἐκ τῆς κόρης γινώσκωμεν τὰ τῶν αἰσθητῶν σώματα καὶ χροιάς. καὶ οἱ μικρὰν ἔχοντες τὴν κόρην ἐκ γενετῆς μὲν ὁράσεως ὀξυωπεστάτης, ὅτι τὸ πνεῦμα ἰσχυρῶς ἐξακοντίζεται διὰ τοῦ στενοῦ τρήματος τῆς κόρης καὶ οὐδαμῶς χεῖται, ὅπερ ἐπὶ τῆς μεγάλης γίνεται κόρης.
>
> The optic *pneuma* in the eyes is carried down from the brain and accomplishes a twofold activity or function – to fill the eye and give it bulk and that, when it is ejected from the pupil, we may know the bodies and colors of perceptible objects. Those who have a small pupil from birth have the keenest vision, because the *pneuma* is ejected with force through the narrow hole of the pupil and in no way is scattered, as happens in the case of a large pupil.[47]

Leo follows Meletios closely, describing the keenness of sight according to the force with which the *pneuma* is ejected from the eye. Such language of the pneumatic efflux might easily have been construed as part of a tactile theory of vision, yet no authors who were in the Galenic tradition and who read and used these works show any inclinations toward this reading, suggesting that they had a nuanced and extensive grasp of Galenic thought, including his sharp critiques of the Stoics.

One of the challenges of teasing out a cogent theory of sight from the medical works is that the stages in which pneumatic sensation unfolds are not given much attention, since these texts are focused on the physiology of sight and particularly its afflictions and treatments. Hippocrates's treatise on sight, for example, says little regarding the function of vision itself. It assumes the existence of the visual *pneuma*, and the reader's familiarity with it, and devotes its attention to the outward qualities of the eye, its diseases, and treatments, such as purging, scraping, and cauterizing.[48]

Paul of Aegina, for example, states for one such affliction, "such a state is wholly incurable, being occasioned by weakness of the optic *pneuma*."[49] Aetios of Amida attributes similar impairments of sight, such as night-blindness or amblyopia, as being caused by the "thickening of the optic *pneuma* " (διὰ... τοῦ ὀπτικοῦ πνεύματος παχυνομένου).[50] The

[47] Leo the Physician, *De natura hominis synopsis*, 38, ed. and trans. Renehan, 32–35.

[48] See Hippocrates, *De visu*, ed. and trans. Elizabeth M. Crank, *Two Hippocratic Treatises:* On Sight *and* On Anatomy (Leiden: Brill, 2006), 38–47.

[49] "ἀνίατος δέ ἐστιν ἡ τοιαύτη διάθεσις ὑπὸ ἀσθενείας γινομένη τοῦ ὀπτικοῦ πνεύματος." Paul of Aegina, *Epitomae medicae libri septem*, III.22.35, ed. Heiberg, I:187; modified trans. Adam, I:422–423.

[50] Aetios of Amida, *Iatricorum liber*, VII.49, ed. Olivieri, 303.

physiological focus on vision led to concerns that sexual intercourse might deplete the optical *pneuma*, as attested to by Aristotle and others in the Greek and Persian worlds.[51]

Leo the Physician, echoing both Meletios and Nemesios, offers a rare glimpse of a Byzantine medical writer's ideas of the physical theory:

> Ὅρασίς ἐστιν ἡ γινομένη διὰ τῶν ὀφθαλμῶν ἐνέργεια. τεσσάρων μάλιστα πρὸς ἐναργῆ διάγνωσιν ἡ ὄψις χρῄζει· ἀβλαβοῦς αἰσθητηρίου, συμμέτρου κινήσεως καὶ διαστήματος ἀέρος καθαροῦ καὶ λαμπροῦ φωτός.
>
> Vision is the activity which takes place by means of the eyes; for clear discernment sight needs four things especially: an undamaged organ of perception, a commensurate motion, an interval of pure air and a shining light.[52]

Leo draws the reader's attention to the importance of light and the "interval of pure air" (διαστήματος ἀέρος καθαροῦ), which allude to the mediation of sight through the pneumatic fusion with light and pure air. We might even consider the "commensurate motion" (συμμέτρου κινήσεως) to refer to the action and activity of the perceptible object, which stages vision as dual action of movement and countermovement by the perceiver and the perceived body, as in Philo's characterization of the eye's rays as countermovement.

The medical authors had widespread influence across many social spheres. Not only do we know that medicine played an important role in imperial circles, as discussed above, but it informed the popular understanding of ophthalmology. Afflictions of the eyes were of particular importance in the Byzantine world, being varied and difficult to cure and leading many to seek remedies medical and divine. In her study of ninth- and tenth-century pilgrims who visited saints' shrines for healing, Alice-Mary Talbot notes that illnesses of the eye came in third (at around 9 percent) as the most common affliction, many of which were specifically related to blindness or more general eye diseases.[53] This was superseded only by demonic possession (about a third of reported cases), and forms of paralysis (17 percent).

The practices undertaken to cure these various ailments varied from gazing upon the relics of saints, to praying to their icons, to merely

51 Touraj Daryaee, "Sight, Semen, and the Brain: Ancient Persian Notions of Physiology in Old and Middle Iranian Texts," *Journal of Indo-European Studies* 30:1/2 (2002): 103–128, esp. 109–113.

52 Leo the Physician, *De natura hominis synopsis*, 43, ed. and modified trans. Renehan, 36–37.

53 Alice-Mary Talbot, "Pilgrimage to Healing Shrines: The Evidence of Miracle Accounts," *Dumbarton Oaks Papers* 56 (2002): 153–173, esp. 158.

invoking their name. In the case of Athanasia of Aegina, as recounted in her *Vita*, pilgrims would approach her open coffin and physically touch her relics to their afflicted eyes. In life, Athanasia cured a man with diseased eyes by placing her hand upon his eyes; and through her relics, an infant who was blind in one eye and partly paralyzed was cured by being placed in her coffin.[54] Another woman who was suffering from a "very great inflamed swelling of the eyes so that they drooped [from the sockets] like apples [from a tree branch]," stood before the coffin and placed her head upon Athanasia's holy relics and was cured within two or three days.[55] Such ophthalmia – inflammation of the eyes for various reasons – was a common problem and was immensely painful, as this account vividly attests.[56]

Because ailments of the eyes could be some of the more painful common afflictions, they were of particular concern to Byzantine doctors. Consider for instance the twelfth-century *typikon* of Emperor John II Komnenos for the Monastery of Christ *Pantokrator* in Constantinople. In establishing a hospital on the grounds, eight of the fifty beds were to be set aside for those suffering from ophthalmia (ὀφθαλμιώντων), sicknesses of the stomach, and "any other very acute and painful illnesses" (ἑτέροις τισὶν ὀξυτάτοις ἐπωδύνοις).[57] As the *typikon* later specifies, those who had to take a purgative were to join those with ophthalmia and bowel diseases so that they would not interrupt the rest of others with their continuous restlessness and pain.[58] Such agonizing eye afflictions occupied much space in the medical handbooks, with some volumes dedicated exclusively to them, discussing everything from the anatomy of the eye to the diseases of

54 Anonymous, *Vita Sanctae Athanasiae*, 19 (11) and 30 (17), ed. L. Carras, "The Life of St. Athanasia of Aegina" in A. Moffatt (ed.), *Maistor: Classical, Byzantine, and Renaissance Studies for Robert Browning* (Canberra: Australian Association for Byzantine Studies, 1984), 212–224; trans. Lee Francis Sherry, "Life of St. Athanasia of Aegina," in Alice-Mary Talbot (ed.), *Holy Women of Byzantium: Ten Saints' Lives in English Translation* (Washington, DC: Dumbarton Oaks Research Library and Collection, 1996), 137–158, esp. 149–150, 156.

55 "φλεγμονὴν μεγίστην περὶ τοὺς ὀφθαλμοὺς παθοῦσα, ὥστε κρέμασθαι αὐτοὺς μήλων δίκην." Anonymous, *Vita Sanctae Athanasiae*, 19 (18), ed. Carras, 223; trans. Sherry, 156.

56 See Emilie Savage-Smith, "Hellenistic and Byzantine Ophthalmology: Trachoma and Sequelae," *Dumbarton Oaks Papers* 38 (1984): 169–186.

57 Anonymous, *Pantokrator Typikon*, 36.908–910, ed. Paul Gautier, "Le typikon du Christ Sauveur Pantocrator," *Revue des Études Byzantines* 32 (1974): 1–145, esp. 83; trans. Robert Jordan, "Pantokrator: *Typikon* of Emperor John II Komnenos for the Monastery of Christ *Pantokrator* in Constantinople" in *Byzantine Monastic Foundation Documents*, vol. II, ed. John Thomas and Angela Constantinides Hero (Washington, DC: Dumbarton Oaks, 2000), 725–781, esp. 757. On hospitals in Byzantium, see Timothy S. Miller, *The Birth of the Hospital in the Byzantine Empire*, 2nd edition (Baltimore, MD: Johns Hopkins University Press, 1997), esp. 12–29.

58 Anonymous, *Pantokrator Typikon*, 51.1172–1175, ed. Gautier, 99; trans. Jordan, 762.

the eyelids and surrounding areas, down to possible treatments ranging from herbs to surgery.

In his autobiographical poems, Gregory of Nazianzos paints a vivid image of one of his own eye injuries. He carelessly lacerated the corner of the eye, where his eyelids meet, with a twig:

> Οὐδ' ὁπότ' ἀφραίνοντι νόῳ λυγὸν ἀμφὶς ἑλίξας,
> Οὔτι ἑκὼν βλεφάρου περιηγέα κανθὸν ἄμυξα,
> Αἱματόεντα δ' ἔθηκα. Φάος δ' ἐμόν, οἷα φονῆος,
> Ὤλετο, ἀμπλακίην δὲ πικρὴν μέγα πένθος ἔτετμεν.
>
> Nor when being foolish in mind, I spun a twig around on all sides,
> And without willing it, tore the surrounding corner of the eyelid
> Making it all covered in blood. My light, just as a murderer,
> Was destroyed, and great pain overtook the prickly error.[59]

Gregory's account here captures not only the great pain and sorrow of his error, but also the fact that his sight, denoted here by "light" (φάος), was temporarily lost, presumably from the damage and blood loss. His erudite language invokes anatomical terms – the "surrounding corner of the eyelid" (βλεφάρου περιηγέα κανθόν), seemingly redundant to readily communicate to those who may not know the specialized medical terminology.[60]

This survey of medical thought demonstrates the consistent, widespread penetration of knowledge of diseases and anatomy of the eyes, so much so that it finds its way into texts as diverse as typika, vitae, and erudite poems. Furthermore, the late antique medical writers discussed here endured as major influences throughout the Middle and Late Byzantine periods.[61] Their work presents some healthy challenges to any notion of haptic vision. Let us grant that some people (e.g. those against whom pseudo-

[59] Gregory of Nazianzos, *Carmina de se ipso*, 1.330–333, ed. PG 37: 995.

[60] The ambiguity of the term is even evidenced by various modern translations, which understand these lines in quite significantly different manners. Denis M. Meehan, on the one hand, understands the first two of the lines as stating, "Nor when I unwittingly drove a thorn into my eyeball, and made it all bloody, while foolishly toying with a twig." On the other hand, Mary E. Keenan has suggested that those two verses refer to Gregory's pain when he "imprudently lanced a swollen eyelid," and which made him lose consciousness. However, as compelling as this reading might be here, I am unable to substantiate this understanding of the text given any contemporaneous medical terminology. See Denis M. Meehan (ed. and trans.), *Saint Gregory of Nazianzus: Three Poems: Concerning His Own Affairs, Concerning Himself and the Bishops, and Concerning His Own Life* (Washington, DC: Catholic University of America Press, 1987), 35–36; Mary Emily Keenan, "St. Gregory of Nazianzus and Early Byzantine Medicine," *Bulletin of the History of Medicine* 9 (1941): 8–30, esp. 31.

[61] For a brief survey of these writers and more, see John Scarborough, "Symposium on Byzantine Medicine: Introduction," *Dumbarton Oaks Papers* 38 (1984): ix-xvi.

Alexander and Plotinos argued, discussed above) with but partial medical knowledge misconstrued the Stoic *pneuma* as a tactile iteration of sight. Nevertheless, its absence from learned texts is significant, and inevitable, since a haptic interpretation is impossible from a proper reading of Galen. Second, medical writers addressed classical theories of vision within the context of Galen, whereas philosophical ones concerned themselves with vision as discussed by Plato and Aristotle. In fact, I would argue that the debate held by medical students in the courtyard of Holy Apostles in the thirteenth century as to whether "we see according to a sending out or a coming in" (κατ' ἐκπομπὴν ὁρῶμεν ἢ εἰσπομπήν)[62] referred to a debate not between followers of Plato against those of Democritus and the Epicureans, but rather more likely between disciples of Galen and those of a Platonic–Aristotelian hybrid. In other words, the primary fault line fell between medical and philosophical knowledge.

Nemesios of Emesa's *De natura hominis*

Nemesios, bishop of Emesa, composed his *De natura hominis* in the fourth century. While at times treated as a medical author, Nemesios truly stands out for his scope, breadth, and goals, since he sought to amalgamate a variety of texts and methods to reflect on the nature of the human being, physiologically, cognitively, and spiritually, focusing on sensual perception, cognition, and the faculties of the soul. This treatise became one of the most influential volumes on the human condition in Byzantium. Not only is it an important source for John of Damascus's *Expositio fidei* and his defense of treatises on icons for matters of human perception and psychology, but it was the key source for many of the Byzantine medical handbooks, including that of Meletios. However, most quotes from the work were left uncredited by Byzantine authors, who preferred to name the more august figures that Nemesios himself credited.

Nemesios's entry on sight in *De natura hominis* provides an extensive summary of theories of vision from antiquity to his present, explicitly citing Hipparchus, the 'Geometricians' (probably Euclid and Ptolemy), the Epicureans, Plato, Aristotle, Galen and the Stoics, and Porphyry. His work is deeply indebted to the writings of Alexander of Aphrodisias and especially Aetios of Antioch, from whom Nemesios partly derives his

[62] Nicholas Mesarites, *Descriptio ecclesiae SS. Apostolorum*, 42.5, ed. and modified trans. Downey, 895, 917.

survey of sight, at times verbatim.[63] Nemesios even appears to have consulted Theon of Alexandria's recension of Euclid's *Optics*, which explains the attention paid to the "Geometricians" in his work. While that attention is not very extensive, it is still quite notable for a treatise focused on neither geometry, mathematics, nor astronomy.[64] Appearing after the entry on the imagination, the discussion of sight is succeeded by entries on touch, taste, hearing, smell, thought, and memory that carefully place the senses within the mental and cognitive processes responsible for processing and preserving sensory impulses and for enabling thought and reason.

Nemesios begins his survey by explaining the views of Hipparchus, who says that "rays extend from the eyes and by their extremities lay hold of the external bodies like the touch of hands (χειρῶν ἐπαφαῖς καθαπτούσας), and thus transmit their apprehension (ἀντίληψιν) to the sense of sight."[65] This statement, nearly identical to that of Aetios in *De placita philosophorum*, is one of the few to explicitly characterize the power of sight or the rays as being tactile or capable of touch.[66] Yet, Nemesios does not place any import on this, and moves immediately to the views of the "Geometricians," explaining the theory of rays emitted from the eye in the shape of a cone, and dwelling on geometrical aspects of vision. He moves on to explain the key theories ranging from the Epicureans to Galen, writing:

> οἱ δὲ Ἐπικούρειοι εἴδωλα τῶν φαινομένων προσπίπτειν τοῖς ὀφθαλμοῖς. Ἀριστοτέλης δὲ οὐκ εἴδωλον σωματικόν, ἀλλὰ ποιότητα δι᾽ ἀλλοιώσεως τοῦ πέριξ ἀέρος ἀπὸ τῶν ὁρατῶν ἄχρι τῆς ὄψεως παραγίνεσθαι. Πλάτων δὲ κατὰ συναύγειαν τοῦ μὲν ἐκ τῶν ὀφθαλμῶν φωτὸς ἐπὶ ποσὸν ἀπορρέοντος εἰς τὸν ὁμογενῆ ἀέρα, τοῦ δὲ ἀπὸ τῶν σωμάτων ἀντιφερομένου, τοῦ δὲ περὶ τὸν μεταξὺ ἀέρα εὐδιάχυτον ὄντα καὶ εὔτρεπτον συνεκτεινομένου τῷ πυροειδεῖ τῆς ὄψεως.
>
> The Epicureans say that [vision occurs by] images (εἴδωλα) of what appears falling upon (προσπίπτειν) our eyes. But Aristotle says that it is not a corporeal image but a quality that comes from the object of sight to [the organ of] sight through an alteration of the surrounding air. Plato, however,

[63] On Nemesios's sources, see R. W. Sharples and P. J. van der Eijk, *On the Nature of Man* (Liverpool University Press, 2008), 94–101. See also Thomas Halton, "The Five Senses in Nemesius, *De Natura Hominis* and Theodoret, *De Providentia*," *Studia Patristica* 20 (1989): 94–101.

[64] Euclid, *Opticorum recensio Theonis*, ed. J. L. Heiberg, *Opera omnia*, vol. VII (Leipzig: Teubner, 1895), 144–246. See Johan L. Heiberg, *Geschichte der Mathematik und Naturwissenschaften im Altertum*, Handbuch der Altertumswissenschaft 5.1.2 (Munich: C. H. Beck, 1926), 75 n. 5.

[65] "Ἵππαρχος δέ φησιν ἀκτῖνας ἀπὸ τῶν ὀφθαλμῶν ἀποτεινομένας τοῖς πέρασιν ἑαυτῶν, καθάπερ χειρῶν ἐπαφαῖς καθαπτούσας τοῖς ἐκτὸς σώμασι, τὴν ἀντίληψιν αὐτῶν πρὸς τὸ ὁρατικὸν ἀναδιδόναι." Nemesios, *De natura hominis*, 7, ed. Morani, 57; modified trans. Sharples and van der Eijk, 104.

[66] Pseudo-Plutarch (Aetios of Antioch), *De placita philosophorum*, IV.13, ed. Mau, 123.

> says that it is a meeting (συναύγειαν) of the light from the eyes, so far as it flows (ἀπορρέοντος) into the air that is homogeneous with it, and of the light travelling in the opposite direction (ἀντιφερομένου) from bodies, while the light in the air between (μεταξὺ ἀέρα), which is easily diffused and changed, extends away to the fire-like [thing] of sight.[67]

Here, we have a comparison of the key theories that circulated within medieval philosophy, particularly that which looked to Plato and Aristotle. As expected, he extols the theory of the union or meeting of rays, which here are described as light (φωτός). This allows Nemesios to transition to the work of Galen, which he initially describes as being in agreement with Plato, at least in how sight perceives size and scale.

However, Nemesios goes on to develop at length the theory of the optic *pneuma* through a series of quotations and paraphrases taken from Galen's *On the Doctrines of Hippocrates and Plato.*[68] Nemesios's summary is worth citing at length:

> ἀλλὰ μὴν καὶ τὸ πνεῦμα τὸ ὀπτικὸν οὐχ οἷόν τε τοσαύτην ἰσχὺν ἐκκρινόμενον λαμβάνειν, ὡς περιχεῖσθαι παντὶ τῷ βλεπομένῳ. ‛λείπεται οὖν τὸν πέριξ ἀέρα τοιοῦτον ὄργανον ἡμῖν γίνεσθαι, καθ' ὃν ὁρῶμεν χρόνον, ὁποῖον ἐν τῷ σώματι τὸ νεῦρον ὑπάρχει τὸ ὀπτικόν. τοιοῦτον γάρ τι πάσχειν ἔοικεν ὁ περιέχων ἡμᾶς ἀήρ· ἥ τε γὰρ αὐγὴ τοῦ ἡλίου, ψαύουσα τοῦ ἄνω πέρατος τοῦ ἀέρος, διαδίδωσιν εἰς ὅλην τὴν δύναμιν, ἥ τε διὰ τῶν ὀπτικῶν νεύρων αὐγὴ φερομένη τὴν μὲν οὐσίαν ἔχει πνευματικήν, ἐμπίπτουσα δὲ τῷ περιέχοντι καὶ τῇ πρώτῃ προσβολῇ, τὴν ἀλλοίωσιν ἐργαζομένη, διαδίδωσιν ἄχρι πλείστου συνεχοῦς ἑαυτῇ, ἄχρις ἂν εἰς ἀντιτυπὲς ἐμπέσῃ σῶμα.' γίνεται γὰρ ὁ ἀὴρ ὄργανον τῷ ὀφθαλμῷ πρὸς τὴν τῶν ὁρωμένων διάγνωσιν τοιοῦτον, οἷόνπερ ἐγκεφάλῳ τὸ νεῦρον, ὥστε ὃν ἔχει λόγον ἐγκέφαλος πρὸς τὸ νεῦρον, τοῦτον ἔχειν τὸν ὀφθαλμὸν πρὸς τὸν ἀέρα ἐψυχωμένον ὑπὸ τῆς ἡλιακῆς αὐγῆς. ὅτι δὲ πέφυκεν ὁ ἀὴρ τοῖς πλησιάζουσι σώμασι συνεξομοιοῦσθαι, δῆλον ἐκ τοῦ καὶ πυρροῦ τινος ἢ κυανοῦ ἢ καὶ ἀργύρου λαμπροῦ διαφερομένου, φωτὸς ὄντος, ὑπὸ τοῦ διενεχθέντος ἀλλοιοῦσθαι τὸν ἀέρα.

> But moreover the optic *pneuma*, when being emitted, also would not be able to gain (λαμβάνειν) so great a strength to fill-up (περιχεῖσθαι) all that was seen. "So it remains that the surrounding air becomes such an instrument for us at the time that we see, as is the optic nerve in the body. For the air that surrounds (περιέχων) us seems to be affected in some such way. For as the ray from the sun which touches the upper limit of the air hands on its power

[67] Nemesios, *De natura hominis*, 7, ed. Morani, 57; trans. Sharples and van der Eijk, 105.

[68] For a comparison, see Bolesław Domański, *Die Psychologie des Nemesius* (Munich: Aschendorff, 1900), 101–102. See Galen, *De placitis Hippocratis et Platonis*, ed. and trans. Phillip De Lacy, *On the Doctrines of Hippocrates and Plato*, Corpus medicorum Graecorum 4.1.2 (Berlin: Akademie Verlag, 1980), esp. 452–460.

to the whole, likewise also the ray as it travels through the optic nerve has the character of *pneuma*, and falls on its surroundings and brings about an alteration as soon as it touches them, and thus passes it (διαδίδωσιν) on to the limit of what is continuous with it until it falls upon a resistant body." For air becomes an instrument for the eye for the recognition of visible objects such as is the nerve for the brain, so that the eye has the same relation to the air that has been given soul power (ἀέρα ἐψυχωμένον) by the sun's ray as the brain has to the nerve. That air naturally becomes like bodies near to it is clear from the fact that air, when something bright, red or blue or even shining silver travels through it, is altered by that which travels through it.[69]

It is within the context of the *pneuma*'s apprehension of the sensible object that we might find the most pronounced association between sight and touch when Nemesios writes that it, to translate it more literally: "falls upon that which surrounds it and the first encounter, causing an alteration" (ἐμπίπτουσα δὲ τῷ περιέχοντι καὶ τῇ πρώτῃ προσβολῇ, τὴν ἀλλοίωσιν ἐργαζομένη). This statement stresses a critical point in these theories of vision: that touch does not necessarily imply haptic contact. In Greek, statements that describe something as encompassing or being encompassed by something, or touching or being touched, regarding the operation of light, air, or *pneuma*, do not imply haptic, tactile contact, but rather are used to denote relative position. This does not suggest sense perception and cannot do so, given that Galen has explicitly stated that air is not sentient, but only can transmit alterations through itself back to the sentient properties. While one might wish to blame this pseudo-haptic language for conflations of sight and touch in modern historiography, it demonstrates that more so than Plato, it is ironically Galen's clarifying words that might be blamed. Yet, given Galen's clear and consistent desire to distance sight from touch, it is difficult to make a clear case for any instance wherein Galen's language could have been misread by a Byzantine author to imply a touching seeing.

Nemesios's phrase is a direct quotation from Galen, and it finds its way into Meletios's *De natura hominis* verbatim.[70] For a moment here, I wish to pause and consider the possible renderings of this phrase, using modern translation as a manner of not only examining our primary sources, but also the expectations with which we as scholars have read these texts. In the block quote above, I have retained R. W. Sharples and P. J. van der Eijk's translation, and then subsequently also rendered the phrase more literally,

[69] Nemesios, *De natura hominis*, 7, ed. Morani, 58–59; trans. Sharples and van der Eijk, 105–106.

[70] Galen, *De placitis Hippocratis et Platonis*, VI.5.7, ed. and trans. De Lacy, 454–455. See Meletios, *De natura hominis*, ed. Cramer, 71.19.

while nevertheless retaining their reading of the phrase. There is a critical problem that emerges here in our understanding of the participle περιέχοντι, understood as surrounding or embracing. It almost seems as if Sharples and van der Eijk have translated this, or at least clarified the statement with "touches them," taking for granted the tactility of sight. There is certainly a degree of tactility that could be read into the use of the verb περιέχειν in order to suggest that there is some form of embrace occurring through the optic *pneuma*, but this would be a modern misreading of the Greek, which is best translated as "scatter about" or "drench" rather than "embrace," if we are to more adequately render its usage.

Yet, it is crucial that this portion of the text begins precisely with "would not be able to gain (λαμβάνειν) so great a strength to scatter about (περιχεῖσθαι) all that was seen." Thus, moving on to demonstrate that in fact the *pneuma* is not able to encompass the object, but rather it is through the air that encompasses us all that the *pneuma* can sense objects: that is, by activating this encompassing medium of air when the *pneuma* first falls upon it (ἐμπίπτουσα δὲ τῷ περιέχοντι καὶ τῇ πρώτῃ προσβολῇ τὴν ἀλλοίωσιν ἐργαζομένη). This clearly indicates that περιέχοντι grammatically refers to the air, as in the earlier "the air that encompasses us" (ὁ περιέχων ἡμᾶς ἀήρ); and, in terms of syntax, the removal of the comma in the phrase makes this reading even clearer, as is the case in Benedict Einarson's revised critical edition of the text.

Therefore, for our purposes here, Nemesios's statement is better rendered in Phillip De Lacy's translation of Galen's original, which Nemesios repeats almost verbatim:

> τοιοῦτον γάρ τι πάσχειν ἔοικεν ὁ περιέχων ἡμᾶς ἀὴρ ὑπὸ τῆς τοῦ πνεύματος ἐκπτώσεως, ὁποῖόν τι καὶ πρὸς τῆς ἡλιακῆς αὐγῆς. ἐκείνη τε γὰρ ψαύουσα τοῦ ἄνω πέρατος αὐτοῦ διαδίδωσιν εἰς ὅλον τὴν δύναμιν, ἥ τε διὰ τῶν ὀπτικῶν νεύρων ὄψις φερομένη τὴν μὲν οὐσίαν ἔχει πνευματικήν, ἐμπίπτουσα δὲ τῷ περιέχοντι καὶ τῇ πρώτῃ προσβολῇ τὴν ἀλλοίωσιν ἐργαζομένη διαδίδωσιν ἄχρι πλείστου, συνεχοῦς αὐτῷ δηλονότι τοῦ πέριξ σώματος ὑπάρχοντος.
>
> It seems that the effect produced on the air around us by the emission of the *pneuma* is of the same sort as the effect produced on it by the light of the sun. For sunlight, touching the upper limit of the air, transmits its power to the whole; and the vision that is carried through the optic nerves has a substance of the nature of *pneuma*, and when it strikes the surrounding air it produces by its first impact an alteration that is transmitted to the furthest distance.[71]

[71] Galen, *De placitis Hippocratis et Platonis*, 7.5.6–7, ed. and trans. De Lacy, 454–455.

Here, we can appreciate the point made by Galen (and Nemesios and Meletios by extension), where, faced with the inability to scatter about (περιχεῖσθαι) the visible object, the *pneuma* activates the air so that it might then transmit (διαδίδωσιν) sensory data. In this model, as Nemesios clearly recounts, the air operates as a sort of prosthetic organ for the eye (ὁ ἀὴρ ὄργανον τῷ ὀφθαλμῷ) by virtue of having "ensouled" the air (ἀέρα ἐψυχωμένον). Hence the faulty metaphor of the mediating walking stick, because while a stick can transmit touch back to the hand that holds it in a crude manner, the air can transmit sense data back to the *pneuma* that activates it. This system works on the notion of similarity. Just as the rays (or lights) in Plato coalesce, like with like, here the air has the ability to "recognize" and be joined to a form of likeness, as Nemesios explains.

Nemesios uses Galen to transition to the thinking of Porphyry, whose theory of sight relies on the notion that the soul simply recognizes fellow, perceptible beings. Yet, Nemesios's report on Porphyry is limited, moving quickly past his theories, and turning to broader matters regarding the operation of vision, such as how it grasps size, shape, and place, and how it relates to the other senses. Nemesios asks how sight (and by extension other senses too) can on its own not only perceive common properties but also infer sensory details not proper to sight, for example, heat from merely looking at a fire. This passage eliminates potential confusion between sight and touch, as Nemesios vehemently argues:

> μὴ ἀπατήσῃ δέ τινας ὡς καὶ τῶν θερμῶν ἡ ἀντίληψις ὄψεώς ἐστιν, ἐπειδὴ τὸ πῦρ ὁρῶντες ἴσμεν εὐθέως ὅτι καὶ θερμόν ἐστιν· ἂν γὰρ ἐπὶ τὴν πρώτην ὄψιν ἀναγάγῃς τὸν λόγον, εὑρήσεις ὡς τότε, ὅτε πρῶτον ἡ ὄψις ἐθεάσατο τὸ πῦρ, μόνου τοῦ χρώματος καὶ σχήματος ἔσχε τὴν ἀντίληψιν, προσελθούσης δὲ καὶ τῆς ἁφῆς ἐγνωρίσαμεν ὅτι καὶ θερμόν ἐστιν, ὅπερ διεφύλαξεν ἡ μνήμη παραλαβοῦσα παρὰ τῆς ἁφῆς. νῦν οὖν, ὅταν θεασώμεθα πῦρ, οὐδὲν ἄλλο βλέπομεν ἢ τὸ σχῆμα καὶ τὸ χρῶμα τοῦ πυρός, ἡ δὲ διάνοια συμπαραλαμβάνει τοῖς ὁρωμένοις διὰ τῆς μνήμης καὶ τὴν θερμότητα.
>
> Nobody should be under the delusion that the grasping of hot things is through sight because when we see fire, we immediately know that it is also hot. For if you reduce the account to the first sight of it you will find that then, when sight first observed fire, it received only its colour and shape; but when touch (ἁφῆς) also approached it we recognized that it is hot, and this was preserved by memory, which took it from touch (μνήμη παραλαβοῦσα παρὰ τῆς ἁφῆς). So now, when we observe fire, we see nothing other than the colour and form of the fire, but thought (διάνοια) adds on heat as well through the memory, in addition to what is seen.[72]

[72] Nemesios, *De natura hominis*, 7, ed. Morani, 61; trans. Sharples and van der Eijk, 108.

Here, Nemesios almost seems to address what is at times described by Aristotle as the common sense, whereby the synthetic faculty of perception stitches together the unique experiences of the various senses so as to comprehend perceptibles as models in the imagination beyond their mere sensual impression, addressed more fully in the next chapter. Nemesios leaves no room for confusion that sight might possess the qualities of touch, even when it recalls the experiences of touch and other senses. These are supplied by thought (διάνοια) and memory (μνήμη), and to which the imagination (φαντασία) also contributes.

Therefore, Nemesios's language permits no confusion of Galenic *pneuma* being constituted as touch, problematic modern translations notwithstanding. Nemesios emphatically leaves no possibility of conflating any two senses, demonstrating the uniqueness and separation of each sense and explaining how forms of synesthetic experience (e.g. when seeing might lead to the experience of touch) are a product of the associative powers of the cognitive faculties, not one sense channeling through another. The only way to construe haptic vision from this passage is to give credence to the opinion he attacks: "Nobody should be under the delusion that the grasping of hot things is through sight" (μὴ ἀπατήσῃ δέ τινας ὡς καὶ τῶν θερμῶν ἡ ἀντίληψις ὄψεώς ἐστιν).

Nemesios's popular text is not only an exemplary survey of classical and late antique theories of sight for Christian audiences, but also a model of discourse written with care and criticality. He does not simply apply loose metaphors to the senses, but he places them in dialogue with each other. He articulates the senses' commonalities and distinctions, not just in the section on sight, but everywhere in his discussion of sensation and cognition.

Symeon Seth and the Arab World

Symeon Seth, a contemporary of Michael Psellos, is another important figure in Byzantine theories of vision, since his work eloquently juxtaposes the mathematical optics of Ptolemy and Euclid with philosophical theories of sensation and perception. Like Nemesios, his erudite and syncretic work evidences a deep knowledge of Plato, Aristotle, the Stoics, Galen, and various late antique commentators whose work enjoyed prominence in Byzantine intellectual circles, as well as the thought of his Arab contemporaries. In his *Conspectus rerum naturalium*, Symeon Seth discusses the operation of sight and various experiments on optics, derived from Ptolemy and even explicitly citing his *Optics*. There, Symeon says that the mathematicians and many of the philosophers believe that rays come out of the eyes and seize (καταλαμβάνουσι) the object of sight. This passing remark demonstrates the problematic use of metaphor in these sources, since this might suggest that Symeon subscribed to a haptic extramission where there was a literal touching and grasping of the object seen. However, a proper grasp of Symeon's point requires context, with careful attention to his sources and other discussions on the senses. His entry "On Seeing" (Περὶ ὁράσεως) reads as follows:

> Περὶ τοῦ πῶς ὁρῶμεν πολλὰς ἔσχον οἱ παλαιοὶ ἀμφιβολίας καὶ ἄμφω τὰ μέρη τοσοῦτον ἕκαστον τὴν ἰδίαν δόξαν ἐκράτυνεν ὡς καὶ τὸν πρῶτον φιλόσοφον ἐπαπορῆσαι τίνι τούτων τὴν νικῶσαν ἐπιψηφίσοιτο. οἱ μὲν οὖν Ἀριστοτελικοὶ ἐδόξασαν ὡς τὰ τῶν ὁρωμένων ἰνδάλματα διαπορθμεύει ὁ ἀὴρ ἐπὶ τὸ κρυσταλλοειδὲς τὸ ἐν τῷ ὀφθαλμῷ καὶ τοῦτο διαφανὲς ὂν δέχεται τὸν τύπον τῶν ὁρατῶν. οἱ δὲ γεωμέτραι καὶ πολλοὶ τῶν φιλοσόφων ἀπεφήναντο ὡς ἀκτῖνές τινες ἐξέρχονται ἐκ τῶν ὀφθαλμῶν καὶ καταλαμβάνουσι τὸ ὁρατόν, παρεικάσαντες ταύτας ταῖς τοῦ ἡλίου ἀκτῖσιν. ὥσπερ γὰρ αὗται ἅμα τῷ τὸν ἥλιον ἀνατεῖλαι ἀχρόνως ἐπιλάμπουσι τοῖς τῇδε, πολλοῦ τοῦ διαστήματος μεταξὺ τῆς γῆς καὶ τοῦ ἡλίου ὄντος, οὕτω καὶ ἡ ἡμετέρα ὄψις ἅμα τῷ ἀναβλέψαι καταλαμβάνει ἀχρόνως τὸν οὐρανόν. καὶ ἔνθα ἂν διαπερῶσιν αἱ ἡλιακαὶ ἀκτῖνες, ἐκεῖσε καὶ ἡ ὄψις, ὡς ἐπὶ τῶν διαφανῶν σωμάτων, καὶ ἔνθα κωλύονται αὗται διελθεῖν, καὶ ἡ ὅρασις οὐ διαβαίνει, καὶ ἔνθα ἀνακλᾶται ἡ ἡλιακὴ ἀκτίς, ἐκεῖθεν ἀντανάκλασιν ποιεῖται καὶ ἡ ὄψις ὡς ἐπὶ τῶν λείων σωμάτων. καὶ γὰρ ὅτε ἐπιλάμπει ὁ ἥλιος ὕδατι καθαρῷ, ὁρῶμεν τὰς ἀκτῖνας ἀπὸ τούτου ἀνακλωμένας καὶ ἡμῶν ὁρώντων τὸ ὕδωρ ἀνακλᾶται ἡ ἡμετέρα ὄψις, ὥστε φαίνεσθαι ἡμῖν τὰ ἄνωθεν τοῦ ὕδατος καὶ τὰ πέριξ ἐν τῷ ὕδατι ὄντα. ὡσαύτως καὶ ἐπὶ τῶν κατόπτρων ἀνακλᾶται προδήλως ἡ ἡλιακὴ ἀκτίς, καὶ τοσοῦτον ὡς ἀπὸ τῶν κοίλων κατόπτρων καὶ καίειν· ἀνακλᾶται δὲ ἡ ὄψις ἐπεὶ ἀντανακλωμένης αὐτῆς πρὸς τὸ ἡμέτερον πρόσωπον, ὁρῶμεν τούτου τὸ ὁμοίωμα καὶ δοκοῦμεν ἐν τῷ κατόπτρῳ ὁρᾶν. καὶ μέχρι τούτου ἰσόρροπα καὶ ἰσοτάλαντα τὰ ἀμφισβητούμενα ἦν. ἐφευρὼν δέ τις τῶν γεωμετρῶν τὴν ἐν τῷ ὕδατι τῆς ὄψεως ἀνάκλασιν, καὶ τοῦτο πιστωσάμενος δι᾽ αὐτῆς τῆς αἰσθήσεως, δυναμικωτέρας τῶν γεωμετρῶν τὰς δείξεις ἀπέδειξεν· εἰ γάρ τις ἔν τινι κοίλῳ ἀγγείῳ θήσει δακτύλιον ἢ νόμισμα ἢ ἕτερόν τι τῶν σωμάτων καὶ ἀπομακρυνεῖ ὥστε μὴ τοῦτο ὁρᾶν, ὕδατος ἐπιχεθέντος τούτῳ, ὀφθήσεται. ἐξ οὗ ἐτεκμήραντο ὡς διὰ μὲν τοῦ ἀέρος ἰθυτενῶς ἡ ὄψις δίεισιν, ἐν δὲ τῷ ὕδατι περικλᾶται. διὰ τοῦτο γὰρ καὶ ἡ κώπη ἐν τῷ ὕδατι περικεκλασμένη φαίνεται, τῆς ὄψεως τοῦτο τὸ πάθος ὑφισταμένης καὶ δοκούσης τοῦ ὁρωμένου τοῦτο εἶναι. διὰ ταύτην οὖν τὴν αἰτίαν, ὡς ἐν τοῖς ὀπτικοῖς δείκνυται, ἐν μὲν τῷ ἀέρι ἐφ᾽ ὅσον τὰ ὁρώμενα πορρώτερον ᾖ, μικρότερα φαίνεται, ἐν δὲ τῷ ὕδατι τὸ ἀνάπαλιν. ὅσον γὰρ ἐστι βαθύτερον τὸ ὕδωρ, τοσοῦτον τὰ ἐν τῷ βάθει αὐτοῦ μείζονα φαίνεται. ἐξ ὧν δῆλον ὡς ὑγιεστέρα ἐστὶν ἡ δόξα ἡ δι᾽ ἀκτίνων ἀπὸ τῶν ὀφθαλμῶν ἐκπεμπομένων πρεσβεύουσα γίνεσθαι τὴν ὅρασιν. φησὶ δὲ ὁ

Πτολεμαῖος ἐν τοῖς ὀπτικοῖς ὅτι αἰθερῶδές τί ἐστι καὶ τῆς πέμπτης οὐσίας τὸ ὀπτικὸν πνεῦμα.

Concerning how we see, the ancients had many quarrels and both sides confirm each their own belief, just as you might question the first philosopher, [one] will question the prevailing [belief] by anyone. On the one hand, the Aristotelians argued that the air transmits the images of the things seen onto the crystalline in the eye and this being transparent (διαφανές) accepts the image (τύπον) of the thing seen. On the other hand, the geometricians and many of the philosophers declared that some rays come out (ἐξέρχονται) of the eyes and seize (καταλαμβάνουσι) the thing seen, these being comparable to the rays of the sun. For just as when the sun gushes them forth they instantaneously shine upon here, despite there being a great interval between the sun and the earth, in this way our sight [can] at the same time immediately seize the heavens being looked at. And whence the rays of the sun traverse, thither also goes sight, as upon diaphanous bodies, and whence they are prevented from penetrating, thither sight does not proceed, and whence the rays of the sun reflect, thither sight also makes a reflection, as on smooth bodies. For when the sun shines forth on pure water we see the rays reflecting from this and when we gaze at the water our sight is reflected, therefore to us is visible both the things that are above the water and those things that are in it. In the same way, upon mirrors the rays of the sun reflect clearly, and even go so far as to burn when coming from concave mirrors. But sight reflects when we see its [i.e. our face's] likeness (ὁμοίωμα), after it is being reflected toward our face, and think we see it in the mirror. And up to there the disputed [images] are equally scaled and equally balanced. But one of the geometers, having discovered the reflection in the water of the face (ὄψεως), and having relied on this through bare sensation, demonstrated proofs more powerful than the other geometers. For if someone were to put in some hollow container a ring or coin or some other object and set the vessel far enough away that they could not see the [object], when water is poured into it, [the object] would become visible. From which they proved that sight passes straight through the air, but is refracted in the water. For this explains the appearance of a handle refracted in the water, when sight succumbs to this experience it thinks [the refraction] to be the thing seen. Thus, on this same basis, just as is demonstrated in the *Optics* (ἐν τοῖς ὀπτικοῖς), in the air the farther the things seen are, the smaller they appear; but in water [it is] the opposite. For the deeper the water is, the larger the thing in its depth seems. Hence, it is evident that the belief advocating that the emissions of rays is the source of sight is more sound. Ptolemy says in the *Optics* (ἐν τοῖς ὀπτικοῖς) that the optical *pneuma* is some ether-like thing of the fifth essence (τῆς πέμπτης οὐσιας).[73]

[73] Symeon Seth, *Conspectus rerum naturalium*, IV.71–74, ed. A. Delatte, *Anecdota Atheniensia et alia*, vol. II (Liège: H. Vaillant-Carmanne, 1939), 17–89, esp. 71–73.

Beginning with a summary of the Aristotelian theory of sight, Symeon soon turns his attention to describing the view espoused by the "mathematicians and many of the philosophers," who describe how rays emerge from the eyes and seize the visual object (ἀκτῖνές τινες ἐξέρχονται ἐκ τῶν ὀφθαλμῶν καὶ καταλαμβάνουσι τὸ ὁρατόν). Here, Symeon is referring to the views of the extramissionists, to which, in contrast to the Aristotelians, the rest of his commentary is devoted. But rather than focusing on Platonic extramission, as might be expected, Symeon is interested in the mathematical aspect of optics.

So, rather than explaining the physiology of seeing, Symeon follows closely in the geometrical tradition. His treatise is deeply indebted to Ptolemy's *Optics* and has a strong interest in Euclidean geometry. Ptolemy – the only figure mentioned by name from the geometers – is the basis for the explanatory excursus of the "floating coin" experiment, quoted above, where a coin is placed in an empty cup placed far enough away that the bottom cannot be seen, and then water is poured in so that the refracted image of the coin appears on the surface of the water.[74] Symeon's text likewise evidences knowledge of other optical geometry (not addressed directly by Ptolemy), such as the study of so-called burning mirrors, attested to in sources from Euclid to Anthemios of Tralles to Arabic thinkers contemporary with Symeon Seth.[75]

The emphasis on Euclid and Ptolemy, and the focus on the experiments of the floating coin and burning mirrors are notable for their shared prominence in contemporary Arabic texts. Symeon Seth was unique. He learned astronomy from Arabic sources, translated the *Kalīlah wa Dimnah* from the Arabic, and spent time in Cairo between 1057 and 1059.[76] His

74 Ptolemy, V.5–7, trans. Smith, 230–232.

75 The study of "burning mirrors" is described in Euclid's *Catoptrics*, 30, at length and with further nuance in Diocles's treatise *On Burning Mirrors* (which comes down to us only as an Arabic translation of the Greek original), in the writings of Anthemios of Tralles on conics, and in the treatise by Ibn Sahl, *Fī al-'āla al-muḥriqa* (the *On Burning Instruments*), written in Baghdad around 984. See Euclid, *Catoptrica*, 30, ed. Heiberg, *Opera omnia*, 285–343, esp. 340–342; Diocles, *On Burning Mirrors*, ed. and trans. G. J. Toomer, *On Burning Mirrors: The Arabic Translation of the Lost Greek Original* (Berlin: Springer, 1976); Anthemios of Tralles, *Περὶ παραδόξων μηχανημάτων*, ed. and trans. G. L. Huxley, *A Study in Later Greek Geometry*, Greek, Roman, and Byzantine Studies Monographs 1 (Cambridge, MA: Harvard University Press, 1959); Roshdi Rashed, "A Pioneer in Anaclastics: Ibn Sahl on Burning Mirrors and Lenses," *Isis* 81.3 (1990): 464–491. See also A. Mark Smith, *Ptolemy and the Foundations of Ancient Mathematical Optics: A Source Based Guided Study* (Philadelphia: American Philosophical Society, 1999), 153–160.

76 On Symeon Seth and his learning, see Paul Magdalino, "The Byzantine Reception of Classical Astrology" in C. Holmes and J. Waring (eds.), *Literacy, Education and Manuscript Transmission in Byzantium and Beyond* (Leiden: Brill, 2002), 33–58, esp. 46–48; Scarborough, "Symposium," xiii-xiv; Delatte, *Anecdota Atheniensia*, 1–2.

purported birthplace, suggested as both Alexandria and Antioch, helps explain his fluency in Arabic and intimate knowledge of Arabic sources. In the mid eleventh century, contemporary Islamic thinkers on medicine were actively interested in critical readings of Galen's work and it is even possible that some of Symeon's work derived from that of his contemporaries, such as Ibn Riḍwān from Giza, who was active in Cairo, and Ibn Buṭlān from Baghdad, who spent time in Cairo before arriving in Constantinople in 1053 during the Great Schism.[77] Both Arabic writers have been suggested by Petros Bouras-Vallianatos and Sophia Xenophontos to have been the basis for Symeon's *Refutation of Galen*, which rebutted some of Galen's physiological ideas.[78] That treatise speaks to Symeon's training and work as a doctor, as well as his critical engagement with Galen's corpus. It also attests to Galen's enduring popularity, for as Symeon Seth writes in his introduction, "you are on almost everyone's lips, considered faultless in every respect and praised as superhuman" (ἐπὶ γλώττης σχεδὸν πάντων κείμενον καὶ ἄπτωτον πάντῃ λογιζόμενον καὶ ὑπὲρ ἄνθρωπον εὐφημούμενον).[79] It would be reasonable to infer that Symeon Seth would have been familiar as well with contemporary Arabic writings on vision, which similarly followed in the tradition of Plato, Galen, Euclid, and Ptolemy.

Many Arabic treatises then in circulation, such as the writings of Ḥunayn ibn 'Isḥāq and Al-Kindī in the ninth century, addressed how emanations from an object enabled sight. The Arabic writers' reliance on either a united rays theory or an optical *pneuma* would have been in keeping with the contemporary Byzantine thinking. But even if clear avenues of translation and exchange cannot be securely determined, Symeon Seth's topics and very selection of examples in his description of geometrical optics is identical to those that drew the interest of Ibn Sahl, Al-Haytham, Ḥunayn ibn 'Isḥāq, and Al-Kindī. While this does not securely prove his familiarity with his contemporaries' works, it nevertheless suggests that Symeon is representative of a certain culture of pan-Mediterranean learning.[80]

77 See Joseph Schacht and Max Meyerhof, *The Medico-Philosophical Controversy between Ibn Butlan of Baghdad and Ibn Ridwan of Cairo* (Cairo: Egyptian University, 1937); Jacques Grand'Henry, *Le livre de la méthode du médecin de 'Ali b. Ridwan (998–1067)*, 2 vols. (Louvain-la-Neuve: Université catholique de Louvain, Institut orientaliste, 1979).

78 Petros Bouras-Vallianatos and Sophia Xenophontos, "Galen's Reception in Byzantium: Symeon Seth and his Refutation of Galenic Theories on Human Physiology," *Greek, Roman, and Byzantine Studies* 55 (2015): 431–469, esp. 447–448.

79 Symeon Seth, *Oratio antirrhetica ad Galenum*, I.12–13, ed. and trans. Bouras-Vallianatos and Xenophontos, "Galen's Reception in Byzantium," 459, 464.

80 Symeon Seth's similarities, skills, and travel should lead us to consider learning in the Mediterranean world as not necessarily having to be rooted on clear-cut buffer zones of encounter and translation, but as a mutually conducted process of converging inquiry around a

Symeon Seth can be placed squarely in the extramissionist camp on account of his attribution of the eyes' effluxes to the *pneuma*, which he credited to Ptolemy but was part of the same second-century milieu that gave Galen the same term. But Symeon's theory roots the optical *pneuma* within a power mediating the organ of sense and the sensible object, a unified efflux that could still be abstracted into individual rays for the sake of mathematics. Symeon is specifically addressing the concerns of the mathematicians, who stressed the visual rays as a mathematical construct for the calculation of reflection and refraction, as the floating coin and burning mirrors experiments suggest. Symeon's Arabic contemporary, Ibn al-Haytham, for instance, despite being an intromissionist who argued that the efflux of rays was not the cause of sight, advocated for the preservation of the extramissive visual rays for the sake of geometric computations. So, it would be hasty to assume that the visual rays here are the basis of a haptic theory of vision, since even intromissionists in the period relied primarily on its mathematical efficacy.

In the subsequent entry on hearing, Symeon opens his discussion with a quick summary of the sense of sight, simply stating, "The creator has made the eye spherical and projecting out, just as the optical *pneuma* is destined to go out (ἐκπέμπειν μέλλοντα) toward the things seen and to seize them (καταλαβεῖν)."[81] While Symeon Seth subscribes to the theory of optical *pneuma*, we should neither overlook nor overemphasize the tactile suggestion of his use of the term seize or grasp (καταλαβεῖν) in both descriptions of the *pneuma*'s emission, which might be interpreted as evidence of a haptic theory of vision. His emphasis on the *pneuma* possibly speaks to his training as a medical doctor and his familiarity with medical treatises, which advocated treating medical ailments by intervening on the optical *pneuma*.

The term is absent altogether from the entries on hearing, smell, and even in the combined one on touch and taste. Yet, it does reappear immediately after them, in the entry on the mind, which additionally

certain body of canonical texts. The fact alone that they came from Greek copies into the Arab- and Persian-speaking worlds attests to existing and thriving avenues of interchange, and should allow us to think that even if they are not canonical enough to include in an encyclopedia for the emperor (particularly when they corroborated elementary ideas in the established literature on vision), their findings would still be known to Byzantine scholars, such as Symeon Seth and his colleagues. See, for example, Paul Magdalino, "The Road to Baghdad in the Thought-World of Ninth-Century Byzantium" in Leslie Brubaker (ed.), *Byzantium in the Ninth Century: Dead or Alive? Papers from the Thirtieth Spring Symposium of Byzantine Studies, Birmingham, March 1996* (Burlington, VT: Ashgate, 1998), 195–213.

81 "Τὸν ὀφθαλμὸν ὁ δημιουργὸς ὡς ἐκπέμπειν μέλλοντα τὸ ὀπτικὸν πνεῦμα πρὸς τὸ καταλαβεῖν τὰ ὁρατὰ σφαιροειδῆ καὶ προεξέχοντα πεποίηκε." Symeon Seth, *Conspectus rerum naturalium*, IV.75, ed. Delatte, 73.

deploys sight and light as metaphors for understanding the action of the five mental faculties: mind, intellect, belief, imagination, and perception (νοῦ, διανοίας, δόξης, φαντασίας καὶ αἰσθήσεως).[82] For instance, Symeon writes that, following Plato and Aristotle, just as vision needs (δεῖται) light to actualize sight, the mind needs meaning for intellection; and, following Gregory of Nazianzos, he says that just as the sun gives off light for seeing, the divine gives meaning for the mind.[83] Then, he continues:

> ταῦτα γὰρ αἱ αἰσθήσεις καταλαμβάνουσι καὶ ἡ φαντασία καὶ ἡ δόξα· ὁ δὲ νοῦς νοεῖ τὰ νοητά· ἕκαστον γὰρ οἰκείᾳ γνώσει διαγινώσκεται· τὰ νοητὰ τῷ νῷ, τὰ δοξαστὰ τῇ δόξῃ, τὰ αἰσθητὰ ταῖς αἰσθήσεσιν.
>
> For the senses seize these things, as do the imagination and belief. And the mind thinks intelligible things. For each attends to the knowledge proper to it: the intelligible by the mind, the believable by belief, the perceptible by perception.[84]

Here, Symeon Seth uses "seize" (καταλαμβάνουσι) not haptically but in a manner that stresses precisely the mental faculties' grasping and comprehension of their proper objects of contemplation.

This language of grasping is common in the context of cognition and perception, where terms like *antilepsis*, *prolepsis*, and *katalepsis* describe various forms of apprehension and comprehension. As shall be discussed in subsequent chapters, such terminology is indebted to Stoic notions of perceptual cognition, namely the *katalēptikē phantasia* that denotes the cognitive impression derived from perception that leads one to belief and assent.[85] While literally denoting seizing or grasping, the term *katalepsis* in perceptual and cognitive terms is to be understood as comphrension and

82 "Αἰσθήσεων πέντε οὐσῶν τῆς ψυχῆς, καθάπερ καὶ τοῦ σώματος, ἤτοι νοῦ, διανοίας, δόξης, φαντασίας καὶ αἰσθήσεως, ὁ νοῦς ἐπέχει τὴν πρώτην τάξιν χρώμενος ταῖς λοιπαῖς ὡς ὑπηρέτισιν." Symeon Seth, IV.83, ed. Delatte, 78–79.

83 "Πλάτων δὲ καὶ Ἀριστοτέλης φασὶν ὅτι ὥσπερ ἡ ὄψις δεῖται φωτὸς πρὸς τὸ διενεργῆσαι τὴν οἰκείαν ἐνέργειαν, οὕτω καὶ ὁ νοῦς δεῖται πρὸς τὸ νοεῖν τοῦ πρώτου νοῦ. ὡς γὰρ ὁράσεως οὔσης καὶ ὁρατοῦ, εἰ μὴ τρίτον προσῇ τὸ φῶς, μάτην ἔσται ταῦτα, οὕτως ἐπὶ τοῦ νοῦ καὶ τοῦ νοητοῦ, εἰ μὴ ἐπιλάμψῃ ὁ πρῶτος νοῦς ὃς ἀνάλογος τῷ ἡλίῳ ἐστίν, οὐ δύναται ὁ νοῦς ἐνεργεῖν τὴν οἰκείαν ἐνέργειαν. συγκατατίθεται δὲ τούτοις καὶ ὁ μέγας Γρηγόριος λέγων ὡς καθάπερ ὁ ἥλιος παρέχει τῇ ὄψει τὸ ὁρᾶν, οὕτω δὴ καὶ τὸ θεῖον τῷ νῷ τὸ νοεῖν." Symeon Seth, IV.83, ed. Delatte, 79.

84 Symeon Seth, IV.84, ed. Delatte, 79–80.

85 See Francis H. Sandbach, "Phantasia Katalēptikē" in A. A. Long (ed.), *The Problems in Stoicism* (London: Athlone, 1971), 9–21; David Sedley, "Zeno's Definition of *phantasia kataleptike*" in Theodore Scaltsas and Andrew S. Mason (eds.), *The Philosophy of Zeno: Zeno of Citium and his Legacy* (Larnaca: Municipality of Larnaca, 2002), 135–154. On Symeon Seth and *phantasia*, albeit Aristotelian *phantasia*, see George Arabatzis, "*Phantasia* et intellect chez Symeon Seth: Sur les sources du *Conspectus rerum naturalium*" in M. C. Pacheco and J. F. Meirinhos (eds.), *Intellect et imagination dans la philosophie médiévale*, vol. 1 (Turnhout: Brepols, 2006), 483–493.

καταλαβεῖν is used to denote all the senses' apprehension of sensory perceptions, as well as the mind's comprehension of these sense perceptions. Thus, while one may claim that Symeon Seth is characterizing extramission haptically in his use of καταλαβεῖν, this seems highly unlikely given that the verb, wholly in line with general usage, denotes merely the act of apprehension.

In Symeon Seth's writing, the term καταλαβεῖν is used consistently, and its absence from the entries on touch and taste, both of which were considered to be reliant on tactility, is telling. The term refers to the perceptual action of mind seizing the object of sensation, and is not a literal indicator of haptic perception. The notion of seizing or taking hold indicates the sensual contact between seer and seen, which makes perception possible, not the confusion or collapse of sight and touch. Consider, for example, the use of the term in the fourteenth-century *Harmonics* by Manuel Bryennios, a treatise derived from Ptolemy's work and its subsequent commentators. Bryennios writes regarding the interplay between vision and cognitive perception: "For just as the eye cannot perceive (ἀντιλαμβάνεσθαι) any colour in the absence of light, no more can the mind grasp (καταλήψεως) what is being said without having a clear idea of the subject-matter; that is to say, without knowing its definition."[86] Here, the seemingly haptic language of seize (ἀντιλαμβάνεσθαι) and grasp (καταλήψεως) is used not just for sight but also for the mental operation. The term denotes apprehension through the senses and the mind, and does not connote any form of physical touch or feeling.

Nikephoros Blemmydes

Syncretic thinkers such as Nemesios and Symeon are important. In addition to surveying various ways to define vision, they also considered sight within human psychology as a whole. They clarify how vision was thought to operate, not only physically and physiologically but also through perception and cognition. This cohesive approach took precedence in Byzantine anthropology, which sought to reconcile the division between Galenic medical circles and Platonic-Aristotelian philosophical circles.

86 "καὶ γὰρ ὥσπερ ἄνευ φωτὸς οὐδὲν τῶν χρωμάτων ἡ ὅρασις ἀντιλαμβάνεσθαι δύναται, οὕτως ἀληθῶς οὐδὲ νοῦς ἄνευ τῆς τοῦ ὑποκειμένου ἀκριβοῦς καταλήψεως, ταὐτὸ δ' εἰπεῖν τῆς τοῦ ὁρισμοῦ γνώσεως κατανοεῖν τὰ περὶ αὐτοῦ λεγόμενα δύναται." Manuel Bryennios, *Harmonica*, II.6, ed. and trans. G. H. Jonker, *The Harmonics of Manuel Bryennius* (Groningen: Wolters-Noordhoff, 1970), 172–173.

Although Galen and his commentators/compilers worked with classical theories of vision, Galen himself rarely figured in the major commentaries on Plato and Aristotle. Robert Todd has claimed that Galen's first overt appearance in a commentary on an Aristotelian work is in the late eleventh-, early twelfth-century commentary on the *Parva Naturalia* by Michael of Ephesus.[87] Earlier, in the sixth century, John Philoponos, in his commentary on *De anima*, when he argues against extramission, seems to target specifically Galen's anatomical understanding of the flow of the optic *pneuma* to the eye and beyond, but he does not name Galen.[88] This corroborates the critical fracture identified throughout this chapter, that between philosophy and medicine. Philosophically inclined authors privileged classical learning, and Galen's *pneuma* features far less prominently until later centuries. Clearly, if Todd's judgment is right, then Michael of Ephesus represents an important turning point, where Galenic thought gained respect in philosophical circles. This trend, as we shall see, was limited to a few key fourteenth-century figures, namely, Nikephoros Blemmydes and those who followed him – his student the emperor Theodore II Doukas Laskaris and the commentator Sophonias, who drew heavily from both Michael of Ephesus and Blemmydes.

Nikephoros Blemmydes privileged Galen's theory of vision in his *Epitome physica*. Trained as a medical doctor, Blemmydes followed Galen closely, yet also engaged Alexander of Aphrodisias's commentary on Aristotle's *Meteorology*, Plotinos's *Enneads*, and possibly also Philoponos's commentary on *De anima*, as Börje Bydén suggests.[89] Sections from his chapter on sight found their way into Sophonias's account of vision in his paraphrase of Aristotle's *De anima*.[90] His influence is attested by a letter from his most illustrious student, the emperor Theodore II Doukas Laskaris, who commits to Galen's views. There, he describes the eyes as

87 Robert B. Todd, "Galenic Medical Ideas in the Greek Aristotelian Commentators," *Symbolae Osloenses* 52:1 (1977): 117–134.

88 John Philoponos, *In Aristotelis libros de anima commentaria*, ed. M. Hayduck, *In Aristotelis libros de anima commentaria*, Commentaria in Aristotelem Graeca 15 (Berlin: Reimer, 1897), 363.31–365.17, 336.24–35.

89 For a useful summary of Blemmydes's argument and sources, see Bydén, *Theodore Metochites'* Stoicheiosis Astronomike, 203–205.

90 Bydén suggests the possibility that these two contemporaries, Sophonias and Blemmydes, might have used a now-lost common source, which of course should not be overlooked. See Bydén, *Theodore Metochites'* Stoicheiosis Astronomike, 205. See also Sophonias, *In Aristotelis libros de anima paraphrasis*, ed. M. Hayduck, *In libros Aristotelis de anima paraphrasis*, Commentaria in Aristotelem Graeca 23.1 (Berlin: Reimer, 1883), 77.2–38. On Sophonias, see Divna Manolova, "Sophonias the Philosopher: A Preface of An Aristotelian Commentary: Structure, Intention, and Audience," unpublished MA thesis, Central European University (2008), esp. 1–38.

being the receptacles for a "sticky and dense spirit" (γλίσχρῳ καὶ παχυτέρῳ θυμῷ), words used throughout Galen's oeuvre to describe the *pneuma* and the humors, and "from which the *pneuma* of sight hurls forth" (ἐξ ὧν τὸ πνεῦμα ἐξακοντίζεται τὸ τῆς ὄψεως).[91]

In summarizing the views of the "mathematicians," Nikephoros Blemmydes interestingly characterizes their views in a manner that appears to be explicitly haptic. Though at first he states that they stress the operation of the air in relation to light, he goes on to claim that some of the mathematicians "say that some rays pour out from the eyes and fall upon the visible object" (ἀκτινάς φασί τινας ἐκχεομένας τῶν ὀφθαλμῶν καὶ προσπιπτούσας τῷ ὁρατῷ), being reflected in various ways toward visible objects.[92] This section derives from Alexander of Aphrodisias's commentary on Aristotle's *Meteorology*,[93] and Blemmydes explains that this theory of vision differs from the views of the natural philosophers:

> τοῖς δὲ φυσικοῖς οὐ δοκεῖ δι' ἀκτίνων ἐκπομπῆς τελεῖσθαι τὴν ὅρασιν, ἀλλὰ τῇ εἰσδοχῇ τῆς τῶν ὁρωμένων ποιότητος. Ὁρᾶσθαι γάρ φασι τὰ μὲν εὐθείας ὁρώμενα, τῷ πάσχειν ὑπὸ τοῦ ὁρωμένου χρώματος τὸ μεταξὺ τοῦ ὁρωμένου καὶ ὄψεως διαφανές, καὶ μεταδιδόναι τῇ ὄψει τοῦ πάθους οὔσῃ καὶ ταύτῃ διαφανεῖ.
>
> But the natural philosophers do not hold that vision is brought about by a sending out of rays, but rather through the reception of the quality of visible objects. They say that what is seen directly is seen by virtue of the fact that the transparent medium between the object and the eye is affected by the color seen, and that it imparts its affection to the eye which is also transparent.[94]

Here, the natural philosophers clearly represent Aristotle, signaled by the term *transparent* (διαφανές). Galenic *pneuma* appears almost as a synthesis or compromise between the mathematicians and Aristotle. Blemmydes suggests that he favors a Galenic understanding of the outflow of the *pneuma* since this is wholly supported by the fact that the optic nerves coming from the brain are hollow so as to conduct the optic *pneuma*.

Several sections later, Blemmydes stresses the limited range of the *pneuma* and its consequent need to unite itself through an alteration with

[91] Theodore II Doukas Laskaris, *Epistula*, 135, ed. N. Festa, *Epistulae CCXVII* (Florence: Istituto di Studi Superiori Pratici e di Perfezionamento, 1898), 190–191, esp. 190.

[92] Nikephoros Blemmydes, *Epitome Physica*, 21.3, ed. PG 142:1192B.

[93] See Alexander of Aphrodisias, *In Aristotelis meteorologicorum libros commentaria*, 141.3–142.20, ed. M. Hayduck, *In Aristotelis meteorologicorum libros commentaria*, Commentaria in Aristotelem Graeca 3.2 (Berlin: Reimer, 1899), 141–142.

[94] Nikephoros Blemmydes, *Epitome Physica*, 21.3, ed. PG 142:1192B; modified trans. Bydén, *Theodore Metochites'* Stoicheiosis Astronomike, 203.

the intervening air, along with light, so as to be able to perceive the visible object. There he works closely with Galen's text on the interaction of the *pneuma* with the encompassing air, a Galenic passage that Nemesios of Emesa also echoed closely, quoted above.[95] But in Blemmydes the description of the encompassing air seems to play with both the Galenic and Aristotelian conceptions of the intervening space. Galen originally wrote, "when it strikes the surrounding [air] (περιέχοντι) it produces by its first impact an alteration that is transmitted to the furthest distance."[96] Yet, Blemmydes renders it:

> ἐμπίπτουσα δὲ τῷ περιέχοντι διαφανεῖ καὶ τῇ πρώτῃ προσβολῇ τὴν ἀλλοίωσιν ἐργαζομένη διαδίδωσιν ἄχρι πλείστου, συνεχοῦς ἑαυτήν, μέχρις ἂν καταλάβοι τὸ ὁρατόν, συνεκτεινομένη τῷ περὶ τὸν μεταξὺ ἀέρα φωτί, καθὰ συγγενεῖ.
>
> When it strikes the surrounding transparent it produces by its first impact an alteration that is transmitted to the furthest distance until it apprehends the visible thing, spreading out around the medium of air with light, like onto like.[97]

Although the phrase is nearly identical to Galen, the "surrounding" is said to be "transparent" (περιέχοντι διαφανεῖ), thus conflating Galen's medium of light-like air to the Aristotelian medium credited to the broad category of the "natural philosophers." Furthermore, the final portion of the sentence is altered to stress that through the union of light with the transparent and the *pneuma* sight happens and the object is apprehended. Following Plato, Blemmydes explains this union through the idea that because like unites with like, so too transparent bodies and light.

Blemmydes's account may remind us of similar tendencies to synthesize the theories of antiquity in Nikephoros Choumnos (see Chapter 1), but with a focus on different precedents and purposes. Nikephoros Choumnos desired to unify Aristotelian and Platonic theories of vision, based on earlier debates and syntheses. And, we can likewise observe that another one of his contemporaries and a staunch critic of Aristotle, Nikephoros Gregoras, would advocate his theory of vision in his *Solutions* for the empress Helena Palaiologina, explaining quite resolutely in his header that

95 Galen, *De placitis Hippocratis et Platonis*, 7.5.6–7, ed. and trans. De Lacy, 454–455. Cf. Nemesios, *De natura hominis*, 7, ed. Morani, 58–59; trans. Sharples and van der Eijk, 105–106.

96 "ἐμπίπτουσα δὲ τῷ περιέχοντι καὶ τῇ πρώτῃ προσβολῇ τὴν ἀλλοίωσιν ἐργαζομένη διαδίδωσιν ἄχρι πλείστου, συνεχοῦς αὐτῷ δηλονότι τοῦ πέριξ σώματος ὑπάρχοντος." Galen, *On the Doctrines of Hippocrates and Plato*, 7.5.7, ed. and trans. P. De Lacy, *Corpus medicorum Graecorum*, 454–455.

97 Blemmydes, *Epitome Physica*, 21.12, ed. PG 142:1193D.

the work is a "Demonstration concerning sight, that it is not that which crosses over to the visible thing, but that it becomes a reception of it."[98] The affiliations and preferred sources of each of these writers were much more fluid than their training or background might lead us to believe, a sign of the broader state of philosophy in late Byzantium and of theories of sight less divided than we might suppose. Not only did late Byzantine authors privilege different theories at different times to suit specific rhetorical and philosophical goals, but likewise they could resolve or ignore the careful nuances and distinctions between these theories.

The survey of sources in this chapter has shown that the Byzantine tradition was dynamic. Rather than observing a polarization or ossification of thought over time in the Byzantine world, we instead witness a desire to synthesize conflicting theories – so as to reconcile groups traditionally at odds, philosophers and medical doctors, theologians and worshippers. Furthermore, we have seen that conflation of touch and vision is not a Byzantine phenomenon. The only glimmers of such a conflation are to be found in the echoes of pseudo-Alexander and Plotinos against anonymous readers who had read a single Stoic metaphor absent from the philosophical parameters that originally framed that metaphor. The best, most articulate sources were in fact quite careful to distinguish between sight and touch. We should follow that example. The history that we have discussed in this chapter is one regarding the purposeful, nuanced, and eloquent distancing of sight from touch, despite whatever metaphors or haptic language might be used to model vision. In the next chapter we turn to the conversation on the commonalities and shared characteristics of the senses. This will allow us to add a further dimension to this line of thought by considering how frequent associations between the various senses were, what these shared characteristics sought to say about the operation of sensation and perception, and how they came to articulate the individuality of each of the senses and each one's unique affordances.

[98] "περὶ τῆς ὄψεως ἀπόδειξις, ὅτι οὐκ αὐτὴ διαβαίνει πρὸς τὰ ὁρατά, ἀλλ' ἐκείνων αὕτη γίνεται παραδοχή." Nikephoros Gregoras, *Solutiones questionum*, 2, ed. P. L. M. Leone, "Nicephori Gregorae 'Antilogia' et 'Solutiones questionum'," *Byzantion* 40 (1970): 488–513, esp. 497–502, at 497. See Rodolphe Guilland, *Essai sur Nicéphore Grégoras, l'homme et l'oeuvre* (Paris: P. Geuthner, 1926), 220–221. See also Divna Manolova, "Discourses of Science and Philosophy in the Letters of Nikephoros Gregoras," unpublished PhD dissertation, Central European University (2014).

CHAPTER 3

The Commonalities of the Senses

The primary sources we have seen so far articulately argue against any conflation of sight and touch. So how did it emerge? Inasmuch as the examples in the last chapter opened up questions regarding the tactility of sight, the ones in this chapter seek to further resolve them by demonstrating that the conflation that developed was not merely sight with touch, but all the senses with each other. That is, ancient, late antique, and Byzantine thinkers alike were occupied with the commonalities of all the senses. When we study vision in a vacuum, it is striking to see the preponderance of statements that juxtapose sight and touch. Thus, one might naturally find them to be unique and indicative in indexing an entanglement between sight and touch. However, such comparisons are embedded within a larger explanatory framework, one that uses cross-sensory comparisons and interrelations as a way of clarifying the uniqueness of each of the senses and how they individually relate to one another. The goal of this chapter is to explore the commonalities of all the senses, to lay a foundation for understanding how touch came to be thought of as particularly common to the senses, and how the mind was thought to synthesize and combine the senses into one common experience.

The notion of common sense is first evident in Aristotle, but it features in four different ways. First, it denotes the commonality of touch across all living beings, since all animals, even if they lack sight, hearing, taste, or smell, are capable of feeling, and, unlike other senses, if touch is removed from a creature then their life too will have perished. Second, it describes the commonality of touch to all the different sense organs: the eye, just like the tongue or the ear, is sensitive to touch. Third, the term refers to qualities of perceptible objects that can be sensed by multiple senses, for example, the size of an object via touch and sight, or movement by all but taste. Fourth, the "common sense" (κοινὴ αἴσθησις) proper can describe the broader sensory capacity of the soul, which instantly takes information from the various senses, compiling and discriminating their properties.

This last aspect, the apex of the other three, will be given its deepest attention in the book's final chapter.

With each definition, touch presents a critical conundrum. Not only is touch shared by all the sense organs and by all living creatures, but it also might be projected as a manifestation of that internal common sense that grasps all sensation. Touch also became a metaphor for all sensation, to describe how the soul senses the perceptible world and how the sensory organ senses the sensible object. These matters are put into relief by texts that sought to summarily categorize the five senses, explain the relation of each to the other, and situate them within the cognitive process. Such synopses appear in the writings of Nemesios of Emesa, Symeon Seth, and Michael Psellos, and they resemble in style and purpose the extracts we have explored in Chapters 1 and 2 from these same authors. But other texts, like the late tenth-century *Souda*, a lexicon of ancient and medieval sources that Paul Lemerle calls "a compilation of compilations," drew from other commentaries, compendia, scholia, and lexica.[1] John of Damascus's eighth-century *Expositio fidei*, another kind of compendium that featured a synopsis of the senses, enjoyed enduring popularity in the Byzantine world, influencing the middle and late Byzantine periods. These compendia were not only used by many authors and thinkers at and after the time of their production, but also reveal the education and reading habits of their authors, and so provide a window into Byzantine society.

The Commonality of Touch, 1: Across All Life

The first kind of common sense states that touch is the only sense common to all living things. Nemesios of Emesa's *De natura hominis* is one of the more eloquent texts to engage with the various articulations of the common sense, clarifying in particular the commonality of touch across all life. Early on in his entry on touch, Nemesios states:

> ἀλλὰ τῶν μὲν πλείστων αἰσθητηρίων διαφθειρομένων οὐδὲν εἰς αὐτὴν τὴν ζωὴν καταβλάπτεται τὸ ζῷον, ἁφῆς δὲ ἀπολλυμένης συναπόλλυται καὶ τὸ ζῷον. ἁφὴ γὰρ μόνη τῶν αἰσθήσεων κοινὴ πάντων ἐστὶ τῶν ζῴων. πᾶν γὰρ ζῷον ἁφὴν ἔχει, τὰς δὲ ἄλλας αἰσθήσεις οὐ πάντα πάσας, ἀλλ᾽ ἐνίας ἔνια, πάσας δὲ ἔχει τὰ τελειότερα.
>
> Yet when most of the sense organs are destroyed an animal is in no way vitally impaired, whereas, if touch perishes the animal perishes with it. For

[1] Paul Lemerle, *Le premier humanisme byzantin: notes et remarques sur enseignement et culture à Byzance des origines au Xe siècle* (Paris: Presses Universitaires de France, 1971), esp. 299.

> touch alone of the senses is common to all animals. For while every animal has touch, not all have all the other senses, but some have some and the higher animals have them all.[2]

While the other senses can be damaged and maimed independently without causing the death of the creature, touch perishes only at the expense of death. The observation here is not only that all living creatures partake in touch, but that there is no way to remove touch from an animal without also killing it. So touch is necessary to life and uniquely common to all living beings.

This point is made by Aristotle in *De anima* when he discusses the fundamental nature of sensation to all living beings and turns his attention first toward touch:

> αἰσθήσεως δὲ πρῶτον ὑπάρχει πᾶσιν ἁφή. ὥσπερ δὲ τὸ θρεπτικὸν δύναται χωρίζεσθαι τῆς ἁφῆς καὶ πάσης αἰσθήσεως, οὕτως ἡ ἁφὴ τῶν ἄλλων αἰσθήσεων.
>
> Touch is the first sense in everything. Just as the merely nutritive faculty may exist apart from touch and from all sensation, so touch may exist apart from all other senses.[3]

Not only does Aristotle characterize the sense of touch here as being a primal factor of human sensation, it is also characterized as being independent of the other senses. Just as the faculty of nutrition is separated from the senses, the same can be said regarding touch. He goes on to clarify soon thereafter, "all animals obviously possess the sense of touch" (τὰ δὲ ζῷα πάντα φαίνεται τὴν ἁπτικὴν αἴσθησιν ἔχοντα), unlike the nutritive faculty, which extends even further, to plants.[4] The matter of touch's commonality across all animals is reiterated throughout Aristotle's work, and, as we shall see in our discussion of "common sense" proper below, became an important point for later authors.

The Commonality of Touch, II: Across the Senses

Second is the commonality of touch across all the sense organs. This type features in the *Souda*'s entry on sight:

> Ὄψις: ὅρασις. καὶ ἡ θεωρία. τὰ μὲν οὖν πρὸ τῆς πόλεως ἀργύρῳ καὶ χρυσῷ περιελάμπετο, καὶ τῆς ὄψεως ἐκείνης οὐδὲν οὔτε τοῖς σφετέροις

[2] Nemesios, *De natura hominis*, 8, ed. Morani, 63; trans. Sharples and van der Eijk, 110–111.
[3] Aristotle, *De anima*, 413b5–7, ed. and modified trans. Hett, 74–75.
[4] Aristotle, *De anima*, 413b9–11, ed. and trans. Hett, 74–75.

ἐπιτερπέστερον οὔτε τοῖς πολεμίοις παρέστη φοβερώτερον. ὅτι ἡ ὄψις τῶν δώδεκα χρωμάτων ἐστὶ κριτική, λευκοῦ καὶ μέλανος καὶ τῶν μεταξύ, ξανθοῦ καὶ φαιοῦ, ὠχροῦ, ἐρυθροῦ, κυανοῦ, ἁλουργοῦ, λαμπροῦ, ὀρφνίου. ἡ δὲ ἀκοὴ κριτική ἐστιν ὀξέος καὶ βαρέος φθόγγου, ἡ δὲ ὄσφρησις κριτικὴ εὐωδῶν καὶ δυσωδῶν ὀσμῶν καὶ τῶν μεταξύ, σηπομένων τε καὶ βρεχομένων ἢ τηκομένων ἢ θυμιωμένων. ἡ δὲ γεῦσις κριτικὴ γλυκέων τε καὶ πικρῶν χυμῶν καὶ τῶν μεταξὺ ε΄· ἑπτὰ [sic] γάρ εἰσι χυμοί, γλυκύς, πικρός, ὀξύς, δριμύς, στυφύς, σομφός, ἁλυκός, στρυφνός. ἡ δὲ ἁφὴ πλειόνων ἐστὶ κριτική· βαρέων, κούφων καὶ τῶν μεταξύ· τραχέων καὶ λείων καὶ τῶν μεταξύ· ξηρῶν καὶ ὑγρῶν. καὶ αἱ μὲν δ΄ αἰσθήσεις ἐν τῇ κεφαλῇ εἰσι καὶ ἰδιάζουσιν καὶ περικλείονται ὀργάνοις, ἡ δὲ ἁφὴ καὶ διὰ τῆς κεφαλῆς καὶ διὰ τοῦ σώματος κεχώρηκε καί ἐστι κοινὴ πάσης αἰσθήσεως.

Sight: Seeing. Also, contemplation. "So the things before the city shone with silver and gold, and nothing was more delightful for them or appeared more terrible to the enemies than that sight."[5] [Note] that sight distinguishes the twelve colors: white and black, and the intermediates, [i.e.] yellow and grey, pale, red, blue, purple, bright, dark. Hearing distinguishes a high and low sound, smelling distinguishes good and bad odors, and the intermediates: rotten and wet [odors], whether melted or flavored with thyme. Taste distinguishes sweet and bitter flavors, and the five intermediates. For there are seven flavors: sweet, bitter, sharp, acrid, sour, porous, salt, sour. Touch distinguishes many things, heavy, light objects, and their intermediates; rough, smooth and their intermediates; dry and wet objects. And there are four senses in the head and they are peculiarly adapted to the organs and are enclosed by them, but touch passes through the head and the body and is common to every sense.[6]

This summary focuses specifically on the issue of the senses' perceptibles, particularly the types of qualities that each of the various senses can detect. Tellingly, the entry concludes its description of sight with a summary that clarifies that "touch… is common to every sense" (ἁφή… ἐστι κοινὴ πάσης αἰσθήσεως). The entry does not draw a direct connection to sight or say that sight is a variant of touch, nor does it deploy any metaphors or language that might suggest this to be the case. It also does not pair touch and taste together, even though taste is often understood as being a species of touch given that the tongue requires direct contact with the object being tasted. Nevertheless, the *Souda*'s author holds that touch is common to all the senses.

[5] Josephus, *Jewish War*, v.351, ed. B. Niese, *Opera*, vol. vi (Berlin: Weidmann, 1895), 351.

[6] Anonymous, *Souda*, "Ὄψις," ed. and modified trans. Marcelo Boeri, *Souda On Line*, www.stoa.org/sol-entries/omicron/1082.

Overall, the *Souda*'s passages on anthropology are often indebted to Aristotle's *De anima*, at least indirectly through the late antique commentators. While the direct sources are often diverse, when it comes to matters of perception, cognition, and general psychology, it tends to rely primarily on John Philoponos's commentaries on Aristotle, as we shall see repeatedly throughout the following chapters. This differs prominently from Byzantine medical works, which tended to prefer Galen as a lens through which to gaze at Aristotle and the other ancient writers. The entry on sight, however, is primarily dependent on a longer passage found in Photios's *Bibliotheca* for the entry on the anonymous *Life of Pythagoras*.[7] There, Photios summarized the theories of classical authors on vision via the figure of Pythagoras. Interestingly, while we would certainly hold the Pythagorean sight theory to be extramissive and one where the efflux of rays from the eyes appears to strike the objects, it lacks much clarity as to how exactly that process works. This was attested in the writings of Alkmaeon of Croton, whose theories are described by Theophrastus in his commentary *De sensibus et sensibilibus*.[8] There, Theophrastus suggests that the "gleaming," transparent quality of the eye reflects the object and thus vision comes about in this manner. This, however, is a problematic assertion given that it does not clarify precisely how the process works, much as we have encountered in the mathematical optics of Ptolemy and Euclid, which are comparably taciturn on the physiological and sensory aspects of vision.

Of particular note in Photios's summary from which the *Souda* is drawn is that when discussing sight directly, Photios discusses the perception of colors as being unique only to the faculty of sight, stating, "According to Pythagoras, Plato, and Aristotle, sight is the judge of the twelve colors."[9] Here, Photios does not differentiate between the theories of these various philosophers, focusing only on the fact that color is exclusive to sight, something which, as the sources in the previous chapters have clarified, is

[7] Photios, *Bibliotheca*, 249, ed. Henry, VII:126–134.

[8] See also Theophrastus, *De sensibus et sensibilibus*, I.26, ed. and trans. Stratton, 88–89.

[9] "Ὅτι ἡ ὄψις, κατὰ Πυθαγόραν καὶ Πλάτωνα καὶ Ἀριστοτέλην, τῶν δώδεκα χρωμάτων ἐστὶ κριτική." Photios, *Bibliotheca*, 249, ed. R. Henry, 8 vols (Paris: Les Belles Lettres, 1959–1974), VII:128. On antique theories of color in Byzantium, see Liz James, *Light and Colour in Byzantine Art* (Oxford University Press, 1996), 47–68. See also Katerina Ierodiakonou, "Hellenistic Philosophers on the Phenomenon of Changing Colours" in B. Holmes and K.-D. Fischer (eds.), *The Frontiers of Ancient Science: Essays in Honor of Heinrich von Staden* (Berlin: De Gruyter, 2015), 227–250; Katerina Ierodiakonou, "Basic and Mixed Colours in Empedocles and in Plato" in Marcello Carastro (ed.), *L'antiquité en couleurs: catégories, pratiques, représentations* (Grenoble: J. Millon, 2009), 119–130; Katerina Ierodiakonou, "Aristotle on Colours" in D. Sfendoni-Mentzou, J. Hattiangadi, and D. M. Johnson (eds.), *Aristotle and Contemporary Science*, vol. II (New York: P. Lang, 2000), 211–225.

incapable of being transmitted through touch. One might presume here that Photios ascribes to Pythagoras an approach to vision that would be commensurate with the views of Plato and Aristotle. Thus, one would be hard pressed to interpret that statement in Photios's *Bibliotheca* and the *Souda* as suggesting that the commonality of touch refers in any taciturn or cryptic manner to an articulation of haptic sight. The interesting complexity of the sentence emerges from the fact that its careful wording refers to a few key points simultaneously.

One can appreciate that Photios's declaration of touch's commonality, repeated in the *Souda*, can refer back to Aristotle's statement, quoted in the first chapter, where he writes that the "other sense organs perceive by contact (ἁφῇ) too, but through a medium; touch alone seems to perceive through itself."[10] This speaks to the commonality of contact of all the senses between perceiver and object, but rather than seeking to make sight tactile, it actually distances touch from the immediacy of unmediated perception. However, the most important aspect of touch's commonality expressed in Photios's passage relates precisely to the fact that touch encompasses the whole of the body. As Galen states in *De usu partium*:

> ὅσα γοῦν τῶν ὀργάνων οὐ κινεῖται μόνον ἁπλῶς καθ' ὁρμήν, ἀλλὰ καὶ περιττοτέραν τὴν αἴσθησιν ἐκτήσατο τῆς κοινῆς ἁπάντων μορίων τῆς ἁπτικῆς, οἷον ὀφθαλμοὶ καὶ ὦτα καὶ γλῶττα, ταῦτα καὶ τὸ μαλακὸν ἔχει καὶ τὸ σκληρὸν γένος τῶν νεύρων.
>
> Accordingly, parts such as the eyes, ears, and tongue, that are not only moved simply in obedience to the will but also have sensation over and above that sense of touch common to all the parts, have both the hard and the soft kinds of nerves.[11]

In describing the physiology behind the individual senses, Galen here states that sense organs have both hard and soft nerves, since they not only submit to the commanding impulse of the body, but also possess the power of sensation above touch, which is accorded to all parts of sensation, but also of the body more generally. Before Galen, the same point was iterated by Aristotle to describe the mediation of the skin.[12]

[10] "πάντα δὲ τῷ δι' ἑτέρου αἰσθάνεσθαι ποιεῖ τὴν αἴσθησιν καὶ διὰ τῶν μεταξύ· ἡ δ' ἁφὴ τῷ αὐτῶν ἅπτεσθαί ἐστιν, διὸ καὶ τοὔνομα τοῦτο ἔχει. καίτοι καὶ τὰ ἄλλα αἰσθητήρια ἁφῇ αἰσθάνεται, ἀλλὰ δι' ἑτέρου· αὕτη δὲ δοκεῖ μόνη δι' αὑτῆς." Aristotle, *De anima*, 435a16–20, ed. and trans. Hett, 198–201.

[11] Galen, *De usu partium*, 16.2 (Khun IV:271), ed. G. Helmreich, *De usu partium*, vol. II (Leipzig: Teubner, 1909), 381; trans. Margaret T. May, *Galen on the Usefulness of the Parts of the Body*, vol. I (Ithaca, NY: Cornell University Press, 1968), 684.

[12] Aristotle, *De anima*, 423a18–19, ed. and trans. Hett, 130–131.

This is the same point being made in Photios's passage since, unlike the other senses that have uniquely adapted organs that contain the receptors for sensation (ἰδιάζουσιν καὶ περικλείονται ὀργάνοις), touch encompasses the entire body. While the other four senses are located in the head, touch passes through the head and permeates the whole of the body (διὰ τοῦ σώματος κεχώρηκε), which according to Galen happens through the nerves of touch that proliferate throughout the body. Thus, every organ of sense also happens to possess touch, since the eye, mouth, nose, and ears are all sensitive to touch. Thus, touch is common to all the senses given that all the senses can experience touch and it permeates the whole of the body. Interestingly, the *Souda* cuts off Photios's final clause in the passage, which ostensibly sought to clarify the broad distribution of touch across the body by stating that touch is common to every sense, "but it is in the hands that it renders most clearly its aptitude for judgment" (μᾶλλον δὲ κατὰ τὰς χεῖρας δηλοτέραν παρέχεται τὴν κρίσιν).[13] In other words, Photios endeavored to stress that while the entire body possesses a sense of touch, including the sensory organs, it is in the hands that touch finds its full potential as a sensory organ whose goal is to judge and distinguish the tactile qualities of objects. The deletion of this line in the *Souda* might be attributed to it being deemed superfluous, particularly when considering the overall remarks of the Aristotelian commentators. Or, quite possibly, it was deleted in order to not delimit the statement's meaning so as to allow any ambiguity to denote the various other definitions tied to the notion of touch's commonality across living bodies and the senses.

The Commonality of Sensible Qualities

The *Souda*'s interests compare well with those of John of Damascus in his entry on the senses in his *Expositio fidei*, a source which many Byzantine authors actively engaged with and which lays the foundation for cognitive theories that John elsewhere incorporated into his theory of the icon. In the *Expositio fidei*, he defines sight:

> Πρώτη αἴσθησις ὅρασις. Αἰσθητήρια δὲ καὶ ὄργανα τῆς ὁράσεως τὰ ἐξ ἐγκεφάλου νεῦρα καὶ οἱ ὀφθαλμοί. Αἰσθάνεται δὲ ἡ ὄψις κατὰ πρῶτον μὲν λόγον τοῦ χρώματος, συνδιαγινώσκει δὲ τῷ χρώματι καὶ τὸ κεχρωσμένον σῶμα καὶ τὸ μέγεθος αὐτοῦ καὶ τὸ σχῆμα καὶ τὸν τόπον, ἔνθα ἐστί, καὶ τὸ διάστημα τὸ μεταξὺ καὶ τὸν ἀριθμὸν κίνησίν τε καὶ στάσιν καὶ τὸ τραχὺ

[13] Photios, *Bibliotheca*, 249, ed. Henry, VII:129.

> καὶ λεῖον καὶ ὁμαλὸν καὶ ἀνώμαλον καὶ τὸ ὀξὺ καὶ τὸ ἀμβλὺ καὶ τὴν σύστασιν, εἴτε ὑδατώδης, εἴτε γεώδης ἤγουν ὑγρὰ ἢ ξηρά.
>
> The first sense is sight: and the sensoria or organs of sight are the nerves of the brain and the eyes. Now sight is primarily perception of color, but along with the color it discriminates the body that has color, and its size and form, and location, and the intervening distance and the number: also whether it is in motion or at rest, rough or smooth, even or uneven, sharp or blunt, and finally whether its composition is watery or earthy, that is, wet or dry.[14]

John describes sight as the first sense, composed of the nerves of the brain and the eyes, and whose primary perception is color, along with other common sensibles, such as size and form, locality, the intervening space, number, motion, texture, and the composition of things. The entry combines philosophical teachings with medical knowledge, indebted to Galen, that notes physiological details such as the relationship of the sense organs with the brain and nerves.

Later, John discusses touch, stating its commonality to all living things, and that its nerves come from the brain and permeate the entire body, much like in the *Souda*'s account. He explains:

> Πέμπτη αἴσθησίς ἐστιν ἡ ἁφή, ἥτις κοινή ἐστι πάντων τῶν ζῴων· ἥτις γίνεται ἐκ τοῦ ἐγκεφάλου πεμπομένων νεύρων εἰς ὅλον τὸ σῶμα. Διὸ καὶ ὅλον τὸ σῶμα, ἀλλὰ καὶ τὰ ἄλλα αἰσθητήρια τὴν τῆς ἁφῆς ἔχουσιν αἴσθησιν. Ὑπόκειται δὲ τῇ ἁφῇ τὸ ψυχρὸν καὶ θερμόν, τό τε μαλακὸν καὶ σκληρὸν καὶ γλίσχρον καὶ κραῦρον, βαρύ τε καὶ κοῦφον· διὰ μόνης γὰρ ἁφῆς ταῦτα γνωρίζεται. Κοινὰ δὲ ἁφῆς καὶ ὄψεως τό τε τραχὺ καὶ λεῖον, τό τε ξηρὸν καὶ ὑγρόν, παχύ τε καὶ λεπτόν, ἄνω τε καὶ κάτω καὶ ὁ τόπος καὶ τὸ μέγεθος, ὅταν εἴη τοιοῦτο ὡς κατὰ μίαν προσβολὴν τῆς ἁφῆς περιλαμβάνεσθαι, καὶ τὸ πυκνόν τε καὶ μανὸν ἤγουν ἀραιὸν καὶ τὸ στρογγύλον, ὅταν εἴη μικρόν, καὶ ἄλλα τινὰ σχήματα. Ὁμοίως δὲ καὶ τοῦ πλησιάζοντος σώματος αἰσθάνεται, σὺν τῇ μνήμῃ δὲ καὶ τῇ διανοίᾳ, ὡσαύτως δὲ καὶ ἀριθμοῦ μέχρι δύο ἢ τριῶν καὶ τούτων μικρῶν καὶ ῥᾳδίως περιλαμβανομένων. Τούτων δὲ μᾶλλον τῆς ἁφῆς ἡ ὅρασις ἀντιλαμβάνεται.
>
> The fifth sense is touch, which is common to all living things. Its organs are nerves which come from the brain and ramify all through the body. Hence the body as a whole, including even the other organs of sense, possesses the sense of touch. Within its scope come heat and cold, softness and hardness, viscosity and brittleness, heaviness and lightness: for it is by touch alone

[14] John of Damascus, *Expositio fidei*, II.18 (32), ed. P. B. Kotter, *Die Schriften des Johannes von Damaskos*, vol. II, Patristische Texte und Studien 12 (Berlin: De Gruyter, 1973), 83–84; modified trans. S. D. F. Salmond, "An Exact Exposition of the Orthodox Faith" in Philip Schaff and Henry Wace (eds.), *Nicene and Post-Nicene Fathers*, second series, vol. IX (New York: Charles Scribner's Sons, 1902), 1–101, at 34.

> that these qualities are discriminated. On the other hand, roughness and smoothness, dryness and wetness, thickness and thinness, up and down, place and size, whenever that is such as to be embraced in a single application of the sense of touch, are all common to touch and sight, as well as denseness and rareness, that is porosity, and rotundity if it is small, and some other shapes. In like manner also by the aid of memory and thought perception of the nearness of a body is possible, and similarly perception of number up to two or three, and such small and easily reckoned figures. But it is by sight rather than touch that these things are perceived.[15]

Here, John begins by pointing out that touch is "common to all living things" (κοινή ἔστι πάντων τῶν ζῴων), before discussing in a far more articulate fashion what the *Souda*'s final line suggests only in passing regarding touch's pervasiveness across all the senses (see above). John details in clear and unambiguous language that the body as a whole possesses a sense of touch and therefore so do the individual organs of sense themselves. He then goes on to another important type of sense commonality, namely, that of sensible qualities.

After discussing the things that touch can uniquely sense, as he has done with each of the other senses, John then explores perceptible qualities that are "common to sight and touch" (Κοινὰ δὲ ἁφῆς καὶ ὄψεως). This refers to the common sensibles, which describes aspects of things that are perceived by various senses, detected through each of the senses' own particular mediums. Just as Philo describes the bold agency of sight, qualified by the intervention of the mind, so John of Damascus here grapples with the fact that, with the aid of memory (μνήμῃ) and thought (διανοίᾳ), we are able to better discern sensory impressions and make better judgments about certain of these common sensibles, such as the nearness of a body. Sight has the place of privilege in the apprehension and comprehension of such matters.

Throughout this work John of Damascus is careful not to indicate that sight operates as a form of touch. Instead, he demonstrates that each sensory organ has itself a sense of touch, and also that despite the fact that the senses share commonalities, which overlap in the qualities they can each detect, they are nevertheless all distinguished from each other. Moreover, he discusses exactly where sight and touch overlap, just as he also does with taste and touch, and smell and hearing in their relevant passages. He typifies Byzantine authors who by and large carefully and precisely explain the differences and similarities between the senses, and do not

[15] John of Damascus, *Expositio fidei*, II.18 (32), ed. Kotter, 85; trans. Salmond, 34.

suggest that any of them were mixed or combined. The key point in John is that all perceptibles appear only to the senses proper to themselves (ἰδίοις αἰσθητοῖς).[16] That is, rather than being mixed or combined, each sense receives only those sensations they are particularly prepared to accept.

John of Damascus was indebted to the *De natura hominis* of Nemesios, which was a key source for John's anthropology. In Nemesios's section on touch, we find one of the more eloquent and concise explanations for the commonality of touch across the various senses, which acknowledges that "every part of the body has a grasp of touch" (πᾶν μόριον τοῦ σώματος τῆς ἁφῆς ἀντιλαμβάνεται).[17] There, Nemesios explains with lucid clarity:

> συμβέβηκεν οὖν ἕκαστον τῶν αἰσθητηρίων διπλᾶς ἔχειν τὰς αἰσθήσεις, τὴν μὲν τῶν ἰδίων αἰσθητῶν, τὴν δὲ τῆς ἁφῆς, ὡς ἡ ὄψις· καὶ γὰρ τὰ χρώματα διακρίνει καὶ τῶν θερμῶν καὶ ψυχρῶν ἀντιλαμβάνεται, ἀλλὰ τούτων μὲν ὡς σῶμα, τῶν δὲ χρωμάτων ὡς ὄψις. ὡσαύτως δὲ καὶ γεῦσις καὶ ὄσφρησις καὶ ἀκοή.
>
> So it happens that each of the sense organs has two kinds of sensation, the one being of its specific sense-objects, the other being touch, as in the case of sight; for it both distinguishes colors and participates in the sensations of hot and cold, but of these latter only *qua* body, of colors only *qua* sight. The same holds of taste, smell and hearing.[18]

Nemesios's words have a striking persuasiveness here, as if responding to allegations of sight being akin to touch. He begins by explaining that all the senses (except touch presumably) have two sensations, because they are subject to the sense of touch as parts of the body and then also receive sensations pertaining to their "specific sense-objects" (ἰδίων αἰσθητῶν). That is to say that in addition to being able to feel touch, each of the sense organs also experiences the sensations of whatever sensibles they are uniquely capable of receiving.

In choosing sight as the example to make his point, he is perhaps playing with associations between the two senses, but he winds up dissociating them. He chooses sight because, unlike say taste, it is not liable to be confused as sharing any property with touch since its "specific sense-object," that is, color, is never transmitted through touch. Thus, he goes

[16] "Χρὴ γινώσκειν, ὅτι ἡ μὲν ὄψις κατ' εὐθείας γραμμὰς ὁρᾷ, ἡ δὲ ὄσφρησις καὶ ἡ ἀκοὴ οὐ κατ' εὐθεῖαν μόνον, ἀλλὰ πανταχόθεν. Ἡ δὲ ἁφὴ καὶ ἡ γεῦσις οὐδὲ κατ' εὐθεῖαν οὐδὲ πανταχόθεν γνωρίζουσιν, ἀλλὰ τότε μόνον, ὅταν αὐτοῖς πλησιάσωσι τοῖς ἰδίοις αἰσθητοῖς." John of Damascus, *Expositio fidei*, II.18 (32), ed. Kotter, 85; trans. Samond, 35.

[17] Nemesios, *De natura hominis*, 8, ed. Morani, 63; trans. Sharples and van der Eijk, 111.

[18] Nemesios, *De natura hominis*, 8, ed. Morani, 63–64; modified trans. Sharples and van der Eijk, 111.

on to explain that while the eye experiences colors, it also experiences temperature. In attempting to explain the difficult commonality of touch to all the senses, rather than collapsing the individual senses into one, Nemesios instead draws an even clearer distinction in the roles of a sense organ, between that of a body and that of its unique sensory function (ὡς σῶμα. . .ὡς ὄψις). Demonstrating that sight is uniquely sight through its perception of colors, Nemesios assures us that its experience of touch is merely its virtue of being part of the body.

What we see in the *Souda* and the *Expositio fidei* is the distillation of earlier elaborations on sense perception. Moving back in time as we have done here, we can unfold layers of thought, whereby a passing line in the *Souda* is the reverberation of a deep and more eloquent voice of an earlier author, such as John of Damascus or, even more so, Nemesios, whose works also reduce and distill earlier texts. At times, their citation, as we witnessed in the case of the *Souda*, can be quite cryptic to the modern reader. Their taciturn and abridged language could possibly lead to confusion. Yet, such pitfalls remind us that none of these tomes were to be read in isolation. The *Souda*, for example, points to various sources and commentaries that a Byzantine reader would have had at hand if they wished to theorize and deepen their understanding of the senses. The *Souda*'s passing reference to the commonality of touch not only indicates in simple terms the sense organ's possession of touch, but speaks to the layered chains and bundles of knowledge that it has distilled.

The Common Sense and the Unity of Sensation

The fourth type of common sense is the culmination of the other three, and will be given full attention in this book's final chapter. In Aristotle, *De anima* III, this fourth definition is featured, to describe the process whereby input of the various senses is brought together into one sensible faculty.[19] This is the common sense proper. Here, however, I wish to consider the broader function. Nemesios, for example, describes this type of common sense in the section on the imagination: "There are five sense-organs but one sense, that of the soul, which recognizes through the sense-organs the sensations (πάθη) that occur in them."[20] This self-conscious

[19] Aristotle, *De anima*, 425a15–425b12, ed. and trans. Hett, 142–145.

[20] "Ἔστι δὲ αἰσθητήρια μὲν πέντε, αἴσθησις δὲ μία ἡ ψυχικὴ ἡ γνωρίζουσα διὰ τῶν αἰσθητηρίων τὰ ἐν αὐτοῖς γινόμενα πάθη." Nemesios, *De natura hominis*, 6, ed. Morani, 56; modified trans. Sharples and van der Eijk, 102.

process not only unites the senses, but also allows the person to recognize perception as it occurs within the bounds of the soul.

Gregory of Nyssa describes the unity of sense perception in his work *De opificio hominis*. Gregory composed this text in 379 as an Eastertide gift for Peter, who would become bishop of Sebasteia, in order to complement the *Hexameron* of their late brother, Basil of Caesarea. In the ninth homily of the *Hexameron*, Basil promised a discussion of human creation and human nature, ostensibly fulfilled in the two additional homilies dubiously attributed to Basil, as discussed Chapter 1.[21] Gregory appears to have taken on the challenge of completing his late brother's work, either because those homilies had been left unfinished or because he actively sought to rework his brother's formulations as he was often known to do. Thus, in *De opificio hominis*, Gregory attempts (among other things) to explicate the action and activity of the soul through the senses. To do this, Gregory compares the mind to an expansive city with many gates receiving visitors, which represents the entry of sensory impressions from the various senses. Yet, these visitors do not congregate in a single place, but rather go on to homes, churches, markets, etc. And thus the "mind investigating and distinguishing each of the ones coming in, stores them in the corresponding places of knowledge."[22] This seems to suggest that at first hand, the mind makes judgments about sense imprints through its initial beliefs and preconceptions, sorting them accordingly into various categories of knowledge.

As Gregory, however, points out, the entrances themselves do not bind or separate the visitors. For example, visitors who are strangers to one another will often enter through the same gate, but then they will disperse and unite with their own kindred, going to the same places and clustering together even if they did not enter as one. This latter point is immensely important to Gregory because it describes how, after providing initial judgments upon sensory impressions, the mental faculty then goes on to collate the information coming from the various senses that describe the same, similar, or related objects. The metaphor of a family entering through various gates but coming together from within exemplifies this point. Gregory explains:

[21] See Basil of Caesarea, *Homiliae in hexameron*, 9.6, ed. S. Giet, *Homélies sur l'hexaéméron*, 2nd edition, Sources chrétiennes 26 bis (Paris: Éditions du Cerf, 1968); trans. Bloomfield Jackson, "The Hexaemeron" in Philip Schaff and Henry Wace (eds.), *Nicene and Post-Nicene Fathers*, second series, vol. VIII (New York: Christian Literature Company, 1895), 51–107, at 107.

[22] "ἕκαστον δὲ τῶν εἰσιόντων φιλοκρινῶν τε καὶ διεξετάζων ὁ νοῦς, τοῖς καταλλήλοις τῆς γνώσεως τόποις ἐναποτίθεται." Gregory of Nyssa, *De opificio hominis*, 10, ed. PG 44:152.

> Πολλάκις γὰρ καὶ ἐκ διαφόρων αἰσθητηρίων μία γνῶσις ἡμῖν συναγείρεται, τοῦ αὐτοῦ πράγματος πολυμερῶς πρὸς τὰς αἰσθήσεις μεριζομένου. Πάλιν δ' αὖ τὸ ἐναντίον, ἔστιν ἐκ μιᾶς τινος τῶν αἰσθήσεων πολλὰ καὶ ποικίλα μαθεῖν, οὐδὲν ἀλλήλοις κατὰ τὴν φύσιν.
>
> For often the knowledge which we gather from the different organs of sense is one, as the same object is divided into several parts in relation to the senses; and again, on the contrary, we may learn from some one sense many and varied things which have no affinity one with another.[23]

Just as sight might simultaneously perceive many objects in the world around the viewer, those individual objects are not necessarily related. And, inversely, just because one's sensation of a single object may enter through various sensory pathways, it does not mean that these sensations are not unified.

In other words, upon seeing an object, hearing its sounds, smelling its odors, touching its textures, tasting its flavors, one is able to recognize when impressions coming in from a particular sense do not all relate to the same object, or when all these stimuli, despite coming from different senses, all belong to the same object. Gregory uses the example of honey:

> Καὶ γὰρ ἰδών τις τὸ μέλι, καὶ τοῦ ὀνόματος ἀκούσας, καὶ τῇ γεύσει λαβών, καὶ τὸν ἀτμὸν διὰ τῆς ὀσφρήσεως ἐπιγνούς, καὶ τῇ ἀφῇ δοκιμάσας, τὸ αὐτὸ πρᾶγμα δι' ἑκάστου τῶν αἰσθητηρίων ἐγνώρισε.
>
> For when one sees honey, hears its name, takes it by taste, recognizes its odor through smell, and tests it by touch, one recognizes the same thing through each of the senses.[24]

Honey was an insightful illustration of the common sense since it engages all five senses and can be used to explain how a syncretic operation can be unified – how the initial judgment of sensory stimuli combines with its later collation through perception and discernment.

Gregory's account depicts the same operation or faculty that Aristotle lays out at the start of Book 3 of *De anima*, but with little of the latter's technical language. Aristotle's language of common sense is more clearly found in the writings of the twelfth-century scholar Michael of Ephesus. Michael was part of Anna Komnene's intellectual circle, alongside figures like Eustratios of Nicaea, and he appears to have been

[23] Gregory of Nyssa, *De opificio hominis*, 10, ed. PG 44:153; trans. H. A. Wilson, "On the Making of Man" in *Nicene and Post-Nicene Fathers*, second series, vol. v, ed. Philip Schaff and Henry Wace (New York: Charles Scribner's Sons, 1902), 387–427, at 306.

[24] Gregory of Nyssa, *De opificio hominis*, 10, ed. PG 44:153; modified trans. Wilson, 306.

commissioned to produce a series of commentaries on Aristotle between 1118 and 1138,[25] that is, after she retired and before she began to compose the *Alexiad*.[26] Michael's commentaries are marked by a nuanced and critical engagement, both with the Aristotelian texts and with the late antique commentators, an engagement that avoids quoting or paraphrasing his predecessors. In his commentary on *De partibus animalium*, Michael of Ephesus explains a passing reference to the common sense[27] that taciturnly describes the parts of the body and their arrangement: "weight hampers the motion of the intellect and of the common sense" (τὸ γὰρ βάρος δυσκίνητον ποιεῖ τὴν διάνοιαν καὶ τὴν κοινὴν αἴσθησιν).[28] Michael interprets this passage as referring to either the commonality of touch across all animals or the union of sensation as one, writing, "And, the common sense [Aristotle] says that it is either touch (for all living things have this), or, as I think, all of the five [senses] together." Pavel Gregorić has argued that Aristotle meant the former explanation.[29] But, as Michael himself states, he prefers the latter, which is not surprising given that this was the primary understanding of the common sense in the Byzantine world.

[25] Robert Browning, "An Unpublished Funeral Oration on Anna Comnena" in Richard Sorabji (ed.), *Aristotle Transformed: The Ancient Commentators and Their Influence* (Ithaca: Cornell University Press, 1990), 393–406, esp. 399. On Michael of Ephesus, see also Peter Frankopan, "The Literary, Cultural and Political Context for the Twelfth-Century Commentary on the *Nicomachean Ethics*" in Charles Barber and David Jenkins (eds.), *Medieval Greek Commentaries on the Nicomachean Ethics* (Leiden: Brill, 2009), 45–62, esp. 47–48; George Arabatzis, *Παιδεία και Επιστήμη στον Μιχαήλ Εφέσιο.* Εις περι ζώων μορίων *A 1,3–2,10* (Athens: Akadēmia Athēnōn, 2006), esp. 17–36.

[26] Michael's toils are recounted in a funeral oration for Anna Komnene, where George Tornikios, metropolitan of Ephesus, states. "I have myself heard the wise man from Ephesus attribute the cause of his blindness to her [i.e. Anna Komnene], because he spent sleepless nights over commentaries on Aristotle at her command, whence came the damage done to his eyes by candles through desiccation" (Ἐγὼ δὲ καὶ τοῦ ἐξ Ἐφεσίων ἠκηκόειν σοφοῦ ταύτῃ τῆς τῶν ὀφθαλμῶν ἀβλεψίας τὴν αἰτίαν προσεπιρρίπτοντος, ὅτι παννύχοις σχολάσειεν ἀϋπνίαις ἐπὶ ταῖς τῶν Ἀριστοτελείων, κελευούσης αὐτῆς, ἐξηγήσεσιν· ὅθεν τὰ ἐλλύχνια τοῖς ὀφθαλμοῖς διὰ ξηρασίαν παθήματα). See George Tornikios, *Orationes*, 14 (283.9–12), ed. J. Darrouzès, *Georges et Dèmètrios Tornikès, lettres et discours* (Paris: Centre National de la Recherche Scientifique, 1970), 283; trans. Browning, "An Unpublished Funeral Oration," 406.

[27] See Pavel Gregorić, *Aristotle on the Common Sense* (Oxford University Press, 2007), esp. 203–214. On Michael of Ephesus's *De partibus animalium*, see George Arabatzis, "Michael of Ephesus and the Philosophy of Living Things (*In De partibus animalium*, 22.25–23.9)" in Börje Bydén and Katerina Ierodiakonou (eds.), *The Many Faces of Byzantine Philosophy* (Athens: Norwegian Institute at Athens, 2012), 51–78.

[28] "κοινὴν δὲ αἴσθησιν ἢ τὴν ἁφὴν λέγει (ταύτην γὰρ ἔχουσι πάντα τὰ ζῷα) ἢ ὅπερ οἶμαι, ὁμοῦ πάσας τὰς πέντε." Michael of Ephesus, *In libros de partibus animalium commentaria*, 84.18–20, ed. M. Hayduck, *In libros de partibus animalium, de animalium motione, de animalium incessu commentaria*, Commentaria in Aristotelem Graeca 22.2 (Berlin: Reimer, 1904), 84.

[29] Gregorić, *Aristotle on the Common Sense*, 96–98.

Commenting on *De somno et vigilia* in his *Parva naturalia* commentary, Michael returns to the association of the common sense with touch, responding to the statement that the common sense is "closely connected with the sense of touch" (τοῦτο δ' ἅμα τῷ ἁπτικῷ μάλισθ' ὑπάρχει), which itself refers to *De anima* 3.13, where Aristotle states the indispensability of touch across the sensory organs.[30] In the specific passage being commented on, Aristotle says that every sense has its own special function as well as one which they share with the rest, namely touch.[31] Since there is "one sense-faculty, and one paramount sense organ" (μία αἴσθησις, καὶ τὸ κύριον αἰσθητήριον ἕν), Aristotle infers that since all the organs also happen to possess touch then it is reasonable that the common sense is somehow related to touch (τοῦτο δ' ἅμα τῷ ἁπτικῷ μάλισθ' ὑπάρχει). To this, Michael states:

> Τοῦτο λέγει τὴν κοινὴν αἴσθησιν· τῆς γὰρ ἁφῆς χωριζομένης τῶν ἄλλων αἰσθήσεων (εἰσὶ γάρ τινα ζῷα, ὡς τὰ ὄστρεια, μόνην ἔχοντα τὴν ἁφήν· ἡ γὰρ γεῦσις ἁφή τίς ἐστι), χωριζομένης οὖν ταύτης τῶν ἄλλων δηλονότι ἅμα τῇ ἁφῇ ἐστιν ἡ κοινὴ αἴσθησις. εἰ δὲ χρὴ τἀληθὲς εἰπεῖν, ἁφὴ καὶ κοινὴ αἴσθησις ταὐτόν ἐστι· ταύτῃ γὰρ κοινωνεῖ πάντα τὰ ζῷα, ἀλλ' οὐ τῇ ὄψει ἢ τῇ ἀκοῇ.
>
> [Aristotle] says this [i.e. touch] is the common sense. Touch is separated from the other senses (for some living things are in possession of touch alone, such as the oyster, since taste is a kind of touch), thus since this [i.e. touch] is separated from the other [senses], clearly the common sense is simultaneously in touch. If it is necessary to speak the truth, touch and the common sense are identical. For all living things share in it, but not in sight or hearing.[32]

Here, touch stands in for a broader notion of commonality that encompasses but also goes beyond touch's pervasiveness across all living things, since it indicates the commonality of touch across the sensory organs and all life, which suggests that in some manner it may be associated with the common sense. We can deduce in Michael of Ephesus's thinking a careful process of thought and speculation as he attempts to make an informed judgment on an ambiguous text. He seems to arrive at the equation of touch and the common sense somewhat syllogistically, reasoning that if

[30] Aristotle, *De somno et vigilia*, 455a23–24, ed. and trans. W. S. Hett, *Parva Naturalia*, Loeb Classical Library 288 (Cambridge, MA: Harvard University Press, 1957), 326–327. See also Aristotle, *De anima*, 435a11–20, ed. and trans. Hett, 198–201.

[31] Aristotle, *De somno et vigilia*, 455a14–26, ed. and trans. Hett, 326–327.

[32] Michael of Ephesus, *In parva naturalia commentaria*, 48.4–9, ed. P. Wendland, *In parva naturalia commentaria*, Commentaria in Aristotelem Graeca 22.1 (Berlin: Reimer, 1903), 48.

touch is separated from the other senses, and the common sense is separate from other senses, then touch must be united or associated with touch (ἅμα τῇ ἁφῇ).

Michael seems uncomfortable with this conclusion, writing, "If it is necessary to speak the truth" (εἰ δὲ χρὴ τἀληθὲς εἰπεῖν). It is as if he has been forced to acknowledge that touch and the common sense are identical, that touch is the common sense. Certainly, touch is the common sense insofar as it is common to all life, but here Michael is also articulating one of the other understandings of the common sense, namely, that as that unitary perceptual faculty it must also be said to be associated with touch. This appears to be the hesitation that Michael shares. We know that the common sense, defined as the synthetic cognitive function that unites all perception, is not a sixth sense, as John Philoponos states.[33] Therefore, the common sense cannot be understood as being "separated from the senses" (χωριζομένης οὖν ταύτης τῶν ἄλλων). Hence, this passage must be referring specifically to the notion that the sense of touch is common to all life, as the example of the oyster suggests. However, Michael's hesitation belies the fact that this hedges too close for comfort to the idea that the perceptive faculty of the soul is touch. Yet, the association of the unitary perceptive faculty with a form of touch would not be unique or unfounded.

Daniel Heller-Roazen has described this construction of the common sense as an "inner touch," which describes the understanding and sensation of sentience, following Aristotle's characterizations along with that of his medieval Greek, Arabic, Hebrew, and Latin commentators.[34] Thus, this idea of the common sense as the action of perceiving that we see and hear, described in *De anima* III.2, taken at times to be a sort of theory of consciousness,[35] could feasibly be understood as a metaphorical "inner touch." This would work well with the haptic language used to describe internal cognition, resonating with Nemesios's self-reflexive comment that every part of the body has an understanding or "grasp of touch" (ἁφῆς ἀντιλαμβάνεται), as cited above.

[33] See John Philoponos, *In Aristotelis libros de anima commentaria*, 477.3–481.30, ed. M. Hayduck, *In Aristotelis de anima libros commentaria*, Commentaria in Aristotelem Graeca 15 (Berlin: Reimer, 1897), 477–481; trans. William Charlton, *On Aristotle's On the Soul 3.1–8* (Ithaca: Cornell University Press, 2000), 52–57.

[34] Daniel Heller-Roazen, *The Inner Touch: Archaeology of a Sensation* (Brooklyn, NY: Zone Books, 2007), esp. 31–42. See also Simon Kemp and Garth J. O. Fletcher, "The Medieval Theory of the Inner Senses," *American Journal of Psychology* 106:4 (1993): 559–576.

[35] See Gregorić, *Aristotle on the Common Sense*, 174–201.

Certainly, as Aristotle explained, all the senses partake in a form of "contact" or touch, between perceptible and perception, but this is not touch proper – unmediated tactile contact – since each sense is necessarily unique. Just as the notion of the optical *pneuma*'s "walking stick of surrounding air" in Galen does not allow for any more unmediated contact or touch than does Plato's united rays theory or Aristotle's theory of the transparent. As we shall explore further in this book's final chapter on synesthesia and sensual remediation, the commonalities of the senses, their common sensibles, and the fact that they all are stitched together through consciousness necessitate the uniqueness of each sense so as to allow for their union, overlap, and crossover. Were the senses not distinguished, there would be no point in their separation or in their individual sensory organs.

Returning to Michael of Ephesus's grappling of the common sense, we should consider how he understands the sense of sight to relate to tactility. Perhaps the most telling instance of this occurs in his commentary on the *Nicomachean Ethics* in a passage dedicated to Aristotle's line "Sight differs from touch in greater purity" (διαφέρει δὲ ἡ ὄψις ἁφῆς καθαρειότητι):[36]

> Ἡ μὲν ὄψις τῶν εἰδῶν ἐστιν αὐτῶν ἀντιληπτικὴ δίχα τῆς ὕλης, ὡς ἐν τῷ δευτέρῳ τῆς Περὶ ψυχῆς πραγματείας δέδεικται. ὥσπερ γὰρ ὁ κηρὸς τὴν σφραγῖδα μόνην τὴν ἐν τῷ χρυσίνῳ δακτυλίῳ ἀπομάττεται χωρὶς τῆς ὑποκειμένης ὕλης (χρυσὸς γὰρ ἂν ἐγίνετο ὁ κηρός, εἰ σὺν τῷ ὑποκειμένῳ ὡς ὕλῃ τῷ χρυσῷ τὴν σφραγῖδα ἐδέχετο), οὕτω καὶ ἡ ὄψις τῶν εἰδῶν ἐστιν ἀντιληπτικὴ χωρὶς τῆς ὑποκειμένης αὐτοῖς ὕλης. ἡ δὲ ἀκοὴ καὶ ἡ ὄσφρησις σωματοειδεῖς οὖσαι καὶ παθητικώτεραι μεθ' ὕλης εἰσδέχονται τὰ αἰσθητά.
>
> Sight is perceptive of forms themselves, without matter, just as Aristotle has shown in the second book of *De anima*. For just as the wax takes the seal alone in the gold ring without the underlying material (for then the wax would become the gold, if the [wax] received the seal together with the underlying gold as matter). Thus, sight of forms is perceptive without the underlying matter with it. But, since hearing and smell are body-like, they are more subject to be affected by matter when they receive sensible things.[37]

Here, Michael extols the greater purity of sight above touch precisely because it is "perceptive without matter" (ἀντιληπτικὴ δίχα τῆς ὕλης),

[36] Aristotle, *Ethica Nicomachea*, 1175b36–1176a1, ed. and modified trans. H. Rackham, *Nicomachean Ethics*, Loeb Classical Library 73 (Cambridge, MA: Harvard University Press, 1926), 602–603.

[37] Michael of Ephesus, *In ethica Nicomachea ix–x commentaria*, 569.8–15, ed. G. Heylbut, *In ethica Nicomachea commentaria*, Commentaria in Aristotelem Graeca 20 (Berlin: Reimer, 1892), 569.

unlike the other sense, which directly interacts with matter. He offers examples in this excerpt of hearing and smell, but then goes on to discuss taste as well, which is even closer to touch, and thus even less pure than hearing and smell, which are yet still inferior to sight. Because the other senses are corporeal or "body-like" (σωματοειδεῖς) and subject to being affected (παθητικώτεραι) by matter, they lack purity.[38]

Michael of Ephesus's use of the wax tablet is a brilliant choice of metaphor for characterizing the purity of sight against touch. The wax-seal metaphor in general, in Plato, Aristotle, and others, is used to describe sense impressions upon the soul, and the memory is often described as wax tablets. The metaphor also suggests a certain degree of tactility in perception since it implies the literal touch of the seal upon the wax. However, Michael of Ephesus uses it here in an almost tongue-in-cheek manner that actively works against any hint of tactility, given that the purity that sight enjoys is manifested in the metaphor because of the fact that the wax can receive a form off the signet ring without taking on its material. This distance from material exchange is precisely what touch lacks in Michael's model, since it is completely on the opposite side of the spectrum, wherein it is the sense most affected by the materiality of the object of its perception. Thus, in playing with this instance of tactility, Michael reminds us of the importance of reading such examples precisely as metaphors, rather than as accurate models for the conceptualization of sense.

Across his various works, Michael demonstrates how articulations of touch are not only pervasive in philosophical thought, but often reflect a metaphorical understanding of noetic perception. In denying even the very corporeality (σωματοειδεῖς) of sight, which he attributes to the lesser senses, Michael unequivocally demonstrates that visual contact, like that between the signet ring and the wax, cannot be understood as being akin to the contact of touch, given that touch commingles with the material it receives its impressions from. As Michael states later in his commentary on the *Nicomachean Ethics*, it is through the articulation of the various forms of purity of each sense, "by which we know the differences and commonalities of sensible-things" (αἷς τὰς διαφορὰς καὶ κοινωνίας τῶν αἰσθητῶν ἔγνωμεν).[39]

38 See George Arabatzis, "Michael of Ephesus on the Empirical Man, the Scientist and the Educated Man (*In Ethica Nicomachea* x and *In Partibus Animalium* i)" in Charles Barber and David Jenkins (eds.), *Medieval Greek Commentaries on the Nicomachean Ethics* (Leiden: Brill, 2009), 163–183, esp. 174–178.

39 "τὸ γὰρ τὴν ζῳώδη καὶ γενεσιουργὸν ἀποπληροῦν χρείαν πῶς ἂν ἐπίσης εἴη καθαρὸν ὁράσει καὶ ἀκοῇ καὶ ταῖς λοιπαῖς δύο, αἷς τὰς διαφορὰς καὶ κοινωνίας τῶν αἰσθητῶν ἔγνωμεν." Michael of Ephesus, *In ethica Nicomachea ix–x commentaria*, 569.26–28, ed. Heylbut, 569.

In other words, since each sensation comes to our attention with some level of contamination with earthly matter, we are able to discern the quality of each sense's impressions. The "inner touch" of the common sense and touch's commonality across living things and the sensory organs might lead to possible confusion. However, as we have seen in this chapter, it is through their specificity that the senses truly operate. While touch can serve as a broader conceptual metaphor for the inherent contact between the sensory organ and its object that sensation depends upon, this is always a contact that occurs through a medium, as when a viewer grasps a visible thing through the intervening medium of water or air.

PART II

Photios and the Unfolding of Perception

Introduction to Part II

[a] Χριστὸς σαρκὶ ἐπεδήμησε καὶ τῆς τεκούσης ὠλέναις ἐφέρετο· [b1] τοῦτο καὶ ταῖς εἰκόσιν ὁρᾶται καὶ πιστοῦται καὶ κηρύσσεται, αὐτοψίας νόμῳ διατρανουμένης τῆς μαθήσεως [b2] καὶ τοὺς θεατὰς ἐφελκομένης εἰς ἀπροφάσιστον συγκατάθεσιν. [c] Μισεῖ τις τὴν διὰ τούτων διδασκαλίαν; πῶς οὐ πρότερον οὗτος τὸ τῶν εὐαγγελίων εἰς μῖσος ἀπώσατο κήρυγμα; [d1] ὥσπερ γὰρ ὁ λόγος ὁ δι᾽ ἀκοῆς, οὕτω δι᾽ ὄψεως ἡ μορφὴ τοῖς τῆς ψυχῆς ἐγχαράσσεται πίναξιν, ὁμόφωνον τῆς εὐσεβείας, [d2] οἷς πονηρῶν δογμάτων ἡ πρόληψις οὐκ ἐνέστικται, τὴν μάθησιν διαγράφουσα. [e] Μάρτυρες δεσποτικῆς ἀγάπης ὑπερήθλησαν, τὸ φίλτρον τοῦ πόθου τοῖς αἵμασι παραστήσαντες καὶ τὴν μνήμην αἱ βίβλοι συνέχουσι· [f1] ταῦτα καὶ ταῖς εἰκόσιν ὁρῶνται διαπραττόμενοι, [f2] ἐναργεστέραν τῆς γραφῆς παρεχομένης τῶν μακαρίων ἐκείνων εἰς γνῶσιν τὴν ἄθλησιν. [g] Ἄλλοις αἱ σάρκες καὶ ζῶσιν ὡλοκαυτώθησαν, προσευχῆς καὶ νηστείας καὶ τῶν ἄλλων πόνων καλλιερούντων τὰ θύματα· [h] ταῦτα καὶ λόγοι καὶ εἰκόνες φέρουσαι μᾶλλον τοὺς ὁρῶντας ἢ τοὺς ἀκούοντας πρὸς μίμησιν ἐπιστρέφουσιν. [i] Ἡ παρθένος τὸν κτίστην χερσὶν ὡς βρέφος βαστάζει. Τίς καθορῶν ἢ τοῖς ὠσὶ ταῦτα βαλλόμενος οὐ μᾶλλον καταπλαγείη τοῦ μυστηρίου τὸ μέγεθος καὶ πρὸς ὕμνον διανασταίη τῆς ἀφάτου καὶ λόγους πάντας νικώσης συγκαταβάσεως; [j] εἰ γὰρ καὶ δι᾽ ἀλλήλων ἑκάτερον συνεισάγεται, ἀλλὰ πολὺ προέχειν ἐπὶ τῶν ἔργων αὐτῶν ἐπιδείκνυται τῆς κατὰ τὴν ἀκοὴν εἰσδυομένης μαθήσεως ἡ διὰ τῆς ὄψεως ἐγγινομένη κατάληψις. [k0] Ἔκλινέ τις τὸ οὖς εἰς διήγημα; [k1] εἵλκυσε [k2] φανταζομένη τὸ ἀκουσθὲν ἡ διάνοια; [k3] νηφούσῃ μελέτῃ τὸ κριθὲν [k4] τῇ μνήμῃ ἐναπέθετο. [l] Οὐδὲν τούτων ἔλαττον, εἰ μὴ καὶ πολὺ μᾶλλον, κρατεῖ τὰ τῆς ὄψεως· [m0] Καὶ γὰρ καὶ αὐτή γε δήπου τῇ προχύσει καὶ ἀπορροῇ τῶν ὀπτικῶν ἀκτίνων [m1] τὸ ὁρατὸν οἱονεί πως ἐπαφωμένη καὶ περιέπουσα [m2] τὸ εἶδος τοῦ ὁραθέντος τῷ ἡγεμονικῷ παραπέμπεται, [m4] ἐκεῖθεν διαπορθμευθῆναι διδοῦσα τῇ μνήμῃ πρὸς ἐπιστήμης ἀπλανεστάτης συνάθροισιν. [n0] Εἶδεν ὁ νοῦς, [n1] ἀντελάβετο, [n2] ἐφαντάσθη, [n3] τοὺς τύπους ἀκόπως [n4] ἐν τῇ μνήμῃ παρεπέμψατο.

> [a] Christ came to us in the flesh, and was borne in the arms of His Mother. [b1] This is seen and confirmed and proclaimed in pictures, the teaching made manifest by means of personal eyewitness, [b2] and impelling the spectators to unhesitating assent. [c] Does a man hate the teaching by means of pictures? Then how could he not have previously rejected and hated the message of the Gospels? [d1] Just as speech is transmitted by hearing, so a form through sight is imprinted upon the tablets of the soul, [d2] giving to those whose apprehension is not soiled by wicked doctrines a representation of knowledge concordant with piety. [e] Martyrs have suffered for their love of God, showing with their blood the ardour of their desire, and their memory is contained in books. [f1] These deeds they are also seen performing in pictures, [f2] as painting presents the martyrdom of those blessed men more vividly to our knowledge. [g] Others have been burnt alive, a sacrifice sanctified by their prayer, fasting and other labours. [h] These things are conveyed both by stories and by pictures, but it is the spectators rather than the hearers who are drawn to emulation. [i] The Virgin is holding the Creator in her arms as an infant. Who is there who would not marvel, more from the sight of it than from the report, at the magnitude of the mystery, and would not rise up to laud the ineffable condescension that surpasses all words? [j] For even if the one introduces the other, yet the comprehension that comes about through sight is shown in very fact to be far superior to the learning that penetrates through the ears. [k0] Has a man lent his ear to a story? [k2] Has his intelligence visualized and [k1] drawn to itself [k2] what he has heard? [k3] Then, after judging it with sober attention, [k4] he deposits it in his memory. [l] No less – indeed much greater – is the power of sight. [m0] For surely, having somehow through the outpouring and effluence of the optical rays [m1] touched and encompassed the object, [m2] it too sends the essence of the thing seen to the mind, [m4] letting it be conveyed from there to the memory for the concentration of unfailing knowledge. [n0] Has the mind seen? [n1] Has it grasped? [n2] Has it visualized? [n3] Then it has effortlessly [n4] transmitted the forms to the memory.[1]

Delivered on Holy Saturday, 29 March 867, before the emperors Michael III and Basil I, Patriarch Photios's Homily 17 commemorates the depiction and unveiling of the image of the Virgin and Child in the apse of Hagia Sophia.[2] There, Photios discusses the process of visual sensation and perception. In praising sight's effectiveness over hearing, he has seemed to

[1] Photios, *Homiliae*, 17, ed. B. Laourdas *ΟΜΙΛΙΑΙ* (Thessaloniki: Hetaireia Makedonikōn Spoudōn, 1959), 170–171; trans. Cyril Mango, *The Homilies of Photius Patriarch of Constantinople* (Cambridge, MA: Harvard University Press, 1958), 293–294. Inserted references are mine, to facilitate discussion in Part II. For the rationale behind references beginning with 0, and for the missing [m3], please see Table 1 and accompanying discussion in Chapter 5.

[2] Mango, *The Homilies*, 279, cf. 286 n. 1.

modern readers to suggest that extramissive rays of vision have haptic and tactile power. But Photios's homily does not advocate the tactility of sight. Rather, as I argue in this part, he provides a heretofore overlooked detailed description of the procedural unfolding of perception – from the sensory contact with the object through its apprehension by the mind, its visualization in the imagination, and its storage in the memory.

An erudite scholar, statesman, and theologian, Photios provides us in this homily with an indispensable window into theories of vision, sensation, and perception in post-iconoclastic Constantinople. Its importance should not be understated: not only does this text come down to us from one of the best and most diversely educated intellectuals of the time, but also, being a homily, it targeted a diverse audience of Constantinopolitan figures, and so can be interpreted as representative of Byzantine thought.

Perhaps this breadth of application seems presumptive. The highly learned figures discussed thus far, like Nemesios of Emesa, John Philoponos, or Michael Psellos, might lead us to question how applicable they are to a broader Byzantine sphere. The same could be said in the case of Photios, whose extensive medical knowledge was said to border on professional. Not only did he have a low opinion of doctors of his time, disagreeing with and criticizing some of their practices, but he also practiced medicine himself, as attested by his letters that evidence both the issuing of prescriptions and the acknowledgement of thanks for cures.[3] Photios's knowledge even elicited the comment of Niketas David Paphlagon, who sarcastically remarked that Photios's knowledge of medicine and other secular arts "not only surpassed all his own generation, but, one might almost say, he even rivaled the ancients."[4]

Nevertheless, while Photios may have been referencing an eruditely nuanced theory of vision, the homily was written to be legible – and plausible – to his congregation. This sermon is perhaps the closest one can get to an articulate popular understanding of vision in the extant written sources. Thus, Photios's thoughts and explanations in this genre and context are different from the philosophical and medical treatises on vision surveyed in the previous three chapters. They help us grasp what a more common and ecumenical theory of vision might have looked like, which by necessity should have been accessible and legible to a relatively educated

[3] Warren T. Treadgold, *The Nature of the Bibliotheca of Photios* (Washington, DC: Dumbarton Oaks, 1980), 97–110, esp. 103.

[4] "ὡς μὴ μόνον σχεδὸν φάναι τῶν κατὰ τὴν αὑτοῦ γενεὰν πάντων διενεγκεῖν, ἤδη δὲ καὶ πρὸς τοὺς παλαιοὺς αὐτὸν διαμιλλᾶσθαι." Niketas David Paphlagon, *Vita patriarchae Ignatii*, ed. and trans. Andrew Smithies, *The Life of Patriarch Ignatius* (Washington, DC: Dumbarton Oaks, 2013), 34–35.

and cosmopolitan audience characteristic of the capital's church. That visual theory also needed to be articulated so as to accomplish its intended theological function, to expound on the icon of the Theotokos and Christ in the apse.

The goal of Part II is to carefully unpack the processes of sensation and perception that Photios outlines in this homily. My explanation is intended to demonstrate how classical and late antique learning contoured his understanding of perception, and how and why the seeming confusion of sight and tactility emerges as a prominent discursive turn in Byzantine texts. The key lines that will be the focal point of these chapters are the final segment of the excerpted homily above that details the actions of perception, [**k, n**]. The chapters that follow are organized around the litany of rhetorical questions that summarize Photios's theory of perception, namely [**n**]: "Has the mind seen? Has it grasped? Has it visualized? Then it has effortlessly transmitted the forms to the memory."

In Chapter 4, we shall turn our attention to the question of sensation itself, considering how the mind sees through the senses – in the case of sight, through the "outpouring and effluence" of optical rays [**m**]. I will carefully consider this language in order to explain in detail Photios's theory of vision, which as I will show followed Plato's theory of united rays. Then I shall turn in Chapter 5 to the second stage of sensory perception, where the mind grasps the physical object and converts its stimuli and impulses into noetic form, transitioning from the realm of sensation to that of perception proper, and its cognitive unfolding. Here, we shall consider likewise how the claim to having "touched and encompassed" [**m**] the visible object relates to earlier theories of the mind's apprehension of the outside world.

Chapters 6 and 7 build on this haptic cognitive language to consider the important role of the imagination in visualizing sensory data. The first of these chapters will focus on the Stoic heritage of the *katalēptikē phantasia*, a cognitive impression from the outside world deemed to have a direct claim to truth. Chapter 7 will consider early Christian suspicions of the imagination; at times *phantasia* was depicted as being fantastical, fictional, and contrafactual – more like our modern notion of fantasy. However, as I shall show in the context of Photios's works and that of other thinkers, while the images of *phantasia* could be suspect, the imaginative faculty itself had overwhelming support as something inherent to human perception, thought, and memory. These two chapters lay the foundation to consider how the mind passes judgment on sensation's impulses and the mind's reception of them.

In this part's final chapter, I consider the role that initial preconceptions and beliefs played in the judgment and discernment of the representations being processed by perception. Looking comparatively at Photios's earlier description of hearing [**k**], I observe that while this echoes sight's unfolding perfectly [**n**], there is one subtle difference between the two, namely, while sight "effortlessly" passes on the imprints to the memory, hearing judges them "with sober attention." This denotes an additional process of judgment that Photios seems to negate or elide in the context of vision, fueling his argument that vision is superior to hearing because it provides us with ready-made images that need not be visualized by the imagination before proceeding to the mind. In the conclusion to this part, I summarize my very close readings, to provide a comprehensive picture of Photios's ordering of perception, from the work of the senses to the preservation of memory. This likewise allows me to shed some further light on Photios's sources for his theory of vision, primarily Nemesios of Emesa and John of Damascus.

Throughout these chapters, I will use Cyril Mango's translation as a foil to my argument, as a way of reading Photios against the grain. My goal is not to demonstrate it to be faulty or lacking, but rather to expose how our preconceptions of Byzantine vision are rooted and reflected in Mango's translation. By responding, resisting, and defending my reading against Mango's own, I wish to grapple with the proliferation of haptic language in this translation to argue that Photios means precisely the opposite. Mango's translation allows me to expose the various haptic eddies that have caught scholars repeatedly, and thus caused us to overlook a perspective far richer than is evident at first glance, one that is not merely an exaltation of haptic vision, but is a compact but detailed theory of cognitive perception.

CHAPTER 4

"Has the Mind Seen?"
The Language of Effluxes

In the first stage of Photios's summary of visual perception, he asks [**n**], "Has the mind seen?" (Εἶδεν ὁ νοῦς;), a moment that suggests the action of sensation itself, when the mind is made aware through the senses of the external world. The stage agrees with the initial portion of the previous statement [**mo–1**], "having somehow through the outpouring (προχύσει) and effluence (ἀπορροῇ) of the optical rays touched and encompassed the object," which is then passed on to the mind. In these words, Photios provides us with a succinct but eloquent characterization of the physiology of sight: vision grasps via the "outpouring" (προχύσει) and "effluence" (ἀπορροῇ) of the optical rays. In this chapter, I wish to focus on the meaning of these two words. My intention here is to use Mango's popular translation, which has served repeatedly as the bedrock for modern understanding of post-iconoclastic visuality, as a rhetorical foil to illustrate the subtle and significant differences between Photios's words and their modern articulation.

If disregarded as being part of an oratorical flourish, the two terms could be generalized and the statement smoothed out to suggest that the outpouring and effluence of rays from the eye grasp the object. But the words are strikingly different. While the term *prochysis* (προχύσει) denotes the notion of an outpouring as in the case of a river or sweat issuing from the body, *aporroe* (ἀπορροῇ) on the other hand denotes a subtler emanation and is often associated in the atomists, first found in Empedocles's theory of vision, with emissions from visible objects. In the homily these two words provide us with a notable pairing of prefixes, προ- and ἀπο-, which may indicate bidirectional motion of rays rather than only that which emerges from the viewer. This pair of terms could be read as a description of rays hitting and reflecting off the object, but their synchronous deployment nevertheless seems to allude to the simultaneity of rays being emitted from the eyes and from the object, as would be expected in a properly Platonic theory of extramission, or, more properly, of united rays.

LSJ defines both words quite similarly, either as emissions pouring out or flowing off, but their usage in late antique and Byzantine texts suggests a differentiation that produces a more nuanced translation of Photios's passage. As one might expect, *prochysis* denotes a flowing outwards or a deposition of materials, such as mud in a creek bed, and is often used in this explicit manner. For example, Oribasios, the fourth-century personal physician of Julian the Apostate, uses the term in his medical writings to describe the ejaculation of sperm from the body, based on Galen's work on the matter; Theodoret of Cyrus in his fifth-century *Commentary on the Psalms* uses it to describe the gushing-out of water from stones in reference to Psalm 114:7–8; and George the Synkellos in his late eighth-, early ninth-century *Chronography* uses "προχύσει" to describe the "sprinkling" (πρόσχυσιν), of lamb's blood upon the doorways during Passover in Exodus 12:13 (via Hebrews 11:28).[1] Such examples are consistent with various other uses, which denote a directional outpouring of a substance toward an object. This is also consistent with what one would expect from *prochysis* in Photios's text as referring to the rays pouring out from the eyes.

However, the crux of the matter here is how we come to understand its counterpart, *aporroe*. The term *aporroe* is attested often and is used in texts relating to vision to describe precisely the rays emanating from the object. In Plato's *Timaeus*, discussed in Chapter 1, the term appears in Plato's discussion of colors emanating from the visible bodies, described as "a flame which issues (ἀπορρέουσαν) from the several bodies."[2] As discussed, Plato here explains how this fiery emanation from the bodies of visible things produces sensation by uniting with the fiery stream outpouring from the eyes. Plato deploys the verb form ἀπορρέουσαν specifically to convey the issuing of rays from the object specifically, a strong precedent for this term's later association with the emanations coming off perceptible things.

[1] See Oribasios, *Collectiones medicae (libri incerti)*, 9.23.6, ed. J. Raeder *Collectionum medicarum reliquiae*, vol. IV (Leipzig: Teubner, 1933), 93; Theodoret of Cyrrhus, *Interpretatio in Psalmos*, ed. PG 80:1792; George the Synkellos, *Ecloga chronographica*, ed. A. A. Mosshammer, *Ecloga chronographica* (Leipzig: Teubner, 1984), 142.

[2] "φλόγα τῶν σωμάτων ἑκάστων ἀπορρέουσαν, ὄψει σύμμετρα μόρια ἔχουσαν πρὸς αἴσθησιν· ὄψεως δ' ἐν τοῖς πρόσθεν αὐτὸ περὶ τῶν αἰτίων τῆς γενέσεως ἐρρήθη." Plato, *Timaeus*, 67c, ed. and trans. Bury, 172–73.

In other treatises on vision and philosophy, the term is used to denote specifically this emanation from the outside perceptible body or object, understood as being external to the subject. Later ἀπορρέοντα will appear in Psellos's commentary on Aristotle, discussed at length in Chapter 1, building upon the term's earlier transmission in Aetios and Alexander of Aphrodisias:

> πάσχουσα γάρ, φησίν, ἡ ὄψις ὑπὸ τῶν ὁρατῶν αἰσθάνεται, ὥσπερ καὶ ἑκάστη τῶν ἄλλων αἰσθήσεων, ἀλλ' οὐχὶ ποιοῦσα καὶ ἐκπέμπουσα, πάσχουσα δὲ οὐ τῷ δέχεσθαι ἀπορρέοντά τινα ἀπὸ τῶν αἰσθητῶν.
>
> For, [Aristotle] says, sight perceives by being affected by the things seen, just as each of the other [senses] perceives, but not by making and sending out, and not by being affected by admitting what flows from the things perceived.[3]

Here Psellos clearly delineates the difference between the rays shooting out of the eyes (ἐκπέμπουσα) and the emanations from the object (ἀπορρέοντά). Elsewhere in this same commentary, he uses the term consistently to describe as well the intromissionists' theory of the entrance of the atomic *eidola* being emanated by the object (εἴδωλά τινα ἀπορρέοντα), as in the case of Democritus; and the notion of the mixing of optical rays to describe the ones coming from the object (ἐφαπλουμένας ἀπορροίας), as in the case of Empedocles.[4] And, furthermore, in his encyclopedic entry on the senses in *De omnifaria doctrina*, Psellos also uses it to describe the operation of sight according to Plato. In his entry "On the Five Senses" (Περὶ τῶν πέντε αἰσθήσεων), Psellos says:

> Ἡ ὅρασις γίνεται ὅταν τὸ ἀπὸ τῶν ὀφθαλμῶν φῶς ἐκφερόμενον συναντήσῃ τῷ ἀπὸ τῶν σωμάτων ἀπορρέοντι, ὅπερ ὁ Πλάτων ὀνομάζει συναύγειαν.

[3] Michael Psellos, *Opuscula psychologica*, 8, ed. D. J. O'Meara, *Philosophica minora*, vol. II (Leipzig: Teubner, 1989), 16–17; trans. Barber, *Contesting the Logic of Painting*, 94.

[4] The term appears therein six times, and only once in this short treatise does it describe the flowing of rays from the eye, following its same use in the Alexander of Aphrodisias source text. This anomalous appearance emerges as part of a line that chooses to stress the exteriority of the outpouring of the rays from the eye, similar to what occurs in a passage from Aetios, discussed below. See Psellos, *Opuscula psychologica*, 8, ed. O'Meara, 16; cf. Alexander of Aphrodisias, *In librum de sensu commentarium*, ed. Wendland, 32:13–15. On Empedocles, see David Sedley, "Empedocles' Theory of Vision and Theophrastus' *De sensibus*" in William W. Fortenbaugh and Dimitri Gutas (eds.), *Theophrastus: His Psychological, Doxographical, and Scientific Writings* (New Brunswick, NJ: Transaction Publishers, 1992), 20–31; A. A. Long, "Thinking and Sense-Perception in Empedocles: Mysticism or Materialism," *Classical Quarterly* 16 (1966): 256–276; D. O'Brien, "The Effects of Simile: Empedocles' Theories of Seeing and Breathing," *Journal of Hellenic Studies* 90 (1970): 140–179.

> The act of sight comes into being when the light sent out from the eyes comes into contact with the [light] emanating from the bodies, the very thing which Plato calls united rays.[5]

In each of the two excerpts presented here, Psellos neatly uses doubled participles to capture the dual process of Platonic extramission. Notably, the pairing of ἐκ- and ἀπο- convey the outpouring and emanation of rays that unite to make vision possible, analogous to the pairing of προ- and ἀπο- verbs in Photios's own homily. And, in *De omnifaria doctrina* this language appears once again in his entry "On Colors" (Περὶ χρωμάτων) to explain that "color is the visible quality of the appearance of bodies."[6] There, Psellos again explicitly states that he is following Plato, and explains that Plato shows that "from the visible body and from the seeing eyes some emanations (ἀπόρροιαι) are sent out, which in the medium of the interval mix to produce the appearance of colors."[7] Here, however, for the sake of conciseness and brevity, Psellos uses only a single verbal term, ἀπόρροιαι, for both the emanations from the eyes and those from the object. Likewise, he uses the preposition ἀπό for both the eyes and for the object so as to stress the efflux of rays going into the intervallic medium (ἐν τῷ μεταξὺ τοῦ διαστήματος), where the two mix and produce the appearance of color. Yet, as he goes on to argue, it is not merely perception that occurs in the interval, but colors vary from person to person given the humors of the eye and their own internal tints and colorations. Thus, we can understand the external passivity of the ἀπόρροιαι, which despite being used at times with technical precision to describe the efflux from objects, can also be applied to both effluxes to stress effectively their exteriority or outward flow.

Psellos's passage "On the Five Senses" in *De omnifaria doctrina* derives closely from a popular text, Aetios of Antioch's *De placita philosophorum*, which summarizes various theories of sight. In the Aetios text, however, the participle form (ἀπορρέοντος) is used to describe the optical rays that emanate from the viewer, rather than the object. In the words of Aetios:

[5] Michael Psellos, *De omnifaria doctrina*, 108, ed. L. G. Westerink, *De omnifaria doctrina* (Nijmegen: Centrale Drukkerij, 1948), 60.

[6] "Χρῶμα ἐστὶν ὁρατὴ ποιότης τῆς τῶν σωμάτων ἐπιφανείας." Michael Psellos, *De omnifaria doctrina*, 89, ed. Westerink, 52.

[7] "λέγει δὲ ὅτι καὶ ἀπὸ τῶν ὑποκειμένων σωμάτων καὶ ἀπὸ τῶν ὁρώντων ὀμμάτων ἀπόρροιαι τινὲς πέμπονται, αἵτινες ἐν τῷ μεταξὺ τοῦ διαστήματος μιγνύμεναι τὰς ἰδέας τῶν χρωμάτων ποιοῦσιν." Michael Psellos, *De omnifaria doctrina*, 89, ed. Westerink.

Δημόκριτος Ἐπίκουρος κατ' εἰδώλων εἰσκρίσεις ᾤοντο τὸ ὁρατικὸν συμβαίνειν, καὶ κατά τινων ἀκτίνων ἔκκρισιν μετὰ τὴν πρὸς τὸ ὑποκείμενον ἔνστασιν πάλιν ὑποστρεφουσῶν πρὸς τὴν ὄψιν. Ἐμπεδοκλῆς τοῖς εἰδώλοις τὰς ἀκτῖνας ἀνέμιξε, προσαγορεύσας τὸ γιγνόμενον ἀκτινείδωλον συνθέτως. Ἵππαρχος ἀκτῖνάς φησιν ἀφ' ἑκατέρου τῶν ὀφθαλμῶν ἀποτεινομένας τοῖς πέρασιν αὐτῶν οἷον χειρῶν ἐπαφαῖς περικαθαπτούσας τοῖς ἐκτὸς σώμασι τὴν ἀντίληψιν αὐτῶν πρὸς τὸ ὁρατικὸν ἀποδιδόναι. Πλάτων κατὰ συναύγειαν, τοῦ μὲν ἐκ τῶν ὀφθαλμῶν φωτὸς ἐπὶ ποσὸν ἀπορρέοντος εἰς τὸν ὁμογενῆ ἀέρα, τοῦ δ' ἀπὸ τῶν σωμάτων φερομένου, τοῦ δὲ περὶ τὸν μεταξὺ ἀέρα εὐδιάχυτον ὄντα καὶ εὔτρεπτον συνεκτεινομένου τῷ πυρώδει τῆς ὄψεως αὕτη λέγεται Πλατωνικὴ συναύγεια.

Democritus [and] Epicurus suppose that sight is caused by the insinuation of little images into the visive organ, and by the exit of certain rays which return to the eye after striking upon the object. Empedocles supposes that images are mixed with the rays of the eye; these he styles the rays of images. Hipparchus, that the visual rays extend from both eyes to the superficies of bodies, and give to the sight the apprehension of those same bodies, after the same manner in which the hand touching the extremity of bodies gives the sense of feeling. Plato, that sight comes from united rays; there is a light which goes some distance from the eyes into a congruous air, and there is likewise a light emitted from bodies, which meets and is joined with the fire of vision in the intermediate air, which is liquid and mutable; and the conjunction of these rays gives the sense of seeing. This is Plato's corradiancy, or splendor of united rays.[8]

In the context of our broader concerns with tactility, it is worth noting that the text of Aetios makes it clear that only the proponents of intromission, such as Democritus and Epicurus, see vision as occurring through the entrance of rays returning from the object after having struck its surface. And even then, Aetios stresses that it is only Hipparchus's theory of vision that explicitly explained sight as touch. This clarification here is critical since not only was Aetios's account selected by the learned Psellos to represent a definitive and comprehensive view of vision and its variants in his encyclopedia, but it is also used

[8] Pseudo-Plutarch (Aetios of Antioch), *De placita philosophorum*, IV.13, ed. Jürgen Mau, *Plutarchi moralia*, vol. V.2.1 (Leipzig: Teubner, 1971), 123; modified trans. William W. Goodwin, *Plutarch's Morals* (Boston: Colonial Press, 1905), 168. For the work's textual history, see David T. Runia, "The Placita Ascribed to Doctors in Aëtius' Doxography on Physics" in J. Mansfeld and D. T. Runia (eds.), *Aëtiana: The Method and Intellectual Context of a Doxographer*, vol. III (Leiden: Brill, 2010), 515–575.

in various other encyclopedic sources, ranging from Nemesios's *De natura hominis* to John Stobaios's *Anthology*.[9]

Curiously, the term ἀπορρέοντος is used in Aetios's text to describe rays emanating from the eye into space, unlike what one might expect, those returning to the eye. The rays are characterized as emanations into the intermediary space. Aetios is not emphasizing the act of the active viewer's sending out of rays, but rather their progression into space. This suggests that the verb, while somewhat flexible in its technical use, nevertheless connotes an inherent exteriority to the viewer. This is consistent with usage in other late antique and Byzantine texts. An example is the term's usage in Plotinos's *Enneads* in his discourse on love. Plotinos describes the emergence of love as born from the Soul through the action of the gaze looking upon the object of contemplation. Plotinos writes:

> Καὶ πρῶτον ἦν ὅραμα αὐτῇ τοῦτο καὶ ἑώρα ὡς πρὸς ἀγαθὸν αὐτῆς καὶ ἔχαιρεν ὁρῶσα, καὶ τὸ ὅραμα τοιοῦτον ἦν, ὡς μὴ πάρεργον ποιεῖσθαι τὴν θέαν τὸ ὁρῶν, ὡς τῇ οἷον ἡδονῇ καὶ τάσει τῇ πρὸς αὐτὸ καὶ σφοδρότητι τῆς θέας γεννῆσαί τι παρ' αὐτῆς ἄξιον αὐτῆς καὶ τοῦ ὁράματος. Ἐξ οὖν τοῦ ἐνεργοῦντος συντόνως περὶ τὸ ὁρώμενον καὶ ἐκ τοῦ οἷον ἀπορρέοντος ἀπὸ τοῦ ὁρωμένου ὄμμα πληρωθέν, οἷον μετ' εἰδώλου ὅρασις, Ἔρως ἐγένετο...
>
> That was the first object of its vision; it looked towards it as towards its good, and it rejoiced in the looking; and the quality of what it saw was such that the contemplation could not be void of effect; in virtue of that quasi-rapture, of its position in regard to its object, of the intensity of its gaze, the Soul conceived and brought forth an offspring worthy of itself and of the vision. Thus, from the strenuous action about the object-being-seen and from some sort of effluence from that object-being-seen, the eye is fulfilled, like a vision with an image, and so Eros is born...[10]

In Plotinos's text, ἀπορρέοντος works double duty. On the one hand, the term suggests the rays issuing forth from the eyes of the other, who is the object of desire; on the other hand, it speaks to a theory that held that vision was produced by these twin effluxes of rays, from both viewer and the object of contemplation. Deploying

[9] Nemesios, *De natura hominis*, 7, ed. M. Morani, *De natura hominis* (Leipzig: Teubner, 1987), 58; John Stobaios, *Anthologium*, I.52.7, ed. O. Hense and C. Wachsmuth, *Anthologium*, vol. I (Berlin: Weidmann, 1884), 484.

[10] Plotinos, *Enneades*, III.5.3, ed. P. Henry and H.-R. Schwyzer, *Opera*, vol. I (Leiden: Brill, 1951), 225–417; modified trans. Stephen Mackenna and B. S. Page, *The Six Enneads* (London: Faber, 1966), 175–176.

a complex and metaphorical iteration of the theory of vision to explain the birth of Eros, Plotinos's use of ἀπορρέοντος seems to suggest that the term can be fruitfully applied to that object of sight – even if it too projects optical rays. This object of sight is inherently exterior to the body, just as it is in the case of Basil of Caesarea's discussion of the evil eye as "some flowing emission (ἀπορρέοντος) of destruction from the jealous eyes."[11] In the case of the evil eye, this efflux comes from another person, who envies the victimized subject, a subject who is harmed by the rays coming towards them. Metaphorically then, Plotinos's Eros emerges from the union of the rays from the viewer's eyes in the "strenuous action about the object-being-seen" and the "some sort of effluence" from the same object. So, the comparison of love to vision is completed and comes full circle by the fact that in this union – of rays and lovers – is born not only Eros, but also vision itself.[12]

Similar rhetoric can be found in an earlier text, the late second-century *Leukippe and Kleitophon* by Achilles Tatius, whose continued popularity well into the Middle Byzantine world is attested in entries in the *Souda* and in Photios's *Bibliotheca*, in a comparative literary analysis by Michael Psellos, and in a discussion of the novel in a commentary on Hermogenes by the early twelfth-century metropolitan of Corinth, Gregory Pardos.[13] Achilles Tatius describes love at first sight like lightning, writing, "Her face flashed on my eyes like lightning" (καταστράπτει μου τοὺς ὀφθαλμοὺς τῷ προσώπῳ), or later when Thersandros glimpses Leukippe's beauty for an instant "like a lightning flash"

[11] "οἷον ῥεύματός τινος ὀλεθρίου ἐκ τῶν φθονερῶν ὀφθαλμῶν ἀπορρέοντος." Basil of Caesarea, *Homilia de invidia* (Homily 11), ed. PG 31:380.

[12] Such a telling metaphorical use of vision appears as well in Nicholas Kabasilas's fourteenth-century liturgical commentary, where he compares the worshiper's union with Christ to the union of the eye with light for the enabling of vision. See Nicholas Kabasilas, *Explicatio divinae liturgiae*, 44, ed. R. Bornert, J. Gouillard, P. Périchon, and S. Salaville, *Explication de la divine liturgie*, Sources chrétiennes 4 bis (Paris: Éditions du Cerf, 1967), 252; trans J. M. Hussey and P. A. McNulty, *Commentary on the Divine Liturgy* (Crestwood, NY: St. Vladimir's Seminary Press, 1960), 100.

[13] *Souda*, "Ἀχιλλεὺς Στάτιος," alpha 4695, ed. and trans. Jennifer Benedict, *Souda-On-Line*, www.stoa.org/sol-entries/alpha/4695; Photios, *Bibliotheca*, 87, ed. Henry, II:12–14; Michael Psellos, *De Heliodoro et Achille Tatio judicium*, ed. and trans. Andrew R. Dyck, *The Essays on Euripides and George of Pisidia and on Heliodorus and Achilles Tatius*, Byzantina Vindobonensia 16 (Vienna: Österreichische Akademie der Wissenschaften, 1986), 39–50; Gregory Pardos, *Commentarium in Hermogenis librum περὶ μεθόδου δεινότητος*, ed. C. Walz, *Rhetores Graeci*, vol. VII.2 (Stuttgart: Cotta, 1834), 1090–1352. See also Nigel G. Wilson, *Scholars of Byzantium* (Baltimore: Johns Hopkins University Press, 1983), 172–177, 184–190.

(ὡς ἁρπαζομένης ἀστραπῆς).[14] Here, it is not merely the lovers' gazes that intertwine, but rather it is the very beauty of the woman's countenance that impinges upon the male viewer's visual stream.

These matters are made explicitly clear in another passage, when Kleinias gives advice to Kleitophon, describing how the lovers look into each other's eyes and form a bodily union in that mutual gaze:

> οὐκ οἶδας οἷόν ἐστιν ἐρωμένη βλεπομένη· μείζονα τῶν ἔργων ἔχει τὴν ἡδονήν. ὀφθαλμοὶ γὰρ ἀλλήλοις ἀντανακλώμενοι ἀπομάττουσιν ὡς ἐν κατόπτρῳ τῶν σωμάτων τὰ εἴδωλα· ἡ δὲ τοῦ κάλλους ἀπορροή, δι' αὐτῶν εἰς τὴν ψυχὴν καταρρέουσα, ἔχει τινὰ μίξιν ἐν ἀποστάσει· καὶ ὀλίγον ἐστὶ τῆς τῶν σωμάτων μίξεως· καινὴ γάρ ἐστι σωμάτων συμπλοκή.
>
> You have no idea how marvelous a thing it is to look on one's beloved. This pleasure is greater than that of consummation, for the eyes receive each others' reflections, and they form therefrom small images of bodies as in a mirror. Such an effluence of beauty flowing down through them into the soul is a kind of copulation at a distance. This is not far removed from the intercourse of bodies – it is in fact a novel form of intimate embrace.[15]

This is a critical passage since it evidences a clear Platonic orientation for Tatius's theory of vision, resorting once again to the problem of an "outpouring" (ἀπορροή) from the beautiful object that streams into the eye and beyond. The language here parallels that found in Plato's *Phaedrus*, namely in his discussion of Eros and vision, which uses the same comparison to mirrors and to this emanation of beauty, as Helen Morales notes in her study of the novel.[16] Regardless of one's interpretation, Tatius describes vision erotically, in terms of mutual effluence.

Through him, we can appreciate how Plotinos's focus on effluxes allows him to construct a theory for the lovers' gazes that nevertheless articulated a non-haptic instance of contact and entanglement. This

[14] Achilles Tatius, *Leucippe et Clitophon*, 1.4.2 and 6.6.3, ed. E. Vilborg, *Leucippe and Clitophon* (Stockholm: Almqvist & Wiksell, 1955), 5–6, 114; trans. John J. Winkler, "Leucippe and Clitophon" in Bryan P. Reardon (ed.), *Collected Ancient Greek Novels* (Berkeley: University of California Press, 1989), 170–284, esp. 179 and 252. On its author and date, see Karl Plepelits, "Achilles Tatius" in Garth L. Schmeling (ed.), *The Novel in the Ancient World* (Leiden: Brill, 1996), 387–416; K. Plepelits, *Achilleus Tatios: Leukippe und Kleitophon* (Stuttgart: A. Hiersemann, 1980), esp. 1–18.

[15] Achilles Tatius, *Leucippe et Clitophon*, 1.9.4–5, ed. Vilborg, 11; trans. Winkler, 183.

[16] Helen Morales, *Vision and Narrative in Achilles Tatius'* Leucippe and Clitophon (Cambridge University Press, 2004), 131–135.

can be ascribed to various factors, all evidencing a coherent and consistent theory of vision following a loose Platonic iteration. Not only does Alexander of Aphrodisias serve as a key source for Plotinos, as evidenced by Porphyry's *Life of Plotinos*,[17] but furthermore Plotinos's theory of visual perception relied precisely on an Aristotelian-infused version of Plato's theory where the sympathy of viewer and object come together to produce vision.[18] In his treatise on sight, Plotinos once again stresses the efflux of light from the eye and its sympathetic (συμπάθεια) union with outside bodies.[19] Elsewhere, he likewise states that not only is the eye a thing of light (ὀφθαλμὸς φωτοειδής), but it throws itself out in search of light (πρὸς τὸ φῶς βαλών) since colors are a form of light (χρόας φῶτα ὄντα).[20] These accounts demonstrate that Plotinos chooses to focus on the activity of the viewer by stressing the extramissive effluxes from the eye, yet still crucially acknowledging the necessity and importance of the light that actively radiates from objects. These emanations are the source of vision via a model of sympathetic union of like with like derived from Plato, where similitude plays a crucial role.[21] The light from the eyes unites with that of colors, since they are both modes of light. However, Plotinos's theory of perception was based not on the mere mechanics of physiological sight, but rather on a concern with the soul's ability to perceive other beings.[22]

[17] Porphyry, *Vita Plotini*, 14.13, ed. and trans. A. H. Armstrong, *Porphyry on Plotinus*, Loeb Classical Library 440 (Cambridge, MA: Harvard University Press, 1969), 40–41.

[18] Eyjólfur Kjalar Emilsson, *Plotinus on Sense-Perception: A Philosophical Study* (Cambridge: Cambridge University Press, 1988), 36–62. See also Gordon H. Clark, "Plotinus' Theory of Sensation," *Philosophical Review* 51 (1942): 357–382; H. J. Blumenthal, "Plotinus' Adaptation of Aristotle's Psychology: Sensation, Imagination, and Memory" in R. Baine Harris (ed.), *The Significance of Neoplatonism* (Norfolk, VA: International Society for Neoplatonic Studies, 1976), 41–58.

[19] Plotinos, *Enneads*, IV.5 (29), ed. and trans. A. H. Armstrong, *Enneads*, 7 vols., Loeb Classical Library 440–445, 468 (Cambridge, MA: Harvard University Press, 1966–1988), II:115–129. See also H. J. Blumental, *Plotinus' Psychology: His Doctrines of the Embodied Soul* (The Hague: Martinus Nijhoff, 1971), esp. 67–79.

[20] "Καὶ ὁ νοῦς λόγος. Διὸ τὸν ἐφ' ἑκάστου λόγον ὁρῶν τὸ κάτω ὡς ὑπὸ τὸ φῶς σκοτεινὸν ἥγηται, ὥσπερ ὀφθαλμὸς φωτοειδὴς ὢν πρὸς τὸ φῶς βαλὼν καὶ χρόας φῶτα ὄντα τὰ ὑπὸ τὰ χρώματα σκοτεινὰ καὶ ὑλικὰ εἶναι λέγει κεκρυμμένα τοῖς χρώμασι." See Plotinos, *Enneads*, II.4.5 (12), ed. and trans. Armstrong, I:169.

[21] On Plotinos's theory of vision, see A. Mark Smith, *From Sight to Light: The Passage from Ancient to Modern Optics* (University of Chicago, 2014), 134–142.

[22] Compare this to Porphyry's theory of visual perception, as paraphrased in Nemesios's *De natura hominis*, which believes that sight is neither the product of an optical cone, *eidola*, or some other cause, but instead it is merely the soul itself encountering fellow visible objects. See Peter Lautner, "Perception and Self-Knowledge: Interpreting Fr. 264 Smith," *Bulletin of the Institute of Classical Studies* 50 (2007): 77–90.

While Plotinos argued against a medium for vision in the Aristotelian sense, stressing instead the immediacy of sight, his reasoning against this was precisely that if we considered sight to be due to impressions on an intermediary, which are passed along, then sight "would be a matter of touch. But there is no touch in acts of vision" (ἁφῇ γὰρ τοῦτο, ἐν δὲ τοῖς ὁράμασιν οὐχ ἁφή).[23] He then goes on to use a well-known Aristotelian example: pressing an object to the eye makes vision impossible. Denying the tactility of sight could serve different ways for articulating the mediation and presence offered by the visible object. Yet all these variants adopted the non-tactility of sight as a basic fact. Even when Plotinos does indeed choose to stress the intimacy of contact between viewer and object, and thus compares seeing to touch, he makes sure to word this as a similarity or *likeness* to touch (ἀλλ' ἁφῇ ἔσται ἐοικὸς τὸ ὁρᾶν). He does not make sight a species, modality, or form of touch, since he explicitly argues against this elsewhere in the same treatise.[24] Hence, Plotinos helps us to understand the conceptual nuance of such metaphors' operations, and cautions us not to confuse sight and touch and treat them as if they were interchangeable.

Returning then to Photios's account, we might see in his doubled terms – *prochysis* and *aporroe* – a similar allusion to hint at that union and contact that occurs in vision. In the late antique commentaries known to Photios, variants of *aporroe* appear throughout with notable consistency and use. For instance, it appears twenty-two times and with sharp technical precision in Theophrastus's *Commentary on the Senses* to describe all forms of emanations that issue from the visible object in the theories of Plato, Democritus, Empedocles, and others.[25] Notably, while Theophrastus's commentary on perception does not make an appearance in Photios's *Bibliotheca*, several excerpts from other commentaries by Theophrastus are included, suggesting that Photios was familiar with his broader corpus.[26] Photios was also very familiar with the work of Themistios[27] and other commentators on Aristotle such as Alexander of Aphrodisias, who deploys the term fifty-four times, mostly

[23] Plotinos, *Enneads*, IV.5.2, ed. and trans. Armstrong, II:119.
[24] Plotinos, *Enneads*, IV.5.4, ed. and trans. Armstrong, II:121.
[25] Theophrastus, "De sensu et sensibilibus" in H. Diels (ed.), *Doxographi Graeci* (Berlin: Reimer, 1879), 499–552; ed. and trans. G. M. Stratton, *Theophrastus and the Greek Physiological Psychology before Aristotle* (New York: Macmillan, 1927), 65–151. On this commentary, see H. Baltussen, "The Purpose of Theophrastus's "de Sensibus" Reconsidered," *Apeiron* 31:2 (1998): 167–199.
[26] See Photios, *Bibliotheca*, 278, ed. Henry, VIII:157–170.
[27] See also H. J. Blumenthal, "Photius on Themistius (Cod. 74): Did Themistius Write Commentaries on Aristotle?," *Hermes* 107 (1979): 168–182.

to refer to emanations from perceptible objects. At times, Alexander uses it to refer to both emanations – from the eyes and perceptibles. Yet, he avoids using any form or variant of the word *aporroe* (in opposition to another term for the emissions from the eye) as a name specifically for the rays from the eyes to the object. There are, however, four anomalous instances.[28] Those diversions are all relegated to a section commenting on Aristotle's refutation of the theory that the eye sees by something that comes out (438a25), and once more in a later citation back to this earlier section. In these cases, we witness a desire to stage a certain passivity of the efflux going from the eye into space. This is often paired with the preposition ἀπό stating the origin of these effluxes, rather than with ἐκ or πρός, which these commentators tend to prefer when emphasizing the forceful action of a movement outward into space from the eyes.

Hence, between Psellos, Theophrastus, Alexander of Aphrodisias, Achilles Tatius, and Plotinos, it is possible to understand the technical dimension and weight that *aporroe* must have had in Photios's homily – a technical importance that was not relegated to the perceptual treatises alone. It stands to reason then that when presented with the task of accurately describing the operation of the five senses in *De omnifaria doctrina*, Michael Psellos precisely altered – or rather, corrected – the language found in the Aetios text, where Aetios distractingly used ἀπορρέοντι to describe the outpouring from the eyes, rather than the emanation from the object. This neatly clarifies the distinction between the two rays, and would have been in keeping with the term's established uses in the rest of the commentaries and sources, as well as its popular connotation of exteriority in religious and philosophical texts. Its application to both emanations in the entry on colors goes against this more common use that places emphasis on the emergence of colors in the external medium. This contrasts with the more anthropocentric narrative that stages seeing from the perspective of the seer, who sends out emissions toward the object and toward whom effluxes come from the object, perfectly denoted in Photios through the doubling of prepositions and terms. As indicated earlier, the fact that Psellos, following Aetios, uses the preposition ἀπό twice for both eyes and object stresses the exteriority of vision, and demonstrates that the precision of any of these technical terms cannot be

[28] Alexander of Aphrodisias, *In librum de sensu commentarium*, ed. Wendland, 27.20–34.21, anomalies at 30.22, 31.21, 32.14, 58.25; trans. Alan Towey, *On Aristotle On Sense Perception* (London: Duckworth, 2000), 37–43.

reduced to mere terminology. But their meanings depend primarily on the grammatical construction of what the author wishes to stress is flowing out, from where, or toward what. Hence, while the uses of these terms at times evidence some misleading flexibility, the outpourings and emanations in Photios gain lucid clarity and cohesion because they are deployed in unison through the pairing of prepositions and prefixes, such as προ- and ἀπο-. Thus, the language of effluxes is structured relationally between the duality of the outpouring and emanation of rays, which together articulate the distinction between those coming from the viewer and off the object, respectively.

CHAPTER 5

"Has It Grasped?"
Apprehending the Object

Next, Photios rhetorically asks of the mind, "Has it grasped?" (ἀντελάβετο;) [**n1**]. Thus, we are immediately made aware that it is not the visual rays or the eyes that grasp, but rather "the mind" (ὁ νοῦς) [**no**]. This grasp delineates a second stage in Photios's perceptual unfolding, whereby the mind comes to apprehend the external physical object through the senses via an immaterial form. In this chapter, my aim is to articulate how a language of touching and encompassing the object through sight describes not sensation, but rather perception. Looking comparatively at how Photios and others segmented and named stages of sense perception, I will contextualize this enigmatic haptic vocabulary and place it in its proper cognitive context. What we see in this stage of grasping [**n1**] or apprehending the object is the process whereby the bodily senses hand over their stimuli to the mind's processes – the faculties of thought, imagination, judgment, and memory.

The doubled technical terms that invoke the two kinds of optical rays have led me to reevaluate Photios's homily as an account of Plato's united rays theory, or of a close permutation of this visual theory, inflected by Aristotle and the late antique commentators (see Chapter 4). Thus, the first stage describes the "outpouring (προχύσει) and effluence (ἀπορροῇ) of the optical rays" [**mo**] understood as shorthand for the outpouring of rays sent out from the eyes and the effluence of rays emanating from the object seen. However, if Photios is indeed citing Plato's theory with a proper and nuanced grasp of its function, then it must be demonstrated how it is that Photios then goes on to say that these rays "touched (ἐπαφωμένη) and encompassed (περιέπουσα) the object." [**m1**] Mango's reading is partly to blame for scholars seeing undeniable evidence for haptic vision within Photios's homily.[1] My argument is that we have thus

[1] For other matters of vision in Photios's Homily 17 in relation to Mango's translation, see Beatrice Daskas, "Nota sulla Theotókos descritta da Fozio, Hom. XVII 2 (p. 167.14–17 Laourdas)," *ACME:*

far overlooked the subtle nuances of Photios's terminology. Note that in his translation, Mango takes the "surely" or "perhaps" of δήπου [**m0**] from the beginning of the sentence and unites it with the "as if somehow" of οἱονεί πως [**m1**] preceding the comparison to touch. Thus, the result is, "For surely, having somehow ... touched and encompassed the object," which turns a careful metaphor into a statement of mere wonder and amazement at the cryptic functions of sight. The passage is more accurately translated as, "For surely, ... as if somehow touching and encompassing the object seen."

This problem is only exacerbated further by his "essence of the thing seen," [**m2**] a translation that suggests an unmediated model of contact where matter itself is seemingly being transmitted through sight, something for which there is no precedent in any of the classical, late antique, or Byzantine texts on vision. An "essence" is not being transmitted through sight – a notion that would require something like φύσις or οὐσία, not εἶδος – but instead it is the *form* proper of the object alone that is transmitted: "form of the thing seen" (εἶδος τοῦ ὁραθέντος) [**m2**]. The terminology suggests that Photios was indeed subscribing to a Platonic notion of sight that stresses the union of rays for the transmission of form. This emphasis on form is stressed earlier, when he opens the broader discussion on hearing and sight, writing [**d1**] "Just as discourse is through hearing, thus through sight form (μορφή) is inscribed (ἐγχαράσσεται) into the tablets of the soul."[2] While the vivid language of inscription (ἐγχαράσσεται) is used to describe the impression made on memory, the language of outward form is nevertheless used so as to stress a homology between sight and hearing – specifying that the object of transmission and inscription is never essential but rather formal.

To guide our reading of these important statements, it is necessary to first view these passages in the context of this section of his homily. The language of impression and tactility is to blame for the modern interpretations of Photios's homily, and likewise was a challenge for classical writers as they attempted to diagram the processes of perception that in some way relied on a form of contact between object and the viewer's consciousness. While relying on haptic metaphors just as we do to characterize our "grasp" or "handle" of a concept, Photios (as the others before him) was

Annali della Facoltà di Lettere e Filosofia dell'Università degli Studi di Milano 64:2 (2011): 339–351. See also Sergei Mariev, "Echi delle teorie ottiche antiche nelle Omelie di Fozio," *Bulgaria Mediaevalis* 2 (2011): 71–80.

[2] "ὥσπερ γὰρ ὁ λόγος ὁ δι' ἀκοῆς, οὕτω δι' ὄψεως ἡ μορφὴ τοῖς τῆς ψυχῆς ἐγχαράσσεται πίναξιν." Photios, *Homiliae*, 17, ed. Laourdas, 170.

not suggesting that sight occurred as touch. Instead, Photios is attempting to outline the psychosomatic stages of perception, seen in the unfolding of his evocative litany of questions [**n**]:

> Εἶδεν ὁ νοῦς, ἀντελάβετο, ἐφαντάσθη, τοὺς τύπους ἀκόπως ἐν τῇ μνήμῃ παρεπέμψατο.
>
> Has the mind seen? Has it grasped? Has it visualized? Then it has effortlessly transmitted the forms to the memory.[3]

This passage carefully marks the procedure behind visual perception, the way the mind processes and understands visual information: the object must first be seen; the mind must grasp and understand what it is seeing, a process that must be given a clear visualization in the mind (which seeing makes quite effortless in comparison to hearing); and then this form must be deposited into memory as a stable source of knowledge.

What is perhaps most significant here is that this process does not differ for the sense of hearing, and one may compare it with sight to see Photios's procedural aspects of perception. Just before his musings on vision, Photios had said [**k**]:

> Ἔκλινέ τις τὸ οὖς εἰς διήγημα; εἵλκυσε φανταζομένη τὸ ἀκουσθὲν ἡ διάνοια; νηφούσῃ μελέτῃ τὸ κριθὲν τῇ μνήμῃ ἐναπέθετο.
>
> Has [a person] inclined their ear to the story? Has their intelligence drawn to itself and visualized the thing heard? Then, judging it with sober attention, they deposit it into the memory.[4]

The two passages [**k** and **n**], a neatly paralleled pair of litanies of questions for both sight and hearing, walk us through the mental processes of perception ranging – in the same order – from initial perception to memorization. They lend themselves to the arrangement set out in Table 1.

This process does not at all address the act of seeing or hearing itself, which are passingly cited in the initial Εἶδεν and Ἔκλινέ of the two respective segments. Hence, we can see that ἀντελάβετο [**n1**] agrees with εἵλκυσε [**k1**], which can be translated as grasping and drawing in to oneself, respectively. This is obscured in Mango's translation, which for stylistic purposes depicts the process out of order in the rhetorical questions on hearing: "Has his intelligence visualized and drawn to itself what he has heard? (εἵλκυσε φανταζομένη τὸ ἀκουσθὲν ἡ διάνοια;)."[5] Hence, we

[3] Photios, *Homiliae*, 17, ed. Laourdas, 171. [4] Photios, *Homiliae*, 17, ed. Laourdas, 170.
[5] Photios, *Homiliae*, 17, ed. Laourdas, 170; trans. Mango, 294.

Table 1. *Photios's theory of perception*

Faculty	Hearing	Sight	Description of sight
		Pre-cognition	
αἴσθησις sensation	[**ko**] Ἔκλινέ τις τὸ οὖς εἰς διήγημα. Has [a person] inclined their ear to the story?	[**no**] Εἶδεν ὁ νοῦς, Has the mind seen?	[**mo**] Καὶ γὰρ καὶ αὐτή γε δήπου τῇ προχύσει καὶ ἀπορροῇ τῶν ὀπτικῶν ἀκτίνων For surely, through the outpouring and emanation of the optical rays
		Cognition	
(1) ἀντίληψις apprehension	[**k1**] εἵλκυσε Has their intelligence drawn to itself, and	[**n1**] ἀντελάβετο, Has it grasped?	[**m1**] τὸ ὁρατὸν οἱονεί πως ἐπαφωμένη καὶ περιέπουσα as if somehow contacting and regarding the visible-object
(2) φαντασία imagination	[**k2**] φανταζομένη τὸ ἀκουσθὲν ἡ διάνοια; imagined the thing heard?	[**n2**] ἐφαντάσθη, Has it imagined?	[**m2**] τὸ εἶδος τοῦ ὁραθέντος τῷ ἡγεμονικῷ παραπέμπεται, the form of the thing that has been seen is sent to the intellect,
(3) κριτική judgment	[**k3**] νηφούσῃ μελέτῃ τὸ κριθὲν Then, judging it with sober attention,	[**n3**] τοὺς τύπους ἀκόπως Then, effortlessly it has	N/A
(4) μνήμη => ἐπιστήμη memory => knowledge	[**k4**] τῇ μνήμῃ ἐναπέθετο they deposit it into the memory.	[**n4**] ἐν τῇ μνήμῃ παρεπέμψατο transmitted the forms to the memory.	[**m4**] ἐκεῖθεν διαπορθμευθῆναι διδοῦσα τῇ μνήμῃ πρὸς ἐπιστήμης ἀπλανεστάτης συνάθροισιν. letting it be conveyed from there to the memory for the concentration of unfailing knowledge.

must understand that, for sight, "Has it grasped? (ἀντελάβετο)" is undoubtedly to be read as an act of perception, better translated as "apprehended" or "perceived." This perceptual reading of ἀντελάβετο as grasping with the mind is a well-established convention, as evidenced by its definition in LSJ, and it is used with various senses including hearing. Furthermore, this is the only possible reading, given that at this point in the process, the optical rays have already done their job – that is, the mind has already seen, as the first question indicates [**k1** and **n2**].

Subsequently, φανταζομένη [**k2**] in hearing is paralleled with ἐφαντάσθη [**n2**], in sight, stressing the importance of imagining perceptual information both in seeing and hearing. As he writes in the sentence preceding his rhetorical questions on hearing [**j**]: "For even if the one introduces the other, yet the comprehension (κατάληψις) that comes about through sight is shown in very fact to be far superior to the learning that penetrates through the ears." In discussing sight and hearing together, Photios makes the argument that seeing is superior to hearing because sight provides the mind with ready-made images. That is, before one can deposit information into the memory for future recollection, it must first be given form in the mind, hence the importance of the imagination as an intellectual faculty that visualizes sensory impulses. Thus, in perception, sound first requires mental vision because one must first visualize spoken words before they can be understood [**k2**]. Hence sight is a superior sense simply because it circumvents this initial step in perceptual cognition. The significant difference in these two perceptual processes appears in hearing's need for "judging with sober attention" (νηφούσῃ μελέτῃ τὸ κριθέν) [**k3**], whereas crucially in sight after imagination the mind can "effortlessly transmit" (ἀκόπως... παρεπέμψατο) these images into the memory [**n3**]. This prerequisite for sober judgment distinguishes sight from hearing, not any form of material tactility or proximity with the object of perception.

This passage of Photios's homily must be understood as a careful and terse treatise on the established Byzantine notions of visual and sonic perception beyond sensation. The homily is not focused on how one senses one's environment, but rather on how one interprets and processes these sensations in the mind. Photios denotes the physiological sensation of the visual object with an enigmatic simile: "as if somehow touching (ἐπαφωμένη) and encompassing (περιέπουσα) the object" [**m1**]. The perceptual nuances of this simile are now more comprehensible. It does not refer to the act of sensation alone but to the entire process of sight from sensation through perception on to memory – a process that is recapitulated in the rhetorical questions that follow. The long sentence [**m**], like the questions [**k** and **n**], is neatly divided into four different parts: the sensation of the visual object through the optical rays, the grasping of the object by the mind, the visualization of that form in the intellect, and the transmission of the images to memory.

By understanding the clause, "as if somehow touching (ἐπαφωμένη) and encompassing (περιέπουσα) the object-seen," [**m1**] as parallel to "Has it perceived? (ἀντελάβετο)," [**n1**] it is then possible to see that both rely on a language of grasping to convey the apprehension of the visual form by the mind. This explains why Photios deploys both ἐπαφωμένη and περιέπουσα to describe the process neatly summarized by the deceptively haptic ἀντελάβετο, with its common understanding as apprehension or perception in the context of various sensory processes, as stated earlier. Furthermore, while one could reason that the light touching of surface implied by ἐπαφωμένη could suggest the alleged tactility of the optical rays, περιέπουσα suggests that both these terms require qualification. The term περιέπουσα denotes a sense of handling or treatment of a thing. It appears in contemporaneous sources as a sense of ill- or well-treatment. For instance, the Patriarch Nikephoros uses it in regard to the "honored and well-regarded" (τιμῶσα καὶ περιέπουσα) teachings of the church.[6] The verb does not imply physical grasping or encompassing, but rather mentally regarding an object, person, or thing. This certainly does not allow for Mango's rendering of the term as "encompassed," unless one understands the word metaphorically, to indicate a mental process.

The same reading is necessary with ἐπαφωμένη [**m1**], which in this context seems to indicate this broader notion of contact between sensible objects and sensing beings. This term appears in a perceptual context in Porphyry's "Commentary on Ptolemy's *Harmonics*,"[7] a text that evidences close similarities with the works of Alexander of Aphrodisias, Sextus Empiricus, and Galen on sense perception.[8] This text would later be used

[6] "ἐπειδὴ δὲ ἐκ τῆς καθολικῆς ἐκκλησίας ἐξοστρακίζειν ὁρίζονται, ὁμολογοῦσιν καὶ ἄκοντες ὡς πᾶσα ἡ καθολικὴ καὶ ἀποστολικὴ ἐκκλησία ταῦτα δέδεκται, τιμῶσα καὶ περιέπουσα." Nikephoros, *Refutatio et eversio definitionis synodalis anni 815*, ed. J. Featherstone, *Refutatio et eversio definitionis synodalis anni 815*, Corpus Christianorum, Series Graeca 33 (Turnhout: Brepols, 1997), 68.24–28.

[7] Porphyry, *Εἰς τὰ ἁρμονικὰ Πτολεμαίου ὑπόμνημα*, ed. and trans. Andrew Barker, *Porphyry's Commentary on Ptolemy's Harmonics: A Greek Text and Annotated Translation* (Cambridge University Press, 2015).

[8] On the relations between Porphyry and others, see Riccardo Chiaradonna, "Platonismo e teoria della conoscenza stoica tra II e III secolo D.C." in M. Bonazzi and C. Helmig (eds.), *Platonic Stoicism – Stoic Platonism: A Dialogue Between Platonism and Stoicism in Antiquity* (Leuven University Press, 2007), 209–241, esp. 228–234; Riccardo Chiaradonna, "Platonist Approaches to Aristotle: From Antiochus of Ascalon to Eudorus of Alexandria (and Beyond)" in Malcolm Schofield (ed.), *Aristotle, Plato and Pythagoreanism in the First Century BC* (Cambridge University Press, 2013), 28–52,

by George Pachymeres for his section on Harmonics and Music in the *Quadrivium* as well as by Manuel Bryennnios for his own commentary on Ptolemy's *Harmonics*.[9] Porphyry's commentary here offers more than mere lexical clarity. The word ἐπαφωμένη is used exactly, to characterize the process of perception:

> τῆς γὰρ ὕλης εἰδοπεποιημένης ὑπὸ τοῦ ῥηθέντος λόγου τὴν ψυχὴν συμβέβηκε τοῖς οὖσιν ἐφισταμένην καὶ οἷον ἀποσπῶσαν αὖθις ἀπὸ τῆς ὕλης τὰ εἴδη καὶ δεχομένην εἰς ἑαυτὴν καὶ τρόπον τινὰ ἀποκαθιστᾶσαν εἰς τὸ ἄϋλον γίνεσθαι τὴν κρίσιν. τὸ μὲν γὰρ πρῶτον ἀπὸ τῆς αἰσθήσεως ἡ ἀντίληψις, οἷον ἐπαφωμένη τοῦ ὄντος ἀναλαμβάνειν αὐτὰ πειρᾶται καὶ οἷον εἰσαγγέλλειν τε καὶ εἰσάγειν εἰς τὴν ψυχὴν ὥσπερ ὁδηγός τις καὶ εἰσαγωγεύς. μετὰ δὲ ταῦτα ἡ δοξαστικὴ ὑπόληψις ὑποδέχεται τὸ εἰσαχθὲν προσαγορεύουσα αὐτὸ καὶ ἀναγράφουσα διὰ λόγου τῇ ψυχῇ, οἷον εἴς τι γραμμάτιον ἐνυπάρχον αὐτῇ. τρίτη δ' ἐστὶ μετὰ ταῦτα δύναμις εἰκονιστὴ τῶν ἰδιωμάτων καὶ ὄντως ζωγραφική τις ἢ πλαστικὴ ἢ φαντασία οὐκ ἀρκουμένη τῷ τῆς προσαγορεύσεως εἴδει καὶ τῷ τῆς ἀναγραφῆς, ἀλλ' ὅνπερ τρόπον οἱ τοὺς καταπλέοντας εἰκονίζοντες ἢ κατὰ τοὺς τοῖς συμβόλοις παρακολουθοῦντας τὴν ἀκρίβειαν τῆς ὁμοιότητος ἐκλογίζονται· οὕτω καὶ αὕτη τοῦ πράγματος ἅπασαν τὴν μορφὴν ἐκλογιζομένη, ὁπόταν τοῦτον τὸν τρόπον ἀκριβώσῃ, τότε ἀπέθετο ἐν τῇ ψυχῇ τὸ εἶδος. καὶ τοῦτο ἦν ἡ ἔννοια, ἧς ἐγγενομένης τε καὶ βεβαιωθείσης ἡ τῆς ἐπιστήμης ἐγγίνεται διάθεσις, ἀφ' ἧς ὥσπερ ἀπὸ πυρὸς πηδήσαντος ἐξαφθὲν φῶς ὁ νοῦς ἀναφαίνεται οἷόν περ ὄψις ἀκριβὴς εἰς τὴν προσβολὴν τὴν ἐπὶ τὸ ὄντως ὄν.
>
> For when matter has been informed by the aforementioned rational principle, it happens that the soul focuses upon entities, and again draws off, as it were, the forms from matter; it receives them into itself (δεχομένην εἰς ἑαυτήν) and in some fashion restitutes them so that the act of discrimination may be non-material. For in the first place apprehension (ἀντίληψις) arising from sense-perception seems to contact (ἐπαφωμένη) the entity and tries to take up (ἀναλαμβάνειν) the forms to make a kind of report on them and introduce them into the soul like some guide or introducer. Next, belief-making assumption receives what has been introduced, names it and describes it through words in the soul as if upon some writing tablet existing in it. Next, the third faculty is one

esp. 36; Richard Sorabji, "Universals Transformed: The First Thousand Years after Plato" in P. F. Strawson and Arindam Chakrabarti (eds.), *Universals, Concepts and Qualities* (Aldershot: Ashgate, 2006), 105–126; Lautner, "Perception and Self-Knowledge," 77–90, esp. 83–84 n 23. On a possible connection to Thrasyllus, see also Harold Tarrant, *Thrasyllan Platonism* (Ithaca, NY: Cornell University Press, 1993), esp. 119–131.

9 See Ingemar Düring, *Ptolemaios und Porphyrios über die Musik*, Göteborgs Högskolas årsskrift 40 (Gothenburg: Elanders Boktryckeri Aktiebolag, 1934), esp. 8–9; Manuel Bryennios, *Harmonica*, ed. and trans. G. H. Jonker, *The Harmonics of Manuel Bryennius* (Groningen: Wolters-Noordhoff, 1970).

> that makes images out of distinctive features and is really like a faculty of painting or sculpting or the *phantasia* (φαντασίας). It is not content with the form produced by naming and describing, but just as those who try to descry persons sailing into port, or by way of those who attend to features to match them, and work out details of similarities, so this faculty works out the whole structural outlook of the object, and when it achieves accuracy in this way, then it stores the form in the soul. This is the mental image. And when it has arisen within and received confirmation, the condition of knowledge comes about. From this, like a light kindled from leaping fire intellect shines forth, just like an accurate vision for a focus on true being.[10]

This is a general account of sensation and perception, not focused on any particular sense, yet it likewise attempts to stress this notion of the contact that occurs between subject and object as perceptual form is drawn out from matter. The passage is as technical and dense as that of Photios, and it lends itself to the arrangement in Table 2.

The outline I provide is not the only way to interpret Porphyry.[11] The exact parsing of the various stages matters mainly as a modern tool for the comparative study of Porphyry's understanding of the cognitive processes, and less as an absolute characterization of his psychology. Throughout the late antique and Byzantine period, there is a great deal of variety regarding the unfolding of this process. For example, it is often common for belief to follow the imagination and some even seek to conflate the imagination with sensation itself. For our purposes, Porphyry's stages parallel Photios's description of perceptual unfolding: sensual perception through contact, the assumption and reception of the sensation introduced, the creation of images through the imagination, and the eventual preservation of this form for the sake of knowledge.

The passage from Porphyry is no anomaly. In his commentary on Hermogenes's *Progymnasmata* from soon after 1025, John Sikeliotes recounts how *phantasia* is derived from sensation (*aisthesis*), which then passes on the impressions formed through imagination onto belief (*doxa*),

[10] Porphyry, *Εἰς τὰ ἁρμονικὰ Πτολεμαίου ὑπόμνημα*, ed. I. Düring, *Kommentar zur Harmonielehre des Ptolemaios* (Göteborg: Elanders, 1932), 13–14; modified trans. Peter Lautner, "Mental Images in Porphyry's Commentary on Ptolemy's *Harmonics*," *Apeiron* 48:2 (2015): 220–250, esp. 225–226.

[11] Peter Lautner ("Mental Images"; see esp. 226–231) has proposed in his study on Porphyry's perceptual processes that Porphyry understands perception as having four stages or faculties involved (αἴσθησις/ἀντίληψις, δοξαστικὴ ὑπόληψις, φαντασία, and ἔννοια). Michael Chase, "Porphyry on the Cognitive Process," *Ancient Philosophy* 30 (2010): 383–405, esp. 385, has proposed a division of six stages (ἀντίληψις, δοξαστικὴ ὑπόληψις, φαντασία, ἔννοια, ἐπιστήμη, and νοῦς). Chiaradonna, "Platonismo e teoria," 229, proposed five (αἴσθησις, ἀντίληψις, δοξαστικὴ ὑπόληψις, φαντασία, and ἔννοια).

Table 2. *Porphyry's theory of perception*

Faculty	Operations	Description of operation	Descriptive similes
(0) αἴσθησις sensation	N/A	N/A	N/A
(1) ἀντίληψις apprehension	(a) εἰσαγγέλλειν go in and announce (b) εἰσάγειν lead in	(a) ἐπαφωμένη contact (b) ἀναλαμβάνειν. . . πειρᾶται tries to take up	(a) ὁδηγός guide (b) εἰσαγωγεύς introducer
(2) δοξαστικὴ ὑπόληψις belief-making assumption	(a) ὑποδέχεται welcomes	(a) προσαγορεύουσα call by name (b) ἀναγράφουσα διὰ λόγου describes through words	(a) γραμμάτιον writing tablet
(3) δύναμις εἰκονιστή image-making faculty	*Operation is here described by analogy*	*After image-making:* τότε ἀπέθετο ἐν τῇ ψυχῇ τὸ εἶδος then the form is deposited in the soul	(a) ζωγραφική (painting) (b) πλαστική (sculpting) (c) φαντασία (imagination) (d) Extended painting simile
(4) ἔννοια => ἐπιστήμης thought => knowledge[a]	(a) ἐγγίνεται διάθεσις becomes set-up	*Emerges when thought (ἔννοια) has:* (a) ἐγγενομένης been created (b) βεβαιωθείσης been confirmed	(a) ἀπὸ πυρὸς πηδήσαντος ἐξαφθέν light kindled from leaping fire (b) ὄψις ἀκριβής accurate vision

[a]The creation and confirmation of "thought" that sets up "knowledge" seems to operate as an intermediary step between (3) and (4), even if Porphyry does not distinguish them as such. On the role of judgment via Photios, see Chapter 8.

before going on to intellect (*dianoia*), and finally on to the mind (*nous*), which "receives and describes" (δεξάμενος καὶ διαρθρώσας) all matters related to knowledge in itself.[12] Sikeliotes summarizes this process with a metaphor of entering a household:

> καὶ ἀναλογοῦσιν ἡ μὲν αἴσθησις πύλῃ, δεχομένῃ τοὺς ἔξωθεν εἰσιόντας, ἡ δὲ φαντασία τῇ μετὰ τὴν εἴσοδον στάσει μέχρις ἂν ἀπαγγείλῃ ὡς πυλωρῷ τῇ δόξῃ τῶν ὧν ἔχει χρείαν· ἡ δὲ διάνοια τῷ θαλαμοφύλακι, ἣ καὶ μουσικαῖς συνεξήρμοσται μεσότησι, καθάπερ ἀφ᾽ ὑπάτης εἰς νήτην διαχωροῦσα, καὶ ἀπὸ νήτης εἰς παρυπάτην πρὸς τὰ ἄκρα διανυστὰ ποιεῖ τὰ ζητούμενα· ὁ δὲ νοῦς τῷ οἰκοδεσπότῃ, ὃς κατὰ τὸ προσῆκον ἕκαστα διατάττει καὶ διοικεῖ.
>
> Sensation is analogous to a gate, which receives those coming in from outside. The imagination is like a pause after the entrance until one is announced to belief, like a door-keeper necessarily holds them. And then intellect is like a servant, which puts together musical intervals, passing from the lowest pitch to the higher pitch, and from the higher one to next toward the highest, producing the things sought. And the mind is like the head of the house, who makes arrangements and administers each thing in a suitable way.[13]

Here Sikeliotes presents a process where *phantasia* occupies a place analogous to the initial apprehension that we witness in Porphyry and Photios, while belief is delayed until after the imagination and operates as a prelude to higher intellectual functions, captured through the metaphor of ascending musical intervals. For Sikeliotes, then, it is clear that perception unfolds as *aisthesis*, *phantasia*, *doxa*, *dianoia*, and *nous*, wherein the goal of the author is to stress the loftier aspects of cognition. Thus, he stresses and divides these higher functions across *doxa*, *dianoia*, and *nous*, relocating any operations associated with *doxa* after *phantasia*, and giving little focus to the earlier stages of perception that are seen as more earthly, material, and passive.

The perceptual processes described by Porphyry and Sikeliotes illumine Photios's passage. The differences between the two show that Photios was closer to Porphyry than he was to Sikeliotes. Both Photios and Porphyry held to a five-stage model for the cognitive process: *aisthesis*, *antilepsis*, *phantasia*, and *episteme*/*mneme* – with Photios notably and self-consciously

[12] John Sikeliotes, *Commentarium in Hermogenis librum περὶ ἰδεῶν*, ed. C. Walz, *Rhetores Graeci*, vol. VI (Stuttgart: Cotta, 1834), 87–88. See also Panagiotis Roilos, "Ancient Greek Rhetorical Theory and Byzantine Discursive Politics: John Sikeliotes on Hermogenes" in Teresa Shawcross and Ida Toth (eds.), *Reading in the Byzantine Empire and Beyond* (Cambridge University Press, 2018).

[13] John Sikeliotes, *Commentarium in Hermogenis librum περὶ ἰδεῶν*, ed. Walz, *Rhetores Graeci*, 88.

lacking the actions of the *doxastike hypolepsis*, a matter to which we shall return at the end of this chapter. Photios's pre-cognitive stage is the act of sensation. In the case of listening, this is depicted by "Has a man lent his ear to a story?" [**ko**] and in seeing it is indicated by "the outpouring and effluence of the optical rays," [**mo**] or, in the ensuing questions, simply, "Has the mind seen?" [**no**]. In Porphyry's commentary sensation (αἴσθησις) comes before the cognitive side of perception, where *antilepsis* is described as the first step (τὸ μὲν γὰρ πρῶτον). As Chase suggests, this is "the transition from sensation to intellection."[14] Sensation in Porphyry is the process whereby the subject received the external object into itself (δεχομένην εἰς ἑαυτήν), occurring before *antilepsis* proper.

In his first cognitive stage, Porphyry uses *antilepsis* and *epaphao* to describe the contact of the mind with the sensible stimuli. As he writes, "apprehension (ἀντίληψις) arising from sense-perception seems to contact (ἐπαφωμένη) the entity." The term *antilepsis* is echoed by Photios's use of the rhetorical question, "Has [the mind] grasped?" (ἀντελάβετο), to reflect this same stage where the visible thing is received into the mind. Or, in the case of hearing, this is when the person has "drawn to itself" (εἵλκυσε) the thing heard. This is the same stage that Photios describes with regards to sight in wonder, saying, "as if somehow touching (ἐπαφωμένη) and encompassing (περιέπουσα) the object-seen," which therefore would be better stated: "as if somehow contacting (ἐπαφωμένη) and regarding (περιέπουσα) the visible thing." This stresses the stage in sense perception where the form of the thing being seen is apprehended, grasped by the mind, and understood, before it can come to generate mental images for its storing in memory. Photios's sense of wonder, conveyed through the οἱονεί πως [**mɪ**], is then to be understood more as an indication of the abstracted and difficult understanding of sensual contact that occurs in this perceptual process, rather than a mere marveling at the tactility or physiology of sight. Instead, οἱονεί πως seems to indicate precisely the obscurity of how the mind initially grapples with sensual data in a stage where the rays have allowed for sight, but just before the imagination has capably produced an image of the forms for their inscription in memory.

The crucial clarification suggesting that apprehension proceeds from sensation comes from Porphyry's "apprehension from sensation" (ἀπὸ τῆς αἰσθήσεως ἡ ἀντίληψις), which suggests that apprehension (ἀντίληψις) is

[14] Chase, "Porphyry on the Cognitive Process," 385–386.

not sensation itself, but emerges from (ἀπό) it.[15] Photios grapples with this same temporal nature of the perceived object when he describes the transition from the stage of apprehension into that of imagination (φαντασία). In the second stage, where he details the mind's act of establishing contact and grappling with the object, Photios describes the visible-thing as τὸ ὁρατόν [**m1**]. Yet, further in the same clause, when he speaks of the transmission and implicit visualization of the form of the object, he describes this entity with an aorist passive participle, that is to say, as "the form of the thing-that-has-been-seen" (τὸ εἶδος τοῦ ὁραθέντος). The temporal marker succinctly points to the preceding perceptual processes and stresses the fact that the images that come into consciousness are not physical realities, but rather noetic forms of what has been already perceived. The image that is transmitted is not just "the thing-that-has-been-seen," but only its form (εἶδος).

The transition from sensation to intellection is highly important. Porphyry appears to capitulate the process as a division between *morphe* and *eidos*.[16] As he writes near the end of the excerpted passage, the imagination "works out the whole structural outlook of the object (ἅπασαν τὴν μορφήν), and when it achieves accuracy in this way, then it stores the form in the soul (ἀπέθετο ἐν τῇ ψυχῇ τὸ εἶδος)." For Photios, it is that form of the thing-that-has-been-seen that is sent to the intellect (τὸ εἶδος τοῦ ὁραθέντος τῷ ἡγεμονικῷ παραπέμπεται) [**m2**]. Earlier, however, when he is describing the transmission of the Gospels through hearing as parallel to vision, Photios uses μορφή to denote the outward appearance: "the form through sight is imprinted upon the tablets of the soul" (δι' ὄψεως ἡ μορφὴ τοῖς τῆς ψυχῆς ἐγχαράσσεται πίναξιν) [**d1**]. Here, imprinting or engraving (ἐγχαράσσεται) suggests a process of transformation or translation, whereby the outward form is carved into the soul's tablets just as the written text is transformed into speech, or speech into written text.

Porphyry deploys the same metaphor in relation to his stage between *antilepsis* and *phantasia*, in the stage described as "belief-making assumption" (δοξαστικὴ ὑπόληψις), which is absent from Photios's unfolding of visual perception, an issue to which I shall return. As Porphyry writes, *doxastike hypolepsis* "describes [the object] through words in the soul as if

[15] On the term *antilepsis* in the commentators, particularly Alexander of Aphrodisias, see Robert B. Todd, "Lexicographical Notes on Alexander of Aphrodisias' Philosophical Terminology," *Glotta* 52:3/4 (1974): 207–215, esp. 208–209.

[16] See Lautner, "Mental Images," 241–243.

upon some writing tablet existing in it" (ἀναγράφουσα διὰ λόγου τῇ ψυχῇ, οἷον εἴς τι γραμμάτιον ἐνυπάρχον αὐτῇ). Here ἀναγράφουσα precisely indicates a process of engraving or registering through logocentric means on some tablet (τι γραμμάτιον), which plays well with Photios's use of the metaphor in relation to the aural transmission of the Gospel text. Thus, *eidos* is the noetic form produced by the entire process, notably by the imagination, which is then simply sent (παρεπέμψατο) or is allowed to be given (διαπορθμευθῆναι διδοῦσα), in Photios's account. Similarly, the same is deposited (ἀπέθετο) in the soul for Porphyry and, in terms of hearing, deposited (ἐναπέθετο) in the memory for Photios. It is by virtue of this act of storage that the final stage in the process of perception comes about, which Porphyry characterizes as the "condition of knowledge" (ἡ τῆς ἐπιστήμης. . .διάθεσις). This is the same "unfailing knowledge" (ἐπιστήμης ἀπλανεστάτης) [**m4**] in Photios that results once the forms from seeing and hearing are deposited into the memory for future emulation.

The stages of Photios and Porphyry may be summarily compared in Table 3.

Thus, we may then responsibly revise Mango's translation of Photios's crucial passage on sight as follows:

> [**m0**] For surely, through the outpouring [i.e. from the eye] and emanation [i.e. from the object] of the optical rays, [**m1**] as if somehow contacting and regarding the visible-object, [**m2**] the form of the thing that has been seen is sent to the intellect, [**m4**] letting it be conveyed from there to the memory for the accumulation of unfailing knowledge. [**n0**] Has the mind seen? [**n1**] Has it apprehended? [**n2**] Has it imagined? [**n3**] Then, effortlessly it has transmitted [**n4**] the images to the memory.

Table 3. *Comparison of Photios's and Porphyry's theories of perception*

Photios, Homily 17		Porphyry, *Harmonics*
	Pre-cognition	
Εἶδεν ὁ νοῦς,		αἴσθησις
	Cognition	
(1) ἀντελάβετο,		(1) ἀντίληψις
		(2) δοξαστικὴ ὑπόληψις
(2) ἐφαντάσθη,		(3) δύναμις εἰκονιστή, or φαντασία
(3) τοὺς τύπους ἀκόπως		(3/4) Judgment of ἔννοια
(4) ἐν τῇ μνήμῃ παρεπέμψατο		(4) ἔννοια/ἐπιστήμης

With this revision, it becomes clear that the evidence for tactile sight is less apparent and therefore less consequential than what most scholars have assumed. For even if visual sensation (up to the second stage) were indeed seen as operating on a model of tactility, at the second stage of perception this tactile contact with the object would be curtailed. The physical touch of the image at this point would be remediated in any case through the cognitive faculties that produce mental images of the outward physical appearance (μορφή) for the form's (εἶδος) deposition into memory. What is effortlessly deposited into the memory by sight are furthermore the typologies (τοὺς τύπους) of images, which then serve as the model for emulation and reproduction. Hence, the impression of the *typos* generated by the imagination is no longer mere form as *eidos*, but a potential form, ready for emulation. This is in accordance with the typological language associated with the relations between the Old and New Testaments as well as the representation of forms in icons, whereby representations – both verbal and visual – are always caught in typological chains that can be ever replicated and reproduced, while always signaling back to their archetypes and prototypes.[17] Hence, vision is always mediated. It is never haptic because the mind does not merely impress material or haptic forms directly taken from the object onto itself. Rather it remediates sensual information into mental images that are then deposited in the soul or memory. In a word, the mind has not "grasped" – it has, rather, apprehended.

[17] See Frances Young, "Typology" in Stanley E. Porter, Paul Joyce, and David E. Orton (eds.), *Crossing the Boundaries: Essays in Biblical Interpretation in Honour of Michael D. Goulder* (Leiden: Brill, 1994), 29–48, esp. 34–39.

CHAPTER 6

"Has It Visualized?" I
The Grasp of the Imagination

The second stage in Photios's cognitive process of perception is *phantasia*, or, for our purposes here, the imagination. This is indicated by his third question "Has it visualized?" (ἐφαντάσθη) [**n2**], or better yet, "Has it imagined?" Here, Photios cues a complex history regarding the role of the imagination in perception. *Phantasia* is notoriously difficult to translate, depending on context. For example, in Stoic epistemology, it is often best rendered as an "impression" or "presentation," stressing the vivid self-evidence of its impressions.[1] While these resonances are not fully abandoned in the medieval world, the *phantasia* that primarily concerns us here is best understood as referring to the imagination and the imaginative capacities of the mind, rather than the mere impressions of perception. As we shall see, this imaginative faculty is understood to take raw sensory data and construct working models of perceived objects, which affords us the ability to think, judge, and remember the world around us. But the mental forms of *phantasia* were not limited simply to the visual, but applied to all the senses. So here I will often translate *phantasia* as "imagination," given that this term today does not simply

[1] For a general survey of *phantasia*, see Alessandra Manieri, *L'immagine poetica nella teoria degli antichi: Phantasia ed enargeia* (Pisa: Istituti editoriali e poligrafici internazionali, 1998); Gerard Watson, *Phantasia in Classical Thought* (Galway University Press, 1998); cf. G. Watson, "The Concept of 'Phantasia' from the Late Hellenistic Period to Early Neoplatonism," *Aufstieg und Niedergang der Römischen Welt* 36:7 (1994): 4766–4810; Shadi Bartsch, "'Wait a Moment, *Phantasia*': Ekphrastic Interference in Seneca and Epictetus," *Classical Philology* 102:1 (2007): 83–95; Jeremy Tanner, *The Invention of Art History in Ancient Greece: Religion, Society and Artistic Rationalisation* (Cambridge University Press, 2006), 293–287; Irmgard Männlein-Robert, "Zum Bild des Phidias in der Antike: Konzepte zur Kreativität des bildenden Künstlers" in Thomas Dewender and Thomas Welt (eds.), *Imagination – Fiktion – Kreation: Das kulturschaffende Vermögen der Phantasie* (Munich: K. G. Saur, 2003), 45–67. See also Mireille Armisen, "La notion d'imagination chez les Anciens: I – Les philosophes," *Pallas* 26 (1979): 11–51; Mireille Armisen, "La notion d'imagination chez les Anciens: II – La rhétorique," *Pallas* 27 (1980): 3–37; Mary-Anne Zagdoun, *La philosophie stoïcienne de l'art* (Paris: Centre National de la Recherche Scientifique, 2000); Anne Sheppard, *The Poetics of Phantasia: Imagination in Ancient Aesthetics* (London: Bloomsbury, 2014).

imply visualizing, but rather conjures the whole gamut of sensory experiences that might be stimulated by a narrative.

The goal in this and the subsequent chapter is to flesh out the place, role, and function of the imagination in perception and thought. This goal will be partly achieved by placing the medieval Greek notions of *phantasia* in dialogue with their ancient counterparts in order to clearly demonstrate how the Byzantine articulation approaches our modern "imagination," while often retaining its accessory language of impressions and imprints from the Stoic past. This conversation will allow us to appreciate the longstanding classical and Stoic heritage of haptic language around cognition, and the conflicted place of the imagination as counterfactual fantasy in early Christian discourse. In this and subsequent chapters, we shall see how the imaginative faculty is responsible for the possibility of judgment, thought, recollection, and the actualization of the common sense.

While Photios's rhetorical question makes it clear that this second phase of cognitive perception is concerned with the imagination, this matter is less explicit in the corresponding section of his previous summary, where he simply states, "the form of the thing that has been seen is sent to the intellect" (τὸ εἶδος τοῦ ὁραθέντος τῷ ἡγεμονικῷ παραπέμπεται) [**m2**]. Of first note in Photios's statement is its location in human cognition, namely in this case, the suggested location in the realm of the intellect (ἡγεμονικόν). Photios's *Lexicon*, the *Synagoge* in the *Lexica Segueriana*, and the *Souda* all equate the term *hegemonikon* to the mind (νοῦς). However, the *Souda* insightfully clarifies that it is the place "where the images and impulses occur, and from where thought is transmitted" (ἐν ᾧ αἱ φαντασίαι καὶ αἱ ὁρμαὶ γίνονται, καὶ ὅθεν ὁ λόγος ἀναπέμπεται), which is located in the heart.[2] This definition of the *hegemonikon* derives verbatim from Diogenes Laertius's book on Zeno from the *Lives of the Eminent Philosophers*, where he describes the nature of semen.[3] That life of Zeno is a critical text for the Stoic understanding of *phantasia*. Therefore, Photios's concordance with Diogenes Laertius's

[2] Anonymous, *Synagoge*, eta 248:22, ed. L. Bachmann, *Anecdota Graeca*, vol. I (Leipzig: Hinrichs, 1828), 248; Photios, *Lexicon*, eta 28, ed. C. Theodoridis, *Lexicon*, vol. II (Berlin: De Gruyter, 1998), 252; Anonymous, *Souda*, "ἡγεμονικόν," eta 41, ed. and trans. Stefano Sanfilippo, *Souda-On-Line*, www.stoa.org/sol-entries/eta/41. On the *hegemonikon* in late antique and Byzantine commentaries, see Robert B. Todd, "Galenic Medical Ideas in the Greek Aristotelian Commentators," *Symbolae Osloenses* 52:1 (1977): 117–134.

[3] Diogenes Laertius, *Vitae philosophorum*, VII.1.159, ed. and trans. R. D. Hicks, *Lives of Eminent Philosophers*, vol. II, Loeb Classical Library 185 (Cambridge, MA: Harvard University Press, 1925), 262–263.

key text on *phantasia* only strengthens our understanding that this phase in the process indicates the imagination, corroborated by the later question "Has [the mind] imagined?" (ἐφαντάσθη) [**n2**]. For Photios, the activity of the *phantasia* in vision occupies this second step in cognitive perception from *antilepsis*, *phantasia*, and on to *episteme*, achieved through *mneme*.

It is necessary to consider the classical and late antique heritage of the *phantasia* to comprehend its tactile implications and associations. The sentences on sight in Homily 17 are part of Photios's defense of images, accusing those who might argue against images of likewise arguing against the teachings passed on by the Gospels, a common defense of the iconophiles.[4] Photios uses the cognitive process as part of his defense, and this process is centered upon the faculty of the imagination. Images present the blessed figures "more graphically to our knowledge" (ἐναργεστέραν... εἰς γνῶσιν) [**f2**], alluding to the fact that while oral rhetoric might cue the imagination to generate images of the wondrous stories, images provide us already with vivid visualizations of these narratives. Here, "more graphically" (ἐναργεστέραν) alludes to the language of the vivid visual quality of rhetorical form. Through discourse, narratives and stories are placed before one's eyes via the imagination, as René Nünlist, Ruth Webb, and others have discussed at length, and to which we shall return in Chapter 9.[5] Immediately, before discussing hearing through his series of rhetorical questions, Photios reflects precisely: "For even if each one [i.e. hearing] introduces (συνεισάγεται) the other [i.e. seeing], yet the comprehension (κατάληψις) that comes about through sight is shown in very fact to be far superior to the learning that penetrates through the ears." [**j**] This notion that the two senses are in a sense "brought in together" (συνεισάγεται) speaks to the power of the imagination to synthetically visualize the narratives in the mind, whether they be heard or seen. Moving on to describe hearing and eventually sight, Photios thus seems to be using the common operation of the imagination in both faculties of sense as support for his argument that they mutually introduce each other.

[4] See Charles Barber, *Figure and Likeness: On the Limits of Representation in Byzantine Iconoclasm* (Princeton University Press, 2002), 135–137.

[5] See René Nünlist, *The Ancient Critic at Work: Terms and Concepts of Literary Criticism in Greek Scholia* (Cambridge University Press, 2009), 194–198. See also the bibliography in Ruth Webb, *Ekphrasis, Imagination, and Persuasion in Ancient Rhetorical Theory and Practice* (Burlington, VT: Ashgate, 2009).

What Photios offers then is an insight into the Byzantine articulation of ancient theories of cognition reliant on the imagination.[6] The imagination (φαντασία) is understood as being one of the faculties of the soul, attested in Aristotle's definition of imagination and encountered throughout late antique and Byzantine sources like the *Doctrina patrum* and Symeon Seth, where the five powers of the soul are usually listed as mind (νοῦς), thought (διάνοια), belief (δόξα), sensation (αἴσθησις), and imagination (φαντασία).[7] In the *Bibliotheca*'s entry on the *Life of Pythagoras* by an anonymous writer, Photios also gives a list of the organs of knowledge, which include sensation, imagination, art, belief, understanding, knowledge, wisdom, and mind.[8] There, the imagination is described as an impression or movement in the soul (φαντασία δὲ κίνησις ἐν ψυχῇ), which follows Aristotle's definition.[9] These short examples show that φαντασία in antiquity and late antiquity was varied and had nuanced degrees of definition. The term may be fairly translated in different contexts as imagination, fantasy, impression, appearance, presentation, or representation.

In the complex history of φαντασία, the Stoic version appears to take precedence, though it is certainly filtered for the Byzantines through late antique rereadings of it, particularly those of the commentators on Aristotle and Galen. Since for the Stoics all knowledge emerges from sense perception, the role of the φαντασία becomes crucial as it is the faculty

[6] For a survey of imagination in Byzantium, see Webb, *Ekphrasis*, 107–130; Nadine Schibille, *Hagia Sophia and the Byzantine Aesthetic Experience* (Burlington, VT: Ashgate, 2014), 27–37; Glenn Peers, "Imagination and Angelic Epiphany," *Byzantine and Modern Greek Studies* 21 (1997): 113–131; Liz James, "Art and Lies: Text, Image, and Imagination in the Medieval World" in Antony Eastmond and Liz James (eds.), *Icon and Word: The Power of Images in Byzantium* (Burlington, VT: Ashgate, 2003), 59–71; Aglae Pizzone, "Late Antique Φαντασία and the Greek Fathers: A Survey" in David Hernández de la Fuente (ed.), *New Perspectives on Late Antiquity* (Newcastle upon Tyne: Cambridge Scholars Publishing, 2014), 416–432; Aglae Pizzone, "Theodore and the Black Man: Imagining (through) the Icon in Byzantium" in Andreas Speer and Philipp Steinkrüger (eds.), *Knotenpunkt Byzanz* (Berlin: De Gruyter, 2012), 47–70. Also, for a brief note on the matter, see Aglae Pizzone, "When Homer met *Phantasia*: Fiction, Epic Poetry, and Entertainment Literature in Byzantium," *British Academy Review* 19 (2012): 42–45, esp. 44. On art and imagination in Byzantium, for example, see John Onians, "Abstraction and Imagination in Late Antiquity," *Art History* 3:1 (1980): 1–24; Leslie Brubaker, "Perception and Conception: Art, Theory, and Culture in Ninth-Century Byzantium," *Word & Image* 5:1 (1989): 19–32.

[7] Symeon Seth, *Conspectus rerum naturalium*, IV.83, ed. Delatte, 78–79; Anonymous, *Doctrina patrum*, ed. F. Diekamp, *Doctrina patrum de incarnatione verbi* (Munich: Aschendorff, 1907), 296.

[8] "Ὅτι γνώσεως ὀκτὼ φασιν ὄργανα, αἴσθησιν, φαντασίαν, τέχνην, δόξαν, φρόνησιν, ἐπιστήμην, σοφίαν, νοῦν." Photios, *Bibliotheca*, 249, ed. Henry, III:132.

[9] Photios, *Bibliotheca*, 249, III:132. On *phantasia* as a movement or alteration, see Christof Rapp, "Intentionalität und *phantasia* bei Aristoteles," *Ancient and Medieval Theories of Intentionality*, ed. Dominik Perler (Leiden: Brill, 2001), 63–96, esp. 83–85. See also Aristotle, *De anima*, 427b11–18, ed. Hett, 160–161.

that mediates sensory input and the intellect.[10] In the Stoic context, φαντασία could be a product of either sensual perception or from the combined acts of reasoning, which were described literally as "graspable imagination" (καταληπτική φαντασία) by figures like Sextus Empiricus, Diogenes Laertius, and Athenaeus.[11] This was an apprehensible imagination that was subject to confirmation by belief and reason, and thus one that could feed into thought and memory. As Diogenes Laertius defines it in his entry on Zeno from his *Lives of Eminent Philosophers*:

> Τὴν δὲ φαντασίαν εἶναι τύπωσιν ἐν ψυχῇ, τοῦ ὀνόματος οἰκείως μετενηνεγμένου ἀπὸ τῶν τύπων <τῶν> ἐν τῷ κηρῷ ὑπὸ τοῦ δακτυλίου γινομένων. τῆς δὲ φαντασίας τὴν μὲν καταληπτικήν, τὴν δὲ ἀκατάληπτον· καταληπτικὴν μέν, ἣν κριτήριον εἶναι τῶν πραγμάτων φασί, τὴν γινομένην ἀπὸ ὑπάρχοντος κατ᾽ αὐτὸ τὸ ὑπάρχον ἐναπεσφραγισμένην καὶ ἐναπομεμαγμένην· ἀκατάληπτον δὲ τὴν μὴ ἀπὸ ὑπάρχοντος, ἢ ἀπὸ ὑπάρχοντος μέν, μὴ κατ᾽ αὐτὸ δὲ τὸ ὑπάρχον· τὴν μὴ τρανῆ μηδὲ ἔκτυπον.
>
> A presentation (φαντασίαν) is an imprint on the soul: the name having been appropriately borrowed from the imprint made by the seal upon the wax. There are two species of presentation, the one apprehending a real object, the other not. The former, which they take to be the test of reality, is defined as that which proceeds from a real object, agrees with that object itself, and has been imprinted seal-fashion and stamped upon the mind: the latter, or non-apprehending, that which does not proceed from any real object, or, if it does, fails to agree with the reality itself, not being clear or distinct.[12]

Here, we observe the language of apprehension (καταληπτικός) and impression (τύπωσις) as that used by Photios and others discussed, whereby the qualities of sensual apprehension are explicitly related to the actions of the mental faculties and reasoning (and not haptic activity).

[10] See Anna Maria Ioppolo, "Presentation and Assent: A Physical and Cognitive Problem in Early Stoicism," *Classical Quarterly* 40:2 (1990): 433–449; Michael Frede, "Stoics and Skeptics on Clear and Distinct Impressions" in Myles Burnyeat (ed.), *The Skeptical Tradition* (Berkeley, CA: University of California Press, 1983), 65–93; Julia Annas, "Truth and Knowledge" in Malcolm Schofield, Myles Burnyeat, and Jonathan Barnes (eds.), *Doubt and Dogmatism: Studies in Hellenistic Epistemology* (Oxford: Clarendon Press, 1980), 84–104; Anthony A. Long, "Stoic Psychology and the Elucidation of Language" in Elizabeth Asmis and Giovanni Manetti (eds.), *Knowledge through Signs: Ancient Semiotic Theories and Practices* (Brussels: Brepols, 1996), 938–950.

[11] For example, see Sextus Empiricus, *Adversus Mathematicos*, VII.151–159, ed. and trans. R. G. Bury, *Against the Logicians*, Loeb Classical Library 291 (Cambridge, MA: Harvard University Press, 1935), 82–87; Diogenes Laertius, *Vitae philosophorum*, VII.1.45–46, ed. and trans. Hicks, 154–157; Athenaeus, *Deipnosophistae*, VIII.354f–355a, ed. and trans. S. Douglas Olson, *The Learned Banqueters*, vol. IV, Loeb Classical Library 235 (Cambridge, MA: Harvard University Press, 2008), 122–125.

[12] Diogenes Laertius, *Vitae philosophorum*, VII.1.45–46, ed. and trans. Hicks, 154–157.

Interpreting the *katalēptikē phantasia* and its *typos* as a "cognitive impression," however, caused problems for Chrysippos, who saw the wax-seal metaphor as a crude articulation of the function of the *phantasia*.[13] As Sextus Empiricus explains, Chrysippos objected on the grounds that this model of impressions would prevent the imagination of multiple forms at once, as if all pressed upon one another and contradicted each other's form.[14] Chrysippos resolves that this "presentation is an alteration of the soul" (φαντασία ἐστὶν ἑτεροίωσις ψυχῆς).[15] This understanding of the imagination or presentation as an "alteration" (ἑτεροίωσις), echoes Aristotle's own articulation of imagination as "some kind of movement" (ἡ δὲ φαντασία κίνησίς τις).[16] As David Sedley has noted, however, it is not in Chrysippos's character to contradict Zeno. Rather he tends to approach Zeno's philosophy by "reinterpreting his *ipsissima verba*."[17] Sedley thus reasons that in understanding this seemingly haptic relation between Zeno's *phantasia* as an "impression in the soul" (τύπωσιν ἐν ψυχῇ) and the external object, "from what is" (ἀπὸ ὑπάρχοντος),[18] one might understand that Chrysippos had an alternative reading of the statement where ἀπό indicates a representational function.[19] This attested function in Hellenistic Greek thus suggests that the phrase should be understood as "representing what is actual." Not only does this open the possibility for non-veridical forms of the imagination, but it also suggests a representational relation between type and prototype. This rereading of Zeno on the part of Chrysippos to reinterpret what he meant is alluded to by Sextus Empiricus when he writes, "He himself [Chrysippos], therefore, suspected (ὑπενόει) that the term 'impression' (τύπωσιν) was used by Zeno in the sense of 'alteration.'"[20] This suspicion (ὑπενόει), recounted by Sextus Empiricus, indicates not that Chrysippos disagreed with Zeno, but rather that he disagreed with the interpretations of his intended meaning.

Despite attempting to distance "impression" (τύπωσις) from a reductive literalness, this language of impression continued to be associated with the

[13] See Baron Reed, "The Stoics' Account of Cognitive Impression," *Oxford Studies in Ancient Philosophy* 23 (2002): 147–180.

[14] Sextus Empiricus, *Adversus mathematicos*, VII.228–230, ed. and trans. Bury, 122–125.

[15] Sextus Empiricus, *Adversus mathematicos*, VII.230, ed. and trans. Bury, 124–125.

[16] Aristotle, *De anima*, 427b12, ed. Hett, 160–161.

[17] David Sedley, "Zeno's Definition of *phantasia kataleptike*" in Theodore Scaltsas and Andrew S. Mason (eds.), *The Philosophy of Zeno: Zeno of Citium and His Legacy* (Larnaca: Municipality of Larnaca, 2002), 135–154, esp. 139.

[18] Diogenes Laertius, *Vitae philosophorum*, VII.1.45, 54, ed. and trans. Hicks, 154–157, 162–163.

[19] Sedley, "Zeno's Definition," 144.

[20] "αὐτὸς οὖν τὴν τύπωσιν εἰρῆσθαι ὑπὸ τοῦ Ζήνωνος ὑπενόει ἀντὶ τῆς ἑτεροιώσεως." Sextus Empiricus, *Adversus mathematicos*, VII.230, ed. and trans. Bury, 124–125.

katalēptikē phantasia in Stoic logic. As Verity Platt has demonstrated, this logic had implications beyond Stoic epistemological discourse, coming to define authority and authenticity through the clear and distinct impressions of seals.[21] In the late antique and Byzantine period, this language of τύπωσις was retained as a manner of regarding noetic images more broadly, as seen for example in Photios, and φαντασία became ever more defined by its cognitive functionality. Photios himself was familiar with Stoic discourse on the *katalēptikē phantasia*, as evidenced by his detailed entry in the *Bibliotheca* on Aenesidemus's now-lost *Pyrrhonian Discourses*.[22] Despite his less than favorable view, Photios describes how Aenesidemus criticizes the Academics for their seemingly haphazard agreements and disagreements with the Stoics. He says that the Academics "firmly determine many other things, while saying that they disagree only about the καταληπτική φαντασία."[23] Despite this fleeting remark on the imagination, Photios's entry is replete with the language of comprehension (κατάληψις) and comprehensibles (καταληπτά), demonstrating his own grasp of the *katalēptikē phantasia* and its implications, which we have already seen in his Homily 17. Likewise, in the *Amphilochia*, Photios devotes attention to the question of God's perceptibility through the "God-fitting imagination" (θεοπρεπής φαντασία), which as Georgi Kapriev has explained, struggles with the unknowability of the divine form and the imagination's emergence outside the viewer.[24]

Photios's interest in the imagination seems to be rooted largely in how the *phantasia* mediates between sensation and the mind, and between various stages of perceptual cognition. Such mediation had long been assigned to *phantasia* in ancient thought and epistemology. As Michael Squire observes, at a basic level even the ancient articulations of "*phantasia* existed in the gap between sense perception and rational knowledge: it refers to an intuitive insight or impression that generates cognitive

[21] Verity Platt, "Making an Impression: Replication and the Ontology of the Graeco-Roman Seal Stone," *Art History* 29:2 (2006): 233–257. See Anthony A. Long and David Sedley, *The Hellenistic Philosophers*, vol. 1 (Cambridge University Press, 1987), 236–241. See also Michael Koortbojian, "Mimesis or Phantasia? Two Representational Modes in Roman Commemorative Art," *Classical Antiquity* 24:2 (2005): 285–306.

[22] Photios, *Bibliotheca*, 212, ed. Henry, III:119–123.

[23] "ἄλλα τε πολλὰ βεβαίως ὁρίζουσι, διαμφισβητεῖν δέ φασι περὶ μόνης τῆς καταληπτικῆς φαντασίας." Photios, *Bibliotheca*, 212, ed. Henry, III:120. On Photios on Aenesidemus, see Charles Brittain, *Philo of Larissa: The Last of the Academic Skeptics* (Oxford University Press, 2001), 83–94, 134.

[24] Georgi Kapriev, "Die gottgebührenden Phantasien und die Gottesschau bei Photios" in *Intellect et Imagination dans la Philosophie Médiévale*, vol. 1 (Turnhout: Brepols, 2006), 473–481, esp. 478.

understanding of the thing perceived."[25] This is in keeping with the interests and outlook of the late antique commentators, who approached the subject matter primarily through Aristotle's *De anima* III.3 and its commentaries. As H. J. Blumenthal notes, Neoplatonist psychology was largely Aristotelian in character, and the same is true of its doctrine of the *phantasia*.[26] That is, whereas late antique terms such as *phantasia* and *typos* and their logic of sensory impression come from Stoic tradition, the roots of those terms and logic came from the peripatetic tradition. For instance, in the case of Proklos, his work changes over time as to what faculty mediates between thought (διάνοια) and sensation. As he argues in the *In Timaeum*, it is belief (δόξα) that occupies this role, but in his later *In Euclidem*, belief appears only marginally, now replaced by the imagination.[27] This intermediary position, like that of Plotinos before him, led him to reason that the *phantasia* was dual, sharing a place within the lower and upper portions of the soul.[28] Likewise, "Simplikios" (possibly Priscian of Lydia)[29] and "Philoponos" (the *De anima* III commentary, probably by Stephanos of Athens/Alexandria)[30] extensively grappled with the relationship of *phantasia* and sensation, both resolving there to be a joint operation in some regard, even though the two are distinct and the imagination is also higher than sensation.[31] Because of the challenges to its clear parsing, Themistios, even while deferring to Aristotle's statement

[25] Michael Squire, "Apparitions Apparent: Ekphrasis and the Parameters of Vision in the Elder Philostratus's *Imagines*," *Helios* 40:1–2 (2013): 97–140, esp. 101.

[26] See H. J. Blumenthal, "Neoplatonic Interpretations of Aristotle on 'Phantasia,'" *Review of Metaphysics* 31:2 (1977): 242–257.

[27] See H. J. Blumenthal, "Plutarch's Exposition of the *De Anima* and the Psychology of Proclus" in Bent Dalsgaard Larsen (ed.), *De Jamblique à Proclus* (Geneva: Fondation Hardt, 1975), 123–147; A. Charles, "L'imagination, miroir de l'âme selon Proclos" in M. P. M. Schuhl and M. P. Hadot (eds.), *Le Néoplatonisme* (Paris: Editions du Centre National de la Recherche Scientifique, 1971), 241–248. See also D. Gregory MacIsaac, "*Phantasia* between Soul and Body in Proclus' Euclid Commentary," *Dionysius* 19 (2001): 125–136.

[28] H. J. Blumenthal, *Plotinus' Psychology*, 80–99.

[29] See Carlos Steel, *On Aristotle on the Soul 2.5–12* (Ithaca, NY: Cornell University Press, 1997), 105–140.

[30] See Wanda Wolska-Conus, "Stephanus d'Athènes et Stephanus d'Alexandrie: essai d'identification et de biographie," *Revue des Études Byzantines* 47 (1989): 5–89; Peter Lautner, "Philoponus, *In de Anima III*: Quest for an Author," *Classical Quarterly* 42 (1992): 510–522; H. J. Blumenthal, "John Philoponus and Stephanus of Alexandria: Two Neoplatonic Christian Commentators on Aristotle?" in D. J. O'Meara (ed.), *Neoplatonism and Christian Thought* (Albany, NY: SUNY Press, 1982), 54–63.

[31] Simplikios, *In Aristotelis libros de anima commentaria*, ed. M. Hayduck, *In libros Aristotelis de anima commentaria* (Berlin: Reimer, 1882), 207.27–215.13. John Philoponos, *In de anima*, 501,10–506,15, ed. M. Hayduck, *In Aristotelis de anima libros commentaria* (Berlin: Reimer, 1897), 501–506; trans. William Charlton, *On Aristotle's On the Soul 3.1–8* (Ithaca, NY: Cornell University Press, 2000), 77–82.

regarding their distinctness, suggested a simultaneity of operation that made their independence and unique identity difficult to understand.[32]

This lowliness of *phantasia* as a cognitive faculty becomes associated in the Neoplatonic tradition with Aristotle's definition of the "passive mind" or "passive intellect" (παθητικὸς νοῦς).[33] In Book 3 of *De anima*, Aristotle states:

> καὶ ἔστιν ὁ μὲν τοιοῦτος νοῦς τῷ πάντα γίνεσθαι, ὁ δὲ τῷ πάντα ποιεῖν, ὡς ἕξις τις, οἷον τὸ φῶς· τρόπον γάρ τινα καὶ τὸ φῶς ποιεῖ τὰ δυνάμει ὄντα χρώματα ἐνεργείᾳ χρώματα.
>
> Mind in the passive sense is such because it becomes all things, but mind has another aspect in that it makes all things; this is a kind of positive state like light; for in a sense light makes potential into actual colours.[34]

Illustrated through a metaphor from visual perception, where the mind is compared to the illuminated transparent that makes sight possible, the active mind here is seen as being able to create things. The passive mind, on the other hand, "becomes all things" (πάντα γίνεσθαι), suggesting that it does not partake in creative action of the "active intellect" (ποιητικός νοῦς), but rather passively receives impressions from sensation alone. As H. J. Blumenthal has noted, while *phantasia* in mathematical contexts had an active function, even before Proklos, in terms of thought and perception *phantasia* retained a primarily passive overtone until after Plotinos.[35]

In later Byzantine periods, these associations between the imagination and the passive intellect were still active. In his eleventh-century commentary on Hermogenes's *Progymnasmata*, John Sikeliotes equated *phantasia* with "passive intellect," affording it a role intimately tied with initial sensation.[36] The same is found in John Galenos's twelfth-century allegorical commentary on Hesiod.[37] And, as we shall soon see, the association between the *phantasia* and the "passive intellect" features in the *Souda*'s entry on *phantasia*. That inclusion is notable because of its indebtedness to the commentaries associated with Philoponos. Michael Psellos, in fact,

[32] Themistios, *In Aristotelis libros de anima paraphrasis*, ed. R. Heinze, *In libros Aristotelis de anima paraphrasis* (Berlin: Reimer, 1899), 92.4–93.31; trans. Robert B. Todd, *On Aristotle On the Soul* (London: Duckworth, 1996), 115–117.

[33] H. J. Blumenthal, *Aristotle and Neoplatonism in Late Antiquity: Interpretations of the De Anima* (London: Duckworth, 1996),

[34] Aristotle, *De anima*, 430a14–20, ed. and trans. Hett, 170–171.

[35] Blumenthal, *Aristotle and Neoplatonism*, 158. See also Dominic J. O'Meara, *Pythagoras Revived: Mathematics and Philosophy in Late Antiquity* (Oxford: Clarendon Press, 1989), 167–168.

[36] John Sikeliotes, *Commentarium in Hermogenis librum περὶ ἰδεῶν*, ed. Walz, 87–88.

[37] See Panagiotis Roilos, "'Unshapely Bodies and Beautifying Embellishments': The Ancient Epics in Byzantium, Allegorical Hermeneutics, and the Case of Ioannes Diakonos Galenos," *Jahrbuch der Österreichischen Byzantinistik* 64 (2014): 231–246, esp. 240.

dedicates much of his work on the allegory of the Sphinx to contemplating the division between *phantasia* and *nous* as a division between the terrestrial and the irrational versus the heavenly and rational.[38] In Psellos's approach, the more attached one is to the sensory affects of the world, coextensive with the realm of *phantasia*, the less steadfast one is in their reason and intellect, which he associates with a loftier existence analogous to the divine.[39] Thus, *phantasia* here serves as a metaphor for the material experiences of humanity by virtue of being understood as intimately connected to those sensory impulses that we could associate with the "passive intellect," whereas *nous* proper is the rational cognition of the loftier, heavenly powers.

In Priscian of Lydia's sixth-century *Metaphrasis* on Theophrastus's work on sense perception, the author provides an entire section devoted precisely to the problem of the imagination in perception.[40] Not only does the text preserve the thinking of Theophrastus from his now-lost commentaries on Aristotle's *De anima*, but it also works closely with Iamblichus's partially preserved commentary.[41] As Priscian, following Iamblichus, mentions, the imagination:

> πάσας ἀποτυποῦται καὶ ἐκμάττεται τὰς τῶν εἰδῶν ὁμοιότητας καὶ ταῖς τῶν ἑτέρων δυνάμεσιν ἐμφάσεις εἰς τὰς ἑτέρας διαπέμπει, τὰς μὲν ἀπὸ τῆς αἰσθήσεως εἰς δόξαν ἀνεγείρουσα, τὰς δὲ ἀπὸ νοῦ δευτέρας τῇ δόξῃ προτείνουσα, ἐν ἑαυτῇ δὲ ἀπὸ τῶν ὅλων τὰ φαντάσματα παραδεχομένη.
>
> represents in itself and moulds all the likenesses of the forms, and transmits representative images connected with the one set of faculties to the others, rousing up the one set [of images] from sense perception to opinion, and presenting to opinion the second, from intellect, and receiving in itself the images from the wholes.[42]

[38] Michael Psellos, *Opuscula logica, physica, allegorica, alia*, 44, ed. J. M. Duffy, *Philosophica minora*, vol. 1 (Leipzig: Teubner, 1992), 158–161. See Graeme Miles, "Living as a Sphinx: Composite Being and Monstrous Interpreter in the 'Middle Life' of Michael Psellos" in Danijela Kambaskovic (ed.), *Conjunctions of Mind, Soul and Body from Plato to the Enlightenment* (New York: Springer, 2014), 11–24, esp. 17–18. For a parallel, though not exact, reading of the enigma of the Sphinx in the *Tabula Cebetis*, see Michael Squire and Jonas Grethlein, "'Counterfeit in Character but Persuasive in Appearance': Reviewing the *Ainigma* of the *Tabula Cebetis*," *Classical Philology* 109:4 (2014): 285–324.

[39] See Panagiotis Roilos, "*Phantasia* and the Ethics of Fictionality in Byzantium: A Cognitive Anthropological Perspective" in Panagiotis Roilos (ed.), *Medieval Greek Storytelling: Fictionality and Narrative in Byzantium* (Wiesbaden: Harrassowitz Verlag, 2014), 9–30, esp. 12–13.

[40] Priscian, *Metaphrasis in Theophrastum*, ed. I. Bywater, *Prisciani Lydi quae extant* (Berlin: Reimer, 1886), esp. 23.1–25.26; trans. Pamela Huby, *On Theophrastus on Sense-Perception* (Ithaca, NY: Cornell University Press, 1997), esp. 32–35.

[41] See Matthias Perkams, "Priscian of Lydia, Commentator on the *De anima* in the Tradition of Iamblichus," *Mnemosyne*, 4th series, 58:4 (2005): 510–530.

[42] Priscian, *Metaphrasis in Theophrastum*, ed. Bywater, 23.14–18; trans. Huby, 32.

Here, Priscian stresses the intermediary function of the *phantasia*, which processes sensory input, passing it to the faculty of belief (δόξα), but which also appears to conduct the input from the intellect to belief as well. As Priscian goes on to explain in his commentary, the imagination is a separate entity from sensation not only because it is roused by sensory stimuli, but also because it can generate its images (φαντάσματα) independently, a point which is often noted by other commentators as well. This not only addresses the problem of hallucinatory imaginings, close to what one might call fantasies, as a counterfactual imagination, but also explains the ability to think with the imagination's images even when the objects are not present. In many regards, Priscian's reading of Theophrastus and Iamblichus presents the imagination in the way it is generally seen by subsequent Byzantine writers such as Photios.

Throughout his section on the imagination, Priscian retains the signature language of impression (ἀποτυποῦν) and molding (ἐκμάττειν) derived through the wax-seal metaphor of imprints. This language of imprints for perception and the imagination may be readily associated with the Stoics, but it certainly goes back to Aristotle in *De anima* and even before to Plato in *Theaetetus*,[43] and is also pervasive in the commentaries of Alexander of Aphrodisias and John Philoponos.[44] Nevertheless, Priscian explicitly states that the imagination is not a "movement" (κίνησις), nor is it "molded (διαπλαττομένη) by receiving something from outside like wax."[45] This disclaimer allows one to appreciate that while *phantasia* is not understood as an "impression," the metaphorical language of sealing and imprinting is retained to explain cognition and representation more generally. This language becomes significant in post-iconoclastic discourses around the icon, as in the work of Theodore of Stoudios, and is used in liturgical commentaries to describe symbolism and representation.[46] Thus, the Stoic uses

[43] See Aristotle, *De anima*, 423b17–21, ed. and trans. Hett, 132–133; Plato, *Theaetetus*, 191C–192C, ed. and trans. Fowler, 184–189.

[44] See William Charlton, *Philoponus: On Aristotle's On the Intellect (de Anima 3.4–8)* (London: Duckworth, 1991), 13–16; Gérard Verbeke, *Jean Philopon: commentaire sur le De anima d'Aristote* (Paris: Éditions Béatrice-Nauwelaerts, 1966), liii-liv.

[45] "οὐκ οὖσα πάθος οὐδὲ κίνησις ἀλλ' ἀμέριστος καὶ ὡρισμένη ἐνέργεια, καὶ οὐ κατὰ παραδοχὴν ἔξωθεν ὡς ὁ κηρὸς διαπλαττομένη ἀλλ' ἔνδοθεν." Priscian, *Metaphrasis in Theophrastum*, ed. Bywater, 23.25–27; trans. Huby, 33.

[46] See Bissera Pentcheva, "Painting or Relief: The Ideal Icon in Iconophile Writing in Byzantium," *Zograf* 31 (2006–2007): 7–14, esp. 10–11. See also Roland Betancourt, "A Liturgical Commentary in Verse: Introduction and Translation," *Orientalia Christiana Periodica* 81:2 (2015): 433–472, esp. 434–437.

of the term were avoided in the Byzantine period in favor of a cognitive interpretation of the functions and operations of the *phantasia*, which offered potential claims to truth via haptic contact or impression.

The *Souda* elaborates on the faculty of the imagination in its entry on "Imagination" (Φαντασία). This entry is derived from John Philoponos's introduction to his commentary on Aristotle's *De anima* with only minor alterations. As a Christian commentator on Aristotle, he is unsurprisingly relied upon for the entries on psychology from Byzantine handbooks.[47]

> Φαντασία: γνωστικαὶ καὶ ζωτικαὶ καὶ ὀρεκτικαὶ δυνάμεις τῆς ψυχῆς: φαντασία καὶ αἴσθησις. διαφέρουσι δὲ ἀλλήλων, ὅτι ἡ μὲν αἴσθησις πρὸς τὸ ἐκτὸς ἀποτείνεται, ἡ δὲ φαντασία ἔνδον ἔχει τὴν γνῶσιν. καὶ ἡ μὲν αἴσθησις τὸ παρὸν μόνον καὶ οὗ ἀντιλαμβάνεται ἔξωθεν τοῦτο οἶδεν, ἡ δὲ φαντασία παρὰ τῆς αἰσθήσεως λαβοῦσα τῶν αἰσθητῶν τοὺς τύπους ἐν ἑαυτῇ τούτους ἀναπλάττει: ὅθεν καὶ παθητικὸν νοῦν Ἀριστοτέλης αὐτὴν καλεῖ, νοῦν μὲν ὡς ἔνδον ἔχουσαν τὸ γνωστὸν καὶ ἁπλῇ προσβολῇ, ὥσπερ ἐκεῖνος, ἐπιβάλλουσαν: παθητικὸν δέ, διότι μετὰ τύπων, καὶ οὐκ ἀσχηματίστως. ἐκλήθη δὲ φαντασία, οἱονεὶ φαοστασία τις οὖσα: φαντασία γάρ ἐστιν ἡ τῶν φανθέντων στάσις: ἵστησι γὰρ ἐν αὐτῇ τὰ ἔξω φανέντα. ἑκατέρα δὲ αὐτῶν περὶ μερικὰ ἀποτείνεται: τὸ γὰρ λευκὸν οἶδεν, οὐ μὴν πᾶν λευκόν. διαφέρει δέ, ὅτι ἡ μὲν τὸ ἔξω, ἡ δὲ τὸ ἐντὸς οἶδε. καὶ ἡ μὲν φαντασία τῶν ε' αἰσθήσεων δέχεται τοὺς τύπους, ἑκάστη δὲ τῶν αἰσθήσεων μόνον τὸ ἴδιον αἰσθητὸν γινώσκει.

> Imagination: The cognitive and vital and appetitive faculties of the soul [are] imagination and perception. But they differ from each other, because perception refers to the external object, whereas imagination deals with knowledge within [the knower]. And while perception knows only what is present (that is, that external object which it apprehends), imagination, receiving from perception the imprints of the perceptible things remodels them in itself. Hence Aristotle also calls it [i.e. imagination] "passive intellect" – "intellect" insofar as it contains what is known within, and attends to pure stimuli, like that [other intellect]. [Imagination] is "passive" because it exists accompanied by imprints and [does] not [operate] without forms. And it was named "fantasy" as if it were a sort of φαοστασία [i.e. something related to light]. For imagination (φαντασία) is the state of the objects which have appeared (φανθέντων), since the things which have appeared (φανέντα) from outside stand in it. And each of them is related to particulars, for it knows that something is white, and that not everything

[47] On human cognition in Philoponos, see G. Verbeke, "Levels of Human Thinking in Philoponus" in C. Laga, J. A. Munitiz, and L. Van Rompay (eds.), *After Chalcedon: Studies in Theology and Church History Offered to Professor Albert Van Roey for His Seventieth Birthday* (Leuven: Peeters, 1985), 451–470.

> is white. There is a difference, since while one knows the external object, the other knows the internal. Imagination also receives the imprints of the five senses, but each of the senses knows only its proper sensation.[48]

The entry stipulates that while perception can perceive only what is at hand, the imagination can bring forth images of things that are absent and once were present. Despite stressing *phantasia*'s association with the "passive intellect," the entry nonetheless articulates the active function that it plays within perception and cognition. Hence, for Philoponos the imagination is a faculty that allows for the mediation of presence and absence by virtue of its ability to remediate sensual data as mental images that may be represented.[49] This language of grasping (ἀντιλαμβάνεται) and taking (λαβοῦσα), as we have observed it before, is demonstrated to be an element of perception's action as the imagination remolds (ἀναπλάττει) the images (τύπους) received from the external outward form of visible objects via sensation into mental forms, since it cannot reason or think without these forms. This involves an act of mediation whereby the external, physical object is transformed into a noetic object.

The faculty of imagination is critical not only for cognition and the processing of sensation, but also for the collection and collation of sensory data that comes in through the fragmented individuality of the five senses. As the entry concludes, "Imagination also receives the imprints of the five senses, but each of the senses knows only its proper sensation" (καὶ ἡ μὲν φαντασία τῶν ε' αἰσθήσεων δέχεται τοὺς τύπους, ἑκάστη δὲ τῶν αἰσθήσεων μόνον τὸ ἴδιον αἰσθητὸν γινώσκει). In receiving the images or imprints (τύπους) from the senses (and note that the term *typos* is used not only for visual forms, but for all sensual data) imagination stitches these sensations together to produce a fluid understanding of the sensible object, whose various properties arrive separately through the five individual and unmixed senses. The stitching power of the imagination amounts to what Aristotle calls the "common sense," where diverse sensory inputs are brought into a cohesive mental understanding of an object that is not exclusively visual, sonic, gustatory, tactile, or olfactory.

Significantly, Aristotle's definition of the common sense is that it mediates the disjuncture of the senses by synaesthetically bringing them

[48] Anonymous, *Souda*, "Φαντασία," phi 84, ed. and modified trans. Marco Boeri, *Souda-On-Line*, www.stoa.org/sol-entries/phi/84. Cf. John Philoponos, *In Aristotelis libros de anima commentaria*, ed. Hayduck, 5.34–6.10; trans. Philip J. van der Eijk, *On Aristotle's On the Soul 1.1–2* (Ithaca, NY: Cornell University Press, 2005), 19–20.

[49] On John Philoponos and the imagination, see Peter Lautner, "Philoponean Accounts of *Phantasia*," *Acta Antiqua Academiae Scientiarum Hungaricae* 34 (1993): 159–170.

together in the mind, at times it would seem through the action of the imagination.[50] In the late antique and Byzantine world, the common sense was at times conflated with the imagination or seen as simultaneous with sensation. The idea that the common sense coincides with sensation and is subsequently used by the imagination appears to take precedence, for example with Philoponos.[51] Naturally, then the *Souda* pairs a statement on the particularity of each of the senses to their own organs with a statement on their eventual communion since this indicates the synthesis during cognition.

The conflation of the *phantasia* with the common sense recalls Porphyry's initial stage of "apprehension" (ἀντίληψις) in the perceptual process, which likewise coincides with the sensation of the object, which, as cited above, "seems to contact (ἐπαφωμένη) the entity and tries to take up (ἀναλαμβάνειν) the forms to make a kind of report on them and introduce them into the soul like some guide or introducer." While Porphyry's initial apprehension (ἀντίληψις) does not entail *phantasia* or the common sense in any way, for Photios the comparison sheds light on the two participles indicating the "contacting (ἐπαφωμένη) and regarding (περιέπουσα)" [**m1**] of the visible object. The fact that περιέπειν connotes a sense of handling and overall treatment of a thing (almost an act of circumscription through the prefix περι-) may even lead one to read the term as suggesting that it is not only the visual morphology of the object that is being apprehended, but rather its totality, that is, with any other synthesized sensory information. In the case of pictorial representation, like the mosaic image of the Theotokos and Christ upon which Photios is reflecting, high above the viewer in the apse, the sight then presumably also would cue a series of associations for the spectators: the touch of her garb and flesh with those rosy cheeks, the sweetly perfumed smell of royalty, the sound of her voice that Photios describes as "not incapable of speaking" (μηδ' ἂν τὸ φθέγξασθαι παραιτήσασθαι),[52] and so on. Here, it is the role of the imagination precisely to stir up every possible sensory imprint, united through the common sense, so as to vividly enliven the apprehension and comprehension of the depiction upon the apse. In a word, it has visualized.

[50] See Pavel Gregorić, *Aristotle on the Common Sense* (Oxford University Press, 2007). See also J. Owens, "Aristotle on Common Sensibles and Incidental Perception," *Phoenix* 36 (1982): 215–36.

[51] See Richard Sorabji, "Aristotle on Demarcating the Five Senses" in J. Barnes, M. Schofield, and R. Sorabji (eds.), *Articles on Aristotle*, vol. IV (London: Duckworth, 1979), 76–92. See also Malcolm Schofield, "Aristotle on the Imagination" in M. C. Nussbaum and A. O. Rorty (eds.), *Essays on Aristotle's* De Anima (Oxford University Press, 1995), 249–278.

[52] Photios, *Homiliae*, 17, ed. Laourdas, 167; trans. Mango, 290.

CHAPTER 7

"Has It Visualized?" II
The Problem of Fantasy

The challenge of the imagination for many Christian writers was that imagination did not always depict truth. This understanding of *phantasia*, however, could often stand alongside other notions, leading to differences of opinion. Were these fictional figments of fantasy or correct cognitive impressions of the imagination? In this chapter, after delineating some of the early Christian conversations on this matter, I wish to turn to Homily 10, which Photios composed for the inauguration of the Pharos Chapel in the palace of Constantinople. This homily is best known and cited for its attention to the sensory effects of architecture and art. The homily provides us with some telling understandings of the relationship between imagination as a cognitive faculty and imagination as a contrafactual phantasm of the mind. My goal here is to contextualize the manner in which perception at times vacillates between its confirmed authenticity and the deceptions of dazzling sensory displays. This sets a foundation for considering how perception casts judgments on the world that can lead toward the confirmation of belief and to one's assent to its teachings – these latter issues being the subject of the subsequent chapter. We shall also observe here that while the noun *phantasia* is commonly used to describe a type of image or impression of the senses that would fall under the jurisdiction of a counterfactual presentation, this had quite a different tenor to *phantasia* as a mental faculty, as found in cognitive and perceptual writings – even if at times a noticeable slippage between the two could be sensed. It is this tension that presents a generative challenge in comprehending the role of the imagination in Photios's approach to perception and art.

For early Christian writers, the imagination posed a significant epistemological challenge. This faculty of the mind had the power to deceive and delude the Christian subject. Approaching the *phantasia* from the literary perspective, Panagiotis Roilos has stressed its problematic presence in Byzantium given its "fictionality," that is, the way in which it situated itself against what could be termed as real, as a verisimilitude to reality.[1] In a Christian context, φαντασία could approximate the modern meaning of "fantasy," yet one relative to pagan myths, idol worship, magic, and enchantments, as in the case of Irenaeus in the second century who describes the "ill-composed wisdom" or "*phantasia*" (κακοσυνθέτῳ σοφίᾳ [φαντασίᾳ]) of heretical doctrines, likened to a recomposed idol who deceives "through the illusions of jewels" (διὰ τῆς τῶν ψηφίδων φαντασίας).[2] Similarly, in Athanasios of Alexandria's fourth-century *Life of St. Antonios*, *phantasia* is associated with the activities of demons who, despite being powerless, are capable of terrorizing with "the illusion of their hordes and figures" (τῇ τῶν ὄχλων φαντασίᾳ καὶ τοῖς σχηματισμοῖς).[3] And, in the late fourth century, Epiphanios of Salamis's *Panarion* likewise uses the term to characterize the delusions and false images conjured by Simon Magus.[4]

For Evagrios and Gregory of Nyssa in the fourth century, as Kevin Corrigan has argued, demons cannot arouse genuine contemplation (θεωρία) in the mind, but they can certainly insert false images to elucidate the mind's erroneous assent.[5] The tension here posed between the images of *phantasia* and proper contemplation via *theoria* is given an eloquent twist by Clement of Alexandria in the *Stromata*, where he articulates the "grasping contemplation" (καταληπτικὴ θεωρία) of the pure at heart in achieving a vision of the divine.[6] In Clement's text, therefore, we see the inklings of early Christian writers attempting to grapple with the danger of the imagination's deceptive *phantasia*, yet acknowledging the imagination as a necessary perceptual faculty with direct implications on contemplation. Thus, here it would seem that Clement has reworded the Stoic language of *katalēptikē phantasia* in terms of contemplation, so as to reorient the function of judgment and sober attention to the processes of

[1] See Roilos, "*Phantasia* and the Ethics of Fictionality in Byzantium," 9–30.

[2] Irenaeus, *Adversus haereses*, I.1:15, ed. W. W. Harvey, *Libri quinque adversus haereses*, vol. I (Cambridge University Press, 1857), 67. Cf. Hippolytus, *Refutatio omnium haeresium*, IV.37:3–4, ed. M. Marcovich, *Refutatio omnium haeresium*, Patristische Texte und Studien 25 (Berlin: De Gruyter, 1986), 124–125.

[3] Athanasios of Alexandria, *Vita Antonii*, 28.9, ed. G. J. M. Bartelink, *Vie d'Antoine*, Sources chrétiennes 400 (Paris: Éditions du Cerf, 2004), 214.

[4] Epiphanios of Salamis, *Panarion*, ed. K. Holl, *Epiphanius*, vol. I, Die griechischen christlichen Schriftsteller 25 (Leipzig: Hinrichs, 1915), 238.

[5] Kevin Corrigan, *Evagrius and Gregory: Mind, Soul and Body in the 4th Century* (Burlington, VT: Ashgate, 2009), 94–95.

[6] Clement of Alexandria, *Stromata*, VII.3.13, cf. VII.15.91, ed. L. Früchtel, O. Stählin, and U. Treu, *Clemens Alexandrinus*, vol. III, 2nd edition (Berlin: Akademie Verlag, 1970), 10, cf. 64.

cognition through the operation of *theoria* rather than through the problematic *phantasia.* As Simon Goldhill has demonstrated, Clement of Alexandria carefully deployed the language of impression (ἐναπομάττω) in connection to the Stoics' *katalēptikē phantasia.* And as is well known, Clement consistently deployed his training in Greek philosophy to engage with his Alexandrine colleagues.[7] Therefore, as Aglae Pizzone has reasoned in her survey of φαντασία in the Church Fathers, the imagination became a "surrogate of contemplation" over the course of late antiquity.[8]

In the work Evagrios, who studied under Basil and Gregory of Nazianzos, we observe a telling censorship of the term *phantasia,* nevertheless retained for its overarching function and operation as a mental faculty. Columba Stewart has lucidly dissected Evagrios's theory of cognition, observing that Evagrios believes that the mind operates on "ideas" or those things that are perceived (νοήματα), through which one creates an inner vision related to things perceived in the outside world.[9] These *noēmata,* which Stewart rightly translates as "depictions," bear the "form" (μορφή) of perceived objects and ideas, and they emerge primarily through sensory inputs, namely sight and hearing.[10] This is caused by their "being impressed" (τυπούμενος) and their "impressions" (τύποι) on the mind, particularly when coming from sight.[11] Those imprints that are unmoved by passion are described as "bare" (ψιλά) *noēmata;*[12] however, those imprints rooted in passionate emotions, such as hatred, lust, suffering, pain, ambition, or greed, are described by Evagrios as *eidola* and *phantasia,* images and fantasies.[13] Thus, we can observe Evagrios replacing the Stoic term *phantasia* with *noēma* in order to distinguish his theory of sensory perception and cognition from the potential implications of what more colloquial uses of *phantasia* imply, namely fantasy. This is a clear stance by

7 Simon Goldhill, "The Erotic Eye: Visual Stimulation and Cultural Conflict" in Simon Goldhill (ed.), *Being Greek Under Rome: Cultural Identity, the Second Sophistic, and the Development of Empire* (Cambridge University Press, 2001), 154–194, esp. 172–180.

8 Pizzone, "Late Antique Φαντασία," 429.

9 Columba Stewart, "Imageless Prayer and the Theological Vision of Evagrius Ponticus," *Journal of Early Christian Studies* 9:2 (2001): 173–204, esp. 187–190. See Evagrios, *De malignis cogitationibus,* 24.24–25, ed. Paul Géhin, Claire Guillaumont, and Antoine Guillaumont, *Sur les pensées,* Sources chrétiennes 438 (Paris: Éditions du Cerf, 1998), 238.

10 Evagrios, *De malignis cogitationibus,* 25, ed. Géhin, Guillaumont, and Guillaumont, 240–244.

11 Evagrios, *De malignis cogitationibus,* 2 and 4, ed. Géhin, Guillaumont, and Guillaumont, 154–156, 162.

12 Evagrios, *De malignis cogitationibus,* 8.20, 40.6, ed. Géhin, Guillaumont, and Guillaumont, 178, 290.

13 Stewart, "Imageless Prayer," 188–189 n 73.

a thinker who was educated in philosophy in a wholly Christian context, standing against the usual Hellenistic Christianity of late antiquity. Evagrios may alter the central terminology here, but he retains the logic that these *noēmata* emerge from sensory perception, most vividly through sight, and that they are the imprints of sensation upon the mind. In his quest for pure prayer, however, Evagrios wishes to find a wholly aniconic approach toward God, and that is premised on the perception of the light of God, but that sets aside lowly cognition steeped in the throes of sense-perception.[14] The goal is to approach God wholly "unimprinted" (ἀτύπωτος).[15] The mind is strengthened in prayer when it does not imagine (μηδὲν. . .φανταζόμενος).[16] Thus, although the function or language of *phantasia* as a cognitive faculty is not abandoned wholesale, the noun *phantasia* refers to the presentations or impressions of the imagination used exclusively to refer to ill fantasies and temptations.

While the Church Fathers argued against the deceitful powers of the images (φαντασίαι) of the imagination, some came to advocate the centrality and importance of the mental faculty of the imagination (τὸ φανταστικόν). The role of the imagination in producing mental images from the stimuli of the senses for the sake of perception lays the foundation for the defense of the icon in the Iconoclastic controversies, particularly on the part of Theodore of Stoudios, as shall be discussed in the following chapter. Maximos the Confessor would likewise place a similar focus on the imagination as a faculty of the soul (φανταστική), despite at other times being quite critical of the φαντασία for its deceptive qualities.[17] Pizzone, who has surveyed many of these same texts, sees the views of Maximos as marking a definitive shift in the conception of φαντασία.[18] Arguably, however, there appears to be less of a decided shift away from or resolution to the concerns raised regarding the imagination's deceptions and phantasms, issues which even the Stoics confronted. Instead, throughout the late antique period there is a marked difference between the spurious images and forms of the imagination denoted by *phantasia*, and *phantasia* denoting the cognitive processing of sensual data. Already in the fourth century, Eusebios readily accepted the function of the *phantasia* as a

[14] See Evagrios, *De oratione*, 71, ed. PG 79:1165A–1200C.

[15] Evagrios, *De malignis cogitationibus*, 41.27, ed. Géhin, Guillaumont, and Guillaumont, 294.

[16] Evagrios, *Practicus*, 65, ed. Antoine Guillaumont and Claire Guillaumont, *Traité pratique ou le moine*, vol. II, Sources chrétiennes 171 (Paris: Éditions du Cerf, 1971), 650.

[17] Maximos the Confessor, *Capita de caritate*, III.32, ed. A. Ceresa-Gastaldo, *Capitoli sulla carita* (Rome: Editrice Studium, 1963), 158.

[18] Pizzone, "Theodore and the Black Man," 62.

mental faculty, comparing the conception of the Trinity to the apprehension involved when looking upon a ruler's portrait. As Eusebios writes, "The person who has seen the image of the king (which has been accurately made to his likeness), engraving the imprints of its form from the painting, forms a mental image (φαντασιοῦται) of the king."[19] Here, Eusebios eloquently deploys the Stoic language of imprint (τύπους) and impression (ἀποματτόμενος) in conjunction with the operation of the imagination, where taking such imprints from the external form enables one to imagine the figure of the king.

Therefore, even writers critical of *phantasia* in some contexts could use it elsewhere as part and parcel of the perceptual process. Gregory of Nyssa, for example, was suspicious of the imagination's fantasies, yet he also advocated the importance and necessity of *phantasia* in Christian thought. In his *Contra Eunomium*, Gregory writes:

> ἀλλὰ καὶ τὸ παρὰ τοῦ ἀποστόλου εἰρημένον οὕτως ἂν μάλιστα νοηθείη καλῶς καὶ προσφυῶς ἔχον, τὸ ἀπαύγασμα εἶναι τῆς δόξης καὶ χαρακτῆρα τῆς ὑποστάσεως τὸν υἱόν. τοῖς γὰρ ἀδυνατοῦσιν ἐπὶ τὸ ὕψος τῆς θεογνωσίας τὸν λογισμὸν ἀγαγεῖν διὰ τῆς τῶν νοητῶν θεωρίας, τούτοις ἐκ τῶν τῇ αἰσθήσει προφαινομένων ἐμποιεῖ τινα τοῖς ἀκούουσιν ὁ ἀπόστολος τοῦ ἀληθοῦς φαντασίαν.
>
> The saying of the Apostle might also best be understood as right and relevant, that the Son is "the radiance of his glory and the stamp of his substantial reality" (Heb 1.3); for where people are unable to raise up their thinking to the height of divine knowledge by mental speculation, the Apostle uses perceptible phenomena to put in their ears an impression of the truth.[20]

Here, the notion or image of truth (ἀληθοῦς φαντασίαν) emerges from sense perception in accordance with the function of the imagination, seeming to stand in for the proper mental contemplation (νοητῶν θεωρίας). The apostle appears as an authorized mediator through which to convey the impressions of the imagination with an understanding of its authenticity, veracity, and authority. In other words, while these images might pose a concern for the early Christian thinker, this concern can

[19] "ὡς γὰρ καὶ ὁ τὴν βασιλικὴν εἰκόνα τὴν ἐπ' ἀκριβὲς ἀφωμοιωμένην αὐτῷ τεθεαμένος τοὺς τῆς μορφῆς τύπους διὰ τῆς γραφῆς ἀποματτόμενος φαντασιοῦται τὸν βασιλέα." Eusebios, *De ecclesiastica theologia*, 3.21, ed. G. C. Hansen and E. Klostermann, *Werke*, vol. IV (Berlin: Akademie Verlag, 1972), 181.

[20] Gregory of Nyssa, *Contra Eunomium*, III.6.12, ed. W. Jaeger, *Gregorii Nysseni opera*, vol. II.2 (Leiden: Brill, 1960), 190; trans. Stuart G. Hall, *Contra Eunomium III*, ed. Johan Leemans and Matthieu Cassin (Leiden: Brill, 2010), 155.

often be qualified by the judgment of a reasonable interpreter. For instance, Origen would qualify his possible misgivings, stating that the "rational animal, however, has something besides its imaginative nature, namely, reason, which judges the images (φαντασίας)."[21] Here, the function of judgment clearly asserts one's ability to assent to images as being accurate.

The problem of a hasty assent comes to the foreground in how *phantasia* could be swayed maliciously. As Pseudo-Dionysios the Areopagite eloquently summarizes the matter:

> Ἄλλως τε τί τὸ ἐν δαίμοσι κακόν; Θυμὸς ἄλογος, ἄνους ἐπιθυμία, φαντασία προπετής. Ἀλλὰ ταῦτα, εἰ καὶ ἔστιν ἐν δαίμοσιν, οὐ πάντη οὐδὲ ἐπὶ πάντων οὐδὲ αὐτὰ καθ' αὑτὰ κακά. Καὶ γὰρ ἐφ' ἑτέρων ζῴων οὐχ ἡ σχέσις τούτων, ἀλλ' ἡ ἀναίρεσίς ἐστι καὶ φθορὰ τῷ ζῴῳ καὶ κακόν. Ἡ δὲ σχέσις σῴζει καὶ εἶναι ποιεῖ τὴν ταῦτα ἔχουσαν τοῦ ζῴου φύσιν.
>
> Or, again, what is this evil in the demons? Unreasoning anger, mindless desire, or rushed mental images. And yet things of this sort, even if they are to be found among demons, are not totally, completely, and innately evil. For in other living beings it is not so much the state of such qualities but rather the loss of them which brings ruin to a creature and is therefore evil. State of them can actually ensure life, can form the nature of the living being which has them.[22]

None of these various faculties are inherently evil, as Pseudo-Dionysios states, in fact they are considered to be essential to the life of the beings that possess them. However, it is their unrestrained and unchecked operations that generate their evil: "unreasoning" (ἄλογος) anger, "mindless" (ἄνους) desire, and "rushed" (προπετής) imagination. The notion of a rushed or hasty imagining is one to which we shall return in Part III, yet here it speaks to how one might assent too readily to the motions and imprints of the imagination. Given Pseudo-Dionysios's interest in the affective states of the body, this suggests a hasty assent specifically to sensory input, which structures one's actions toward the world via, for example, desire and anger. Hence, this concept of the hasty imagination suggests a tenuous and potentially dangerous form of perception.

[21] "Τὸ μέντοι λογικὸν ζῷον καὶ λόγον ἔχει πρὸς τῇ φανταστικῇ φύσει, τὸν κρίνοντα τὰς φαντασίας." Origen, *De Principiis*, 3.1.3, ed. H. Görgemanns and H. Karpp, *Origenes vier Bücher von den Prinzipien* (Darmstadt: Wissenschaftliche Buchgesellschaft, 1976), 466; modified trans. G. W. Butterworth, *On First Principles* (New York: Harper & Row, 1966), 159.

[22] Pseudo-Dionysios the Areopagite, *De divinis nominibus*, IV.23, ed. B.R. Suchla, *De divinis nominibus*, Patristische Texte und Studien 33 (Berlin: De Gruyter, 1990), 171; modified trans. Colm Luibheid, *Pseudo-Dionysius: The Complete Works* (New York: Paulist Press, 1987), 91.

Photios himself eloquently elaborates on the prowess and dangers of the imagination's function in relation to speed and variegation. In his well-known *ekphrasis* of the Pharos Chapel in Homily 10, he states:

> Δοκεῖ δὲ λοιπὸν ἐντεῦθεν τά τε ἄλλα ἐν ἐκστάσει εἶναι καὶ αὐτὸ περιδινεῖσθαι τὸ τέμενος· ταῖς γὰρ οἰκείαις καὶ παντοδαπαῖς περιστροφαῖς καὶ συνεχέσι κινήσεσιν, ἃ πάντως παθεῖν τὸν θεατὴν ἡ πανταχόθεν ποικιλία βιάζεται τοῦ θεάματος, εἰς αὐτὸ τὸ ὁρώμενον τὸ οἰκεῖον φαντάζεται πάθημα.
>
> Then everything else seems to be in ecstasy and the church itself seems to whirl around; for the viewer, with his twistings and turnings in every direction and his constant movements that the variety of the spectacle imposes on him, the experience is imagined (φαντάζεται) onto the very thing being seen, the church.[23]

In describing the psychic disposition of the viewer in the richly decorated and variegated chapel, Photios demonstrates that, beyond perpetual processing and cognitive reasoning, the imagination may affect as well the way in which a person comes to understand one's sensual surroundings. Here, the vertigo-inducing space causes the viewer to twirl around, but the imagination understands this as if it were the space itself that was spinning rather than the viewer. Since the Pharos Chapel presents the viewer with a unique sensual space that induces such ecstatic movement, it would seem that the mind has no common frame of reference to understand the spatial movement and accurately process the sensory data. Hence, after losing itself in intense concentration on the nearby variegated details, the mind begins to project the motion onto the surroundings themselves. The viewer experiences a unique and heightened experience of perception in which the usual standards of looking no longer apply, so the imagination is tricked into processing this sensual data as part of its understanding of the perceptible world. Thus, the operation of imagination can not only be deceptive, but it can also affect how one comes to perceive sensation and one's very place in the world.

So Photios's words may be read not so much as a commentary on affective sensual experience, as Bissera Pentcheva and others have done, but rather as a reflection on the very bounds of perception and the imagination's role in it.[24] Photios is confronting the limits of sensual

[23] Photios, *Homiliae*, 10, ed. Laourda, 101, modified trans. Ruth Webb, "The Aesthetics of Sacred Space: Narrative, Metaphor, and Motion in *Ekphraseis* of Church Buildings," *Dumbarton Oaks Papers* 53 (1999): 59–74, at 68.

[24] See Bissera Pentcheva, *The Sensual Icon: Space, Ritual, and the Senses in Byzantium* (University Park, PA: Pennsylvania State University Press, 2010), esp. 141–143; Liz James, "Senses and Sensibility in Byzantium," *Art History* 27:4 (2004): 523–537, esp. 528; Webb, "The Aesthetics of Sacred Space," 67–69.

experience in a state where the critical, analytic reasoning that accompanies perception is overridden. Thus, in remodeling sensation into a mental image the imagination is misled. Such an experience subverts the apprehensibility (καταληπτική) of the imprints that comes from the real object of the imagination. That which, as Diogenes Laertius describes, as cited above, "proceeds from a real object, agrees with that object itself, and has been imprinted seal-fashion and stamped upon the mind." Implicitly, then, Photios's description of the experience of the Pharos Chapel is an account that works against the neat perceptual process that occurred in his Homily 17 on the Virgin in Hagia Sophia. There, sensation, apprehension, imagination, and memorization all unfolded in neat, orderly progression, both sonically and visually. Yet, in the Pharos Chapel, the experience confounds this orderly flow, not allowing for the proper and invested perception that the contemplation of the Gospel narratives or the image of the Virgin might demand. In fact, Photios draws attention to this very problem when he goes on to say soon after this description:

> Οὕτω πάντα πεπλήρωται θαύματος. Ἑνὶ δέ μοι δοκεῖ μόνον ὁ τοῦ τεμένους ἀρχιτέκτων διημαρτηκέναι, ὅτι περ εἰς ἕνα καὶ τὸν αὐτὸν χῶρον ἅμα πάντα συλλέξας τὰ κάλλη οὐκ ἐᾷ τὸν θεατὴν καθαρῶς ἐντρυφῆσαι τῷ θεάματι, ἄλλων ἀπ' ἄλλων ἐφελκόντων τε καὶ μετασπώντων καὶ μὴ παραχωρούντων ὅσον τις ἐθέλει τοῦ ὁρωμένου κορέννυσθαι.
>
> So full of wonder is everything. In one respect only do I consider the architect of the church to have erred, namely that having gathered into one and the same spot all kinds of beauty, he does not allow the spectator to enjoy the sight in its purity, since he is carried and pulled away from one thing by another, and is unable to satiate himself with the spectacle as much as he may desire.[25]

In its variegated spectacle, the chapel denies the spectator the ability to properly and purely (καθαρῶς) enjoy the sights all around since they are constantly being jolted in various directions and thus may not clearly satisfy themselves with the desired quality of contemplation. Thus, Photios presents us with a foil to the perceptual processes evidenced in his other homily, allowing us to understand precisely the important problem of truth and verisimilitude in perception.

In these passages the familiar and expected language of tactile apprehension is wholly missing from the perceptual processes. If gazing upon an icon or listening to the Gospels is characterized by a clearly apprehensible and confirmable perception that results in a Byzantine equivalent to the

[25] Photios, *Homiliae*, 10, ed. Laourda, 102; modified trans. Mango, 187.

Stoics' *katalēptikē phantasia*, denoted by Photios with the term *katalepsis*,[26] then certainly the experience of the Pharos borders on *akatalēptikē phantasia*. As Diogenes Laertius defines it, this non-apprehending imagination either "does not proceed from any real object, or, if it does, fails to agree with the reality itself, not being clear or distinct" (ἀκατάληπτον δὲ τὴν μὴ ἀπὸ ὑπάρχοντος, ἢ ἀπὸ ὑπάρχοντος μέν, μὴ κατ' αὐτὸ δὲ τὸ ὑπάρχον· τὴν μὴ τρανῆ μηδὲ ἔκτυπον). A similar language of κατάληπτον/ἀκατάληπτον, as mentioned previously, permeates Photios's summary of Aenesidemus's *Pyrrhonian Discourses* in the *Bibliotheca*. The failure to agree with reality is precisely what has occurred in Photios's perceptual model when he says that the viewer "imagines" (φαντάζεται) that their personal experience has been transferred to the chapel. At times, such deceits may be caused or sustained through magic. Take, for example, an account in the *Life of St. Leo of Catania* where women mistook a road for a river, and so lifted their skirts and exposed themselves as they crossed it.[27] Such mistakes illustrate the early Christian discomfort with *phantasia*. In this narrative, while the mirage was enabled through the malicious deeds of the magician Heliodorus, the account alludes to the conceivable mistaking of a roadway, composed of stone slabs with their trodden polish and shine in the sunlight, giving the impression of a coursing stream. This mental image or fantasy is derived from what is present, but fails to agree with actual reality. It is for this reason that the workings of the imagination and its *phantasiai* must always be judged and assessed by the higher faculties of thought, be it *doxa*, *dianoia*, *nous*, or all three, before one can properly assent to their truth.

In order to convey the deceit of perception in his homily, Photios is careful to translate this experience of the *akatalēptikē phantasia* into a rhetorical conceit for his description of the Pharos Chapel's sensual experience. The language of perception is one where sensation comes upon the viewer without their consent, almost abrasively as the flashing of spinning lights impinges upon the eyes. Immediately before Photios voices his concern with the architect's work, he tellingly alludes to Democritus's theory of intromission, saying:

> Δημόκριτος εἶπεν ἄν, οἶμαι, τὴν τοῦ ἐδάφους λεπτουργίαν ἐνιδὼν καὶ ταύτῃ τεκμηρίῳ χρώμενος, μὴ ἂν πόρρω εἶναι τοῦ μὴ οὐχὶ καὶ τὰς ἀτόμους αὐτοῦ καὶ ὑπ' ὄψιν πιπτούσας ἀνευρῆσθαι.

[26] On *katalepsis* in the commentators, see Todd, "Lexicographical Notes," 209–210.

[27] Anonymous, *Vita Leonis Cataniae*, 14.1–9, 17.15–18, ed. Alexander G. Alexakis and trans. Susan Wessel, *The Greek Life of St. Leo Bishop of Catania* (Brussels: Société des Bollandistes, 2011), 156–159, 160–161.

> Democritus would have said, I think, seeing the fine workmanship of the pavement and proclaiming this [to be] proof, that his atoms were not far from being discovered falling upon sight.[28]

Perception here is not to be characterized by the disciplined and procedural states of subsequent forms of grasping, apprehending, and comprehending definite perceptible objects. Instead, it impinges upon the sight (ὑπ' ὄψιν πιπτούσας) within the context of a polymorphous dazzle of rushed mental images. For our purposes, this text with Photios's knowledgeable allusion to the theory of atomic intromission certainly offers an insight into the endurance of classical theories of vision in the period. The text also aptly plays with the notion that the functions of sensation and perception break with the proper and expected standards, so that one might even say that Democritus's atoms were present in the fine filigree. While this is not to say that Photios does not admire or enjoy the chapel's aesthetic experience, it is clear that he sees it as posing a challenge to proper perception. This vision is caustic, lacking the comprehensive tactility of the mind's grasp through disciplined acts of perception in exchange for the impinging strikes (πιπτούσας) of external images.

Ironically, to convey this overwhelming experience, which not only seems to violate the norm but also transgresses the limits of perception, Photios rhetorically constructs a model of visual sensation that is akin to touch. That is to say, he resorts to ancient theories of intromission as the most apt metaphor to convey this bewildering experience. As Aetios's popular summary recounted, "Democritus and Epicurus suppose that sight is caused by the insinuation of little images (εἰδώλων) into the visive organ, and by the entrance of certain rays which return to the eye after striking upon the object."[29] For the Byzantines, following the commentators, it was often the intromissionists, such as Hipparchus, who subscribed to a tactile theory of optical sensation, whereby external images or rays from the object touched the visive organs. In Aetios's summary of Democritus's and Epicurus's theories we see this precisely. Images coming into the eyes, or the shooting out of rays that touched the object and returned to the eye, are both intromissive models that approached tactility.

[28] Photios, *Homiliae*, 10, ed. Laourdas, 102; modified trans. Mango, 187.

[29] "Δημόκριτος Ἐπίκουρος κατ' εἰδώλων εἰσκρίσεις ᾤοντο τὸ ὁρατικὸν συμβαίνειν, καὶ κατά τινων ἀκτίνων ἔκκρισιν μετὰ τὴν πρὸς τὸ ὑποκείμενον ἔνστασιν πάλιν ὑποστρεφουσῶν πρὸς τὴν ὄψιν." Pseudo-Plutarch (Aetios of Antioch), *De placita philosophorum*, IV.13, ed. Mau, 123; trans. Goodwin, 168.

Therefore, when Photios suggests that Democritus "would have said, I think" (εἶπεν ἄν, οἶμαι) that the proof for intromission lay within the decorations of the Pharos, he was precisely alluding to a non-normative theory of vision so as to capture the ways in which the Pharos exceeded the established limits and bounds of common perception. Photios does not concede to Democritus's theory, making clear to mark this as a hypothetical musing, but he deploys it to demonstrate how the effect of the chapel on the cognitive side of perception resulted in an image or imagination that did not accord with physical reality. Hence, the seemingly haptic language of grasping and apprehension used for cognition suggests forms of sensation that were not really seen as sensual variants of touch. Rather, when perception became discordant with reality – in this case, a non-graspable fantasy (ἀκαταληπτική φαντασία) –it took on an explanation that aligned it with touch, to illustrate the fallacy and the overstepping of established norms. By stressing the abrasive flashes of sensual stimuli on the viewer and citing the theories of Democritus, Photios was able to capture the complexity of the imagination in processing sensual data, offering us a glimpse into an instance where this proper process goes awry and thus must be conveyed through the language of intromissive impingement.

Before entering the sanctuary of the Pharos Chapel, the viewer adheres to the expected practices of perception, gazing upon the white-marble walls of the atrium, which are so neatly fitted together that their joints appear like rulings on stone rather than seams. There, Photios describes:

> Προπύλαια μὲν γὰρ τῷ ναῷ περικαλλῆ διεσκεύασται· καὶ γὰρ μαρμάρων διαλεύκων πλάκες λαμπρόν τι καὶ χάριεν ἀποστίλβουσαι, ὅλην ἐναπολαβοῦσαι τὴν πρόσοψιν καὶ τὴν πρὸς ἀλλήλας θέσιν καὶ τὴν τῶν περάτων συνάφειαν τῷ ὁμαλῷ καὶ λείῳ καὶ τῷ προσηρμόσθαι λίαν ἀποκρύψασαι εἰς ἑνὸς λίθου συνέχειαν, γραμμαῖς εὐθείαις ὥσπερ ἐπιγεγραμμένου, θαῦμα καινὸν ὁραθῆναι καὶ ἥδιστον, τὴν τῶν ὁρώντων φαντασίαν ὑπάγουσιν· ἐξ οὗπερ τὰς ὄψεις συνέχουσαι καὶ πρὸς ἑαυτὰς ἐπιστρέφουσαι οὐκ ἐθέλειν ποιοῦσι τὸν θεατὴν μεταχωρῆσαι πρὸς τὰ ἐνδότερα, ἀλλ᾽ ἐν αὐτῷ τῷ προτεμενίσματι τοῦ καλοῦ θεάματος ὁ προσιὼν ἐμπιπλάμενος καὶ τοῖς ὁρωμένοις ἐρείδων τὰ ὄμματα, ὥσπερ τις ἐρριζωμένος τῷ θαύματι ἕστηκεν.
>
> The entrance to the shrine is very beautifully composed: for the slabs of very-white marble, bright and graceful, shine brightly, and encompass the whole facade and are fitted against each other [so that] the union of their ends by their evenness, smoothness, and very close attachment conceals the connections as if it were a single stone with straight lines inscribed on it, a remarkable wonder and most-pleasant to see; [thus] they lead on the beholder's imagination: from where, restraining sight and turning it back

> toward themselves, they make the spectator not desire to go into the things inside, but in it, the one cleaving to this vestibule takes their fill of the beautiful spectacle, pressing their eyes to the sights, they stand-firm just as someone fixed with wonder.[30]

These marble revetments "lead on the beholder's imagination" (τὴν τῶν ὁρώντων φαντασίαν ὑπάγουσιν) to believe that they are crafted from a single piece of marble. Yet the word "lead on" or "suggest" (ὑπάγουσιν) qualifies the deceit of the imagination here, acknowledging the effective trickery of the marbles while still acknowledging that the viewer can perceive the laudable deceit. In his translation, Cyril Mango chose to stress the full trickery of the marble revetments, seeing the imagination as having been led to believe in the work's seamlessness.[31] Yet the text argues instead that it is not just the act of trickery but also its detection by the viewer that captures the sustained attention of the imagination.

The single stone comes from the imagination's conception. Because the intellect judges that this is not true, the marble slabs restrain (συνέχουσαι) sights and turn them back toward themselves (πρὸς ἑαυτὰς ἐπιστρέφουσαι), causing the viewer to remain there and subject them to closer examination until they can satiate their curiosity and come to understand what is occurring. This perception is steadfast: the viewer is described as being petrified, planted like a tree, while carefully observing the marble revetments. The spectator does not wish to move, a sharp contrast to what will be their ecstatic and ceaseless motion within the chapel. Moreover, this language of self-reflexivity seems to suggest that there is also a cognitive action of self-judgment and discernment as the imagination critically reevaluates sensory imprints against what it knows and suspects to be true. This process is alluded to by durational language of contemplation: the spectator is the one who "cleaves" to the beautiful sight (τοῦ καλοῦ θεάματος ὁ προσιὼν), and takes their fill (ἐμπιπλάμενος) with eyes "fixed" upon or, rather, "pressed" (ἐρείδων) to the sights. If we characterized the perception of the Pharos's interior as dominated by an *akataleptikē phantasia*, which evidenced a disjuncture between the perceived reality and

30 Photios, *Homiliae*, 10, ed. Laourdas, 100–101.

31 "The atrium of the church is splendidly fashioned: for slabs of white marble, gleaming bright and cheerful, occupy the whole façade, and by their evenness and smoothness and close fitting they conceal the setting of one to another and the juncture of their edges, so that they suggest to the beholder's imagination the continuousness of a single piece of stone with, as it were, straight lines ruled on it – a new miracle and a joy to see. Wherefore, arresting and turning towards themselves the spectator's gaze, they make him unwilling to move further in; but taking his fill of the fair spectacle in the very atrium, and fixing his eyes on the sight before him, the visitor stands as if rooted to the ground with wonder." Photios, *Homiliae*, 10, trans. Mango, 185–186.

the world, then certainly the perception of its atrium is regulated by a *katalēptikē phantasia*, where the trickery of sight and its attempts to lead on the imagination are undermined by the viewer's desire to remain and contemplate the scene and discern what is truly being seen. The architect lamentably hampers this process in the interior of the church. That is, Photios's seemingly tongue-in-cheek complaint regarding the architect's erring should be taken more seriously. He means to make us think about the struggles of the faculties of discernment and assent that fall late in the processes of sensation, imagination, and perception.

CHAPTER 8

"Then It Has Effortlessly. . ." *Judgment and Assent*

The final stage in Photios's process of perception entails the depositing of the imagination's images produced from sensation into the storehouses of memory, as Photios states in his summary: "for the concentration of unfailing knowledge" (πρὸς ἐπιστήμης ἀπλανεστάτης συνάθροισιν) [**m4**]. This is summarized in his rhetorical questions with the declaration "Then it has effortlessly transmitted the forms to the memory" (τοὺς τύπους ἀκόπως ἐν τῇ μνήμῃ παρεπέμψατο) [**n3–4**]. At this point, the journey of perception has been completed. Yet, a critical stage of perception, found in other sources, has been elided here, namely, the judgment about the veracity of these imprints, assents to their truths. This notion is crucial to the Stoic articulation of the *katalēptikē phantasia*. And it appears earlier in the Photios quotation in the case of hearing, where he states, "Then, after judging it with sober attention, he deposits it in his memory" (νηφούσῃ μελέτῃ τὸ κριθὲν τῇ μνήμῃ ἐναπέθετο) [**k3–4**]. The "sober attention" (νηφούσῃ μελέτῃ) of hearing's judgment is contrasted to sight's curt "effortlessly" (ἀκόπως). So the difference deserves our attention, particularly after we become intimately aware of the potential delusions and derailments that are possible with a fantastical operation of the imagination.

The question of truth and perception is embedded deeply in the logic describing the imprints and cognitive operations of the φαντασία. As Michael Frede has elucidated, in pursuing an epistemology geared toward enabling the attainability of truth, the Stoics argued that nature has provided us with, in Frede's words, "clear and distinct impressions" through the actions of the *katalēptikē phantasia*, which inherently cannot diverge from the truth and thus enable an infallible approach toward truth.[1] For this reason, within the Stoic context, *phantasia* is more accurately translated as an impression or presentation, where impression renders

[1] Frede, "Stoics and Skeptics," 65–93, esp. 66.

Zeno's conception of *phantasia* as an imprint (τύπος), while on the other hand presentation speaks best to Chrysippos's articulation of it as an alteration or wave in the soul.[2] Despite their differences, however, what *phantasia* communicates is that, as Diogenes Laertius recounts, "the criterion of truth (Κριτήριον... τῆς ἀληθείας) is the apprehending impression, that is to say [the impression] representing what is."[3] *Phantasia* is for the Stoics synonymous with the sensory impulse itself rather than with the mental faculty that processes sensory impulses to generate an image. That is why it is a cognitive presentation or imprint, rather than the byproduct of an action or mental faculty.

The Stoics debated precisely to what extent the impression could represent the object fully and truly, and a great deal of disparity remained within their positions, which Julia Annas has carefully parsed from sources that are difficult to interpret.[4] Common to every Stoic position, as best we can tell, is that when the *phantasia* manifests an existing external object in a manner in which it cannot be confused with another object, existing or non-existent, it is truthful. This is to say, when the *phantasia* is properly *katalēptikē*, it is a true and accurate impression of the sense object. Charged with truth value, *phantasia* for the Stoics is a presentation of truth warranting assent and allowing for intellection and wisdom to arise from it.

In some regards, the *katalēptikē phantasia* is a tautological concept. It is the criterion of truth because it is true whenever it is true. Sextus Empiricus admits the challenges and discrepancies in the articulation of *phantasia* in Stoic thought,[5] writing:

> Ἀλλὰ γὰρ οἱ μὲν ἀρχαιότεροι τῶν Στωικῶν κριτήριόν φασιν εἶναι τῆς ἀληθείας τὴν καταληπτικὴν ταύτην φαντασίαν, οἱ δὲ νεώτεροι προσετίθεσαν καὶ τὸ μηδὲν ἔχουσαν ἔνστημα. ἔσθ᾽ ὅτε γὰρ καταληπτικὴ μὲν προσπίπτει φαντασία, ἄπιστος δὲ διὰ τὴν ἔξωθεν περίστασιν... ἐνθένδε οὐχ ἁπλῶς κριτήριον γίνεται τῆς ἀληθείας ἡ καταληπτικὴ φαντασία, ἀλλ᾽ ὅταν μηδὲν ἔνστημα ἔχῃ. αὕτη γὰρ ἐναργὴς οὖσα καὶ πληκτικὴ μόνον οὐχὶ τῶν τριχῶν, φασί, λαμβάνεται, κατασπῶσα ἡμᾶς εἰς συγκατάθεσιν, καὶ ἄλλου μηδενὸς δεομένη εἰς τὸ τοιαύτη προσπίπτειν ἢ εἰς τὸ τὴν πρὸς τὰς ἄλλας διαφορὰν ὑποβάλλειν.

2 Karl Thein, "Gods and Painters: Philostratus the Elder, Stoic Phantasia, and the Strategy of Describing," *Ramus* 31 (2002): 136–145, esp. 144 n 9.

3 "Κριτήριον δὲ τῆς ἀληθείας φασὶ τυγχάνειν τὴν καταληπτικὴν φαντασίαν, τουτέστι τὴν ἀπὸ ὑπάρχοντος." Diogenes Laertius, *Vitae philosophorum*, VII.1.154, ed. and trans. Hicks, 162–163.

4 Annas, "Truth and Knowledge," 84–104.

5 Sextus Empiricus, *Adversus mathematicos*, VII.241, ed. and trans. Bury, 130–131.

> While the older Stoics say that this cognitive impression is the criterion of truth, the later ones added the words "one that has no impediment." For there are times when a cognitive impression occurs, but it is incredible owing to the external circumstances... Therefore the cognitive impression is not the criterion of truth unconditionally, but when it has no impediment. This impression, being self-evident and striking, all but seizes us by the hair, they say, and pulls us to assent, needing nothing else to achieve this effect or to establish its difference from other impressions.[6]

Sextus Empiricus states that the *phantasia*, provided it is free from impediments (μηδὲν ἔνστημα), operates as the criterion of truth by virtue of the fact that ontologically the imprint itself is self-evident and has a striking nature (ἐναργὴς οὖσα καὶ πληκτική). Thus, the adjectival *katalēptikē* stands as a reassurance of the impression's verisimilitude and truth, naturally dragging the person to assent. Hence, *phantasia* in Stoic thought is shown not to be an intermediary in perception *per se*, though it certainly operates in that sense when it comes to thinking, memory, or other cognitive acts that deploy its imprints.

Therefore, it is not surprising that Stoic theories of vision articulated a closeness between the soul and the external object, akin to the touch of a walking stick, criticized by Galen, whereby the imprints are transmitted directly and with veracity to the perceiver.[7] Stoic visuality desired a haptic theory of sight that enabled the definition of the *phantasia* as an imprint, following Zeno, to transmit form seamlessly and without impediment from the exterior object into the mind.[8] As discussed in Chapter 2, Galen criticizes this explanation of the optic *pneuma* for the fallacy that since touch cannot perceive colors, sight cannot be described as a species of touch. Yet, Galen's response makes evident that it would be just as much of a fallacy to think reductively that the Stoics might have perceived sight as touch. Rather, the touch metaphor associated with sight allowed one to articulate the immediate and non-mediated perception upon which Stoic epistemology was founded. This approach comports with the development from later Stoic thought described by Sextus Empiricus where the veracity of the *katalēptikē phantasia* required a lack of impediment. Even then, in Sextus Empiricus's account, the qualification of the *phantasia* goes only so

6 Sextus Empiricus, *Adversus mathematicos*, VII.253–254, 257–258, ed. J. Mau and H. Mutschmann, *Sexti Empirici opera*, vol. II (Leipzig: Teubner, 1914), 59–60; trans. Long and Sedley, *The Hellenistic Philosophers*, 246.

7 See Galen, *De placitis Hippocratis et Platonis*, VII.7:19–20, ed. and trans. De Lacy, 474–475.

8 On Stoic visuality and *phantasia*, see Simon Goldhill, "Refracting Classical Vision: Changing Cultures of Viewing" in T. Brennan and M. Jay (eds.), *Vision in Context: Historical and Contemporary Perspectives on Sight* (New York: Routledge, 1996), 17–28.

far as to recognize that its veracity must be confirmed, that it must be confirmed that it is in fact *katalēptikē*.

Over the course of late antiquity and Byzantium, *phantasia* is redefined in Neoplatonic thought as imagination, often through its teachings via Aristotle and the later commentators, whose thinking was inflected by the language of the Stoics. Those commentators nevertheless securely set the imagination's function to be one of mediation, after sensation.[9] Even the work of Galen fell subject to similar refractions in commentaries, such as in the eleventh-century commentary on the *Parva Naturalia* by Michael of Ephesus.[10] Thus, in the late antique and Byzantine period the imagination emerged from a sensory impulse to become a mental faculty that required the robust discretion and attention of other adjoining cognitive faculties such as preconception (πρόληψις and ὑπόληψις), belief (δόξα), thought (διάνοια), and memory (μνήμη). One could no longer just accept or confirm the veracity of the *phantasia*. Its imprints necessitated corroboration and consideration by reason. As discussed previously, early Christian writers struggled with the potential deceits and fallacies of the imagination's images, but their concern was precisely in its unchecked operation.

John Scholastikos, in his scholia on Pseudo-Dionysios's *On Divine Names*, expands on the problem of a hasty assent to the images of the imagination, cited already. John clarifies, paraphrasing Pseudo-Dionysios:

> Ὁ θυμὸς ὁ ἄλογος καθ' ἑαυτὸν καὶ ἡ ἄνους ἐπιθυμία καθ' ἑαυτὴν καὶ ἡ προπετὴς φαντασία οὐ πάντως καὶ ἐπὶ πάντων κακὸν καθ' ἑαυτὰ σκοπούμενα... οὕτως ἔστι καὶ φαντασία οὐ προπετής, καθ' ἣν ἐφαπτόμεθα τῶν νοητῶν τῷ νῷ ἡμῶν τοῦτο ποιοῦντες.
>
> Anger, which is in itself irrational, as well as desire, which is unreasonable, and imagination which is hasty [in assenting], [all these faculties], considered in themselves, are not completely evil in every way... Thus, the imagination according to which we grasp the intelligibles – and we do that with our mind – is not hasty.[11]

9 On the later articulations of the φαντασία, see Watson, *Phantasia in Classical Thought*, esp. 59–161; Blumenthal, "Neoplatonic Interpretations," 242–257; Sheppard, *The Poetics of Phantasia*, esp. 47–70. See also Marcia Colish, *The Stoic Tradition from Antiquity to the Early Middle Ages*, vol. I (Leiden: Brill, 1985), esp. 51–53.

10 Todd, "Galenic Medical Ideas," 117–134.

11 John Scholastikos (= John Scythopolis), *Prologus et scholia in Dionysii Areopagitae librum De divinis nominibus*, 292C (*DN* 171,17), ed. B. R. Suchla, *Prologus et scholia in Dionysii Areopagitae librum "De divinis nominibus" cum additamentis interpretum aliorum*, Corpus Dionysiacum IV.1, Patristische Texte und Studien 62 (Berlin: De Gruyter, 2011), 289; modified trans. Aglae Pizzone, "Toward a Self-Determined and Emotional Gaze: Agathias and the Icon of the Archangel Michael" in Sergei Mariev and Wiebke-Marie Stock (eds.), *Aesthetics and Theurgy in Byzantium* (Berlin: De Gruyter, 2013), 75–103.

This sentiment, which seeks to articulate the dangers of the hasty imagination (προπετὴς φαντασία) and the virtues of its opposite, is reiterated by Maximos the Confessor and again in the twelfth century by Isaac Komnenos, following Pseudo-Dionysios.[12] Pseudo-Dionysios's comments appear to be influenced by Plotinos's struggle with the haste of the imagination, recounted in his treatise on evil:

> Εἰ δέ τις ἀσθένειαν ψυχῆς τὴν κακίαν λέγοι – εὐπαθῆ γοῦν καὶ εὐκίνητον εἶναι τὴν κακὴν ἀπὸ παντὸς εἰς ἅπαν κακὸν φερομένην, εὐκίνητον μὲν εἰς ἐπιθυμίας, εὐερέθιστον δὲ εἰς ὀργάς, προπετῆ δὲ εἰς συγκαταθέσεις, καὶ ταῖς ἀμυδραῖς φαντασίαις εἴκουσαν ῥᾳδίως, οἷα τὰ ἀσθενέστατα τῶν τέχνῃ ἢ φύσει πεποιημένων, ἃ ῥᾳδίαν ἔχει ὑπό τε πνευμάτων ὑπό τε εἱλήσεων τὴν φθοράν – ἄξιον ἂν εἴη ζητεῖν, τίς καὶ πόθεν ἡ ἀσθένεια τῇ ψυχῇ.
>
> But if someone says that vice is a weakness of the soul – pointing out that the bad soul is easily affected and easily stirred, carried about from one evil to another, easily stirred to lust, easily roused to anger, hasty in its assents, giving way freely to confused imaginations, like the weakest of the products of art or nature, which the winds or the sun's heat so easily destroy – it will be worth enquiring what this weakness is and where the soul gets it from.[13]

Here, we observe again the pairing of a hasty assent (προπετῆ. . . εἰς συγκαταθέσεις) with indistinct imaginations (ἀμυδραῖς φαντασίαις), suggesting that the assent to imprints is motivated by evil – a direct retort to the Stoics' criterion of "clear and distinct impressions," which Frede has eloquently detailed. Like the hasty, vertigo-inducing imagination of Photios's viewer in the Pharos Chapel, perception may be deceived into creating false images through a hasty assent. While Sextus Empiricus states that the cognitive impression "all but seizes us by the hair, they say, and pulls us to assent," it is precisely this potential for rushed perception that troubled Christian thinkers, even if the Stoics insisted that these images had to be generated from distinct impressions of the objects. For example, Sextus Empiricus says that a person wishing to perceive something keenly, when receiving only an "indistinct impression" (ἀμυδρὰν. . . φαντασίαν), must strain their sight and approach the object closely, "until they are able to draw in a clear and striking impression of what they are judging"

[12] See Maximos the Confessor, *Capita de caritate*, III.5, ed. A. Ceresa-Gastaldo, *Capitoli sulla carita* (Rome: Editrice Studium, 1963), 146; Isaac Komnenos, *De malorum subsistentia*, 9.1, ed. James John Rizzo, *Περὶ τῆς τῶν κακῶν ὑποστάσεως (De malorum subsistentia)*, Beiträge zur klassischen Philologie 42 (Meisenheim am Glan: Hain, 1971), 30.

[13] Plotinos, *Enneads*, I.8.14, ed. and trans. Armstrong, 310–311.

(μέχρις ἂν τρανὴν καὶ πληκτικὴν σπάσῃ τοῦ κρινομένου φαντασίαν).[14] Thus, despite the Stoics' assertion about the veracity of the cognitive impressions, they also envisioned a hermeneutic cycle where people could judge that they have adequately perceived the object at hand. Yet, this process appears to be simultaneous with perception – it was not a cognitive process.

In the Stoics, especially early on, the model of the criterion of truth sets *phantasia* as being simultaneous with the act of sensation and as being thoroughly unmediated. Francis Sandbach has reasonably suggested that the Stoics treated *phantasia* as simultaneous with sensation in that it requires no further check, being immediately acceptable.[15] For late antique and Byzantine commentators on Aristotle and Galen, however, as discussed previously, even if the role of the imagination appears to be simultaneous to sensation, as Themistios pondered, their distinctness is consistently stressed and the separation between them grows over time. Imagination mediated intellection and sensation, but with time we encounter other faculties introduced to explain the approval or assent (συγκατάθεσις) that *katalēptikē phantasia* so readily could achieve in Stoic epistemology.

In Photios's Homily 17, what is lacking in the description of sight perception is the "belief-making assumption" (δοξαστικὴ ὑπόληψις) that occurs between *antilepsis* and *phantasia* in Porphyry (Table 2, no. 2).[16] As Porphyry describes it, "belief-making assumption receives what has been introduced, names it and describes it through words in the soul as if upon some writing tablet existing in it" (see above). Photios's omission of this step is noticeable, given that for earlier Christian writers, this stage would seem to allow one to slow down the process of assenting to the cognitive impressions instigated by sensation. But it is not missing completely from Photios's homily. He has this preemptory stage in mind when he writes earlier, "so a form through sight is imprinted upon the tablets of the soul, giving to those whose apprehension (πρόληψις) is not soiled by wicked doctrines a representation of knowledge concordant with piety" [**d2**]. Here the term *prolepsis*, translatable as either apprehension or preconception, operates similarly to "assumption" (ὑπόληψις) in Porphyry. In fact, both use the same metaphor of tablets to explain the impressions on the soul. Both *prolepsis* and *hypolepsis* are closely connected to the notion of belief in

[14] Sextus Empiricus, *Adversus mathematicos*, VII.258, ed. Mau and Mutschmann, 60; trans. Long and Sedley, 246.

[15] Francis H. Sandbach, "Phantasia Katalēptikē" in A. A. Long (ed.), *The Problems in Stoicism* (London: Athlone, 1971), 9–21, esp. 19.

[16] On *doxastike hypolepsis* in Porphyry, see Chase, "Porphyry on the Cognitive Process," 386–391.

the *Souda*. Not only does the *Souda* quote Alexander of Aphrodisias's commentary on Aristotle's *Topics* to assert that "belief is a form of assumption" (εἶδος γὰρ ὑπολήψεως ἡ δόξα),[17] but the entry on *prolepsis* also offers *doxa* and *hypolepsis* as synonyms:

> Περὶ προλήψεως: διαφορά ἐστιν ἀληθείας καὶ συνηθείας: ἡ μὲν γὰρ ἀλήθεια γνησίως ζητουμένη πάντως εὑρίσκεται. τὸ δὲ ἔθος, ὁποῖον ἂν παραληφθείη, εἴτε ἀληθὲς εἴθε ψευδές, ἀκρίτως ὑφ' ἑαυτοῦ κρατύνεται: καὶ οὔτε ἀληθεῖ ὄντι αὐτῷ ἥδεται ὁ παραλαβὼν οὔτε ψευδεῖ ἄχθεται. οὐ γὰρ κρίσει, ἀλλὰ προλήψει ὁ τοιοῦτος πεπίστευκε, γνώμῃ τῶν πρὸ αὐτοῦ ἐπ' ἀδήλῳ τύχῃ τὴν ἰδίαν ἐπιδεδωκὼς ἐλπίδα. καὶ οὐκ ἔστι ῥᾳδίως ἀποδύσασθαι τὴν πάτριον περιβολήν, κἂν πάνυ αὐτῷ δείκνυται λυπρὰ καὶ καταγέλαστος οὖσα. ὁ δὲ Χρύσιππος ἐν τῷ ιβ' φυσικῷ λόγῳ δύο φησὶν εἶναι κριτήρια, αἴσθησιν, γνῶσιν καὶ πρόληψιν. ἔστι δὲ πρόληψις ἔννοια φυσικὴ τῶν καθόλου. Ἐπίκουρος δὲ ὁ φιλόσοφος πρόληψιν λέγει, οἱονεὶ κατάληψιν ἢ δόξαν ὀρθὴν ἢ ἔννοιαν ἢ καθολικὴν νόησιν ἐναποκειμένην, τουτέστι μνήμην τοῦ πολλάκις ἔξωθεν φανέντος. οἷον τὸ τοιοῦτόν ἐστιν ἄνθρωπος: ἅμα γὰρ τῷ ῥηθῆναι ἄνθρωπος, εὐθὺς κατὰ πρόληψιν καὶ ὁ τύπος αὐτοῦ νοεῖται, προηγουμένων τῶν αἰσθήσεων. παντὶ οὖν ὀνόματι τὸ πρώτως ἐπιτετευγμένον ἐναργές ἐστι: καὶ οὐκ ἂν ἐζητήσαμεν τὸ ζητούμενον, εἰ μὴ πρότερον ἐγνώκειμεν αὐτό. οἷον τὸ πόρρω ἑστώς, βοῦς ἐστιν ἢ ἵππος; δεῖ γὰρ κατὰ πρόληψιν ἐγνωκέναι ποτὲ ἵππου καὶ βοὸς μορφήν. οὐδ' ἂν οὖν ὠνομάσαμέν τι, μὴ πρότερον αὐτοῦ κατὰ πρόληψιν τὸν τύπον μαθόντες. Ἐναργεῖς οὖν εἰσὶν αἱ προλήψεις: καὶ τὸ δοξαστὸν ἀπὸ προτέρου τινὸς ἐναργοῦς ἤρτηται: ἐφ' ὃ ἀναφέροντες λέγομεν, οἷον πόθεν ἴσμεν εἰ τοῦτό ἐστιν ἄνθρωπος. τὴν δὲ δόξαν καὶ ὑπόληψιν λέγουσι.
>
> On prolepsis. There is a difference between truth and habit. For the truth is discovered by making a genuine and complete search. But habituation, which has been admitted, whether it is true or false, is uncritically confirmed by itself. But the person admitting such a habituation is neither delighted at what is true nor grieved by falsehood. For such a person has not trusted his judgment but his preconception, and has bestowed a peculiar expectation on the understanding of what is before him, making use of a non-evident fortune. And it is impossible easily to get rid of the investiture belonging to one's father, even though such investiture appears to him as being extremely distressful and ridiculous. Chrysippos holds in the twelfth book of his *Physics* that the two criteria are sense-perception, knowledge, and preconception. Furthermore, he says that preconception is a natural conception of universals. The philosopher Epicurus says that preconception is as if it were a comprehension, or correct opinion, or a conception that is a

[17] Anonymous, *Souda*, "ὑπόληψις," upsilon 537, ed. and trans. Oliver Phillips, *Souda-On-Line*, www.stoa.org/sol-entries/upsilon/537.

> universal notion stored in us, i.e. a memory of what often appears to us from the external world. For instance, "such a thing is a man"; for as soon as the word "man" is pronounced its distinguishing feature is immediately conceived of due to preconceptions, since the senses serve as guidance. Therefore, what primarily falls upon each name is something evident. And we would not have investigated what is an object of investigation if we had not known it first. For example, what is standing in front of me, is it a cow or a horse? For by means of preconception we should have known at some time the shape (τύπον) of horse or cow. We would not have given a name to anything if we had not first learned its distinguishing feature by means of preconception. Consequently, preconceptions are evident. And what is an object of opinion depends upon something prior and evident, by referring to which we ask: "how can we know if this thing is a man?" They also call this opinion (δόξα) and supposition.[18]

This entry is one of the more eloquent in the *Souda*, giving an insightful survey of previous writers' views, and likewise referring the reader to the relevant entries on *doxa* and *hypolepsis*. *Prolepsis* has a semiological force in that it allows the perceiving mind to readily associate what is being sensed with a certain discernment about its identity and its effects. Thus it operates as a form of instant comprehension based on the preconceptions and assumptions of the viewer, associated with what the Stoics at times refer to as "common notions" (κοιναὶ ἔννοιαι).[19] As Henry Dyson has argued, *prolepsis* operates as a criterion of truth in the process of perception, intimately tied to the eventual assent to the cognitive imprints.[20] In the second-century *Didaskalikos* of Alcinous, opinion (δόξα) is defined quite similarly as a "combination of memory and sensation" (Δόξα δέ ἐστι συμπλοκὴ μνήμης καὶ αἰσθήσεως), where the preexisting memory of a similar encounter allows the perceiver to immediately distinguish between, say, a man, a horse, or a fire (quite like the *Souda* quotation above).[21] These various functions are being connected with the judgment and discernment of sensory inputs.

As Photios writes, those whose preconceptions are not spoiled by "wicked doctrines" (πονηρῶν δογμάτων) [**d2**] are able to engrave proper

[18] Anonymous, *Souda*, "πρόληψις," pi 2495, ed. and modified trans. Oliver Phillips, *Souda-On-Line*, www.stoa.org/sol-entries/pi/2495.

[19] On preconceptions and common notions, see Henry Dyson, *Prolepsis and Ennoia in the Early Stoa* (Berlin: De Gruyter, 2009). See also Robert B. Todd, "The Stoic Common Notions: A Re-examination and Reinterpretation," *Symbolae Osloenses* 48:1 (1973): 47–75; Werner Theobald, *Hypolepsis: Mythische Spuren bei Aristoteles* (Sankt Augustin, Germany: Academia Verlag, 1999); Frede, "Stoics and Skeptics," esp. 68–70; Ioppolo, "Presentation and Assent," 441–449.

[20] Dyson, *Prolepsis and Ennoia*, 23–47.

[21] Alcinous, *Didaskalikos*, IV.5, ed. Pierre Louis, *Épitomé* (Paris: Les Belles Lettres, 1945), 15.

teachings onto the soul. Wicked doctrines would alter the righteous preconceptions of the subject, corrupting the way the information is understood and judged, and hence potentially leading one to false understandings and to improper assent. What is at stake is assent, given that Photios has been led to consider preconceptions in order to argue that sight, and not hearing, compels spectators "to unhesitating assent" (εἰς ἀπροφάσιστον συγκατάθεσιν) [**b2**]. In Photios's homily, this assent should not be read as hasty, but it certainly does occur readily and without excuse. The privative alpha in ἀπροφάσιστον suggests that it does not allow for pretexts, for any varied plausibility, or for, as compared to εὐπροφάσιστος, disguise or evasion.[22] Hence, Photios's "unhesitating assent" assures one of the veracity of the cognitive impression, which here likewise stresses the superiority of sight over hearing, made more explicit later in the homily ([**k**] versus [**n**]).

Across several of his homilies, Photios stresses the importance of assent in the context of divine comprehension and the imagination. In Homily 11 on Holy Saturday, for example, Photios describes the disciples' impetus to become heralds of the things they witnessed and suggests that it was "indeed the unerring apprehension of things seen, the manifest and indubitable character of the resurrection" (Ἡ μέντοι γε τῶν ὁραθέντων ἀψευδεστάτη κατάληψις καὶ τὸ ἐναργὲς καὶ ἀναμφίβολον τῆς ἐγέρσεως).[23] This "unerring apprehension" (ἀψευδεστάτη κατάληψις) reflects Photios's perceptual language to describe the acquisition of knowledge through sense perception, which results in comprehension (κατάληψις). A comparable instance occurs in Homily 8 on Palm Sunday, describing the virtuous "apprehension of truth" (ἀληθείας κατάληψις).[24] However, the accurate apprehension of truth is eloquently challenged in Homily 4 on the Departure of the Russians, after their attack on the city of Constantinople in 890. Here, Photios recounts the fear of the city's inhabitants upon thinking that the city had been taken:

> Τὸ γὰρ τοῦ συνενεχθέντος ἀπροσδόκητον καὶ τῆς ἐφόδου τὸ ἀνέλπιστον τοιαῦτα φαντάζεσθαι καὶ ἀκούειν ὥσπερ ἀνέπειθεν ἅπαντας, εἰωθὸς μάλιστα τὸ πάθημα τοῦτο τοῖς ἀνθρώποις καὶ ἄλλως ἐπιφοιτᾶν· ὃ μὲν γὰρ ἐκτόπως δεδίασιν, ἀκρίτῳ δόξῃ καὶ μὴ παρὸν ἑστάναι νομίζουσιν, οὗ δ' ὑπόνοιαν οὐ προέλαβον, αὐτεξουσίῳ γνώμῃ καὶ φθάσαντος πόρρω πέμπουσιν.

[22] LSJ, s.v. [23] Photios, *Homiliae*, 11, ed. Laourda, 117, trans. Mango, 207.
[24] Photios, *Homiliae*, 8, ed. Laourda, 87; trans. Mango, 159.

> For the unexpectedness of the event and the unlooked-for attack induced, so to speak, everybody to imagine and hear such things – a symptom that is indeed common among men in such cases: for what they fear excessively they will believe without verification to have happened even when it has not; whereas that of which they have had no previous apprehension they will reject by the arbitrary power of their judgment even when it has come upon them.[25]

In this account, the imagination (φαντάζεσθαι) in light of fear appears to become unchecked, and thus runs rampant with beliefs of terrible deeds occurring all around. In this state, where imagination becomes fantasy, thought comes to rest on uncritical belief (ἀκρίτῳ δόξῃ. . . νομίζουσιν) based on one's unchecked assumptions (ὑπόνοιαν οὐ προέλαβον) as to what is occurring. Here, Photios deals with the counterfactual workings of the imagination, whereby it is resolutely not processing sense information but rather comes to believe false images due to the impetus of fear and a failed operation of *doxa*.

This description of the Russian siege contrasts sharply, however, with Photios's Homily on the Annunciation. Early on in the homily, Photios marvels precisely at the bewildering and inapprehensible nature of the Annunciation following Gabriel's greeting, writing:

> ὢ παράδοξον μυστήριον, πάντα νοῦν καὶ πᾶσαν διαδιδρᾶσκον κατάληψιν· μυστήριον πιστευόμενον, οὐ πολυπραγμονούμενον· θαυμαζόμενον, οὐκ ἐρευνώμενον· μυστήριον προσκυνούμενον, ἀλλ' οὐκ ἐξεταζόμενον· ὑμνολογούμενον, οὐκ ἐξιχνιαζόμενον· τιμώμενον, οὐ ζυγοστατούμενον· ποθούμενον, οὐ καταλαμβανόμενον.
>
> O paradoxical mystery, that escapes every mind and every apprehension! A mystery which is believed, not investigated; wondered at, not searched into; a mystery which is worshipped, but not examined; praised in song, not scrutinized; honored, not weighed; longed for, not apprehended.[26]

Like the counterfactual imagination of the Russian siege, this opening contemplates the ways in which the Annunciation exceeds the bounds and operations of cognition and perception. The mystery eludes one's grasp (διαδιδρᾶσκον κατάληψιν) and cannot be apprehended (οὐ καταλαμβανόμενον), and the series of paradoxes demonstrates that it cannot be accessed through traditional forms of investigation, inquiry, examination, scrutiny, and judgment. Yet, this is exactly the same behavior that the virtuous maiden exhibits when the angel comes to her and speaks words of

[25] Photios, *Homiliae*, 4, ed. Laourda, 44; trans. Mango, 101.
[26] Photios, *Homiliae*, 5, ed. Laourda, 54; modified trans. Mango, 114.

divine promise. The Virgin retorts to herself: "His words, however, which give the impression of being those of a suitor, prompt me to refuse assent (συγκατάθεσιν)."[27] The Virgin here evinces her inability to discern whether the angel is a suitor or part of the heavenly host, since his words (ῥήματα) of conception and birth would appear to be those of a suitor. Taken aback, she is beckoned not to assent, lest she be deceived by this false image. Throughout, the homily is filled with sentences that begin with vision – the verb ὁράω, like Ὁρᾷς or Ἰδοῦσα, stressing the Virgin's perceptual confrontation with the Angel and his message.

In the section that follows, Gabriel assures the Virgin that he does not come to deceive her or to violate her virginity. He goes on to explain at length the incarnation of the Lord. Gabriel explains the relationship between Christ and the Father, through the language of cognition and comprehension to which we have become accustomed, stating, "He of whom the Son was ineffably and timelessly begotten (γέννησις), possesses also the unerring apprehension (ἀψευδῆ κατάληψιν) of Him."[28] Photios then comes to reflect once more on the Virgin's dutiful assent, asking: "Was she immediately softened by these words, and having opened her ears wide with pleasure, did she allow her thoughts to give assent without scrutiny (ἀνεξετάστως συγκατάθεσιν)? Not at all."[29] Thus, the Virgin has not simply assented, readily and without scrutiny, but rather she reflects on the statement of the angel.

Then, the Virgin indicates that *now* she understands that he is "describing the conception, pregnancy, and birth of a son" (σύλληψιν καὶ κυοφορίαν καὶ υἱόν μοι καὶ τόκον λογογραφεῖς). Nevertheless she is all the more perplexed as to how this could possibly occur without intercourse. Here Photios plays with the Virgin's cognitive assent and sexual consent (συγκατάθεσιν) as well as her mental and physical conception (σύλληψιν). Conception (σύλληψις) here resonates with comprehension (κατάληψις), just as it does in English. Thus, one is led to consider that just as God the Father who begets the Son, has a comprehension (κατάληψιν) of him, so also the Virgin has a bodily comprehension (σύλληψιν) of him. She goes on to describe the conception not only with σύλληψιν,

27 "τὰ ῥήματα δὲ μνηστῆρος ἔμφασιν ὑποσημαίνοντα ἀνανεύειν πείθει τὴν συγκατάθεσιν." Photios, *Homiliae*, 5, ed. Laourda, 55; trans. Mango, 115.

28 "ἐξ οὗ γὰρ ἡ τοῦ υἱοῦ ἄρρητός τε καὶ ὑπεραιώνιος γέννησις, οὗτος αὐτοῦ καὶ τὴν ἀψευδῆ κατάληψιν κέκτηται." Photios, *Homiliae*, 5, ed. Laourda, 58; trans. Mango, 118.

29 "Μὴ παραυτίκα κατεμαλακίσθη τοῖς λόγοις ἢ τὰς ἀκοὰς ἡδέως ἀναπετάσασα τοῖς λογισμοῖς ἐφῆκε φέρειν ἀνεξετάστως τὴν συγκατάθεσιν; οὐδαμῶς." Photios, *Homiliae*, 5, ed. Laourda, 58; trans. Mango, 118.

but also with "pregnancy" (κυοφορίαν) and "birth of a son" (υἱόν… τόκον) as if clarifying to herself the bodily implications of the angel's message. Over the course of the Annunciation, the perceptual process of comprehending the angel's words and assenting to them result in her conception and impregnation. She goes on to say that in her abstinence she is not permitted "even to hear of conception" (οὐδὲ τὴν ἀρχὴν εἰς ἀκοὴν κατέρχεται τῆς συλλήψεως),[30] almost suggesting that to hear about conception would be to sully her purity, just listening to the angel's message would result in pregnancy. Therefore, the processes of sense perception and cognition become analogous to the processes of sexual intercourse and conception. But unlike those deceived by the Russian onslaught, subject to fallacy through hasty assent, the Virgin ponders and her pondering brings about assent to true sense perception, just as her pure conception was the result of pure, divine insemination.

Nicholas Mesarites's characterization of the Annunciation in his twelfth-century description of the Church of Holy Apostles deploys this same logic, drawing a connection between the Virgin's perception and conception.

> φθάνει τὸ ῥῆμα ἐπὶ τὴν τῆς παρθένου ἀκουστικήν, εἰσδύνει δι᾽ αὐτῆς ἐπὶ τὸν ἐγκέφαλον, ἐπιδράττεται τοῦ εἰσελθόντος εὐθὺς ὁ τῷ ἐγκεφάλῳ ἐπικαθήμενος νοῦς, γνωρίζει τὸ πρᾶγμα τῇ διανοίᾳ, ἀλλὰ καὶ αὐτῇ τῇ καρδίᾳ κοινοῦται τὸ γνωρισθέν· ἡ δὲ ταράττεται παρευθύ, καὶ διαλογισμοὶ ἐπὶ τὴν τῆς παρθένου καρδίαν ἀναβαίνειν ἀπάρχονται διαλογιζομένης καλῶς, ποταπὸς ἂν εἴη ὁ ἀσπασμός. καὶ πρὸς τὴν τοῦ ἀσπασμοῦ κατεξέτασιν ἤδη χωρεῖ· παρθένος γὰρ ἦν ἀληθῶς ἡ παρθένος, οὐ μέχρι σώματος ἀλλὰ καὶ αὐτῶν λογισμῶν· καὶ τὸν τῆς συλλήψεως τρόπον σαφῶς εἰπεῖν τὸν εὐαγγελιστὴν ἐκπυνθάνεται· 'πῶς γάρ' φησι 'παρθένος οὖσα συλλήψομαί τε καὶ τέξομαι, πεῖραν ἀνδρὸς μὴ γνοῦσα τὸ σύνολον'; ἀκούει γοῦν τὸ ὑπὲρ τὸ πῶς, ὅπερ ἡ τοῦ παναγίου ἐπέλευσις πνεύματος καὶ ἡ τῆς τοῦ ὑψίστου δυνάμεως ἐπισκίασις παρεδήλου σαφέστατα. ὅλην ἑαυτὴν πρὸς τὴν ὑπὲρ νοῦν καὶ λόγον τοῦ λόγου σύλληψιν ὑποτίθησι—δούλην γὰρ ἀποκαλεῖ ἑαυτήν, ἀντίξουν μὴ κεκτημένη τὸ θέλημα πρὸς τὴν τοῦ δεσποτικοῦ θελήματος ἀποπλήρωσιν—καὶ ὁ τοῦ θεοῦ λόγος εὐθὺς ἔργον ἐνανθρωπήσεως ὡς ἂν τις εἴποι ἐγένετο.

> The word comes to the hearing of the Virgin, and enters through it to the brain; the intelligence which is seated in the brain at once lays hold upon what comes to it, recognizes the matter by its perception, and then communicates to the heart itself what it had understood. The heart is immediately agitated, and debates begin to rise up in the maiden's heart

[30] Photios, *Homiliae*, 5, ed. Laourda, 58–59; trans. Mango, 118–119.

> as she debates, in virtuous fashion, what the greeting means. And she already turns to a careful examination of the greeting; for the Virgin was truly maidenly, not merely in her person but even in her very thoughts, and she requests the messenger to describe clearly the manner of the conception. "For how," she says, "shall I, a maiden, conceive and give birth, having no experience at all of a man?" Then she hears something that surpasses wonder, which the coming of the All Holy Spirit and the overshadowing of the power of the Most High intimate in clearest fashion. She yields herself wholly to the conception of the Word, which is beyond thought and speech – for she calls herself a servant, possessing no desire which opposes the fulfillment of the wish of the Lord – and the Word of God at once, as one might say, undergoes the act of incarnation.[31]

While Mesarites's text lacks Photios's terminology, the process he outlines matches Photios's understanding of cognitive perception: sensory input enters the brain, the intelligence lays hold of it, this material is recognized by the intellect, and then it is passed on to the heart for critical debate before she eventually yields (ὑποτίθησι) to assent.

Critical to both Photios and Mesarites's text is that the Virgin must first consider and soberly debate the angel's words, before at last affirming their realities. For Photios, the Virgin's assent, the resolution of the homily, comes in the homily's penultimate passage, which begins by acknowledging the angel, the Virgin's intercourse, and her contemplation: "Such things the archangel was saying, drawing the spotless maiden to assent (εἰς συγκατάθεσιν ἕλκων)."[32] The use of the participle "drawing to" (ἕλκων) here resonates with Photios's use of the same verb in Homily 17 when he asks if the mind has "drawn to itself" (εἵλκυσε) [**kI**] and imagined the things heard, which further intertwines the process of perceptual cognition to the Virgin's assent. Here, there is a clear sense that this is a slow and careful process of drawing the Virgin toward assent, respecting her desire not to be deceived and to properly consider the matter at hand, despite its paradoxical nature, as expressed earlier in the text. Even in a circumstance of perception that bewilders all expectations, the Virgin's righteousness is demonstrated by her ability to soberly scrutinize her perceptions, while still accepting the mysteries of the divine.

Returning then to Photios's sequence of perception in Homily 17, one observes the omission of any form of judgment in the unfolding of sight, be it Porphyry's "belief-making assumption" (δοξαστικὴ ὑπόληψις) or

[31] Nicholas Mesarites, *Descriptio ecclesiae SS. Apostolorum*, XXII.7–9, ed. and trans. Downey, 877, 906.

[32] "Ἀλλὰ ταῦτα μὲν ὁ ἀρχάγγελος εἰς συγκατάθεσιν ἕλκων τὴν ἄσπιλον κόρην ἐφθέγγετο." Photios, *Homiliae*, 5, ed. Laourda, 60; trans. Mango, 120.

some other operation of *doxa*, *hypolepsis*, or *prolepsis*. While preconception (πρόληψις) is noted by Photios early on in comparing sight and hearing, it is likewise missing from his rhetorical questions diagramming the processes of sight and hearing later in the passage.

Ἔκλινέ τις τὸ οὖς εἰς διήγημα;	Εἶδεν ὁ νοῦς,
εἵλκυσε	ἀντελάβετο,
φανταζομένη τὸ ἀκουσθὲν ἡ διάνοια;	ἐφαντάσθη,
νηφούσῃ μελέτῃ τὸ κριθὲν	τοὺς τύπους ἀκόπως
τῇ μνήμῃ ἐναπέθετο	ἐν τῇ μνήμῃ παρεπέμψατο

Has [a person] inclined their ear to the story?	Has the mind seen?
Has their intelligence drawn to itself, and imagined the thing heard?	Has it grasped? Has it imagined?
Then, judging it with sober attention, they deposit it into the memory.	Then, effortlessly it has transmitted the forms to the memory.

As this comparative table demonstrates, the processes of sight and hearing parallel each other with careful precision in Photios's homily with one notable exception: the act of "judging with sober attention" (νηφούσῃ μελέτῃ τὸ κριθέν) in hearing. This step is strongly contrasted in sight through its complement, where the form is "effortlessly" (ἀκόπως) sent on to memory.

Why has Photios neglected the stage in the detailing of sight, both in his rhetorical questions and in the description that preceded it? The answer may be found in Photios's understanding of the function and necessity of assent within vision. It is clear that Photios wishes to stress, as cited earlier, that it is vision which "drags the spectators to unhesitating assent" (τοὺς θεατὰς ἐφελκομένης εἰς ἀπροφάσιστον συγκατάθεσιν) [**b2**]. This language of "dragging" (ἐφελκομένης) resonates with Sextus Empiricus's own characterization of the cognitive impressions "pulling" (κατασπῶσα) one to assent, which speaks to the impressions' striking vividness, "being self-evident" (ἐναργὴς οὖσα), as Sextus describes it. Photios's reasoning is the same, since sight presents things more vividly, more self-evidently (ἐναργεστέραν) than words. In defense of images, such reasoning suggests precisely that images, by virtue of presenting vivid scenes of narratives, impress the mind more strikingly than would those same narratives told through words. Photios offers his descriptions of the perceptual process because he regards sight as superior to hearing, but since both processes are identical except for the effortlessness of sight's imprints engraving in the imagination, it is reasonable to consider that sight itself could be said to bypass the judgment necessitated by hearing. At the very least, the process proceeds far more "effortlessly" (ἀκόπως); that is, without any added exertion, force, or toil.

One implication of this difference between sight and hearing is that "it is the spectators rather than the hearers who are turned toward emulation" (μᾶλλον τοὺς ὁρῶντας ἢ τοὺς ἀκούοντας πρὸς μίμησιν ἐπιστρέφουσιν) **[h]** of the icon. This notion, that an icon impressed upon the soul leads a viewer to emulation, has precedence in the ancient world, and it recurs in post-iconoclastic Byzantine writers such as Michael Psellos.[33] It would be fallacious to read Photios's statement as indicating that hearers are not compelled at all to emulation. After all, Photios's entire defense of images in this passage, delivered orally, compares the teachings of the Gospels (which were known orally) to the work of images, stating that one cannot deny one without denying the other. In the section following the passage on perception, Photios goes on to clarify:

> Ἀθετεῖ τις τοὺς περὶ τούτων λόγους ἱεροὺς καί, οἷς ἅπαν ἐλαύνεται ψεῦδος, οὐ κρείσσους ἄγει ἀμφισβητήσεως; οὗτος πολὺ πρότερον καὶ τῶν σεπτῶν εἰκονισμάτων ἐξορχεῖσθαι τὸ σέβας πεπλάνηται. Ἀλλὰ τιμᾷ ταῦτα καὶ ταῖς προσηκούσαις γεραίρει σεβασμιότησι; τὴν ὁμοίαν ἕξει καὶ περὶ τοὺς λόγους διάθεσιν· ὁποτέρῳ γὰρ ἄν τις τούτων τὰ τῆς τιμῆς ἢ τὰ τῆς ἐξουδενώσεως προσάψῃ, τῶν ἴσων ἀνάγκη καὶ τῷ ἑτέρῳ μεταδιδόναι, εἰ μή τις μετὰ τῆς ἀσεβείας καὶ τοῦ φρονεῖν ὅλως παρενήνεκται, οὐ δυσσεβῶν μόνον, ἀλλὰ καὶ ἑαυτῷ πρεσβεύων μαχόμενα.

> Is there one who rejects the holy writings on these matters and, in spite of the fact that all lies are dispelled by them, considers them to be not above dispute? Then this man has long since transgressed by scorning the veneration of holy images. Does he, on the contrary, reverence the latter, and honor them with proper respect? Then he will be disposed likewise towards the writings. If he treats either one with reverence or with contempt, he necessarily bestows the same on the other, unless, in addition to being impious, he has also abandoned reason, and is not only irreverent, but also preaches things which are in conflict with his own position.[34]

If Photios believed that only a viewer, and not a hearer, is drawn to mimesis, then he himself would oppose the teachings of the Gospel, and

[33] See Verity Platt, "Virtual Visions: *Phantasia* and the Perception of the Divine in *The Life of Apollonius of Tyana*" in Ewen Bowie and Jaś Elsner (eds.), *Philostratus* (Cambridge University Press, 2009), 131–154, esp. 134. Cf. Philostratus, *The Life of Apollonius of Tyana*, II.22.4, ed. and modified trans. Jones, I:184–185. See also Karl Thein, "Gods and Painters: Philostratus the Elder, Stoic Phantasia and the Strategy of Describing," *Ramus* 31:1/2 (2002): 136–145, esp. 142.

[34] Photios, *Homiliae*, 17, ed. Laourda, 171; modified trans. Mango, 294–295.

contradict himself about what sight and sound held in common. To say that "it is the spectators *rather* (μᾶλλον) than the hearers who are turned toward emulation" might suggest that the spoken Gospel is ineffectual at drawing its listeners to assent and emulation. But this statement is a confirmation of the vividness and actuality (i.e. the ἐνάργεια) of images, which make them a communication medium superior to anything heard, since the goal of classical rhetoric is precisely to "place before the eyes" the narratives as if they were occurring before the listeners, as Ruth Webb and others have elucidated. That is, the phrase merely stresses the impact and effect of the visual medium as a more vivid, more striking impression on the mind. Sight is not immune to the necessity of perceptual judgment before its storage in memory for future emulation. Rather it occurs "effortlessly" (ἀκόπως) – images perceived through the proper means are self-evident.

In the writings of the leading ninth-century iconoclast, Patriarch John VII Grammatikos, one may observe the problem of verbal versus visual judgment. John argues that it is wrong to judge the worthiness of a man by means of an image, given that it is words instead that provide us with a man's lineage, histories, deeds, and narratives.[35] This problem concerned iconophile thinkers before and after him, thinkers such as Patriarch Nikephoros and Theodore of Stoudios, and Photios's own response appears to be derived from their defenses, particularly Nikephoros's. In his *Antirrheticus*, Nikephoros writes:

> Ἀλλὰ καὶ οἱ λόγοι αὐτοὶ εἰκόνες εἰσὶ τῶν πραγμάτων, καὶ ἕπονται αὐτοῖς ὡς αἰτίοις. Καὶ πρώτως μὲν τὴν ἀκοὴν εἰσδύονται· πρότερον γὰρ τὰ ἀπηχήματα τῶν λεγομένων τοῖς ἐνηχουμένοις προσπίπτουσι· δευτέρως δὲ δι᾽ ἀναλογισμοῦ ἐπὶ τὴν τῶν δηλουμένων πραγμάτων κατανόησιν ὁ ἠκουτισμένος ἔρχεται. Ἡ δὲ πρώτως καὶ ἀμέσως ἐπ᾽ αὐτὰ τὰ πράγματα, ὡς παρόντα ἤδη, τὸν νοῦν τῶν ἐνορώντων προσάγει, καὶ ἐκ πρώτης θέας καὶ ἐντεύξεως τρανὴν καὶ ἀπεξεσμένην τὴν γνῶσιν τούτων παρέχεται.

> But words themselves are the icons of things, and follow on from them as from their causes. To begin with, they enter hearing; for first the sounds of the things spoken encounter those listening, then, second, the listener achieves understanding of the given facts through analogy. [Images] directly and immediately lead the mind of the viewers to the facts themselves, as if they were present already, and from the first sight and encounter, a clear and perfect knowledge of these is gained.[36]

[35] John VII Grammatikos, *Antirrheticus*, frag. II, ed. Jean Gouillard, "Fragments inédits d'un antirrhétique de Jean le Grammairien," *Revue des Études Byzantines* 24 (1966): 171–181, esp. 174.

[36] Nikephoros, *Antirrheticus*, III.5, ed. PG 100:381C–D; trans. Barber, *Figure and Likeness*, 128.

Just as Photios somewhat cryptically suggests that "the one introduces the other" (δι' ἀλλήλων ἑκάτερον συνεισάγεται) but that the comprehension (κατάληψις) that comes about through sight is superior to that through hearing, so Nikephoros points out that hearing first is received and then must be reasoned with, that is, understood through analogy or, rather, reasoning (ἀναλογισμοῦ).[37] Yet, things coming about through sight, on the other hand, are processed "directly and immediately" (πρώτως καὶ ἀμέσως), achieving a "clear and perfect knowledge" (τρανὴν καὶ ἀπεξεσμένην τὴν γνῶσιν). Nikephoros suggests that images do not require any preliminary judgment for analogical reasoning (ἀναλογισμός), because they are already "as if they were present" (ὡς παρόντα ἤδη). In a sense, we can see John the Grammarian in the mid ninth century implicitly attacking Nikephoros's reasoning by attempting to reiterate the need for reasoning and judgment in all perception. John regards the image as lacking, given that it is believed to circumvent these processes. Iconophile writers responded that the image was superior because it readily and effectively passes the critical stages of sense perception through its vividness and clarity.

Compare Nikephoros's statements above to an earlier comparison of sight and hearing:

> Ὅπερ γὰρ πολλάκις ὁ νοῦς οὐχ εἷλε διὰ τῆς τῶν λόγων ἀκοῆς, ἡ ὄψις ἀπλανῶς παραλαβοῦσα, σαφέστερον ἐφηρμήνευσεν. Εἰς ἀνάμνησιν λοιπὸν ὧν δι' ἡμᾶς Χριστὸς ἔδρασέ τι καὶ ἔπαθε ῥᾳδίως προάγονται, καὶ θᾶττον ἢ διὰ τῆς τῶν λόγων ὑφηγήσεως, ὅσῳ καὶ ὄψις ἀκοῆς θᾶττον εἰς γνῶσιν καὶ πρὸς πίστιν ἑτοιμοτέρα.
>
> For often what the mind has not grasped while listening to a discourse, the sight seizes (παραλαβοῦσα) without risk of error, [that is, it] has interpreted it more clearly. Finally, it is the memory of all that Christ did and of what He suffered for us that [the icons] produce easily and more quickly in fact than the discursive account, as vision is more prompt to knowledge than hearing.[38]

Here, Nikephoros explains not only that sight is a remediation of hearing, but also that images can communicate things that most discourses simply cannot. What the mind does not grasp (εἷλε), sight immediately apprehends (παραλαβοῦσα). When sight comprehends things, it does so "unerringly" (ἀπλανῶς), just as Photios states that sight results in

[37] On Photios and the division of words and images in *Homiliae*, 17, see Barber, *Figure and Likeness*, 135–136; Brubaker, "Perception and Conception," 28–30; James, "Art and Lies," 62–64.

[38] Nikephoros, *Antirrheticus*, III.3, ed. PG 100:380D–381A; trans. Barber, *Figure and Likeness*, 129.

"unerring" (ἀπλανεστάτης) knowledge. Furthermore, it not only seizes things unerringly but it also manages to reason and interpret them more clearly and distinctly (σαφέστερον ἐφηρμήνευσεν). As Charles Barber has noted concerning Nikephoros's thinking on words and images, "words are marked by a potential for ambiguity, giving rise to disputes and doubts. In contrast, the image is direct and immediate."[39] Like Photios's ease (ἀκό-πως) in the transmission to memory of seen objects, sight memory here likewise is easily produced (ῥᾳδίως προάγονται) and its speed is twice remarked upon as being swifter (θᾶττον) than hearing. Liz James has credited this swiftness and ease to the fact that sight "presented the object of memory in a visual form and this was how memory worked."[40] However, the reality of this dynamic is far more nuanced. It is not simply that icons present the perceiver with ready-made images, but rather that the forms that they communicate are more vivid and more striking than those communicated through discourse. Sight not only speeds up the imagination's visualization but also produces more accurate and unerring visualizations for the sake of knowledge. While memory stores these truths, their accuracy and veracity depend upon the faculties of imagination and intellect, which have judged and assented to the results of the *phantasia.*

[39] Barber, *Figure and Likeness*, 129. [40] James, "Art and Lies," 67.

Conclusion to Part II

> Et hoc quidem Zeno gestu conficiebat: nam cum extensis digitis adversam manum ostenderat, 'visum' inquiebat 'huius modi est'; dein cum paulum digitos contraxerat, 'adsensus huius modi'; tum cum plane compresserat pugnumque fecerat, comprensionem illam esse dicebat (qua ex similitudine etiam nomen ei rei, quod ante non fuerat, κατάληψιν imposuit); cum autem laevam manum admoverat et illum pugnum arte vehementerque compresserat, scientiam talem esse dicebat, cuius compotem nisi sapientem esse neminem.
>
> And this Zeno used to demonstrate by a gesture: for he would display his hand in front of one with the fingers stretched out and say "A presentation is like this." Then, he closed his fingers a little and said, "An act of assent is like this." Then, he pressed his fingers closely together and made a fist, and said that that was comprehension, and from this illustration he gave to that process the actual name of κατάληψιν, which it had not had before. But then he used to apply his left hand to his right fist and squeeze it tightly and forcibly, and then say that such was knowledge, which was within the power of nobody save the wise man.[41]

In this account, Cicero explains how Zeno used to explain the unfolding of cognition toward knowledge, following the general stages that have been considered here, such as presentation (*visum*), assent (*adsensus*), comprehension (*comprensionem*), and knowledge (*scientiam*), which correspond with the Greek *phantasia*, *synkatathesis*, *katalepsis*, and *episteme*, respectively. As Cicero notes, it is from this visual explanation through hand gestures that the language of comprehension as grasping or seizing emerges through and around the term *katalepsis*. Sextus Empiricus recounts a similar story in *Against the Rhetoricians* where Zeno uses the clenched fist to explain the succinctness of dialectic and the open hand to

[41] Cicero, *Academica*, II.47 (145), ed. and modified trans. H. Rackham, *Academics*, Loeb Classical Library 268 (Cambridge, MA: Harvard University Press, 1933), 652–655.

suggest the breadth of rhetoric.[42] And, in both *Against the Logicians* and *Outlines of Pyrrhonism*, Sextus also states that "all knowledge is a 'particular state of the regent part,' just as the fist is conceived as a particular state of the hand."[43] These gestural metaphors therefore help contour the understanding of knowledge through a language of grasping, whereby various degrees of grip denote the progressive states of cognition. This language progresses well into the Byzantine world and features prominently around discussions of perception in the various senses, particularly in sight and hearing where nuances and levels of understanding become most pertinent. Likewise, in the Byzantine world we can observe the progression from sensation, apprehension, imagination, judgment, leading to eventual assent, memory, and knowledge, based on Stoic and late antique models with their own adaptations to the individual writer's own focus and desire.

By way of a summary of the teachings garnered from Photios's homily on sensation and perception, I wish to briefly survey the stages of perception as indicated in the homily, and present some considerations regarding their intellectual history and transmission. Earlier, Porphyry's succinct detailing of the various stages of perception offered up an eloquent treatment of each phase for comparison. To summarize, Photios's process of perception proceeds: (0) sensation (αἴσθησις) as a pre-cognitive stage, followed by the cognitive unfolding of perception: (1) apprehension (ἀντίληψις), (2) imagination (φαντασία), (3) judgment (κριτικός), and (4) memory (μνήμη). And, as the narrative description of sight tells us, this results in knowledge (ἐπιστήμη), or, as Photios states, "unerring knowledge" (ἐπιστήμης ἀπλανεστάτης), which is understood as bringing about comprehension (κατάληψις) proper, a sort of additional fifth stage.

From the language used and the order of this progression, it becomes evident that Photios's primary source for cognition comes from John of Damascus's *Expositio Fidei*, particularly as described in the entry on memory, given that in this final stage of perception authors often choose to succinctly summarize the processes that they have detailed

42 Sextus Empiricus, *Adversus mathematicos*, II.7, ed. and trans. Bury, 192–193.

43 "πᾶσα δὲ ἐπιστήμη πῶς ἔχον ἐστὶν ἡγεμονικόν, ὥσπερ καὶ ἡ πῶς ἔχουσα χεὶρ πυγμὴ νοεῖται." Sextus Empiricus, *Adversus mathematicos*, VII.39, ed. and trans. Bury, 20–21. Cf. "ἡ δὲ ἐπιστήμη πῶς ἔχον ἡγεμονικὸν ὥσπερ καὶ ἡ πῶς ἔχουσα χεὶρ πυγμή." Sextus Empiricus, *Pyrrhoniae hypotyposes*, II.8 (81–82), ed. and trans. R. G. Bury, *Outlines of Pyrrhonism*, Loeb Classical Library 273 (Cambridge, MA: Harvard University Press, 1933), 202–203.

beforehand.[44] The relevant passage retains the same order and deploys very similar language to Photios's own description:

> [2] Τὸ μὲν οὖν φανταστικὸν [0] διὰ τῶν αἰσθήσεων [1] ἀντιλαμβανόμενον τῶν ὑλῶν παραδίδωσι [3] τῷ διανοητικῷ ἢ διαλογιστικῷ (ταὐτὸν γὰρ ἀμφότερα)· ὃ παραλαβὸν καὶ κρῖναν παραπέμπει [4] τῷ μνημονευτικῷ.
>
> [0] Through the senses, [1] material objects are apprehended, and [2] the faculty of the imagination (φανταστικὸν) then transmits this to [3] the faculty of thought or reason (for they are both the same), which upon receiving and passing judgment on it, [4] passes it on to the faculty of memory.[45]

Here, I have revised the translation to articulate clearly the procedural unfolding of the various stages, numbered in accordance with Photios's progression. Notably, sensation is somewhat distinguished from the act of apprehension itself, demonstrating the senses as an instrument through which the perceptibles are apprehended (διὰ τῶν αἰσθήσεων ἀντιλαμβανόμενον). Furthermore, while the imaginative faculty is given little attention, discernment is denoted by either the faculties of thought or reason (διανοητικῷ ἢ διαλογιστικῷ), with the critical action of judgment (κρῖναν) being described as its key activity before the information is sent on to the memory.

Compare this description to the similar one that John of Damascus uses to describe sensation overall in his treatise on icons. Here, the passage is not meant to summarize sensation within the context of memory alone, but rather is used to describe the process of perception overall in order to defend the operation of images, a context similar therefore to Photios. There, John of Damascus writes:

> [0] Διὰ γὰρ τῆς αἰσθήσεως [2] φαντασία τις συνίσταται ἐν τῇ ἐμπροσθίῳ κοιλίᾳ τοῦ ἐγκεφάλου καὶ οὕτω [3] τῷ κριτικῷ παραπέμπεται καὶ [4] τῇ μνήμῃ ἐνθησαυρίζεται.
>
> [0] For through the senses a certain [2] imaginative image is constituted in the front part of the brain and thus conveyed to the [3] faculty of discernment and stored in the [4] memory.[46]

[44] For a limited and superficial account of John of Damascus on human cognition, see James R. Payton, "John of Damascus on Human Cognition: An Element in His Apologetic for Icons," *Church History* 65:2 (1996): 173–183.

[45] John of Damascus, *Expositio fidei*, II.20 (34), ed. P. B. Kotter, *Die Schriften des Johannes von Damaskos*, vol. II, Patristische Texte und Studien 12 (Berlin: De Gruyter, 1973), 87.

[46] John of Damascus, *Orationes de imaginibus tres*, I.11, ed. P. B. Kotter, *Die Schriften des Johannes von Damaskos*, vol. III, Patristische Texte und Studien 17 (Berlin: De Gruyter, 1975), 85; trans. Andrew Louth, *Three Treatises on the Divine Images* (Crestwood, NY: St. Vladimir's Seminary Press, 2003), 26.

This features the same progression of processes, yet the act of apprehension is given less force and essentially overlooked, given perhaps only a passing allusion through the verb "constituted" (συνίσταται). What is notable though is that rather than addressing the faculty of the imagination (φανταστικόν), it is the act of constituting a certain cognitive impression itself (φαντασία τις) from the senses. Thus, it would seem that Photios relied more precisely on the *Expositio Fidei*, while still being knowledgeable as to John of Damascus's other description in the icon treatise, which neatly summarizes the critical faculty simply with the act of judgment (κριτικῷ) and understands that the imagination creates images from the senses, which are then the objects conveyed for the unfolding of the process. This is lacking in the *Expositio Fidei*, which makes it uncertain what exactly is being passed on throughout these various stages, beyond that which is "being apprehended from material things" (ἀντιλαμβανόμενον τῶν ὑλῶν) through the senses.

In turn, John of Damascus derived his summary from another highly influential author that has continuously drawn our attention, Nemesios of Emesa's *De natura hominis*. The treatise was cited extensively not only in John of Damascus's *Expositio Fidei*, but also in Meletios's popular medical treatise, the works of Maximos the Confessor, and others.[47] Nemesios's work itself derived greatly from Stoic and Neoplatonic teachings coming from figures as diverse as Porphyry, Posidonius, Diogenes Laertius, Aetios, and Galen.[48] Thus, Nemesios's text not only provides a further chain of antecedents and sources for Photios's cognitive theories, but also evidences the changes and alterations made to the process of perception over the late antique period by Christian thinkers. Consider for instance Nemesios's summary in the passage on memory, where he writes:

> ἡ γὰρ ψυχὴ τῶν μὲν αἰσθητῶν διὰ τῶν αἰσθητηρίων ἀντιλαμβάνεται καὶ γίνεται δόξα, τῶν δὲ νοητῶν διὰ τοῦ νοῦ καὶ γίνεται νόησις· ὅταν οὖν τοὺς τύπους ὧν τε ἐδόξασεν ὧν τε ἐνόησε διασῴζῃ, μνημονεύειν λέγεται.
>
> For the soul grasps objects of sense through the sense-organs and an opinion (δόξα) is formed, while it grasps things intelligible through the intellect and intellection comes about: when the soul preserves the imprints of what it has experienced and of what it has conceived it is said to remember.[49]

[47] Sharples and van der Eijk, *On the Nature of Man*, 4.

[48] Sharples and van der Eijk, *On the Nature of Man*, 18–32. See also Thomas Halton, "The Five Senses in Nemesius, *De Natura Hominis* and Theodoret, *De Providentia*," *Studia Patristica* 20 (1989): 94–101.

[49] Nemesios, *De natura hominis*, 8, ed. Morani, 68; trans. Sharples and van der Eijk, 120.

In this summary, following Plato, we can observe that while the language of imprints (τύπους) is present, the function of the imagination is wholly absent, yet the function of *doxa* is present following sensation and apprehension. Nemesios himself remarks on such matters, as when he accuses Plato of confusing intellection (νοητός) for thought (διανοητός). Nevertheless, he retains the operation of *doxa* alongside apprehension, similar to Porphyry, without a specific disclaimer about it being an error.

Curiously, this function of initial judgment seems to be subsumed by the workings of the imagination, since the entry on the imagination states that, following the Stoics, "for when we see white an affection occurs in the soul through its reception,"[50] which is an example often used to describe the discerning actions of *doxa*. For example, this appears in the opening of the *Souda*'s definition of *doxa*, derived from John Philoponos's commentary on *De anima*, where it explains that opinion (δόξα) is concerned with that which is "universal in sensible things" (τὸ καθόλου ἐν τοῖς αἰσθητοῖς) through which opinion "knows that 'white' is discerned by means of sight, and that all men are bipeds."[51] This is connected to the intuitive discernments of preconception (προλήψις), which likewise instantly are able to allow one to understand what it is one sees but precedes thought proper. In the late antique period, the operations of *doxa* and *phantasia* often became muddled, *doxa* being caught somewhere in between the actions of apprehension (ἀντίληψις) and the imaginative workings of *phantasia*.[52] Despite Aristotle's clear separation of *doxa* and *phantasia*, the division of the *phantasia* into higher and lower registers, almost doubling its functions, often helped to justify these confusions for the Neoplatonic interpreters, as H. J. Blumenthal has argued.[53]

Soon thereafter, Nemesios gives yet another definition of perceptual cognition, which stands alone and is not attached to a citation to any one thinker or school of thought. Here, we find a clear concordance with Photios's own process:

[50] "ὅταν γὰρ λευκὸν ἴδωμεν, ἐγγίνεταί τι πάθος τῇ ψυχῇ ἐκ τῆς λήψεως αὐτοῦ." Nemesios, *De natura hominis*, 6, ed. Morani, 55; trans. Sharples and van der Eijk, 100.

[51] "οἶδε γὰρ ὅτι τὸ λευκὸν διακριτικὸν ὄψεως, καὶ ὅτι πᾶς ἄνθρωπος δίπους." Anonymous, *Souda*, delta 1363, "δόξα," ed. and trans. Marcelo Boeri, *Souda On Line*, www.stoa.org/sol-entries/delta/1363.

[52] See, for example, Peter Lautner, "The Distinction between Φαντασία and Δόξα in Proclus' *In Timaeum*," *Classical Quarterly* 52:1 (2002): 257–269.

[53] Blumental, "Neoplatonic Interpretations," 249–250.

> τὸ μὲν οὖν φανταστικὸν παραδίδωσι τῷ διανοητικῷ τὰ φαινόμενα, τὸ δὲ διανοητικὸν ἢ διαλογιστικὸν παραλαβὸν καὶ κρῖναν παραπέμπει τῷ μνημονευτικῷ.
>
> Thus the faculty of imagination hands on things imagined to the faculty of thought, while thought or reasoning, when it has received and judged them, passes them on to the faculty of memory.[54]

While this summary neglects the entwined initial stages of sensation and apprehension, it does go on to detail the process from the imagination, on to the faculty of thought (διανοητικῷ), which receives and judges (κρῖναν) this and passes it on to memory. This is directly excerpted nearly verbatim by John of Damascus in his summary in the *Expositio Fidei*. Yet, John of Damascus has made the effort to elucidate this process by acknowledging the senses, and alluding to the imagination. In both Nemesios and John, the imagination has already been given extensive treatment in its own sections, hence both authors seem compelled to merely state that the imagination "hands on things imagined" without explaining precisely what it does. And, in Photios, the operations of thought are summarily categorized by judgment alone.

Thus, we may appreciate that the function of initial judgments was still operant in cognition as Porphyry delineated it, comprised of *doxa*, *prolepsis*, or *hypolepsis*, or a combination thereof. Yet, in later periods, these first stages of discernment appear to lose a clear distinction, being usually surmised to be part and parcel of the mind's initial apprehension or subsequent imagination of the object. This, however, is certainly different from the faculty of thought (διανοητικός) proper, indicated in Nemesios, John of Damascus, and Photios, that is defined by the act of judging (κρῖναν). In other words, thought is the action that, through judgment (κρίσις), is ultimately responsible for assent (συγκατάθεσις). This is why Photios's and Nicholas Mesarites's exegeses on the Annunciation, which deploy perceptual cognition as a metaphor for the incarnation, choose to give a great deal of attention to the Virgin's debate and judgment of the angel's proposition before she eventually assents to his message and the incarnation occurs. As Nemesios clearly states, the "different kinds of thought are judgement, assent, avoidance, and impulse," which is in itself similar to a Stoic division of thought found in Sextus Empiricus.[55] In other

[54] Nemesios, *De natura hominis*, 8, ed. Morani, 69; trans. Sharples and van der Eijk, 121.

[55] "τοῦ δὲ διανοητικοῦ εἰσι γενικῶς μὲν αἵ τε κρίσεις καὶ αἱ συγκαταθέσεις καὶ ἀποφυγαὶ καὶ ὁρμαί." Nemesios, *De natura hominis*, 7, ed. Morani, 68; trans. Sharples and van der Eijk, 117. Cf. Sextus Empiricus, *Pyrrhoniae hypotyposes*, 1.65, ed. and trans. Bury, 38–41.

words, beyond the action of discernment, thought enacts the results of said judgment, either by assenting to the sensory inputs or avoiding them (ἀποφυγαί), and acting upon the impulse (ὁρμαί) of thought and its decisions. Thus, the importance of judgment in late antique and Byzantine thinkers can be seen to emerge not only from anxieties regarding the necessity to confirm and accept the generations of the imagination, but also more simply to serve as the process that enables action and emulation, which is preserved in the memory for future and repeated activity. Not only is there initial discernment through opinion, but also thought proper that reasonably weighs in on sensory perceptions and results in true comprehension or "unerring knowledge."

Thus, in Photios, the acts of initial judgment that populated the first stages of perception are neglected, ascribed to apprehension and the imagination, and they are replaced by the final judgment of thought or reasoning (διανοητικὸν ἢ διαλογιστικόν). These are responsible for assent, and it would appear that in visual perception this stage occurs quite "effortlessly" (ἀκόπως), as has been argued. In each of the authors considered herein, the various stages of cognition exhibit contractions and dilations in accordance with the portions or aspects of the process which the authors wish to stress. Photios's learned precision and deployment of technical terms demonstrates his intentionality in using perception to argue for the affordances of the icon. However, it is evident from the overall passages of his homily that he very much acknowledges other aspects of perception that do not fit into his neat litany of questions or descriptions, such as his allusion in passing to preconception (πρόληψις) or assent (συγκατάθεσις). The inclusion of these processes speaks to Photios's keen reading of texts like John of Damascus's *Expositio Fidei* and Nemesios's *De natura hominis*. However, the lack of a slavish citation or an exact cropping of their teachings suggests that Photios was carefully wishing to describe perception for his own goals and purposes. In particular, this meant that he chose to give emphasis to the stages that he deemed important, while also making sure to round off the portions that some of these writers might have left off. Furthermore, Photios's use of Stoic language to describe perception and cognition demonstrates that there was an endurance of Stoic terms and that their models persisted as structures for thinking about thought. However, these terms were heavily refracted through late antique writers, early Christian thought, and the contemporary uses of language around images. Thus, Photios's language has guided us through the various and nestled theories of cognition from the classical to the Byzantine world, demonstrating an unbroken continuity of thought, yet one that developed with great diversity and versatility.

* * *

Therefore, it may be resolved then that the pairing of tactility and sight in the philosophical, medical, and theological sources speaks not to a haptic understanding of visual sensation, but rather to the overlooked cognitive processes of perception. Haptic language is used as a metaphor to understand how the mind is able to receive and make sense of sensory information, converting physical matter into noetic forms for the sake of thought and memory. To suggest then that sight is a species of touch would be to ignore the intricate complexities of Byzantine perceptual practices as they are affiliated with thought, cognition, memory, and action. In my close reading of Photios against the grain of Cyril Mango's translation, my goal has not been to produce a new normative theory of sight for Byzantium. Nor has it been to produce a comprehensive survey of all accounts of sight, but rather to elucidate how writers deployed and also struggled with the haptic language around vision while nevertheless conveying an avowed separation of touch and sight, where tactility was used to characterize the ineffable experience of contact between a viewer and a visible body.

It is necessary to remember that while Photios may have placed emphasis on a Platonic iteration of extramission as a union of rays in his homily on the Theotokos in Hagia Sophia, in his homily on the Pharos Chapel, he described the intricacy of the decoration suggesting that the effects of atomic intromission *could be said* to be occurring as he viewed the fine work of the pavement: as if the atoms of intromissive sight were "falling upon the eyes" (ὑπ' ὄψιν πιπτούσας). Photios does not necessarily subscribe to Democritus's theory of vision, suggested by the hypothetical tone of the sentence. Nevertheless, Photios exhibits his familiarity with what was one elementary understanding of vision, one among many in the Byzantine world. And, he uses it here to stress a perceptual state contrary to the supposed norm. The reference to atomic matter can even be found centuries later in the memorable account of Nikephoros Gregoras, following the coronation of Emperor John VI Kantakouzenos in 1347, that one would find nothing left in the imperial treasury, "but air and dust and, as they say, the atoms of Epicurus (τῶν Ἐπικουρείων εἰπεῖν ἀτόμων)."[56] Such passing comments not only speak to a longstanding popular familiarity with intromissive and atomic theory, but also index the preservation of the teachings of Democritus, Epicurus, and Leucippus, who were discussed by the key commentators on the senses and perception known to the Byzantine world, such as Theophrastus, Aetios, and Alexander of Aphrodisias.

As to any potential hierarchy of visual theories in the expansive Byzantine Empire, the close readings of this literature have already suggested that while the theory of the optical *pneuma* was indeed favored by medical practitioners, it could also be harmoniously combined with proper Platonic iterations of extramission, and even placed in dialogue with the Aristotelian iteration of intromission. Nicholas Mesarites's comment on the debate on sight in the colleges of Constantinople, as to "whether our power of sight directs itself outward or whether images are received by us" (εἰ κατ' ἐκπομπὴν ὁρῶμεν ἢ εἰσπομπήν),[57] should caution us against hastily ascribing any visual theory beyond that offered within our particular and specific objects of study.

While it is possible to infer or deduce theories of vision from the techniques and practices of artists and writers, these should each be treated as independent cases. Each case must be approached with great nuance and care as to the language, its sources, and the eclectic compositions of their theories. This careful approach would adequately acknowledge the embarrassment of riches offered by the tradition of philosophical thought and critical debate in an empire spanning over a thousand years in radically diverse spaces and cultural contexts. If we understand that vision is a lot more flexible and deployed far more self-consciously than previously believed, then we can seek out clues in citations of technical language and metaphor to decipher the theory to which the individual author subscribes. In doing this, however, we must also understand that sensation is not neatly separate from perception in our sources, and that perception itself is comprised primarily of a series of processes that are cognitive. Thus, they must be appraised with a deep understanding of the technical language and linguistic conventions of the optical treatise, as well as with a consideration of the metaphors used to characterize the mental, cognitive, and memory practices of the Byzantine viewer.

[56] "καὶ πλὴν ἀέρος καὶ κόνεως καὶ τῶν Ἐπικουρείων εἰπεῖν ἀτόμων, εὕρηται πλέον οὐδὲν ἐν αὐτοῖς." For a discussion of this text, see Cecily Hillsdale, *Byzantine Art and Diplomacy in an Age of Decline* (Cambridge University Press, 2014), 1. Cf. Nikephoros Gregoras, *Historia Romana*, ed. L. Schopen, *Historiae Byzantinae*, vol. II (Bonn: Weber, 1830), 790.

[57] Nicholas Mesarites, *Descriptio ecclesiae SS. Apostolorum*, XLII.5, ed. and trans. Downey, 895, 917.

PART III

Mediation, Veneration, Remediation

CHAPTER 9

Medium and Mediation

> Medieval optical theory poses similar questions centered on the role of the mediator, whether the species or the diaphanous medium. The same questions persist: does the mediator provide access to knowledge, or does the imperfection of its transmission make it a barrier to clear and perfect knowledge?[1]
>
> – Suzanne Conklin Akbari

In her magisterial analysis of optical theory and Western medieval allegory, Suzanne Conklin Akbari alerts us to the importance of understanding the medium of sight as a critical space for investigation given its role in contouring levels and orders of mediation and representation. In other words, her observations confront the fact that how one defines a sensory medium not only alters how communication through that medium is structured, but also to what extent said medium can enable degrees of presence, vividness, and actuality with regard to its representational capabilities. As Liz James noted, regarding the crucial impact of haptic sight in Byzantine Studies, "Once sight is conceived as tangible, then one's view of the world changes. One touches the world, grasps it, carries it back to the mind. Touching an icon becomes a form of seeing and vice versa."[2] In establishing a medium for sight, we are then not simply constructing a site of transparent transmission, but rather a condition of possibility for representation that consequently comes to determine the limits and boundaries of that representation, contouring the difference between representation and presence, presence and absence.

Medium, as discussed in this volume, is both sensory and perceptual. Thus, it is consequently also artistic, given that the bounds of perception delimit the possibilities of artistic mediums and what their represented

[1] Suzanne Conklin Akbari, *Seeing Through the Veil: Optical Theory and Medieval Allegory* (University of Toronto Press, 2004), 44.

[2] James, "Senses and Sensibility," 528.

forms can achieve. In this chapter, I wish to return to this foundational concern with medium. By building on the dynamics of the medium of sight articulated in Chapter 1 and developed across the intervening chapters, we can expand this conversation to consider then how said medium functions in terms of its artistic and linguistic mediation. In order to do this, I shall begin by asking how it is that we can define an artistic medium for the premodern world, and then move on to consider how theories of representation were articulated through the late antique *Progymnasmata*, ekphrases, and similar texts. Picking up where we left off in Part II, this discussion will further our thinking on the imagination and its imprints in terms of their claims to truth and presence in the context of representation's inherent dependence on a medium and its mediation.

This opens up questions regarding the possible mediation of the divine through ritual and images, and consequently the contingent dynamics of presence and absence. I shall consider these matters in the last two sections of this chapter using liturgical commentaries, *ekphrases* of religious images, and articulate defenses of image-veneration. The aim here is to flesh out the continuation of the classical and late antique conversation around mediation in theories of rhetorical representation; and likewise to understand how sight itself is deployed as a metaphor or rhetorical tool for conceptualizing the mediation of the divine. This not only allows me to demonstrate how sight relates to such issues of artistic mediation in religious spheres, but also to emphasize the important role that mediated theories of vision played in the defenses and articulations of religious veneration. It is not that vision or the icon allows for unmediated presence that makes them acceptable avenues of worship, but rather precisely the opposite. It is by virtue of its mediated representation that one is able to approach images for veneration, understanding the cautious caveats of their limits and limitations, which allow one to avoid problems such as idolatry or the misdirected worship of mere matter.

Medium

In attempting to articulate a theory of the artistic medium for the premodern world, John Guillory struggled precisely with the problem of selecting a term for this concept, as the premodern world may not have had the clear-cut, well-articulated language to encompass its definitions or effects. He observes, for example, that there is no Greek term for "medium" in Aristotle's *Poetics*, even though there is an articulation of the various arts occurring "in different things" (τῷ ἐν ἑτέροις μιμεῖσθαι).[3] In Aristotle's texts on perception, however, the notion of the medium does have an articulated and coherent term, used uniformly throughout the text and readily translatable into English as "the medium" (τὸ μεταξύ) – literally translated as a middle, an in-between, or an interval difference. This *metaxy* or *mesos* is the interval through which sight perception comes about, as when Gregory of Nazianzos states that sight cannot unite with the visible objects without the medium of light and air (ἐν μέσῳ φωτὸς καὶ ἀέρος).[4] When used as a substantive noun, the term comes to articulate a notion of mediation that is constructed as the intervallic space between perception and its perceptible. In touch's dissolution of the interval in haptic sight, the medium collapses both as a physical and cognitive space of separation. The Aristotelian medium of the transparent (διαφανής), which transmits form from the object to the eye through the impressions and conductions of colors upon this receptive body, consequently shares in the properties of the artistic medium as the condition that makes possible the perceptibility of the image. Neither the term *metaxy* nor its resulting logic could be said by any means to be limited to the perceptual theory of Aristotle. Instead, this problem of the middle and its mediation was pervasive in the thinking of antique and late antique theories of perception, which sought precisely to mediate the senses' levels of mediation, depending on the intellectual work they sought to undertake.

Figures like John Philoponos, a committed intromissionist, saw the eye as being affected by the things seen, yet attempted to stress the immediacy of perception as a way of collapsing and negating the mediating action of the intervening medium.[5] Despite this desire to see the immediacy of perception, John nevertheless deploys his language cautiously, clarifying for instance the language of tactility as being specific to cognition, whereas the sense organ itself is merely passively affected. In his commentary on Aristotle, for instance, he writes: "As a sense, it grasps and cognizes them (ἀντιλαμβάνεται αὐτῶν καὶ γινώσκει αὐτά), whereas as a natural body, it

[3] John Guillory, "Genesis of the Media Concept," *Critical Inquiry* 36 (2010), 321–362, esp. 322–333. Cf. Aristotle, *Poetica*, 1447a, ed. Rudolf Kassel, *De arte poetica liber*, corrected edition (Oxford: Clarendon Press, 1966), 8.

[4] "…ἢ τοῖς ὁρατοῖς πλησιάσαι τὴν ὄψιν δίχα τοῦ ἐν μέσῳ φωτὸς καὶ ἀέρος." Gregory of Nazianzos, *Orat.* 28.12 (*De theologia*), ed. J. Barbel, *Die fünf theologischen Reden* (Düsseldorf: Patmos-Verlag, 1963), 86.

[5] See Jean Christensen De Groot, "Philoponus on *De Anima* II.5, *Physics* III.3, and the Propagation of Light," *Phronesis* 28:2 (1983): 177–96. On his theory of vision, see Leslie MacCoull, "Notes on Philoponus' Theory of Vision," *Byzantion* 67:2 (1997): 558–62.

is materially affected (πάσχει) by them."[6] Here, John clearly draws a distinction between perception (αἴσθησις) and the material sensual organ (φυσικὸν σῶμα), utilizing this technical cognitive language of grasping to capture these active perceptual processes and juxtapose them against the passive sensual reception of information by the organs of sensation. Therefore, beyond deploying levels of mediation within specific rhetorical contexts, authors and commentators also used such synesthetic language to distinguish between different states and stages in the perceptual process. The result, then, is that even a passive intromissionist could then go on to emphasize cognition as actively invested, whereby sensual data comes in with immediacy as if unmediated, yet then must be actively processed and remediated by the mind. For example, we find telling variations within the works of figures like Plotinos and Porphyry as these authors attempted to grapple with their own theories of sensation and perception, and in different contexts.[7]

Throughout the antique, late antique, and Byzantine texts, we have witnessed the potentially harmful effects of haptic metaphors in the intellectual history of vision. Writers such as Aristotle, Galen, and Ptolemy often had to grapple with the pitfalls of this metaphorical language, making sure to qualify the explanatory comparisons between all the senses that might be gleaned from earlier sources – a process of clarification that becomes crucial in the work of late antique and medieval commentators. In a similar vein, as we have discussed, Galen attempted to think through the problems of the medium by virtue of the extension of the optical *pneuma*, which issued forth from the eye and united with the intervening air and light to conduct the colors from the object back to the eye. As discussed previously, Galen explicitly criticized the form of mediation enabled by the walking stick metaphor in the Stoics' theory, which likened sight to touch. Galen clarified that since touch cannot transmit color, size, and position this metaphor is wholly in error.

For Galen, the walking stick analogy makes the metaphor of touch too literal for comfort, potentially suggesting that sight is simply a species of

[6] "ὡς μὲν αἴσθησις ἀντιλαμβάνεται αὐτῶν καὶ γινώσκει αὐτά, ὡς μέντοι φυσικὸν σῶμα πάσχει ὑλικῶς ὑπ' αὐτῶν." John Philoponos, *In Aristotelis de anima libros commentaria*, ed. Hayduck, 433.

[7] See Miira Tuominen, "On Activity and Passivity in Perception: Aristotle, Philoponus, and Pseudo-Simplicius," in J. F. Silva and M. Yrjönsuuri (eds.), *Active Perception in the History of Philosophy: From Plato to Modern Philosophy* (London: Springer, 2014), 55–78. See also A. Mark Smith, *From Sight to Light: The Passage from Ancient to Modern Optics* (University of Chicago Press, 2014), esp. 130–50. On Plotinos, see Gordon H. Clark, "Plotinus' Theory of Sensation," *The Philosophical Review* 51:4 (1942): 357–82.

touch at a distance. And in this regard, the optical *pneuma* would then be nothing more than a conduit through which pressures are sent back to the body. This is a fallacy for Galen given that touch cannot grasp a colored body. While Galen fully submitted to the concept of the medium in this regard, he tried to reason a theory that posited this "middle" (μέσου) as an instantaneous sensory conduit for the body, attempting to approximate an unmediated model for the mediation for sight. Galen's revisions of Classical and Stoic theories of vision appear to be consistent with the aesthetic goals of his contemporaries, who, while acknowledging the necessity of mediation, attempted to forge rhetorical and structural models of representation outside the bounds of a medium condition. This section, by undertaking a very brief historiography of the intertwined histories of representation and sensual mediation, demonstrates how the rhetoric of sight's mediation directly informed and was informed by its contemporary theories of representation and that representation's own contingent mediation.

In literary theory, the concept that perhaps best grapples with the limitations and expanses of representation is the concept of *ekphrasis* (ἔκφρασις) and its critical operative *enargeia* (ἐνάργεια), which may be variously defined as the "visibility," "visual vividness," or "graphical quality" of a narrative.[8] *Enargeia* defines the qualities through which a verbal narrative can bring about an "inner vision" or *phantasia* of the story in the minds of its listeners as they imagine events unfolding before them. This would likewise also come to refer to descriptions of visual artifacts, as the term is often characterized in modern literature. This is a turn that is exploited in the Second Sophistic with works such as Philostratus the Elder's *Imagines* that surveyed the paintings in a (real or hypothetical) gallery. Rather than being a mere genre of its own, i.e. of verbal descriptions of visual objects, Jaś Elsner and others have carefully parsed out the various genres of *ekphrasis* itself, such as rhetorical descriptions of an object in a story that interrupt the pace and make an intervention in the narrative, epigrams on objects that contour how they have been seen and how they should be seen, and collections of descriptions of artworks that make up their own tomes.[9] On a basic level, as Elsner writes, *ekphrasis* "was a meditation on the viewing and reception of art, and even an education in how to look."[10]

[8] For one of the more comprehensive surveys and bibliographies on *ekphrasis*, see Michael Squire, "Ecphrasis: Visual and Verbal Interactions in Ancient Greek and Latin Literature," *Oxford Handbooks Online* (Oxford University Press, 2015), www.oxfordhandbooks.com/view/10.1093/oxfordhb/9780199935390.001.0001/oxfordhb-9780199935390-e-58.

[9] Jaś Elsner, "Introduction: The Genres of Ekphrasis," *Ramus* 31 (2002): 1–18.

[10] Elsner, "Genres of Ekphrasis," 15.

Ruth Webb has even argued quite provocatively that *ekphrasis* is not a genre, but rather a technique.[11] That is to say, as Simon Goldhill concisely puts it, "it is not an indicator of content."[12] Beyond being associated with description of buildings, objects, artworks, and so on, *ekphrasis* is simply about "the use of language to try to make an audience imagine a scene."[13] Hence, it is a technique that attempts to exceed the prescribed bounds of representation by stimulating the imagination in a manner that blurs whether the viewer is being told a story or witnessing that story unfold before them. However, not all have followed Webb's argument, which focuses primarily on *Progymnasmata*, or rhetorical handbooks. Michael Squire, for example, has cautiously noted that Webb's perspective has "more to do with the peculiar genre of the *Progymnasmata* than with ancient ideas about ecphrasis," citing that Philostratus had no problem describing his *Imagines* as being composed of *ekphrases* and that Homeric scholia list the description of the shield in the *Iliad* as *ekphrasis*.[14] I, nevertheless, follow here Webb's judgment on this matter given that the *Progymnasmata*, while certainly their own genre, are attempting to classify *ekphrasis* from a broad vantage point in contrast to other techniques and genres of rhetoric. In their definitions, they leave no doubt that this is a rhetorical technique rather than a genre of writing denoted through content. This is similar to the uses of *phantasia*. In that case, we observed that it can denote the mental faculty of the imagination or some form of inner vision, but it also refers to the images or impressions generated by and through this faculty. At certain times in its history, the two are kept strictly apart, if not almost disassociated from each other, while in other periods they are seen to be intimately tied and dependent on each other. The same seems to be the case with *ekphrasis*. While it is certainly a technique of description that prizes clarity and visual vividness in its exposition, the term can likewise be used to describe works, excerpts, or epigrams that foreground this technique. In other words, *ekphrasis* functions more as a literary medium, which in itself can contain various genres within it, rather than as a genre in and

[11] Ruth Webb, "*Ekphrasis* Ancient and Modern: The Invention of a Genre," *Word & Image* 15:1 (1999): 7–18.

[12] Simon Goldhill, "Review of Ruth Webb, *Ekphrasis, Imagination and Persuasion in Ancient Rhetorical Theory and Practice*," *Bryn Mawr Classical Review* 2009.10.03 (2009), http://bmcr.brynmawr.edu/2009/2009–10-03.html.

[13] Webb, *Ekphrasis*, 3.

[14] Michael Squire, *The Iliad in a Nutshell: Visualizing Epic on the Tabulae Iliacae* (Oxford University Press, 2011), 326–327. See also Michael Squire, "Making Myron's Cow Moo? Ecphrastic Epigram and the Poetics of Simulation," *American Journal of Philology* 131:4 (2010): 589–634.

of itself.[15] While there is little at stake in expanding *ekphrasis* to the level of technique, there is much to lose in taking *ekphrasis* to be solely synonymous with a particular genre, which limits its power and operation, particularly within the context of late antiquity and the Byzantine period.

In Byzantium, the concept of *ekphrasis* featured as part of the basic education on rhetoric in the form of students' composition exercises known as *Progymnasmata*, four versions of which survive from late antiquity: by Aelius Theon probably from the first century; Hermogenes of Tarsus probably from the second century; Aphthonios from the later fourth century; and Nicholas of Myra (also known as Nicholas Rhetor), who taught in Constantinople in the third quarter of the fifth century.[16] These earlier works would continue to attract commentaries by various thinkers and educators in the Byzantine world. For example, Theon's work was translated into Armenian and another composition handbook based on Aphthonios's work was produced;[17] and, in the twelfth century, Nikephoros Basilakes, an imperial notary and educator, produced his own *Progymnasmata*, which focused heavily on biblical themes, serving almost as a Christian expansion pack *per se* for the other *Progymnasmata*.[18]

Of particular influence, however, appears to be a work of Aphthonios from the late fourth century that attracted several commentaries. Aphthonios himself was a student of another influential teacher of rhetoric, Libanius of Antioch, who produced his own series of rhetorical exercises, which were compiled posthumously.[19] In his *Progymnasmata*, Aphthonios

[15] On my distinction between medium and genre, see Roland Betancourt, "Stop Crossing Mediums: Grace Helbig and the YouTuber's Medium," *Journal of Popular Culture* 49:1 (2016): 196–223.

[16] For an introduction and partial translations of these four *progymnasmata*, see George A. Kennedy, *Progymnasmata: Greek Textbooks of Prose Composition and Rhetoric* (Atlanta, GA: Society of Biblical Literature, 2003). For the full Greek texts, see Aelius Theon, *Progymnasmata*, ed. L. Spengel, *Rhetores Graeci*, vol. II (Leipzig, 1854), 59–130; Hermogenes, *Progymnasmata*, ed. H. Rabe, *Opera* (Leipzig: Teubner, 1913), 1–27, Aphthonios, *Progymnasmata*, ed. H. Rabe, *Progymnasmata*, Rhetores Graeci 10 (Leipzig: Teubner, 1926), 1 51; Nicholas Rhetor, *Progymnasmata*, ed. J. Felten, *Progymnasmata*, Rhetores Graeci 11 (Leipzig: Teubner, 1913), 1–79. See also Webb, *Ekphrasis*, 39–59.

[17] G. Muradyan, "The Rhetorical Exercises (*Progymnasmata*) in the Old Armenian 'Book of Chreia' (Girk 'Pitoyic'): Translation or Original Composition?," *Revue des Études Arméniennes* 27 (1998–2000): 399–415. See also Kennedy, *Progymnasmata*, xii.

[18] For an introduction and translation, see Jeffrey Beneker and Craig A. Gibson, *The Rhetorical Exercises of Nikephoros Basilakes:* Progymnasmata *from Twelfth-Century Byzantium*, Dumbarton Oaks Medieval Library 43 (Cambridge, MA: Harvard University Press, 2016).

[19] On Libanius and a translation of his *Progymnasmata*, see Craig A. Gibson, *Libanius's* Progymnasmata*: Model Exercises in Greek Prose Composition and Rhetoric* (Atlanta, GA: Society of Biblical Literature, 2008).

defines the concept of *ekphrasis* as follows: "Ekphrasis is a descriptive language, bringing what is shown vividly before the eyes" (Ἔκφρασίς ἐστι λόγος περιηγηματικὸς ὑπ᾽ ὄψιν ἄγων ἐναργῶς τὸ δηλούμενον).[20] The key to *ekphrasis* as a technique was a desire to overstep the boundaries of linguistic representation and "bring vividly before the eyes" (ὑπ᾽ ὄψιν ἄγων ἐναργῶς) the things being signified. Yet, as Nicholas carefully explains further, *ekphrasis* "tries to turn hearers into spectators" (ἣ δὲ πειρᾶται θεατὰς τοὺς ἀκούοντας ἐργάζεσθαι).[21] This suggests a certain failure in that attempt that "all but" (πειρᾶται) brings about sight through hearing, which is well attested in other *Progymnasmata*, such as that of Hermogenes. Simon Goldhill has noted that this qualification is important, for the power of *ekphrasis* is "a technique of illusion, semblance, of making to appear," and in doing so Goldhill suggests *ekphrasis* creates a particular viewer who is meant to see and respond in a certain way.[22] This is to say that *ekphrasis* is not simply a process of literary depiction, but also a crucial structure for the enculturation of the late antique and medieval subject, who not only learns to see through the description of seeing, but also learns to see when they hear. Hence we can begin to appreciate the effects that rhetorical training had on the manner in which late antiquity and Byzantium conceived the bounds of representation in relation to words and images.

For late antique writers, the training in *ekphrasis* encouraged a conception of rhetorical composition that aimed to transmute words into images, but respecting the inherent ability of said medium to do just that. The desire to intermix the two leads writers such as Asterios of Amaseia to boast, "For we, the children of the muses, have in no way less satisfying colors (φάρμακα) than painters."[23] The use of *pharmaka* for colors suggests the fracture of pigment from raw materials akin not only to drug making, but also to the understanding of colors as a treatment or

[20] Aphthonios, *Progymnasmata*, 216.20–24, ed. Rabe, 36–37; modified trans. Kennedy, 117.

[21] Nicholas of Myra, *Progymnasmata*, ed. Felten, 68; trans. Kennedy, 166.

[22] Simon Goldhill, "What Is Ekphrasis for?," *Classical Philology* 102:1 (2007): 1–19, esp. 3 and 18. See also Michael Squire, "Apparitions Apparent: Ekphrasis and the Parameters of Vision in the Elder Philostratus's *Imagines*," *Helios* 40:1–2 (2013): 97–140, esp. 100–101.

[23] "οὐδὲ γὰρ φαυλότερα πάντως τῶν ζωγράφων οἱ μουσῶν παῖδες ἔχομεν φάρμακα." Asterios of Amaseia, *Descriptio picturae sanctae Euphemiae*, 1.15–16, ed. F. Halkin, *Euphémie de Chalédoine: Légendes byzantines*, Subsidia hagiographica 41 (Brussels: Société des Bollandistes, 1965), 5; modified trans. J. Leemans et al., *"Let Us Die that We May Live": Greek Homilies on Christian Martyrs from Asia Minor, Palestine and Syria* (London: Routledge, 2003), 175. See also Ruth Webb, "Accomplishing the Picture: Ekphrasis, Mimesis, and Martyrdom in Asterios of Amaseia" in Liz James (ed.), *Art and Text in Byzantine Culture* (Cambridge University Press, 2007), 13–32.

intervention.[24] In Gorgias of Leonti's *Encomium of Helen*, from the late fifth century BCE, the author writes:

> τὸν αὐτὸν δὲ λόγον ἔχει ἥ τε τοῦ λόγου δύναμις πρὸς τὴν τῆς ψυχῆς τάξιν ἥ τε τῶν φαρμάκων τάξις πρὸς τὴν τῶν σωμάτων φύσιν. ὥσπερ γὰρ τῶν φαρμάκων ἄλλους ἄλλα χυμοὺς ἐκ τοῦ σώματος ἐξάγει, καὶ τὰ μὲν νόσου τὰ δὲ βίου παύει, οὕτω καὶ τῶν λόγων οἱ μὲν ἐλύπησαν, οἱ δὲ ἔτερψαν, οἱ δὲ ἐφόβησαν, οἱ δὲ εἰς θάρσος κατέστησαν τοὺς ἀκούοντας, οἱ δὲ πειθοῖ τινι κακῆι τὴν ψυχὴν ἐφαρμάκευσαν καὶ ἐξεγοήτευσαν.
>
> The effect of speech upon the structure of the soul is as the structure of drugs over the nature of bodies; for just as different drugs dispel different secretions from the body, and some bring an end to disease, and others to life, so also in the case of speeches some distress, others delight, some cause fear, other embolden their hearers, and some drug and bewitch the soul with a kind of evil persuasion.[25]

This use here precisely compares the use of "drugs" (φάρμακα) to different styles of rhetoric that similarly have effects on the soul, just as in Asterios the comparison of rhetoric to painting through this term is stirred up by the intense emotional effects of the painting of St. Euphemia's life and trial that astonishes the author.[26] Similarly, the use of *chromata* for colors also evidences these double meanings across painting and rhetoric, since the term can refer to painted color as well as rhetorical ornamentation and style, as when Plato discusses in the *Republic* the "poetical colors" (μουσικῆς χρωμάτων) of the poet's words.[27]

Furthermore, the use of *pharmaka* for painted colors is likewisc well attested, as in Lucian's first-century *Imagines* where Polystratus states:

> ὁπόσα γὰρ ἢ ποιηταὶ μέτροις διακοσμήσαντες ἢ ῥήτορες δεινότητι κρατύναντες ἐξενηνόχασιν ἢ συγγραφεῖς ἱστορήκασιν ἢ φιλόσοφοι

[24] On *pharmaka* and rhetoric, see Jeffrey Walker, "*Pathos* and *Katharsis* in 'Aristotelian' Rhetoric: Some Implications" in Alan Gross and Arthur Walzer (eds.), *Reading Aristotle's* Rhetoric (Carbondale, LA: Southern Illinois University Press, 2000), 74–92; Michael A. Rinella, *Pharmakon: Plato, Drug Culture, and Identity in Ancient Athens* (Lanham, MD: Lexington Books, 2010), esp. 205–231.

[25] Gorgias of Leonti, *Encomium in Helenam* (Frag. 11:14.87–93), ed. H. Diels and W. Kranz, *Die Fragmente der Vorsokratiker*, vol. II, 6th edition (Berlin: Weidmann, 1952), 292–293; trans. John Dillon and Tania Gergel, *The Greek Sophists* (London: Penguin Books, 2003), 81–82.

[26] See Webb, "Accomplishing the Picture," esp. 23–27. On *ekphrasis* and emotional response, see Ruth Webb, "Imagination and the Arousal of Emotions in Greco-Roman Rhetoric" in Susanna Morton Braund and Christopher Gill (eds.), *The Passions in Roman Thought and Literature* (Cambridge University Press, 1997), 112–127.

[27] Plato, *Respublica*, 601b, ed. and trans. Christopher Emlyn-Jones and William Preddy, *Republic*, vol. II, Loeb Classical Library 276 (Cambridge, MA: Harvard University Press, 2013), 414–415.

> παρηνέκασι, πᾶσι τούτοις ἡ εἰκὼν κεκοσμήσθω, οὐκ ἄχρι τοῦ ἐπικεχρῶσθαι μόνον, ἀλλ' εἰς βάθος δευσοποιοῖς τισι φαρμάκοις εἰς κόρον καταβαφεῖσα.
>
> For all that poets have set forth with the embellishment of meter or orators with the might of eloquence, all that historians have related or philosophers recommended shall give beauty to our picture, not simply to the extent of coloring its surface, but staining it all deeply with some tinting colors till it will take no more.[28]

Lucian elucidates here with concise clarity the interrelation between the rhetoric of poets, philosophers, and historians and the art of painters, explaining the use of "some *pharmakois* " (δευσοποιοῖς τισι φαρμάκοις) for their coloring or tinting effects upon the surface of the painting. Lucian's use demonstrates the almost agentic power of the color as some potion or concoction, while also stressing the commonality of the medium across the arts. This articulates color as a technical term that suggests that its operation is categorized here not by tincture alone, but rather by a shared intervention upon the viewer or listener, hence why it can be seen to operate across visual and rhetorical arts.

Asterios even uses Lucian's same language of tincture or coloring (ἐπικεχρῶσθαι) when he describes the blood of the martyr's torture as they cut out her teeth, saying "the painter has indeed applied (ἐπέχρωσεν) the drops of blood with such realism that you might say they really (ἀληθῶς) stream from her lips."[29] The understanding of colors (φάρμακα) here as a physical mixture, like the drugs from an apothecary, dabbed or smeared upon the painting allows the tincture to transmute into that earthly blood shed by the martyr. It is thus as if the artist has applied (ἐπέχρωσεν) blood itself to the panel, rather than the materials for representation, so much so that one "might say" (εἴποις) that it really is blood. The effect of this dynamic vividness makes it seem precisely as if that which is being represented is actually present, so much so that, as Ruth Webb notes, the viewers "may even lose sight of the fact that they are responding to an illusion and react as if they were indeed present."[30] This is precisely the operation and effect of the vividness (ἐνάργεια) praised by the rhetoricians in *ekphrasis*. Asterios's learned and self-reflexive language,

[28] Lucian, *Imagines*, 16, ed. and modified trans. A. M. Harmon, *Lucian*, vol. IV (Cambridge, MA: Harvard University Press, 1925), 284–285.

[29] "τὰς γὰρ τοῦ αἵματος σταγόνας οὕτως ἐναργῶς ἐπέχρωσεν ὁ γραφεὺς ὥστε εἴποις ἂν προχεῖσθαι τῶν χειλέων ἀληθῶς." Asterios of Amaseia, *Descriptio picturae sanctae Euphemiae* (*Homilia* 11), 4.7–10, ed. Halkin, 7; modified trans. Leemans et al., 176.

[30] Webb, "Accomplishing the Picture," 23.

blurring the teaching and training of painter/writer (ὁ γραφεὺς), provides a deep historical legacy for Asterios's words. He demonstrates both the shared skill or technical knowledge of these "children of the muses," and also his own knowledge of ancient art, *ekphrasis*, and terminology that would have been familiar to Christian and pagan audiences alike.[31]

The question of whether Byzantine artists might have consciously borrowed from ancient rhetoric is debatable. In his 1981 book on the interrelation between art and rhetoric, Henry Maguire weighed in on this matter, writing that: "Byzantine artists borrowed several techniques from ancient rhetoric, but I wish to stress that I do not believe that the artists themselves were conscious of the rhetorical origins of these methods."[32] Going on to herald Leonardo da Vinci as the next artist who would self-consciously deploy art and rhetoric together, Maguire's conclusion is not persuasive, given that he has precisely been able to elaborate on the many ways in which artists did indeed borrow from ancient rhetorical techniques. Many of the concepts he discusses, including *ekphrasis*, were part of the most basic educational training, along with the recitation and copying of canonical texts and followed later by a proper education in composition and rhetoric, deploying often the *Progymnasmata* discussed herein.[33] Jaś Elsner, writing about the interrelation between art and rhetoric in late antiquity, notes rhetorical education was a cultural phenomenon that spanned all aspects of education, including everyone from slaves to freedmen to the elites, from the time of the Republic and well into the Christian empire.[34] While this training may have varied for artists themselves in Byzantium, it is certainly unnecessary to discredit or argue against Byzantine artists' self-reflexive use of ancient rhetorical training. Even if artists themselves were not consciously deploying these ideas, this

[31] On Asterios and his training, see Elizabeth Castelli, "Asterius of Amasea: Ekphrasis on the Holy Martyr Euphemia" in Richard Valantasis (ed.), *Religions of Late Antiquity in Practice* (Princeton University Press, 2000), 464–468, W. Speyer, "Asterios von Amaseia," *Reallexikon für Antike und Christentum*. Supplement 4 (1986), 629–639; Speyer, "Die Euphemia-Reden des Asterios von Amaseia: Eine Missionsschrift für gebildete Heiden," *Jahrbuch für Antike und Christentum* 14 (1971): 39–47.

[32] Henry Maguire, *Art and Eloquence in Byzantium* (Princeton University Press, 1981), 110.

[33] On rhetorical education in Byzantium, see Herbert Hunger, *Die hochsprachliche profane Literatur der Byzantiner*, vol. 1 (Munich: Beck, 1978), esp. 92–120. For earlier periods of education, see George A. Kennedy, *Greek Rhetoric under Christian Emperors* (Princeton University Press, 1983); George A. Kennedy, *A New History of Classical Rhetoric* (Princeton University Press, 1994); Stanley Porter, *Handbook of Classical Rhetoric in the Hellenistic Age, 330 BC – AD 400* (Leiden: Brill, 1997).

[34] Jaś Elsner, "Introduction" in Jaś Elsner and Michel Meyer (eds.), *Art and Rhetoric in Roman Culture* (Cambridge University Press, 2014), 1–32, esp. 2.

learned rhetoric was clear in the texts defending and articulating the contours of the icon. As we have seen in the iconophiles' defenses for images, to attack the image would be to attack the Gospel as well since both result in images in the mind, pictures merely presenting viewers with readymade images for the mind to grasp – the very issues that led Photios to compose his comparison between hearing and seeing in his Homily 17 and carefully detail the processes of vision and hearing as a defense.

The problem that emerges here, for the ancient and modern thinker alike, is to what extent can language actually transgress the bounds of its traditional medium? That is to say, to what extent can we believe or take at face value the notion that readers and listeners could be made into spectators, as Nicholas's fifth-century rhetorical treatise phrases it?[35] While *ekphrasis* had the desire and intention to exceed language, it was the notion of "vividness" (ἐνάργεια) that enabled this transmutation across the boundaries of an individual sense or medium. *Enargeia* is achieved, as Roos Meijering has argued, through two main classes: first, by the "completeness of depiction," that is through the clarity of careful use of detail, both in quantity and specificity; and, secondly, by *translatio temporum*, that is through the temporal translation of past events into the present tense, so as to appear as if they are occurring in the present.[36] Like *phantasia*, the term *enargeia* is difficult to translate since it encompasses a host of qualities and mental functions within its use. Often rendered as "vividness," the term is meant to indicate the generation of mental images that manifest the content of the literary or artistic work being perceived. Thus, as René Nünlist has exceptionally demonstrated, *enargeia* is less indicative of the liveliness or vivacity of a narrative, as it "designates the graphic description that enthralls the audience," which goes on to render the "reader into a spectator," as the rhetorical textbooks recount.[37] There, Nünlist suggests that *enargeia* be translated as the "graphic quality" (or, adverbially, "graphically") so as to properly denote that this operation was a vividness understood in terms of its visual impact.

Inviting connections and comparisons between literary and visual art, this function of *enargeia* allowed one scholar, for example, to state that Homer's description of Ajax was "more graphic even than painting"

[35] Nicholas of Myra, *Progymnasmata*, ed. Felten, 67–71.

[36] Roos Meijering, *Literary and Rhetorical Theories in Greek Scholia* (Groningen: Egbert Forsten, 1987), 39–44.

[37] Nünlist, *The Ancient Critic at Work*, 194.

(καὶ ζωγραφίας ἐναργέστερον).[38] The "mental image" generated by *enargeia*'s graphic aspect was precisely *phantasia*, as evidenced in one scholium praising Homer's ability to project "the entire mental image so graphically that the listeners are captured no less than the spectators" (πᾶσαν φαντασίαν ἐναργῶς προβέβληται ὡς μηδὲν ἧττον τῶν θεατῶν ἐσχηκέναι τοὺς ἀκροατάς).[39] Nina Otto has traced out the history of *enargeia* to show its origins, associated with Epicurus on sense perception, and its development in the Stoa with associations to the operations of the *katalēptikē phantasia* that stress the impact and reliability of sensory impressions.[40] Alessandra Manieri before her has surveyed the interrelation between *phantasia* and *enargeia*, in a study which demonstrates the relationship between the two terms in Greek rhetorical theory and philosophy, wherein *phantasia* thrives and is fully and vividly achieved through the action of *enargeia* itself.[41] Critically, however, it is necessary to understand that *enargeia* did not merely rely on or deploy visual details for the generation of vivid images for the mind, but also included other sensory details. This included sound effects in the act of reading that encompassed mimetic sounds and onomatopoeia, as Meijering notes.[42] Thus, while Nünlist's focus is on the "graphic" quality of the term and its visualization of the narrative in the minds of readers, it is necessary to keep in mind that this process is not solely visual. For this reason, a qualified use of "vividness" might still be more apt for the translation of *enargeia*, just as is the rendering of *phantasia* as the "imagination" or a "mental form," when describing the products of the imagination. Both vividness and imagination do not circumscribe these operations to the visual alone, but speak to a system whereby the various senses are woven together, almost through a form of perceptual synesthesia.

In the Byzantine period, as Stratis Papaioannou's study of *enargeia* demonstrates, the use of this term in rhetoric and theories of representation suggests "the self-evidence of truth beyond mediation" in the tradition

[38] Anonymous, *Scholia in Iliadem*, 16.107–11 ex, ed. H. Erbse, *Scholia Graeca in Homeri Iliadem (scholia vetera)*, 5 vols. (Berlin: De Gruyter, 1969–1977), IV:187.

[39] *Scholia in Iliadem*, 23.362–72, ed. Erbse, *Scholia Graeca.*

[40] Nina Otto, *Enargeia: Untersuchung zur Charakteristik alexandrinischer Dichtung* (Stuttgart: Franz Steiner Verlag, 2009), esp. 37–66. See also Anthony A. Long, "Aisthesis, Prolepsis and Linguistic Theory in Epicurus," *Bulletin of the Institute of Classical Studies* 18:1 (1971): 114–133; Graham Zanker, "*Enargeia* in the Ancient Criticism of Poetry," *Rheinisches Museum für Philologie* 124 (1981): 297–311.

[41] See Alessandra Manieri, *L'immagine poetica nella teoria degli antichi: phantasia ed enargeia* (Pisa: Istituti editoriali e poligrafici internazionali, 1998).

[42] Meijering, *Literary and Rhetorical Theories*, 42–43.

of Stoic and late antique articulations.[43] Papaioannou demonstrates that this is attested to in the works of many significant writers and theologians, such as Philo of Alexandria, Basil of Caesarea, Gregory of Nyssa, and Proklos. Maximos the Confessor, for example, uses the term to describe "the true and active presence" of the divinity (ἐναργῆ τε καὶ ἐνεργὸν παρουσίαν) in his *Mystagogia*.[44] In this text on the liturgy, where Christ is indeed made present in "truth and activity," this statement is correctly metaphysical, exceeding the bounds of an aesthetic, rhetorical theory since the Eucharist alone offers a form of presence not possible from the icon given its consubstantial nature with the body of Christ in *essence* but not *form*. But as Papaioannou goes on to demonstrate, in the post-Iconoclastic period there is a shift in emphasis onto the inability of signification to enable full presence.[45] Thus, we may see Galen's theory of an interactionist extramission as a model of and model for the late antique desire for a notion of re-presentation, where discourse and images could lead to vivid, nearly unmediated realities; yet one that nevertheless repeatedly comes to terms with this impossibility. By the Middle Byzantine period, a definitive shift has occurred that should be attributed to the lessons of the iconomachies and their cultivation of a robustly nuanced semiotic program.

In the middle of the ninth century, John, bishop of Sardis, attests to a move away from desires for unmediated presence in his commentary on Aphthonios's *Rhetorical Exercises*. John writes:

> τὸ 'ὕπ' ὄψιν ἄγων' οἱονεὶ ἀμυδρότερον· τὸ μὲν γὰρ καθαρὰν θέαν δίδωσι τὸ δὲ τύπον ψιλὸν καὶ φαντασίαν τοῦ πράγματος· κἂν γὰρ μυριάκις ἐναργὴς εἴη ὁ λόγος, ἀδύνατον αὐτὸ κατ' ὄψιν ἀγαγεῖν 'τὸ δηλούμενον' ἤτοι ἐκφραζόμενον.
>
> The "under one's sight (ὕπ' ὄψιν)" [is said] as if to mean "less distinctly." For [to bring in front of (κατ') one's sight] gives a pure vision, while [to bring under one's sight] gives a mere impression (τύπον ψιλὸν) and imagination (φαντασίαν) of the thing. For even if the speech were ten thousand times vivid, it would be impossible for it to bring in front of one's sight "the thing shown," namely the thing itself.[46]

[43] Stratis Papaioannou, "Byzantine *Enargeia* and Theories of Representation," *Byzantinoslavica* 3 (2011): 48–60, esp. 52. For the use of the term in antiquity and late antiquity, see Webb, *Ekphrasis*, 87–130.

[44] Maximos the Confessor, *Mystagogia*, 7:35–36 (685C), ed. Raffaele Cantarella, *La mistagogia ed altri scritti* (Florence: Testi Cristiani, 1931), 164.

[45] See also Barber, *Figure and Likeness*.

[46] John of Sardis, *Commentarium in Aphthonii progymnasmata*, ed. H. Rabe, *Commentarium in Aphthonii progymnasmata*, Rhetores Graeci 15 (Leipzig: Teubner, 1928), 216.22–24; modified trans. Papaioannou, "Byzantine *Enargeia*," 55.

Here, we can see John of Sardis elaborating on the precision and clarity of Aphthonios's language, which, while stressing the action of *ekphrasis* and *enargeia*, still is careful to articulate the mediated nature of its effects. By setting deeds under sight (ὑπ' ὄψιν), rather than in front of sight (κατ' ὄψιν), Aphthonios has made sure to stress that this is almost an idiomatic use of the phrase, which suggests something being brought to the mind's eye or to our consciousness, rather than an actual manifestation of presence. John clarifies this by explaining sight as being precisely articulated by cognitive actors, namely the "mere impression and imagination" (τύπον ψιλὸν καὶ φαντασίαν) of the thing being represented through the imagination.

This impression (τύπον) is not something derived as a by-product of sense perception as it was understood in the Stoic articulation of *phantasia katalēptikē*, but rather the impression of the images that are being communicated through a medium, in this case – rhetoric. The qualification of the adjective "mere" (ψιλόν) specifically implies something that is naked or bare, in order to articulate that one perceives here the "stripped impression." More to the point, what one perceives here might be better translated as "the imagination of and impression unaccompanied by the perceptible object" (τύπον ψιλὸν καὶ φαντασίαν τοῦ πράγματος). For Verity Platt, the conceptual logic of *phantasia katalēptikē* produces a model of medium and mediation whereby these "clear and distinct impressions," in Michael Frede's words,[47] foreground communication with others as reliant upon one's assent to these unquestionable impressions of the imagination; so, as Platt writes, "the authority and authenticity of personal and legal relationships is dependent upon the impression of a personal seal."[48] In the Graeco Roman world, not only did the language of impressions speak to the logic of cognition spilling out into legal practices, but also undergirding this metaphorical terminology of cognition and communication was the conceptualization of the importance of stamping and impressing.

What we observe in John of Sardis's commentary is arguably the opposite, whereby the philosophical alterations to the *phantasia* and an image theory rooted on the *typos* over the late antique and Byzantine period have enabled this separation. This is an argument that Simon

[47] Michael Frede, "Stoic and Skeptics on Clear and Distinct Impressions" in Myles Burnyeat (ed.), *The Skeptical Tradition* (Berkeley: University of California Press, 1983), 65–93.

[48] Verity Platt, "Making an Impression: Replication and the Ontology of the Graeco-Roman Seal Stone," *Art History* 29:2 (2006): 233–257, esp. 249.

Goldhill sees as having been brewing since Longinus's *On the Sublime* between what he calls *phantasia* in the "fashionable sense," following Longinus, versus the Stoic philosophical sense: that is, *phantasia* conceived as an image-making function for visualizations or "image productions" (εἰδωλοποιΐας),[49] rather than as a cognitive impression.[50] Hence, John of Sardis's articulation of this "bare imprint" (τύπον ψιλὸν) is notable for clarifying earlier commentaries according to eleventh-century approaches to the *phantasia* and its affordances in mediation. His articulate emphasis on the impossibility (ἀδύνατον) of bringing before sight proper the "thing-being-described itself" (ἤτοι ἐκφραζόμενον) ensures that the reader understands precisely the limits of representation and that a properly Stoic definition of imprints and imagination does not deter the understanding of rhetoric's affordances.

Ruth Webb, however, has suggested that the late antique rhetoricians used *phantasia* in a "looser way" than the Stoics, and therefore these mental images had no direct bearing on them being a judgment of truth, which she positions against Jaś Elsner's understanding of the truth value in *ekphrasis*' deployment of the imagination.[51] Yet, as I have shown already in previous chapters, the composition of *phantasia* in the late antique and Byzantine world was syncretic and eclectic, retaining the language of imprints, assent, and comprehension from Stoic *phantasia* and its Aristotelian precedents, but fused with a more general understanding of *phantasia* as an image-generating faculty of the mind. Thus, rather than seeing Longinus's (possibly) disparaging comment on the "fashionable" use of the term as a violation of the norm, we must see this as evidence of the permutation of *phantasia* into a much broader articulation of mental images. This definition could nevertheless still deploy its Stoic precedents as a form of validation for its imprints' veracity and power, understanding that *ekphrasis* did seek the vividness of truth in mental representation despite its inability to make present in the physical, perceptible world those objects.

Later in the commentary, John of Sardis once again returns to the matter at hand, this time rephrasing and explaining Aelius Theon's perspectives on the virtues of *ekphrasis* and *phantasia*. Responding to Theon's

[49] Longinus, *De sublimitate*, 15.1, ed. and trans. W. H. Fyfe, *On the Sublime*, revised Donald Russell, Loeb Classical Library 199 (Cambridge, MA: Harvard University Press, 1995), 214–215.

[50] Goldhill, "What Is Ekphrasis for?," 6.

[51] Webb, *Ekphrasis*, 107–130, esp. 117. See also Jaś Elsner, *Art and the Roman Viewer: The Transformation of Art from the Pagan World to Christianity* (Cambridge University Press, 1995), esp. 26–28.

statement on the virtues of *ekphrasis*, notably the "vividness of nearly seeing" (ἐνάργεια τοῦ σχεδὸν ὁρᾶσθαι),[52] John writes:

> Θέων δὲ ἀρετὰς λέγει ἐκφράσεως σαφήνειαν καὶ ἐνάργειαν τοῦ σχεδὸν ὁρᾶσθαι τὰ ἀπαγγελλόμενα· ἐναργὲς γὰρ τὸ λίαν φανερὸν καὶ τοῖς ὀφθαλμοῖς ὑποπῖπτον· εἰ γὰρ σαφὴς καὶ ἐναργὴς εἴη ὁ λόγος, ἀπὸ τῆς ἀκοῆς εἰς τοὺς ὀφθαλμοὺς σχεδὸν τὰ λεγόμενα μεθίστησιν· ὁ γὰρ λόγος τὰ δηλούμενα θεωρῶν τούτων τοῖς ὀφθαλμοῖς ὑπογράφει τὸν τύπον καὶ τῇ φαντασίᾳ ζωγραφεῖ τὴν ἀλήθειαν.
>
> Theon says that the "virtues of *ekphrasis* are clarity and the vividness of nearly seeing what is described." Vividness is what is very evident and presents itself to the eyes. For if the language is clear and vivid, what is said is nearly transferred from hearing to the eyes; for discourse, contemplating the things shown, inscribes the imprint of these things in the eyes and paints the truth in the imagination.[53]

John's elaboration here almost seems to respond to the concerns regarding the truth value of *ekphrasis* in this "looser" or "fashionable" definition of *phantasia*. He operates once again on an understanding of *phantasia* as a visualizing faculty that retains the Stoic language of imprints. It is as if John is self-reflexively even playing with this language, given that he not only discusses the imprint (τύπον), but precisely describes it as being traced or inscribed (ὑπογράφει), which alludes to the seal-like properties of the *phantasia*'s clear and distinct impressions. Yet, this term also is used commonly to denote the act of imagining, visualizing, or calling up a mental picture as attested to by Lampe.[54] In *De Sacerdotio*, for example, John Chrysostom uses the same construction that we see here as "inscribes in the eyes" (τοῖς ὀφθαλμοῖς ὑπογράφει), whereby the verb takes the dative "in/with/by the eyes" (ὀφθαλμοῖς) to indicate the act of mental visualization when he commands, "Visualize in the eyes Elijah for me, and the crowd standing around him," and so on.[55] This is worth noting since it helps us better interpret John of Sardis's words when he says to "visualize in the eyes the imprint" (τοῖς ὀφθαλμοῖς ὑπογράφει τὸν τύπον), allowing us to understand this as an idiomatic construction referring to calling up a

52 "ἀρεταὶ δὲ ἐκφράσεως αἵδε, σαφήνεια μὲν μάλιστα καὶ ἐνάργεια τοῦ σχεδὸν ὁρᾶσθαι τὰ ἀπαγγελλόμενα." Aelius Theon, *Progymnasmata*, 11, ed. Spengel, 119; modified trans. Kennedy, 47.

53 John of Sardis, *Commentarium in Aphthonii progymnasmata*, ed. Rabe, 224.21–225.4; modified trans. Kennedy, 219.

54 G. W. H. Lampe, *A Patristic Greek Lexicon* (Oxford University Press, 1961), 1146.

55 "Ὑπόγραψόν μοι τὸν Ἠλίαν τοῖς ὀφθαλμοῖς καὶ τὸν ἄπειρον ὄχλον περιεστῶτα..." John Chrysostom, *De Sacerdotio*, III.4.33–34, ed. A.-M. Malingrey, *Sur le sacerdoce*, Sources chrétiennes 272 (Paris: Éditions du Cerf, 1980), 144–146.

mental image, rather than suggesting that the imprints are somehow visible to the corporeal eyes themselves.

When he turns his attention to the *phantasia*, however, here we see that the imprint of the visualization does indeed carry a value of truth by virtue of its clarity and vividness, for the impression that *ekphrasis* visualizes then goes on to "paint the truth" (ζωγραφεῖ τὴν ἀλήθειαν) in the imagination. The inclusion of "in the eyes" (τοῖς ὀφθαλμοῖς) in this process almost seems to mime the process of sense perception, as if those imprints had themselves been received from sight itself. Thus, going against Ruth Webb's comment that the rhetorician's use of *phantasia* had "no automatic judgment as to the truth of that image," here we see that John of Sardis is suggesting precisely the opposite, given that the imprints appear to be traced out by the eyes and in the imagination go on to depict the truth. While some of this discrepancy might be attributable to a difference as to how these modern scholars are using "truth," the disjuncture echoes the philosophical debate between a proper Stoic definition of the *phantasia* or "cognitive impressions," which lead us inexorably to assent, and the articulations of *phantasia* that nevertheless necessitate that one soberly judge the imprints of perception before assenting to their forms. Thus, this "looser" definition of *phantasia* is by no means ignorant of its history, but rather we see that John of Sardis is working with that to update and revise these earlier texts upon which he is commenting.

Therefore, Jaś Elsner was quite correct in asserting the truth value of *ekphrasis*. As he succinctly characterizes the matter: "Ekphrasis is therefore always concerned with Truth. It is true, however, not to the material reality of a particular image or sculpture, but to the *phantasia* which the speaker experiences in seeing the work of art and which he experiences in delivering his description."[56] Yet, this truth value is naturally not to be taken for granted, nor uncritically. Simon Goldhill has grappled with the implications of this claim to truth, which almost seems to bypass the safeguards of sound perception and cognition which Stoic *phantasia* sought so hard to nuance. As Goldhill argues, this is a particularly poignant conundrum for the conceptualization of the "psychological affect" of *phantasia* as visualization, since in its ability to astonish, dazzle, and amaze, it has the power to perhaps even conceal facts and blind the viewer with its brilliance as a "*weapon* of rhetoric."[57] The danger of this rhetorical weaponry is that *phantasia*, by virtue of its cooperation with *enargeia*, has the power to, in Goldhill's words, "get around the censor of the intellect, to

[56] Elsner, *Art and the Roman Viewer*, 27. [57] Goldhill, "What Is Ekphrasis for?," 5.

cut the listener off from the facts, to leave him not just 'as if a viewer at events,' but with the destabilizing emotions of that event."[58] This was the clout that early Christian figures like Evagrios feared, yet it is also the factor which enabled iconophiles like John of Damascus and Theodore of Stoudios to foreground the icon upon the imaginative investments of the viewer in the image and the viewer's own cognitive predisposition towards images. For if the *phantasia* were not a crucial image-making device with an intimate relation to the revelation of truth, the veneration of icons never could have been supported on the grounds of the imagination, nor could so many Byzantine texts herald the operation of the imagination in sacred sight. Yet, as we have seen, this is by no means new or unique to Byzantium, but in fact can be traced back to the acts of viewing that dominated ancient Greek art, including but not limited to religious images, wherein *phantasia* plays a central role in the viewing of art, as Irmgard Männlein-Robert has shown around the figure of Phidias and his works.[59] Thus, it is unwise to presume that *phantasia* had clearly distinct uses in the late antique and Byzantine periods that followed Stoic definitions precisely, even if they did retain some of their accessory terms and vocabulary. Instead, it is necessary to understand that commentators were actively clarifying earlier writings so as to reflect their own contemporary and more syncretic definitions of *phantasia* as an image-making visualization, that was nevertheless rooted on its Stoic (as well as Aristotelian) intellectual heritage and terminology.

It is not surprising that these clarifications would appear in the ninth century, after the iconomachy, and in keeping with the formalist distancing of the icon from presence as evidenced in the writings of Patriarch Nikephoros and Theodore of Stoudios.[60] Thus, we see rhetorical theory moving in lockstep with visual, artistic theory so as to clarify and qualify alleged claims to full presence in earlier texts and sources. Elsewhere, Papaioannou has characterized this emphasis on mediated presence and the impossibility of actual presence reflected in various aspects of Middle Byzantine thinking and literature as attested, for example, in the epistolary tradition.[61] Thus, I argue that what can be surmised from the late antique

[58] Goldhill, "What Is Ekphrasis for?," 6.

[59] Irmgard Männlein-Robert, "Zum Bild des Phidias in der Antike: Konzepte zur Kreativität des bildenden Künstlers" in Thomas Dewender and Thomas Welt (eds.), *Imagination – Fiktion – Kreation: Das kulturschaffende Vermögen der Phantasie* (Munich: K. G. Saur, 2003), 45–67.

[60] Charles Barber, "From Image into Art: Art after Byzantine Iconoclasm," *Gesta* 34:1 (1995): 5–10.

[61] See Stratis Papaioannou, "Der Glasort des Textes: Selbstheit und Ontotypologie im byzantinischen Briefschreiben (10. und 11. Jh.)" in W. Hörandner, J. Koder, and M. Stassinopoulou (eds.), *Wiener*

and Byzantine history of *ekphrasis* and *enargeia* is precisely the deep cultural shifts occurring throughout the earlier centuries of the Byzantine Empire, where the challenges of the iconoclastic debates forced an articulation and enforcement of a robust theory of mediation, which seems to endure well into the fourteenth century at least. As may be witnessed in the case of Galen in late antiquity and in the work of Michael Psellos in the eleventh century,[62] the interrelation between the ontology of vision and the image may be located in the epistemological question of how we come to know the presence of the image, or representation altogether.

In relation to icons, their treatment "as if" they were living remains a matter of crucial importance because it betrays a deep indebtedness to the representational logic of *ekphrasis* and its aspirations to vividness, even if actual presence is nevertheless impossible, as John of Sardis reminds us. In Evagrios's *Vita* of Pankratios of Taormina, a disciple of St. Peter, the author recounts how after the saint's death he built a church to his memory and decorated it with icons:

> ἐκτύπωσα δὲ καὶ τὸν χαρακτῆρα τοῦ κυρίου μου Παγκρατίου ἐν εἰκόνι, ὡν ἦν ἅπας ἀπαράλλακτον. ὅτε δὲ βλέπω τὸν τίμιον αὐτοῦ χαρακτῆρα ἐν τῇ εἰκόνι, οὕτων δοκῶ ὅτι ἐν σαρκὶ μετ᾽ αὐτοῦ ἐντυγχάνω καὶ βλέπω αὐτόν.
>
> I also depicted the appearance of my lord Pankratios in an image, precisely as he once was. And when I see his reverend appearance in the image, it seems to me then that I meet with him in the flesh and look upon him.[63]

This style of vividness that seems to betray unmediated presence is always qualified and delimited by its very language, such as in Evagrios's disclaimer that because of the icon "it seems" (δοκῶ) as if he is meeting and greeting the one represented in the image. This echoes John Chrysostom's *Vita*, discussed in the next chapter, where Chrysostom venerating his icon of Paul is "imagining (φανταζόμενος) that he was speaking with him in his contemplation."[64] This trope of the illusory presence of the icon nevertheless always goes on to stress the mediation of the mind, which is actively

Byzantinistik und Neogräzistik: Beiträge zum Symposion Vierzig Jahre Institut für Byzantinistik und Neogräzistik der Universität Wien. Im Gedenken an Herbert Hunger (Wien, 4.–7. Dezember 2002) (Vienna: Österreichische Akademie der Wissenschaften, 2004), 324–336.

62 On the relation between Michael Psellos's theories on sensual perception and his image theory, see Barber, *Contesting the Logic of Painting*, 61–98.

63 Evagrios, *Life of St. Pankratios of Taormina*, ed. C. J. Stallman, "The *Life* of St. Pancratius of Taormina," unpublished DPhil dissertation, vol. I, Oxford University (1986), 442; modified trans. Mango, *The Art of the Byzantine Empire*, 137.

64 "φανταζόμενος καὶ διὰ τῆς θεωρίας αὐτῷ ὁμιλῶν." John of Damascus, *Orationes de imaginibus tres*, I.61 (cf. II.57, III.54), ed. Kotter, 162; trans. Louth, 55.

invested in the icon, but still can perceive that this is a state of heightened perceptual activity, rather than the proper presence of what is being depicted.

Ruth Webb concisely describes the "'as if'-ness of both ekphrasis and *enargeia*," as a condition whereby an "audience both sees (metaphorically) and fails to see (literally) the subject matter."[65] It is this mediation between vividness and its inexorable failure that defines Byzantine image and rhetorical theory. The function of the "as if" critically defined how Photios described sight occurring, "as if somehow" (οἱονεί πως) touching and encompassing the object seen, and it described how John Chrysostom attended to that icon of Paul "as if he were alive" (ὡς ἐπὶ ζῶντος αὐτοῦ οὕτω προσεῖχεν αὐτῷ). Likewise, Charles Barber emphasizes that in Photios's ecstatic description of the Pharos Chapel, while Photios praises the heavenly qualities of the space, when he enters he is clear to say, however, that it is "as if one has entered Heaven" (Ὡς εἰς αὐτὸν γὰρ τὸν οὐρανὸν). This stresses the function of the imagination in mediating the experience between lived reality and the perception that the space conjures up in the mind.[66] Hence, the operation of the "as if" is a crucial element that permeates the conceptual logic of Byzantine representation, particularly around questions of divine representation. The "as if" confronts the limits of *ekphrasis*, and structures the longing that drives the experience of the viewer toward the person depicted in word and image.

In the twelfth-century *Progymnasmata* of Nikephoros Basilakes, his encomium of the dog urges the viewers to consider his description of a hunt as if looking upon a painting. There, Basilakes endeavors to "describe in words" (τῷ λόγῳ... διαγράψω) the practice of hunting "as if in a painting" (ὡς ἐν πίνακι).[67] Basilakes goes on to encourage their reader to contemplate this painting that he has set up for them through words:

> ἀλλὰ γὰρ ὅλοις μοι τοῖς ὀφθαλμοῖς πρὸς τὴν γραφὴν ἐνατένιζε· εἰ δὲ καὶ μὴ Ἀπελλῆς τις ἐστὶν ἢ καὶ Πραξιτέλης ὁ ζωγραφῶν, κἀντεῦθεν αὐτὸς μηδέν τι τῶν χρωμάτων καὶ τῆς γραφῆς ὡς ἀτέχνου πρὸς εὐφροσύνην ἀπόναιο. τὸ γοῦν τῆς ἱστορίας σοι γλαφυρὸν οὐκ ἀμυδρὰν εὑρήσεις τὴν ἡδονὴν προτεινόμενον.
>
> Now look upon the painting with your full attention. Even if the artist is not an Apelles or a Praxiteles, and you would therefore derive no happiness

[65] Webb, *Ekphrasis*, 10.

[66] Charles Barber, "From Transformation to Desire: Art and Worship after Byzantine Iconoclasm," *The Art Bulletin* 75:1 (1993): 7–16, esp. 14–15.

[67] Nikephoros Basilakes, *Progymnasmata*, 29.3, ed. and trans. Beneker and Gibson, 132–133.

> at all from the colors and the painting because they are amateurish, at least you will find that the elegant depiction of the story offers you no faint pleasure.[68]

The graphic vividness of the *ekphrasis* here is stressed by the fact that the author tells his reader to literally "fix the eyes upon painting" (τοῖς ὀφθαλμοῖς πρὸς τὴν γραφὴν ἐνατένιζε), not simply to imagine or use their inner eye to perceive the images being stirred up by his description of the hunt. From there on, this segment is filled with imperative commands to "see, then, the dogs attacking the deer" (ὅρα λοιπὸν τοὺς μὲν ἐλάφοις ἐπεισπεσόντας), or "see the very pleasant spectacle" (ὅρα τὸ τηνικαῦτα θέαμα ἥδιστον).[69] The imperative "see" (ὅρα) stresses the visual vividness of the scene, where one is not merely commanded to visualize or imagine the hunt, but rather to witness it unfolding before them.

While his *Progymnasmata* lacks a lesson on *ekphrasis*, Basilakes is not ignorant of the matter. In his *ethopoeiae* on a blind man regaining his sight, Basilakes plays with the operation of *ekphrasis* to turn listeners and readers into spectators by playing on this adage to depict the blind man's strife, stating: "I saw entirely from what I heard, and I translated my hearing into vision" (ὅλως ἑώρων οἷς ἤκουον καὶ τὴν ἀκοὴν εἰς ὅρασιν μετελάμβανον).[70] This suggests that even he who was blind since birth was not deprived of the wonders of *ekphrasis* and *enargeia*. However, it is likewise clear that despite the vivid and graphic power of words to sway into seeing even the blind, words nevertheless cannot truly substitute vision itself. This is captured in the story of the blind man, given that this allusion to an ekphrastic form of mental visualization precisely sets up the miraculous wonder of the blind man's acquisition of actual sight. So much so does hearing pale in comparison to sight, that the blind man upon gaining sight for the first time exclaims: "I will not need my hearing, now that I have vision" (τῆς ἀκοῆς ὡς τά γε τῆς ὄψεως οὐ δεήσομαι).[71] His praises place emphasis on the fact that now his vision is unmediated, he is capable of seeing for himself, not through the eyes of others; namely, of seeing God and the beauties of his temple for himself.[72]

[68] Nikephoros Basilakes, *Progymnasmata*, 29.4, ed. and trans. Beneker and Gibson, 134–135.

[69] Nikephoros Basilakes, *Progymnasmata*, 29.4, ed. and modified trans. Beneker and Gibson, 134–135.

[70] Nikephoros Basilakes, *Progymnasmata*, 38.2, ed. and trans. Beneker and Gibson, 188–191.

[71] Nikephoros Basilakes, *Progymnasmata*, 38.5, ed. and trans. Beneker and Gibson, 192–193.

[72] Nikephoros Basilakes, *Progymnasmata*, 38.2, ed. and modified trans. Beneker and Gibson, 188–191.

In a textbook on rhetoric, it is somewhat surprising if not ironic to see that hearing is not all that it is purported to be. And this tension is not relegated to the narrative of the blind man alone, but present already in the encomium of the dog. In the excerpt above from the encomium, we can also observe a certain tongue-in-cheek reflection on the failures of *ekphrasis*. In their introduction to the text, Jeffrey Beneker and Craig A. Gibson suggest that Nikephoros Basilakes has "self-deprecatingly" asked his listeners to enjoy the amateurish nature of his depiction, which pales in comparison to the skill of an Apelles or a Praxiteles.[73] However, this would seem to be more a self-deprecating comment on *ekphrasis* than on the author himself, given that pleasure ultimately derived from the piece comes from the "elegant" (γλαφυρόν) construction of the story. Yet, it is the painting which the listener sees that is amateurish, or, rather, "as if unskilled" (ὡς ἀτέχνου). This implies then that an ekphrastic representation, when compared to a painting, is analogous to a crude one at best, not that the author's text itself is crude.

Nikephoros Basilakes is not extolling his failures, but rather confronting the limits of *ekphrasis*. This suggests that in some way the encomium's *ekphrasis* fails. As he tells us, if not "joy" (εὐφροσύνην) from the colors and the painting, one will at least find some amateurish "pleasure" (εὐφροσύνην) in the *ekphrasis* given the eloquence of the prose, which of course is that of the author, Basilakes himself.

> οὕτω πάντα λαμπρὰ καὶ χαρίεντα, κἂν μόνον εἰς ἀκοὴν μὴ καὶ τὴν ὄψιν αὐτὴν γενήσονταί τινι τὰ τοῦ πράγματος· ὅλας ἱμέρου πηγὰς καὶ μόνον ἀναστομοῖ τὸ διήγημα.
>
> Thus everything is splendid and lovely, even if the details of the event can only be heard with one's ears and not actually seen. May even this narrative on its own open up all the fountains of your longing.[74]

In the same paragraph as the excerpt above, which was followed by the emphatic imperatives to "see" (ὅρα) the unfolding events, also comes this unabashed qualification of the power of *ekphrasis*. Here, Basilakes acknowledges that indeed the narrative can only be seen, keeping with his predecessors' desires to clarify and curtail any notion that *ekphrasis* and the imagination could actually make manifest the stories being recounted. It is as if each of those imperatives to see only confronted the viewer with their

[73] Beneker and Gibson, *The Rhetorical Exercises*, xv.

[74] Nikephoros Basilakes, *Progymnasmata*, 29.4, ed. and modified trans. Beneker and Gibson, 134–135.

inability to actually see the hunt being vividly detailed for them. Yet, not all is lost: what is established in the *ekphrasis* is perhaps more powerful and vivid than the imprints generated by the *phantasia* as the listener visualized and imagined the hunt unfolding before them. What *ekphrasis* does and what is critical to Byzantine theories of representation is that it opens "all the fountains of longing" (ὅλας ἱμέρου πηγάς). In seeing with the imagination, the person is not satiated, but wishes then to see with their physical eyes, and fully experience the events.

If we are to delineate a concept that undergirds the artistic medium in Byzantium by modern standards, that is as an epistemic condition of possibility for representation, it would be the operation of the "as if." The "as if" manifests a literal medium, a *metaxy* or *mesos*, between that which is physically perceptible and conceptually imaginable. It mediates the interval of representation through which the function of *enargeia* in *ekphrasis* shuttles back and forth between. It is a deception of the viewer's senses whose trickery can do nothing more than signal the fact that the presence of the desired or imagined world has indeed not been represented at all. Then, if the "as if" serves as the Byzantine's medium condition, it is the function of *enargeia*, which serves as the artist's key tactic for subverting and transgressing the boundaries of this medium. The goal then of Byzantine art and rhetoric was to suspend the viewer between physical reality and the mental imaginary. This is not to say that everything operated under the rubrics of a modern notion of symbolism, allusion, allegory, or metaphor, but rather that the perception of art and rhetoric was a tense state of suspension – a buildup of the desire one brings to the icon, as so many reflections on icons note – of being caught in between two possibilities of perception, one potential and the other in actuality.

Mediation

In Michael the Deacon's twelfth-century description of Hagia Sophia, we can admire the deep interrelation between visuality, cognition, and mediation when he describes the sensual effects of the gold in the narthex, stating:

> καὶ ἡτοῦ κρυσοῦ στιλπνότης ἐγγὺς εἶναι τοῦ καταστάζειν ποιεῖ νομίζεσθαι τὸν χρυσόν. τοὺς γὰρ ὑγροὺς ὀφθαλμοὺς τῇ ἀνταυγείᾳ ὥσπερ κυμαίνουσα, τὰς ἐκείνων νοτίδας εἰς τὸν χρυσὸν ἐφάντασε τὸν ὁρώμενον, καὶ δοκεῖ ῥευσεῖσθαι [*sic*] τηκόμενος.

The brightness of the gold almost makes the gold appear to drip down for by its refulgence making waves arise, as it were, in eyes that are moist, it

> causes their moisture to appear in the gold which is seen, and it seems to be flowing in a molten stream.[75]

Michael consistently develops an "as if" language where no verb describing the scene is left unqualified by a degree of mediation, everything is seeming, appearing, imagining. As in Photios's description of the Pharos Chapel, the person's personal condition, here the moisture of the eyes, is projected by the imagination (ἐφάντασε) onto the environment. This detail confronts the physicality of the body in the act of perception, whereby the moisture (νοτίδας) of the eyes, presumably the tears and humors of the eye, affects the reception of sensual data as if everything was made to appear (ποιεῖ νομίζεσθαι) to be dripping and flowing. In *De omnifaria doctrina*, for example, Michael Psellos makes a similar statement in his entry "On Colors" (Περὶ χρωμάτων), clarifying that everyone sees colors somewhat differently as the external sense data is altered by the colors of the individual's humors as they are being apprehended by perception.[76] As such, while Michael demonstrates a careful desire to articulate the physical and physiological aspects of bodily sight, we nevertheless can observe the crucial importance that mediation plays in the description of perception.

One of the more self-reflexive comments on the interrelation of representation, mediation, and visuality is attested in a passing comment made in the closing paragraph of Nicholas and Theodore of Andida's late eleventh-century *Protheoria*. In this liturgical commentary, the Andidan authors at its close liken the commentary's descriptive function in helping the clergy understand the divine mysteries to the actions of a tourist who wishes to perceive the extraordinary and unseeable beauty of a city. The commentary is likened then to the experience of acquiring a tour guide who leads one around by the hand, which is "as if looking through a window" (ὡς διά τινος θυρίδος) from a tower and looking down upon the majesty of the city from there. In the authors' words:

> ὁ δέ γε τούτοις ἐντυγχάνων, μήπω πάσης ἐξηγήσεως τῶν σεπτῶν μυστηρίων ἐν τούτοις τελεῖσθαι νομίσῃ: ἀλλὰ τοιούτῳ τινὶ οἰέσθω

[75] Michael the Deacon, *Ekphrasis of Hagia Sophia*, ed. and trans. Cyril Mango and John Parker, "A Twelfth-Century Description of St. Sophia," *Dumbarton Oaks Papers* 14 (1960): 233–245, at 237.

[76] "οὐ πᾶσι δέ φασιν ὁμοίως τὰ αὐτὰ φαίνεσθαι χρώματα, ἀλλὰ πρὸς τὰς διαφόρους κράσεις τῶν ὁρώντων καὶ ὁρωμένων τὰς χρόας μεταχρωννύεσθαι. τοῖς μὲν γὰρ ἰκτεριῶσιν ὠχρὰ τὰ ὁρώμενα φαίνεται· τοῖς δὲ τὴν κόρην ξανθὴν ἔχουσι ξανθά· τοῖς δὲ φλέγμα περὶ τοὺς ὑμένας τῶν ὀφθαλμῶν ἔχουσιν ὑπόλευκα· καὶ ἄλλοις ἄλλως, κατὰ τὴν διαφορὰν τῶν κράσεων." Psellos, *De omnifaria doctrina*, 89, ed. Westerink, 52.

> ἐντετυχηκέναι, ὡς ἐάν τις τῶν ποθούντων ἰδεῖν κάλλη πόλεως ὑπερφυῆ καὶ ἀθέατα, τύχοι τινὸς ὁδηγοῦντος, ὑφ' οὗ χειραγωγούμενος, ὡς διά τινος θυρίδος ἰσχύσει κατιδεῖν τὴν ἐκεῖθεν ἐκπεμπομένων ἀκτινων αἴγλην τε καὶ λαμπρότητα: οὐ μὴ αὐτὴν τὴν φύσιν τῶν ἔνδον ἀποκειμένων ἀγαθῶν.
>
> He who obtains these things should not expect to complete all the narratives of the august mysteries with things said herein. But, he must imagine that to obtain any kind of such things is as if wanting to see the extraordinary and unseeable beauty of a city. In which case, one obtains a guide by which one is led by the hand, so that as if looking through a window one may be able to look down upon the radiance and splendor of the rays being sent out from that place, [but] not the nature itself of the good things stored there below.[77]

In this text, the Andidan authors notably place emphasis on the agency of the things at a distance as the origins of the rays of vision (ἐκπεμπομένων ἀκτινων), rather than allowing any agency to be projected onto the viewer. This passivity is surprising, given that the activity of extramission is an aspect that has played crucially in extramission's historiographic popularization, promoted as a visuality that heightens and exploits the involved processes of image veneration in the Byzantine world. In this passing allusion to sight though, it is the wondrous mysteries of the city (serving as an analogy for the mysteries of the Divine Liturgy) that possess the agency, while the viewer is merely a receptacle for their sensations. This viewer is not passive though, for it is as if they have acquired a guide and taken the steps to gaze upon this city. The point being made here is that while the liturgical commentary is an effort to comprehend and contemplate, it alone cannot be said to encompass the full realities of those mysteries or how they are revealed to the viewer. The agency of these latter matters must be left in the realms of the divine. Looking at an icon or the liturgy, one is initiated through the somatic and kinesthetic participation in the rites and customs of worship, but one cannot be hubristically ascribed the agency of the perception that is imparted upon them by the divine. As an active participant in the rite or the icon, one is nevertheless a mere mortal who vies for access into the divinely concealed mysteries, which the divine alone chooses to disclose.

What we encounter in the Andidans' closing statements then is a contemplation on the limits of *ekphrasis* and *enargeia* in their liturgical commentary, which is conveyed through a metaphor on sight. Despite reading their narrative, the reader cannot expect to comprehend all the

[77] Nicholas and Theodore of Andida, *Προθεωρία Κεφαλαιώδης*, ed. PG 140:418–468.

mysteries, but this is not merely a question of partiality, as if not all the options have been explicated alone. As the metaphor demonstrates, even when this hypothetical tourist is led to a bird's eye view of the city, they may not believe they have apprehended the nature itself of the city below (οὐ μὴ αὐτὴν τὴν φύσιν τῶν ἔνδον ἀποκειμένων ἀγαθῶν). For sight, like the Andidans' commentary, or "Contemplation" (προθεωρία) as it is titled, can only confront one with that which lies beyond them. The same is said about *ekphrasis*, which, as John of Sardis tellingly explained, presents things "below" the eyes, meaning that it produces a less distinct vision, only a mere image or fodder for the imagination. What lies below the Andidans' tourist, though mediated by sight, still denies full unmediated access; neither is it to be taken lightly that even when they are immersed in the city on the ground, they are struggling with their inability to perceive that "extraordinary and unseeable beauty of the city" (κάλλη πόλεως ὑπερφυῆ καὶ ἀθέατα). Thus, the Andidans offer us a theory of perception where that which is desired is always and successively denied, frustratingly subverted by the limits of human perception. This is an understanding of perception as a theory of perpetual failure and distance, where the perceptive body is always suspended between what they can immediately perceive and that which lies beyond perception, but which they know to be there and true.

In his descriptions of the Chora Monastery, the learned fourteenth-century scholar and statesman Theodore Metochites, similarly conveys a sense of divinely enacted intromissive agency in his ecstatic characterization of the church's fineries and mosaics.[78] There, Metochites describes these as "dazzling the eyes as with brilliant fire" (ἐπαστράπτοντ᾽ ὄμματ᾽ ἠΰτε λαμπετόοντι πυρί) and "sending forth a kind of enchanting glow unto the eyes" (τιν᾽ ὄσσοισι χαρίεσσαν αἴγλαν ἱέντα δὴν μελιηδέα).[79] Likewise, the image of Christ "sends forth grace and great, ineffable wonder" (χάριν, ἠδέ τε θῶυμα μάλ᾽ ἄρρητον, πολλὸν ἱεῖσαν αὐτόθεν),[80] and the liturgical vessels "send forth an all-radiant beauty and grace"

[78] Jeffrey Michael Featherstone, "Metochites's Poems and the Chora" in Holger A. Klein, Robert G. Ousterhout, and Brigitte Pitarakis (eds.), *The Kariye Camii Reconsidered* (Istanbul: Istanbul Research Institute, 2011), 213–237, esp. 218–219. See also Ihor Ševčenko, "Theodore Metochites, the Chora, and the Intellectual Trends of His Time" in Paul A. Underwood (ed.), *The Kariye Djami*, vol. IV (Princeton University Press, 1975), 17–92.

[79] Theodore Metochites, *Poems*, I:1003–1007, ed. Ioannis Polemis, *Carmina*, Corpus Christianorum, Series Graeca 83 (Turnhout: Brepols, 2015), 40; trans. Featherstone, "Metochites's Poems and the Chora," 218.

[80] Theodore Metochites, *Poems*, II:168–169, ed. Polemis, 58; trans. Featherstone, 219.

(παμφανόεσαν ἱέντων ἀγλαΐην τε χάριν).[81] This is a striking depiction, whereby the eyes are merely directed at objects that emit their qualities toward the eyes, but unlike the rays of the Andidan authors, here Metochites deploys a language of fiery effluxes being sent forth from the things seen in Platonic extramission, where ocular fire fuses with objective fire to enable sight. Thus, it is the religious objects here that are given precedence in the agency of sight, where the human viewer can do nothing more than direct the eyes to the image, but it is the image that enables the condition for sight.

Theodore Metochites and his contemporaries believed in iterations of sight that often adhered most closely to the Platonic model of unification and that stressed the medium as either a site of fiery fusion or Galenic *pneuma*. As Börje Bydén has eloquently demonstrated, Palaiologan intellectuals, including Nikephoros Blemmydes, Theodore Doukas Laskaris, Nikephoros Choumnos, and Theodore Metochites all had subtle variations in their theories on sight, yet they all struggled with unifying Platonic and Aristotelian models, often resorting to Galen as an intermediary for thinking through potential unifications.[82] Nikephoros Gregoras, who favored a resolutely Aristotelian model despite being one of the most vocal critics of Aristotle, was the one notable exception in the period.[83] In the case of Metochites, specifically, we find his most eloquent articulation of sight in an account of gazing up at the night sky in the *Semeioseis Gnomikai*, where he writes:

> καὶ τοίνυν ἀνατεινόμενος οὕτω δὴ τοῖς ὀφθαλμοῖς ἄνθρωπος ἄνω, καὶ τῶν οὐρανίων ἐφαπτόμενος, μάλιστα δῆλός ἐστιν ἥδιστα τῇ τοιαύτῃ τῆς αἰσθήσεως ἐνεργείᾳ χρώμενος. καὶ προσορῶμεν γε ἅπαντες ποθοῦντες ἐν τέρψει τοὺς οὐρανίους κόσμους καὶ τὰς τῶν ἄστρων χορείας καὶ τὰς ἀστραπτούσας ἐκεῖθεν χάριτας, καὶ μάλιστ᾽ ἔχοντες ταῖς νυξὶν ἀπροσκόπτως χρῆσθαι, καὶ δίχα παντὸς ἀχλυώδους καὶ ζοφώδους ἐπιτειχίσματος, τὰς ἐκ τῶν ὀμμάτων ἀναπέμπειν ἀκτῖνας καὶ ἀντιπέμπειν ταῖς φερομέναις ἄνωθεν ἐς ἡμᾶς, ἐκ τῆς τῶν ἀστέρων καὶ τῆς τῶν οὐρανίων αἴγλης τε καὶ λαμπρότητος...
>
> And consequently, turning their eyes upwards in this way, and perceiving the heavenly bodies, human beings clearly employ this service of the sense with utmost joy. And we all gaze with sweet desire at the heavenly arrangements and the ring-dancers of the stars and the charms that radiate from there, especially when we are able to employ [the sense]

[81] Theodore Metochites, *Poems*, 1:1026–1027, ed. Polemis, 41; trans. Featherstone, 224.
[82] Bydén, *Theodore Metochites'* Stoicheiosis Astronomike, 199–210.
[83] Bydén, *Theodore Metochites'* Stoicheiosis Astronomike, 208–209.

> unobstructedly during the night and without any misty or murky obstacle, and to send up the rays issuing from the eyes against those [rays] which are carried from above toward us, from the luminescence and the radiance of the stars and the heavenly bodies. . .[84]

While his language in the Chora poems alluded to an intromissive theory of vision, Metochites clearly follows an extramissionist model in his scientific work. Here, Metochites clearly explicates the process whereby rays are sent up and from (ἀναπέμπειν and ἀντιπέμπειν) the eyes and the celestial bodies alike from a distance, allowing us to comprehend that for him sight emerged from a fusion of internal and external rays in some capacity and in keeping with the majority of his learned contemporaries. Thus, contrasting his ekphrastic poems on the Chora with his theory of sight, we can grasp how Byzantine writers would often choose to stress either the emissions from the objects or the emissions from the eyes depending on the agency and rhetorical force that they wished to impart either on the things seen or on the seer, respectively, despite adhering to a model of united rays in keeping with Plato, as in the case of Metochites.

For ancient rhetoricians, as Ruth Webb concludes, the problem of what *ekphrasis* and *enargeia* could actually come to manifest was side-stepped by "assuming that words appeal instead to the mental impressions left by perception."[85] Most of the Byzantine writers surveyed here certainly concur with this conclusion, coming to understand that the ecstatic, heavenly, and inspirited experiences with religious spaces and icons were defined by the viewer's perceptual and cognitive investment in their bodily sensations. Yet for some, desiring to create a more vivid experience with the divine, this at times could seem inadequate. We might compare then the Andidans' comments to Nicholas Kabasilas's own allusion to vision in his fourteenth-century liturgical commentary, where he attempts to convey Christ's intercessory role in section 44 of his treatise, tellingly titled, "On the Mediation of Christ" (Περὶ τῆς τοῦ Χριστοῦ μεσιτείας). There, Nicholas carefully details how Christ serves an intermediary role on our behalf, yet Christ does not merely intercede through words and pleas, but through actions themselves. Christ is able to undertake this, Nicholas writes, because "he unites (συνάπτειν) us to himself, and makes us each. . . sharers

[84] Theodore Metochites, *Γνωμικαί σημειώσεις*, 42, ed. G. Müller and T. Kiessling, *Miscellanea philosophica et historica* (Leipzig: Teubner, 1821), 267; trans. Bydén, *Theodore Metochites'* Stoicheiosis Astronomike, 205.

[85] Webb, "*Ekphrasis* Ancient and Modern," 18.

through him."[86] He goes on to elaborate on this matter through an analogy to sight, writing:

> Καὶ καθάπερ τὸ φῶς δι' ἑαυτοῦ τὸ ὁρᾶν τοῖς ὁρῶσι παρέχον, οἷς ἂν ἐπιλίποι, καὶ τὸ ὁρᾶν ἐπιλείπει, οὕτω καὶ τὴν μετὰ τοῦ Χριστοῦ συνουσίαν ἀνάγκη διηνεκῆ ταῖς ψυχαῖς εἶναι, εἴγε μέλλοιεν ζῆν ὅλως καὶ ἀναπαύεσθαι. Οὔτε γὰρ χωρὶς φωτὸς ὀφθαλμὸς δύναται βλέπειν, οὔτε χωρὶς τοῦ Χριστοῦ ζωὴν ἀληθινὴν καὶ εἰρήνην ἐνεῖναι ταῖς ψυχαῖς δυνατόν, ὅτι αὐτός ἐστι ὁ τῷ Θεῷ καταλλάττων μόνος, ὁ τὴν εἰρήνην ταύτην ποιῶν· ἧς χωρὶς ἐχθροὺς ὄντας τοῦ Θεοῦ τῶν ἀγαθῶν τῶν αὐτοῦ μετέχειν ὁπωσοῦν οὐδεμία ἐστὶν ἐλπίς.
>
> And just as light allows sight through itself [i.e. light] to see, if [light] fails then seeing fails, thus continual union with Christ is necessary for the soul, if it is to live fully and be at rest. For neither without light is it possible for the eye to see, nor is it possible for the soul to have true life and peace without Christ; for he alone reconciles us to God and is the Author of that peace, without which we would be God's enemies, without hope of partaking in his good things.[87]

Here, Nicholas parallels the union with Christ to the eye's union with light, echoing the language used by Byzantine intellectuals to describe the process of Platonic extramission and deploy the "as if" to stress the degree of mediation to which the metaphor itself is subjected.[88] Just as one partakes in sight by virtue of the union of external and internal rays, one partakes in the goodness of Christ through a similar union. Nicholas Kabasilas's text is difficult to comprehend in terms of its visual theory, given that the author does not provide us with any indication of how the rays going out and/or coming in operate in his discussion of sight; in fact he leaves out wholly the language of "rays" (ἀκτῖνες), such as that which we saw in the Andidan commentary and in the majority of the texts discussed thus far. Furthermore, it is somewhat uncertain as to whether he ascribes sight to extramission or intromission, given that the substitution of rays with light strikes the modern reader as a natural and expected language for describing our own intromissive understanding of sight as a product of reflected light. Such theories were being popularized in the Islamic world since the eleventh century by Ibn al-Haytham and Ibn Sīnā, and thus it

[86] "Τὸ συνάπτειν ἑαυτῷ καὶ δι' ἑαυτοῦ τῶν οἰκείων μεταδιδόναι χαρίτων κατὰ τὴν ἀξίαν ἑκάστου καὶ τὸ τῆς καθάρσεως μέτρον." Nicholas Kabasilas, *Explicatio divinae liturgiae*, 44, ed. Bornert et al., 252; trans. Hussey and McNulty, 100.

[87] Nicholas Kabasilas, *Explicatio divinae liturgiae*, 44, ed. Bornert et al., 252; mod. trans. Hussey and McNulty, 100.

[88] See Plato, *Timaeus*, 45c and 46a–b, ed. and trans. Bury, 100–105.

might well be possible to characterize Nicholas Kabasilas's own views under such theories that seem all too familiar to our present-day eye. Yet, Nicholas's text does speak a great deal and quite eloquently about his theory of sight obliquely through the language used to categorize Christ in this analogy. Such language of the union of lights does not manifest itself in the intromissive teachings of Byzantium's Islamic contemporaries, yet it does occur prominently to articulate Platonic extramission. There fiery lights fuse in that intermediate air, just as in Nicholas's text Christ is the mediator for the fusion of the heavenly God and the terrestrial souls.

Most crucial to understanding Nicholas Kabasilas's theory of vision, however, is how the text speaks of the particular nature of that union between Christ and the souls. Here, the inexorability of the mediating interval is reified in the very figure of Christ. Christ has a complex operation here: on the one hand, he is the medium, yet, on the other hand, he also unites the faithful to himself as that medium. Thus, we understand that there is a union that occurs between the mediator and the viewer, an outpouring toward Christ if one were to describe it in visual terms. Hence, we can understand that the union of the faithful comes to that intermediary point manifested by Christ the mediator. However, this is not the end of this process for it is Christ that "reconciles us to God" (τῷ Θεῷ καταλλάττων). Thus, God unites with the faithful by virtue of their union with Christ in that intermediary role. Yet the nature of this union is not by virtue of Christ's intercessory actions alone, or rather, this alone is not the reason why God unifies himself with humanity, for as Nicholas writes:

> Τί γὰρ τὸ καταλλάξαν τῇ φύσει τῶν ἀνθρώπων τὸν Θεόν; Πάντως ὅτι ἄνθρωπον εἶδε τὸν Υἱὸν αὐτοῦ τὸν ἀγαπητόν· οὕτω καὶ ἑκάστῳ σπένδεται τῶν ἀνθρώπων, εἴ τις τὴν μορφὴν κομίζει τοῦ Μονογενοῦς καὶ τὸ ἐκείνου φορεῖ σῶμα, καὶ ἓν πνεῦμα μετ' αὐτοῦ φαίνεται.
>
> What reconciled God to humankind? Simply this, that he saw his beloved son become man. Likewise, he is reconciled personally to every man who wears the stamp of the Only-begotten, and bears his Body, and shows himself to be one spirit with him.[89]

Thus, God's effluence of love (specifically, ἀγάπη) for his son leads him then to become united with humanity by virtue of humanity's own union in spirit with Christ. Hence, what we come to observe here is that there are two vectors of union that make humanity's partaking in the benefits of

[89] Nicholas Kabasilas, *Explicatio divinae liturgiae*, 44, ed. Bornert et al., 252–254; trans. Hussey and McNulty, 101.

God: one emerging from the righteous toward Christ, and another emerging from God toward Christ in his love for him. Thus, it is in Christ, the mediator, that both these vectors are united so as to enable humanity to partake of God's image.

When we consider Nicholas Kabasilas's text diagrammatically, we can quite neatly see at play the theory of Platonic extramission, whereby we can say that God emits the light emanating from the visible objects and the people emit the light issuing from the eyes, which fuse in that intermediary point, here manifested by Christ. This reading is only possible, however, if one is familiar with the language of union and the processes of vision being obliquely indexed by the text. Thus, Nicholas Kabasilas's text can attest to the enduring belief in and nuanced understanding of Platonic extramission up through the fourteenth century. The impenetrability and lack of explicit clarification of this process of extramission in his text though should not be understood as an underdeveloped theory of vision, but instead a telling shorthand that speaks to the manner in which this particular theory may have been taken for granted by Nicholas. Above all, Nicholas's discussion eloquently and crucially reasserts the primacy of mediated experience in the process of sight, even if one were to ascribe his theory to another variant of extramission or intromission. For if he were citing a theory of vision that did not privilege mediation then his whole analogy to the mediation of Christ would be rendered illegible and simply would not work. Yet this is not a mediation of lacking or inaccurate transmission of form; it is a theory of mediation that resolves forms in which mediation can access the divine, even if lacking the direct and immediate contact between human and God – here resolved through the perceptibility of the Christ.

In terms of the limits of perception and representation, however, Nicholas Kabasilas's insightful comparison between sight and a union with Christ, in resorting to perception as a model, in no way breaks with the bounds or limitations of visual perception. While the viewer is allowed a heightened state of contact with God, this happens only through the mediation of the Christ who serves as the interval of perception here. In his desire to explain the union between the human and the divine, Kabasilas contours his theory around human perception, rather than vice versa. Hence, here we see a writer explicitly constructing a theory of communion with God that is quite literally based on the bounds of visual perception. And while the union with Christ might seem to offer up an unmediated contact with the divine, what is made clear is that Christ is merely that diaphanous interval between seer and object, a mediating middle point where the rays unite. He is the external light that allows the Platonic fire of the eye to fuse with it and extend outward, like onto like, but Christ is not the faculty of sight itself, nor does he represent himself as such to the viewer.

In these regards, we come to understand that Byzantine representational and perceptual theories are founded precisely on the failure of grasping what is set before them, always falling short of that grasp. If sight is as if grasping, just as hearing is as if seeing, both are united by a shared inability to perceive in actuality that which they are allegedly promised. It is this paradox of representation that I have argued here comprises the Byzantine medium condition as that condition of perceived possibility, which is ultimately denied, but within which the perceiver is ultimately suspended. This is the cruel optimism of *enargeia*, the sense that one might actually be able to transgress the bounds of the senses and their respective traditional mediums – sight and visual art, hearing and words, and other possible permutations. While this problem is certainly one that emerges from the Byzantine episteme, it is also one birthed in the modern imaginary from the conditions of the artwork and the limits of sight and touch for the contemporary art historian.

When we discuss an artistic medium in Byzantium we should avoid talking about painting, tempera, or wood, but instead endeavor to understand medium as an expanded field of cultural thought and artistic generativity, as an intervallic state of suspended adjacency between the potentiality and actuality of form, its being-in-the-world and its perception. In other words, we must endeavor to think of "medium" as a term that indicates and articulates mediation as a coming to terms with the failure of representation, rather than a reduction of that system to mere material substrates. This is the construction of a condition of possibility around the inherent "as if" function that undergirds the very possibility of representation in relation to that which is absent and unable to be made present. As such, it is necessary to understand how artists – across all traditional media – operate along this medium condition to precisely attempt acts of self-reflexive transgression, acts that vividly (ἐναργῶς) attempt to make seeing into touch, or hearing into seeing, all while orienting themselves around that fundamental failure and inability. Hence, this requires that we reevaluate the slew of synesthetic language and metaphors in Byzantium, which actively grappled with these limitations and provide us with a cultural manifesto of the strategies and tactics that this medium condition enabled and which artists deployed.

Presence/Absence

In her assessment of Asterios of Amaseia's *ekphrasis* on the icon of Euphemia's trial and martyrdom, Ruth Webb underlines the manner in which the painter's *pharmaka* transcend mere colors, but seem to become blood smeared upon the canvas.[90] This, as Webb notes, is not unique to Asterios, but stresses vividness of the painting that appears to delude viewers into thinking that the thing being depicted is actually present. A similar event occurs in Philostratus's *Imagines*, as Jaś Elsner has discussed, where the figures in the narrative stretch out their cloaks to collect the blood dripping from the depiction of the dying Menoikeus.[91] Reflecting on this instance, Webb writes:

> At moments like this there is no separation between viewer and image; the frame dissolves and the viewer is able to interact physically with the figures depicted. This is a response to the figurative arts that we find throughout the tradition of ekphrasis, often expressed in the idea that the figures can be (or can almost be) heard to speak. As Robert Nelson has pointed out, the idea of the haptic gaze, that the eye sends out rays which physically caress the painting as the view contemplates it, provides some 'scientific' backing for this idea of visual contact.[92]

Here, Webb's statements almost operate on the level of metaphor, seeing the contact enabled through the notion of haptic sight as embodying the presentation of the referent that rhetorical and painterly vividness accomplishes. Yet the fact that cannot be ignored here is that just as vividness makes it appear "as if" Paul were alive in John Chrysostom's icon, it also makes it appear "as if" the optical rays somehow touched and encompassed the icon. In a sense, Webb's comparison between *ekphrasis* and Nelson's understanding of Byzantine visuality is quite adept, since both operate "as if" they enabled the contact with unmediated presence. However, both fall precisely short of accomplishing that goal, and it is that failure that enables this entire system of representation whereby the illusion of presence is sustained to make it all but real, yet without overstepping the bounds of physics, psychology, or divine providence.

This issue of presence in art and rhetoric or a tactile contact with objects of devotion through sight is not really so much a matter of presence or

[90] Webb, "Accomplishing the Picture," 22.

[91] Philostratus the Elder, *Imagines*, 1.4, ed. and trans. Fairbanks, 18–19. See Elsner, *Art and the Roman Viewer*, 23; Jaś Elsner, "Naturalism and the Erotics of the Gaze" in N. B. Kampen (ed.), *Sexuality in Ancient Art* (Cambridge University Press, 1996), 247–261.

[92] Webb, "Accomplishing the Picture," 22.

mediation, it is clearly laid out by the defenders of icons throughout Byzantium's history that to forego a system of signification would be heretical. This is to say, that it would be heretical to believe in any capacity that God inheres in the material form or essence of the icon, or that one should ever treat the icon as if that were the case, offering it worship rather than veneration. Hence, what is truly at stake in the question of presence in Byzantine art and rhetoric is precisely the affective state of the viewers that draw themselves to images or texts with all their investment and desire as if that which was being depicted were truly there. Following Longinus, Ruth Webb similarly suggests that while poetry is said to cause surprise or fear (ἔκπληξις), in rhetoric the focus on vividness is nevertheless a desire to result in emotion (πάθος) and excitement (κίνησις).[93] This affecting force emerging from *phantasia* in rhetoric is what comes to "enslave" (δουλοῦν-ται) in Longinus,[94] which Goldhill in particular has stressed as being responsible for the forceful effects of *ekphrasis* to enculturate forms of viewing and seeing.[95]

In these terms, the historical problem of presence becomes more a question of affect than one of mediation or representation. As Jaś Elsner has noted, in the context of Byzantine iconoclasm, despite the fact that both iconophiles and iconoclasts decried the claim that divine presence inhered in the icon in any capacity, "some element of real presence inhered in the image – not fully theorized, to be sure, but potent enough to be worthy of attack through iconoclasm."[96] Yet it is precisely "not fully theorized" and seems to operate more in earlier Roman imperial practices of *damnatio memoriae* than in any specific Byzantine fear or concern with the presence inherent in images. Perhaps the most eloquent articulation of the presence of the icon in the Byzantine sphere came in the eleventh century from Leo of Chalcedon in his critique of Emperor Alexios I Komnenos and his melting down of church property to fund his military campaigns against the Normans. In the destruction of objects that bore the image of holy figures, Leo argued that matter had in some capacity been divinized by virtue of bearing forth the image of the holy ones, and thus it was worthy of absolute veneration, not just the relative veneration that Theodore of Stoudios had ascribed

93 Webb, "Imagination and the Arousal of Emotions," 118.

94 Longinus, *De sublimitate*, ed. and trans. Fyfe, 222–223.

95 Goldhill, "What Is Ekphrasis for?," 4–7.

96 Jaś Elsner, "Iconoclasm as Discourse: From Antiquity to Byzantium," *The Art Bulletin* 94:3 (2012): 368–394, esp. 383.

to icons.[97] Yet these claims were carefully and succinctly demolished by imperial and religious authorities in a synod convened at the Blachernae in 1094/95.[98] There, it was reiterated that icons were the likeness appearing in the material, not the material itself, and that said material could not be worshipped, neither could the icon, since its likeness can only be venerated relatively.[99]

Writing in the last decade of the eleventh century, precisely in the wake of the conflict with Leo of Chalcedon, Eustratios of Nicaea provides us with a nuanced and complex approach to the icon in his *Syllogistic Demonstrations*, which grapples with both the question of presence and the operation of the imagination.[100] There, Eustratios writes regarding the approach toward the icon:

> Καὶ διαστήμασιν ἐσόμεθα καὶ σχήματι λατρεύοντες, τῷ ὑποκειμένῳ μόνῳ διαφέρουσι τῶν προτέρων· ὅπερ ἐκεῖνα μὲν χρυσῷ ἢ ἀργύρῳ ἢ χρώμασιν ἐντετύπωτο, ταῦτα δὲ φαντασίᾳ ἢ μνήμῃ ἢ νῷ.
>
> We will first arrive at the intervals [i.e. surface] and then worship the form, for the latter differs from the former only in its underlying substance. For the former are imprinted in gold or silver or colors, while the latter are imprinted on the imagination or in the memory or in the mind.[101]

Eustratios here brilliantly describes the outward, surface form of the icon as the intervals (διαστήμασιν), understanding the material form of the icon to be an interval of mediation like an aperture or radius. This is what the viewer first confronts, then going on to worship the ineffable shape or schema of the figure (σχήματι λατρεύοντες). Here, the use of adoration or

97 Leo of Chalcedon, *Epistle to Alexios I Komnenos*, ed. Alexander Lauriotes, *Ekklesiastike Aletheias* 20 (1900), 403AB. See Apostolos Glavinas, *Ἡ ἐπὶ Ἀλεξίου Κομνηνοῦ (1081–1118) περὶ ἱερῶν σκευῶν, κειμηλίων καὶ Ἁγίων εἰκόνων ἔρις (1081–1095)* (Thessaloniki: Kentron Vyzantinōn Ereunōn, 1972); John P. Thomas, *Private Religious Foundations in the Byzantine Empire* (Washington, DC: Dumbarton Oaks, 1987), 192–207; Pelopidas Stephanou, "La Doctrine de Léon de Chalcédoine et de ses adversaires sur les images," *Orientalia Christiana Periodica* 12 (1946): 177–99.

98 Barber, *Contesting the Logic of Painting*, 131–157; Annemarie Weyl Carr, "Leo of Chalcedon and the Icons" in Doula Mouriki et al. (eds.), *Byzantine East, Latin West: Art-Historical Studies in Honor of Kurt Weitzmann* (Department of Art and Archaeology, Princeton University, 1995), 579–584.

99 Anonymous, *Blachernae Synod*, ed. PG 127:980–981.

100 On Eustratios of Nicaea and image theory, see Charles Barber, "Eustratios of Nicaea on the Separation of Art and Theology" in Charles Barber and David Jenkins (eds.), *Medieval Greek Commentaries on the Nicomachean Ethics* (Leiden: Brill, 2009), 131–143. See also Barber, *Contesting the Logic of Painting*, 99–130.

101 Eustratios of Nicaea, *Expositio de imaginibus*, ed. Andronikos Demetrakopoulos, *Bibliotheca Ecclesiastica*, vol. 1 (Leipzig: Othonos Vigandou, 1866), 156; modified trans. Pentcheva, *The Sensual Icon*, 186.

worship (λατρεύοντες) reflects the fact that worship is reserved for the divine referent alone, whereas veneration is what is due to icons alone. Hence, we are led to understand that the noetic worship of the form depicted in the icon partakes in proper worship, since there we adore the one represented in the icon, rather than the icon itself which is the mere mediator in this process. Playing with the shared language of imprinting for both physical and mental images, he says that the interval is impressed in matter (i.e. gold or silver or colors), yet the schema of the figure is impressed upon the imagination or the memory or the mind. Thus, what Eustratios is communicating here is that the icon itself as a material body is an interval, whereas the process of veneration is the action of imprinting that form in the psychic faculties.

In his commentary on Aristotle's *Nicomachean Ethics*, Eustratios is quite eloquent about the role of the imagination regarding mathematical objects (μαθηματικὰ φανταστικά) serving as mediators between ideas and sensible particulars. Echoing Ammonius's commentary on the *Isagoge* and Alexander of Aphrodisias on the material intellect, as Michele Trizio has elucidated, Eustratios there describes the relationship between the imagination, sense perception, and thought: describing how the three are held in unison by the imagination and intellect, "being impressed in the imagination, and scrutinized under the intellect (διανοίας)"; then explaining that they have been "impressed immaterially and bodilessly in the imagination, which received these things from sensation," and using the metaphor of "a certain tablet" (τις πίναξ) to describe the impression of noetic forms in the soul.[102] This passage helps us understand Eustratios's careful approach to the icon since it allows us to understand that the forms, having been taken from the senses that are impressed upon the imagination, are precisely "without matter and without body" (ἀύλως καὶ ἀσωμάτως). Thus, it stresses the contrast between the material icon that is a mere interval of mediation, yet

[102] "ἔστι δὲ τὸ λεγόμενον τοιοῦτον, ὅτι τὰ μὲν μαθηματικὰ φανταστικὰ καὶ διανοητὰ καὶ ἐν φαντασίᾳ καὶ διανοίᾳ ἔχει τὴν σύστασιν, ἐν μὲν τῇ φαντασίᾳ τυπούμενα, ὑπὸ δὲ τῆς διανοίας ἐξεταζόμενα· τὰ δὲ τοιαῦτα ἐξ ἀφαιρέσεως λαμβάνομεν, ἤτοι ἐκσπῶντες αὐτὰ καὶ ἀφαιροῦντες ἀπὸ τῶν ὑποκειμένων ἐν οἷς ὑφεστήκασι καὶ τυποῦντες αὐτὰ ἐν τῇ φαντασίᾳ ἀύλως καὶ ἀσωμάτως δεχομένῃ ταῦτα ἀπὸ τῆς αἰσθήσεως, ἣ ἀμέσως ὡς αἰσθητοῖς οὖσι καὶ ἐν ὑποκειμένῳ αὐτοῖς ἐπιβάλλουσα, παραπέμπει τῇ φαντασίᾳ τοὺς τύπους αὐτῆς, ἡ δ' ἐν ἑαυτῇ τούτους δεχομένη καὶ συντηροῦσα ὥσπερ τις πίναξ ὑποτίθησι τῇ διανοίᾳ αὐτούς, ἡ δ' ὡς ὄμμα τῆς ψυχῆς αὐτοῖς ἐπιβάλλουσα ἐξετάζει περὶ αὐτῶν καὶ τῶν ἑπομένων αὐτοῖς καὶ εὑρίσκει ταῦτα καὶ τοὺς λόγους αὐτῶν ἀποδίδωσι." Eustratios of Nicaea, *In Aristotelis ethica Nicomachea vi commentaria*, ed. G. Heylbut, *Eustratii et Michaelis et anonyma in ethica Nicomachea commentaria*, Commentaria in Aristotelem Graeca 20 (Berlin: Reimer, 1892), 344.30–345.6.

nevertheless embodied and material, against the immaterial and bodiless schema that is imprinted in the mind through the action of veneration.

Elsewhere, in his dialogue on icons, Eustratios builds on these ideas to further argue for the role of the icon in veneration, stating that while the image can lead us toward contemplation of the divine through veneration, adoration and worship are reserved for the divine alone, which follows in the doctrines established in the Iconoclastic period.[103] From there, Eustratios states carefully:

> Καὶ ἄλλως δὲ τῷ τρόπῳ διαφέρειν φαίνεται ἐναργῶς ἡ Χριστοῦ προσκύνησις καθ᾽ ἑαυτόν τε κἂν τῇ εἰκόνι· ἐκεῖ μὲν γὰρ ἀμέσως προσκυνοῦμεν αὐτῷ, ἐνταῦθα δὲ διὰ μέσης μιμήσεως· ὥστ᾽ ἐκεῖ μὲν ἔσται λατρευτικὴ ὡς Θεῷ προσφερομένη αὐτῷ, καὶ κοινὴ τῶν ὁμοφυῶν αὐτῷ θείων ὑποστάσεων· ἐνταῦθα δὲ σχετική, ὡς διὰ μέσου προσφερομένη τοῦ πρὸς αὐτὸν σχέσιν ἔχοντος.
>
> Alternatively, the veneration of Christ is vividly shown to differ in character regarding [Christ] himself and the icon. For in the former we venerate him in an unmediated fashion, but in the latter [we venerate] the imitation [of him] through a medium. Thus, on the one hand, in the former adorational (λατρευτική) [*proskynesis*] will be offered to him, as if to God, common to the divine hypostasis, which is similar in nature to him. On the other hand, in the latter relative (σχετική) [veneration] is offered [to the icon] towards him, as if through a medium, which holds a relationship to him.[104]

Eustratios here uses the term *proskynesis*, usually translated as "veneration" (προσκύνησις), to indicate the entire process of veneration, rather than the expected λάτρευσις for worship. In the context of this dialogue on images and their veneration, this use suggests that Eustratios wishes to deploy προσκύνησις to indicate the entire process of veneration, leading one on to the worship of the divine referent in the mind. For this reason, he distinguishes between veneration as adoration (λατρευτική) and as relative (σχετική), so as to clearly distinguish the order in which one approaches the divine through the icon. Thus, allowing him to state that while our icons may be mediated in representation, as in the passage from the *Syllogistic Demonstrations*, their use through veneration moves on to adoration in the mind, which allows for a loftier communion with the divine than is possible with the material flesh of the icon itself. Therefore,

[103] On this matter, see Barber, "Eustratios of Nicaea on the Separation of Art and Theology," esp. 138–139.

[104] Eustratios of Nicaea, *Dialogus de sacris imaginibus*, 148.18–25, ed. Andronikos Demetrakopoulos, *Bibliotheca Ecclesiastica*, vol. 1 (Leipzig: Othonos Vigandou, 1866), 148.

the icon is rightly defined as an interval, which unfolds through the process of veneration, moving toward the adoration of the divine. Yet the presence inherent in the icon is once again explicitly denied, stating precisely that we venerate his "imitation through a medium" (διὰ μέσης μιμήσεως).

Thus, we are left with little but speculation on the so-called "real presence" of the icon, beyond the affective stirrings that images caused, leading viewers to treat them precisely as if they were those whom they represented. This is perhaps best manifested through the arousal of the imagination and what Aglae Pizzone has referred to as the "emotional gaze" in her discussion of the Agathias's inscription upon the icon of the Archangel Michael at Plate.[105] There, as discussed in the next chapter, Agathias writes that when the mortal gazes upon the icon they "direct their passion to a higher imagination" (θυμὸν ἀπιθύνει κρέσσονι φαντασίῃ), leading them to "tremble as if [the Archangel] were present" (ὡς παρεόντα τρέμει).[106] The illusion or *enargeia* of presence is achieved in the icon, as in rhetoric, through the viewer's investment in the image by virtue of the imagination. It is in the imagination that the semblance of presence is constructed as the viewer imagines the animate, living person with whom they can interact and embrace via that so-called eye of faith.

An eloquent articulation of this yearning in the icon, which actively struggles with the dynamics of presence and absence, is found in an epigram upon an icon of St. Theodosios the Cenobiarch composed by Klement the Monk, preserved in the late thirteenth-century *Anthologia Marciana.*[107] The epigram preserved there reads as follows:

> Τῆς πίστεως κάλαμον ἐμβάψας πόθῳ
> καὶ καρδίας γράψας σε πλαξὶ σαρκίναις
> ψυχῆς, Θεοδόσιε, ταῖς κόραις βλέπω.
> Ὁρᾶν δέ, πάτερ, καὶ κατ' αἴσθησιν θέλων,
> γράφω τύπον σὸν ἐξ ἐνύλων χρωμάτων,

[105] Aglae Pizzone, "Toward a Self-Determined and Emotional Gaze: Agathias and the Icon of the Archangel Michael" in Sergei Mariev and Wiebke Marie Stock (eds.), *Aesthetics and Theurgy in Byzantium* (Berlin: De Gruyter, 2013), 75–103.

[106] Anonymous, *Anthologia Graeca*, 1.34, ed. and modified trans. W. R. Paton, *Greek Anthology*, vol. 1, revised Michael A. Tueller, Loeb Classical Library 67 (Cambridge, MA: Harvard University Press, 2014), 32–33.

[107] On the *Anthologia Marciana*, see Foteini Spingou, "Words and Artworks in the Twelfth Century and Beyond: The Thirteenth-Century Manuscript Marcianus gr. 524 and the Twelfth-Century Dedicatory Epigrams on Works of Art," unpublished DPhil dissertation, Oxford University (2012). See also Foteini Spingou, "Byzantine Collections and Anthologies of Poetry" in *Brill Companion on Byzantine Poetry*, ed. Wolfram Hoerandner, Andreas Rhoby, and Nikolaos Zagklas (Leiden: Brill, forthcoming). I wish to thank Foteini Spingou for sharing this forthcoming article with me.

ὦ τύπε καὶ πρώταρχε τοῦ κοινοῦ βίου.
Σὺ δ᾽ ἀλλὰ γράψον ἐν βίβλῳ σεσωσμένων,
ὅταν ἀνοιχθῇ κρίσεως τὸ βιβλίον,
μονότροπον Κλήμεντα, τὸν σὸν οἰκέτην,
πιστὸν μαθητὴν ἐνθέου διδασκάλου,
τοῦ πατριάρχου τῆς Σιὼν Ἰωάννου.

Dipping the pen of faith in yearning
And drawing you in the fleshy tablet of the heart
I gaze on you, Theodosios, with the pupils of the soul.
But, desiring to see you, Father, also with the senses
I inscribe your image out of material colors,
O image and founder of the cenobitic life.
But will you inscribe in the book of the saved,
When the book of judgment is opened,
The unwavering Klement, your servant,
Faithful student of the divinely inspired teacher
Patriarch John of Jerusalem.[108]

Klement's icon is written in the ink of desire in which the "pen of faith" is dipped (Τῆς πίστεως κάλαμον ἐμβάψας πόθῳ). However, this pen paints not the material icon itself upon which this poem was inscribed, but rather the image that is inscribed in the "fleshy tablet of the heart" (καρδίας... πλαξὶ σαρκίναις), resonating back to the metaphor of the tablets of the imagination. This structures in the poem the contradiction between spiritual and earthly sensation, which speaks to the referent's presence in the mind, but absence in the flesh. For Klement writes that while he sees Theodosios with the "pupils of the soul" (ψυχῆς... ταῖς κόραις), he yearns and desires to see him also through the senses (Ὁρᾶν δὲ, πάτερ, καὶ κατ᾽ αἴσθησιν θέλων). This eloquently demonstrates that the presence that eyes of faith enable through the visualization of the person in the imagination, rather than satiating or fulfilling the vision of that person, only amplifies the desire for their presence to earthly, physical sensation itself. By gazing upon the icon of Theodosios, Klement does not commune with him, but rather confronts the unbearable absence of his presence.

It is this conflict that generates the central contradiction in the icon, which, rather than satisfying or reminding you of the presence of the

[108] Klement the Monk, "Εἷς εἰκόνα τοῦ ἁγίου Θεοδοσίου τοῦ κοινοβιάρχου" in Anonymous, *Anthologia Marciana*, 362, ed. Spyridon P. Lambros, "Ὁ μαρκιανὸς κῶδιξ 524," *Νέος Ελληνομνήμων* 8:2 (1911), 123–192, esp. 184; modified trans. Annemarie Weyl Carr, "Reflections on the Medium of the Miraculous" in Sharon E. J. Gerstel (ed.), *Viewing Greece: Cultural and Political Agency in the Medieval and Early Modern Mediterranean* (Turnhout: Brepols, 2016), 33–52, esp. 43.

person, is meant to almost perversely confront you with their absence so as to generate a longing for that presence. This longing for presence is eloquently structured here as a temporal problem, for in the icon there is always the promise of the eventual re-presentation of that person in the heavenly salvation of the final judgment. Thus, the poem rhetorically shifts in the sixth line, notably the central line in this eleven-line poem: "O image and founder of the cenobitic life" (ὦ τύπε καὶ πρώταρχε τοῦ κοινοῦ βίου), which evocatively addresses both the image and the referent themselves. This line manifests precisely the conflict between the image (τύπε) and the referent (πρώταρχε), which in naming Theodosios using "founder" (πρώταρχε) almost seems to play with the proper term for referent in Byzantine image theory, "prototype" (πρωτότυπος), which is also nearly synonymous with "founder" (πρώταρχε), both indicating a primal original.

From there on the poem is essentially a response to the fifth line, "I inscribe your image out of material colors" (γράφω τύπον σὸν ἐξ ἐνύλων χρωμάτων). The last five lines function almost as the apodosis, or main clause, of a conditional statement, set up by the fifth line as a protasis, or dependent clause, since Klement then goes on to ask if his inscription of Theodosios in his heart will lead to his own inscription in the book of the saved. That is to say, if Klement has inscribed Theodosios, will Theodosios then inscribe Klement, which structures a conditional statement with an uncertain future. Yet, in this future possibility lies the possibility of presence itself, understanding that the impression of the icon in the heart may lead to the future manifestation of that image, one that given the final judgment will allow for that future image to be the presence of the referent itself, not just the representation of the image as if through memory. Thus, Klement provides us here with a nuanced and complex meditation on the icon, which aptly diagrams not only the yearning and longing that the icon enacts and its contradiction with sensory perception, but also how it is that presence might come to be achieved in end times through the inscription of the icon, in matter and in the soul alike.

At the behest of Emperor Alexios I Komnenos, following the Blachernai Synod in the wake of the Leo of Chalcedon affair, the court theologian Euthymios Zigabenos composed the *Panoplia Dogmatica*, a lengthy summary of dogmatic theology, accounting for the various heresies against Orthodoxy.[109] In the *Panoplia Dogmatica*, composed around 1111,

[109] See Jacob Wickert, "Die Panoplia dogmatica des Euthymios Zigabenos: Untersuchung ihrer Anlage und ihrer Quellen, ihres Inhaltes und ihrer Bedeutung," *Oriens Christianus* 8 (1908): 278–388.

Zigabenos was at one point tasked with defining the icon and its purpose, writing poignantly:

> τὰς ἱερὰς εἰκόνας ἀναστηλοῦμεν ἵνα ταύτας ὁρῶντες δι' αὐτῶν δοκῶμεν ὁρᾷν τὰ τούτων ἀρχέτυπα, καὶ γίνονται ἡμῖν ὑπομνήματα, καὶ παρηγορία τοῦ πόθου τοῦ πρὸς τὰ τούτων πρωτότυπα.
>
> What the sacred icons make visible are those visible aspects of their archetypes that become the memory and desire for the prototypes for us.[110]

Here his words demonstrate in no uncertain terms that the icon is intended to serve as a reminder (ὑπομνήματα) of the prototype and generate "longing" (πόθου) for it, the same expressions for yearning and desire as used in the Marciana epigram above. The icon is succinctly a site of desire, as has been noted previously by myself and others, but how it sustains, perpetuates, and incites that desire is through the mental investment: both by generating representations in the imagination that make it seem as if the figure is present, though it is not, and by preserving these visually vivid experiences in the memory for the continued endurance of this longing and yearning.

While there may be no "scientific" backing for haptic sight and the direct contact it is said to have enabled, the imagination and the passions it stirs do make it seem as if sight were overstepping its physical and sensual limitations, almost turning spectators into touchers. This is particularly true in states of heightened psychological emotion as notions of the evil eye's origin in passionate envy attest, as discussed by Basil of Caesarea, or, moreover, as associations between haptic sight and the erotic gaze indicate.[111] For example, in a homily on adultery, John Chrysostom cautions, "Even if you have not touched them with your hand (ἥψω τῇ χειρὶ), you have caressed them with your eyes (ἐψηλάφησας τοῖς ὀφθαλμοῖς)."[112] Here, warning against the lustful gaze, which stirs up sexual passions, John Chrysostom implies that this gaze is precisely an act of "viewing with the imagination," to use Verity Platt's useful phrasing once again, whereby the viewer not only gazes upon their objects of desire but copulates with them as their imagination generates erotic images of sexual intimacy.

110 Euthymios Zigabenos, *Panoplia Dogmatica*, ed. PG 130:1168C; trans. Barber, *Contesting the Logic of Painting*, 154.

111 See Elsner, "Naturalism and the Erotics of the Gaze," 247–261; Goldhill, "The Erotic Eye," 154–194; Bartsch, *The Mirror of the Self*, esp. 57–114; Blake Leyerle, "John Chrysostom on the Gaze," *Journal of Early Christian Studies* 1:2 (1993): 159–174.

112 "Εἰ γὰρ καὶ μὴ ἥψω τῇ χειρὶ, ἀλλ' ἐψηλάφησας τοῖς ὀφθαλμοῖς." John Chrysostom, *In Matthaeum*, 17, ed. PG 57:257.

Gregory of Nyssa explicitly draws this connection between erotic love and the desire for divine beauty, articulating the desire of the eyes to commune with their objects of desire. In his first homily, commenting on Song 1:1–4, Gregory writes:

> τίς γὰρ τοῦ τοιούτου κάλλους ἀνέραστος γίνεται, εἰ μόνον ὀφθαλμὸν ἔχοι τὸν ἐνατενίσαι τῇ ὥρᾳ δυνάμενον, οὗ πολὺ μὲν τὸ καταλαμβανόμενον κάλλος, ἀπειροπλάσιον δὲ τὸ διὰ τοῦ φαινομένου στοχαστικῶς εἰκαζόμενον;
>
> For who is there without desire for such beauty, if only they have an eye capable of gazing upon its time [of beauty]? And while beauty so discerned is great, that which [perception] skillfully imagines through the visible is infinitely greater.[113]

From there, Gregory goes on to compare this process to that of earthly erotic love (ὑλικὸς ἔρως), a matter that is reflected in the word chosen to reflect the impossibility of anyone "not desiring" (ἀνέραστος) the divine beauty. As this meditation on sight reflects, we see here that the investment in that divine beauty being discerned or grasped (καταλαμβανόμενον) occurs through the skillful imagination (στοχαστικῶς εἰκαζόμενον) coming through a contemplation of the perceptible world (διὰ τοῦ φαινομένου). The use of καταλαμβανόμενον for discernment works well with the focus on mental visualization here, given that, as we have seen repeatedly in this volume, it is used to denote the cognitive grasp of perception once it has assented to its imaginations. This segment has at times been used as a support for haptic sight, due to an earlier translation of "capable of gazing upon its beauty" (ἔχοι τὸν ἐνατενίσαι τῇ ὥρᾳ δυνάμενον) as "capable of reaching out to its loveliness."[114] However, not only is the language here consistent with the limits of perceptual mediation undertaken by sight and imagination, Gregory of Nyssa is also clear to state that even the vision that the mind conceives through perceptibles is infinitely lesser (ἀπειροπλάσιον) than the reality of that divine beauty. Hence, while this eroticized model of the divine gaze almost seems to transgress the established boundaries of sight, imagination, and

[113] Gregory of Nyssa, *In Canticum canticorum*, 1, ed. H. Langerbeck, *Opera*, vol. 6 (Leiden: Brill, 1960), 38; modified trans. Richard A. Norris, *Homilies on the Song of Songs* (Atlanta, GA: Society of Biblical Literature, 2012), 41.

[114] Gregory of Nyssa, *In Canticum canticorum*, 1, ed. Langerbeck, 38; trans. Casimir McCambley, *Commentary on the Song of Songs* (Brookline, MA: Hellenic College Press, 1987), 53–54. See Georgia Frank, *The Memory of the Eyes: Pilgrims to Living Saints in Christian Late Antiquity* (Berkeley: University of California Press, 2000), 124.

presence, even this heightened state is qualified as paling in comparison to the ineffable reality of the divinity.

Thus, while in John Chrysostom the puerile thoughts of the lustful gaze produce the illusion that one has stroked their desired lover, the perception of the divine through perceptible things in Gregory of Nyssa seems to suggest the same. However, just as the adulterous gaze surely knows that imaginative erotic desire can never compare nor satiate the consummation of those urges, for the viewer of the icon no matter how vivid or clear the illusion of presence may be in their mind's eye that imagination can surely not compare to the real presence of the divine. While some may have succumbed to the belief that their communion with an icon might have satiated that desire, it seems to have been the precise goal of Byzantine thinkers, particularly iconophiles, to remind viewers that this was in fact not the case. Deploying perhaps a popular and "not fully theorized" belief in the real presence of the icon, but making sure to carefully deny it theologically (despite the passions its imaginative viewing might have aroused), Byzantine image theorists like John of Damascus, Nikephoros of Constantinople, and Theodore of Stoudios were able to sustain the icon's place as such a vivid and thriving site of unbridled yearning and desire.[115] It is when we deny the medium and mediation of the icon in exchange for real presence that the icon loses its nuance and power.[116] This is because in doing so we forsake signification altogether for the bare presence of the referent, and in that act also make the Divine Liturgy and the miracle of the Eucharist redundant, if not wholly unnecessary.

[115] Barber, "From Image into Art," 5–10.

[116] On the problem of presence in medieval art historiography, see Roland Betancourt, "Medieval Art after Duchamp: Hans Belting's *Likeness and Presence* at 25," *Gesta* 55:1 (2016): 5–17. On this problem in the Protestant tradition and its implication for art history, see James Simpson, *Under the Hammer: Iconoclasm in the Anglo-American Tradition* (Oxford University Press, 2011). See also Edward James Martin, *A History of the Iconoclastic Controversy* (London: Society for Promoting Christian Knowledge, 1930). On the ancient groundings of this conversation, see Jeremy Tanner, *The Invention of Art History in Ancient Greece: Religion, Society, and Artistic Rationalisation* (Cambridge University Press, 2006); Verity Platt and Michael Squire (eds.), "The Art of Art History in Greco-Roman Antiquity," *Arethusa* 43:2 (2010), 133–369.

CHAPTER 10

Tactility and Veneration

ὅσῳ γὰρ συνεχῶς δι' εἰκονικῆς ἀνατυπώσεως ὁρῶνται, τοσοῦτον καὶ οἱ ταύτας θεώμενοι διανίστανται πρὸς τὴν τῶν πρωτοτύπων μνήμην τε καὶ ἐπιπόθησιν, καὶ ταύταις ἀσπασμὸν καὶ τιμητικὴν προσκύνησιν ἀπονέμειν, οὐ μὴν τὴν κατὰ πίστιν ἡμῶν ἀληθινὴν λατρείαν, ἣ πρέπει μόνῃ τῇ θείᾳ φύσει. ἀλλ' ὃν τρόπον τῷ τύπῳ τοῦ τιμίου καὶ ζωοποιοῦ σταυροῦ καὶ τοῖς ἁγίοις εὐαγγελίοις καὶ τοῖς λοιποῖς ἱεροῖς ἀναθήμασι, καὶ θυμιασμάτων καὶ φώτων προσαγωγὴν πρὸς τὴν τούτων τιμὴν ποιεῖσθαι, καθὼς καὶ τοῖς ἀρχαίοις εὐσεβῶς εἴθισται. Ἡ γὰρ τῆς εἰκόνος τιμὴ ἐπὶ τὸ πρωτότυπον διαβαίνει· καὶ ὁ προσκυνῶν τὴν εἰκόνα, προσκυνεῖ ἐν αὐτῇ τοῦ ἐγγραφομένου τὴν ὑπόστασιν.

For the more continually [the saints] are seen through their iconic representation, the more those who contemplate them are roused to remember and long for the prototypes, and to render to them greeting and honorable veneration: though not true worship, which according to our faith is fitting to the divine nature alone, but the same kind [of veneration] as is offered to the imprint of the precious and life-giving cross, to the holy gospels, and to the other holy dedicated items. Also [we declare] that one may honor these by bringing to them incense and light, as was the pious custom of the ancients; for the honor to the icon is conveyed to the prototype. Thus, the one who venerates the icon venerates the hypostasis of the person depicted on it.[1] – Definition of the Seventh Ecumenical Council (787)

As decreed by the definition of the Seventh Ecumenical Council in 787, responding to the First Iconoclastic Council of Hagia Sophia in 754, the icon's purpose is to elicit a remembrance of and longing (ἐπιπόθησιν) for the divine prototypes, not through worship (λατρείαν) proper, which is reserved for the divinity itself, but through veneration (προσκύνησιν). This veneration is not only rooted on a condition of longing, but it follows

[1] Anonymous, *Definition of the Second Council of Nicea*, ed. J. D. Mansi, *Sacrorum conciliorum nova et amplissima collectio*, vol. XIII (Paris: H. Welter, 1902), 377D–E; modified trans. Daniel J. Sahas, *Icon and Logos: Sources in Eighth-Century Iconoclasm* (University of Toronto Press, 1986), 179.

the ancient customs of bringing incense and lights to the images. As Jaś Elsner has suggested, the Iconoclastic controversy marked in this period a transition from ontology, questioning what the icon was, toward the question of how it is that the icon or any material object allows us to know and access the divinity.[2] In other words, the iconoclastic discourse of the period understood that what was at stake was not the validity of the icon as an ontological possibility, but rather the possibility of using the icon to access or perceive God through the practices of which veneration is comprised.

Throughout the previous chapters, we have seen how the deceptive tactility adjacent to sight has often referred not to vision itself, but to cognitive perception and the workings of the imagination. Thus, we have focused on the various confusions and distinctions of the sensory and perceptual processes. Nevertheless, it is impossible to ignore the incessant pairing of sight with haptic practices, such as the touching, embracing, and kissing of icons. As such, I wish to contextualize these findings by considering the ritual practices of icon and relic veneration. My goal in this chapter is to consider how touch and sight go hand in hand as separate yet simultaneously deployed senses, such as in the Eucharistic rites or in the veneration of relics.

Given the importance of veneration in iconoclastic discourse, it is necessary to contextualize how the matters of sight and tactility that have been discussed so far contributed to the conversations of the eighth and ninth centuries, and built upon preexisting practices of veneration. In other words, we must take seriously that what was at stake in the iconoclastic debates was not the icon, but rather an investigation as to what human perception and cognition allowed for, and which thus contoured what was possible to expect from the icon. Hence, we shall observe how writers, particularly John of Damascus and Theodore of Stoudios, often turned to contemporary discourses on sight, perception, cognition, and comprehension to articulate and justify the operation of the icon. This approach will not only help broaden our perspective on the Iconoclastic controversy, but also reveals a deeply nuanced language regarding the faculties of the mind, particularly regarding the operation of imagination, that has often been overlooked.

In the context of John of Damascus and others, we shall also see how touch manifests itself in the imaginative practices of viewers, who imagine

[2] Jaś Elsner, "Iconoclasm as Discourse: From Antiquity to Byzantium," *The Art Bulletin* 94:3 (2012): 368–394.

themselves to be dialoguing with and embracing the ones represented in icons. These considerations open up a broader conversation in the final sections of this chapter regarding the role that the imagination plays in enlivening and animating the icon in the minds of the faithful. The faithful are able not only to witness past events unfolding before them at holy sites, but also to generate the imprints in the imagination of those who are absent to the senses. This is an articulation of the imagination's counterfactual potentials that are deployed to generate an experience of the divine in the soul, despite that experience being absent to the senses themselves. This process rehabilitates the early Christian suspicions regarding the potential psychological deception of the imagination, put here to good and righteous use. In the viewer's inexorable desire for the holy figure represented in the icon, they are capable of impressing that form into their soul, perceiving not through the bodily senses but through the imaginative faculties of the mind. This mediates between that dynamic of presence and absence discussed in the last chapter, whereby the icon serves as a condensation of desire through which a viewer can generate imprints of the divine in the soul *as if* they had sensed them in the flesh, while acknowledging that inherent lack.

Seeing and Touching in Veneration

The crucial figure in the shift from ontology to epistemology in the iconoclastic discourse is John of Damascus, who early on in the controversy, in Patrick Henry's own words, "realized that the fundamental issue was not whether icons were idols and the worship of them idolatry, but whether any material aids were permissible in Christian worship."[3] While John of Damascus is careful to clearly articulate what an image is and what various types of images exist or are possible, the focus of his treatises on the holy images is on the problem of veneration and what is due or undue to the holy images; understanding the Incarnation and the Christological discourses before him foregrounded the ontological possibility of the icon. It is unsurprising then that many of the texts cited in Nelson's eloquent argument for haptic sight come from around the Iconoclastic period, when a clarification regarding the operation of the image and its perception was theorized and clarified, particularly as embodied in the figure of John of

[3] Patrick Henry, "What Was the Iconoclastic Controversy about?," *Church History* 45:1 (1976): 16–31, esp. 25.

Damascus. As Nelson recounts, observing the importance of the iconomachy on these considerations:

> During and after Iconoclasm, many arguments in support of religious images assumed extramission. Visual theory provided a seemingly natural or scientific defense of the sense of sight and the concomitant legitimacy of the object seen, that is, the icon. For example, John of Damascus takes for granted that his audience shares the notion of extramission, when he asks about an icon, "Shall I not embrace (περιπτύξομαι) with my eyes and lips that which is a wonder to the angels. . .?" And again, when he advocates regarding icons that we should "Kiss (ἀσπάζεσθι) [*sic*] them with the eyes, the lips, the heart."[4]

A challenge is presented here, for not only is John of Damascus an important turning point away from the notion of presence in the image, but also the focus on John of Damascus himself bears witness to a very particular moment that does not give due credit to the work of Theodore of Stoudios and the Patriarch Nikephoros. For instance, it should not be overlooked that Nikephoros was an Aristotelian and used Aristotelian physics to support his image theory, a point that already challenges the claim for haptic extramission's prevalence in Byzantium.[5] For Aristotle was not only close to the intromissionist camp, but also a staunch proponent of the mediation of the transparent (διαφανής) between seer and object. Likewise, Theodore of Stoudios is careful to deploy sight and touch together throughout his treatises on icons, but without confusion so as to stress the intertwined (but never confused) haptic and optic practices of veneration.

In order to carefully deconstruct our assumptions regarding the work of John of Damascus, I wish to begin by looking closely at the two excerpts from John of Damascus's treatises on icons offered by Nelson, which have provided some of the most eloquent justifications for an application of haptic sight to Byzantine visuality in the Iconoclastic and post-Iconoclastic periods. Firstly, we find that in Nelson's citation the text's translation has rendered its meaning quite distinct from that found in the original text. The imperative that one should, "Kiss them with the eyes, the lips, the heart," appears four times in John's text: twice in Treatise II.10, and repeated in III.9. The full passage in which they appear reads as follows:

4 Nelson, "To Say and to See," 153.

5 On the image theory and education of Nikephoros, see Paul J. Alexander, *The Patriarch Nicephorus of Constantinople: Ecclesiastical Policy and Image Worship in the Byzantine Empire* (Oxford University Press, 1958). Cf. Nelson, "To Say and to See," 154.

Ἀλλ᾿ ὅμως ἐπειδὴ πάντα τὰ περὶ αὐτῶν λεγόμενα ἀληθῆ εἰσι καὶ ὁ σκοπὸς πρὸς δόξαν θεοῦ ἐστι καὶ τῶν ὑπ᾿ αὐτοῦ δοξαζομένων ἁγίων καὶ σωτηρίαν ἡμῶν καὶ καθαίρεσιν καὶ αἰσχύνην τοῦ διαβόλου καὶ τῶν δαιμόνων αὐτοῦ, προσκυνοῦμεν καὶ περιπτυσσόμεθα καὶ καταφιλοῦμεν καὶ ὀφθαλμοῖς καὶ χείλεσι καὶ καρδίᾳ ἀσπαζόμεθα, ὁμοίως καὶ πᾶσαν τὴν παλαιὰν καὶ καινὴν διαθήκην τούς τε λόγους τῶν ἁγίων καὶ ἐκκρίτων πατέρων, τὴν δὲ αἰσχρὰν καὶ μυσαρὰν καὶ ἀκάθαρτον γραφὴν τῶν καταράτων Μανιχαίων τε καὶ Ἑλλήνων καὶ τῶν λοιπῶν αἱρέσεων ὡς ψευδῆ καὶ μάταια περιέχουσαν καὶ πρὸς δόξαν τοῦ διαβόλου καὶ τῶν δαιμόνων αὐτοῦ καὶ χαρὰν αὐτῶν ἐφευρεθεῖσαν ἀποπτύομεν καὶ ἀποβαλλόμεθα καίτοι γε καὶ ὄνομα θεοῦ περιέχουσαν. Οὕτως καὶ ἐν τῷ πράγματι τῶν εἰκόνων χρὴ ἐρευνᾶν τήν τε ἀλήθειαν καὶ τὸν σκοπὸν τῶν ποιούντων καί, εἰ μὲν ἀληθὴς καὶ ὀρθὸς καὶ πρὸς δόξαν θεοῦ καὶ τῶν ἁγίων αὐτοῦ καὶ πρὸς ζῆλον ἀρετῆς καὶ ἀποφυγὴν κακίας καὶ σωτηρίαν ψυχῶν γίνονται, ἀποδέχεσθαι καὶ τιμᾶν ὡς εἰκόνας καὶ μιμήματα καὶ ὁμοιώματα καὶ βίβλους τῶν ἀγραμμάτων καὶ προσκυνεῖν καὶ καταφιλεῖν καὶ ὀφθαλμοῖς καὶ χείλεσι καὶ καρδίᾳ ἀσπάζεσθαι ὡς σεσαρκωμένου θεοῦ ὁμοίωμα ἢ τῆς τούτου μητρὸς ἢ τῶν ἁγίων τῶν κοινωνῶν τῶν παθημάτων καὶ τῆς δόξης τοῦ Χριστοῦ καὶ νικητῶν καὶ καθαιρετῶν τοῦ διαβόλου καὶ τῶν δαιμόνων καὶ τῆς πλάνης αὐτῶν, εἰ δὲ θεότητος τῆς ἀΰλου καὶ ἀσωμάτου καὶ ἀοράτου καὶ ἀσχηματίστου καὶ ἀχρωματίστου εἰκόνα τις τολμήσει ποιῆσαι, ὡς ψευδῆ ἀποβαλλόμεθα.

But since everything said about them is true and the purpose is the glory of God and of the saints glorified by him, and our salvation and the overthrow and disgrace of the devil and his demons, all these we venerate and embrace and kiss with eyes and lips and cleave to in the heart, likewise the whole of the Old and New Testaments and the words of the holy and select Fathers, but the shameful and filthy and unclean writing of the accursed Manichees and Greeks and of the rest of the heresies we spit out and reject as containing lies and emptiness, devised for the glory of the devil and his demons and their delight, even though they contain the name of God. So also in the matter of images, it is necessary to search out the truth and the purpose of those who make them, and, if they turn out to be true and upright, promoting the glory of God and his saints, and inspiring virtue and driving away vice, and leading to the salvation of souls, then to accept and honor them as images and copies and likenesses and books for the illiterate, and to venerate and kiss them with eyes and lips, and cleave to them in the heart, as a likeness of God incarnate, or of his Mother, or of saints who shared in the sufferings and the glory of Christ and were conquerors who overthrew the devil and the demons and their error, but if anyone dare to make an image of the immaterial and incorporeal and invisible and formless and colorless divinity, we reject them as false.[6]

[6] John of Damascus, *Orationes de imaginibus tres*, II.10, cf. III.9, ed. Kotter, 98–100; modified trans. Louth, 66–67, cf. 90–91.

The quotation is cropped from the second appearance of the phrase in the passage, which more accurately, following Andrew Louth's 2003 translation, reads as:

> . . .and to venerate (προσκυνεῖν) and kiss (καταφιλεῖν) them with eyes and lips, and cleave (ἀσπάζεσθαι) to them in the heart. . .

However, at the time, Nelson only had available the 1980 translation of the text by David Anderson from which his translation is taken, though no citation is given in the relevant footnotes.[7] Anderson translated the first and second appearances of the phrase in question as follow: (1) "we bow before them, we embrace and love them, we kiss them with our eyes, our lips, our hearts"; and, (2) "Embrace them with the eyes, the lips, the heart; bow before them; love them."[8] From the latter, Nelson took the first portion that attributes the action of the embrace as relating to the eyes, lips, and heart, rather than understanding the more unified operation of the various subsequent parts of the sentence with one another.

Therefore, it becomes increasingly evident that this translation renders incorrectly the term to "cleave" or "greet" (ἀσπάζεσθαι) as to "kiss," which is possible and also the modern Greek meaning of the word. However, here to "kiss" is denoted by καταφιλεῖν properly. On the other hand, the term ἀσπάζεσθαι goes with "in the heart" (καρδίᾳ). Thus, this should not be translated merely as "kiss," since it would be redundant to express the act of kissing in the same sentence with two different terms, without having nuanced differences. In Nelson's case, the result is that the text consequently drops completely the καταφιλεῖν in the translation. This is only exacerbated by the misleading cropping of the quote, cited simply as, "Kiss them with the eyes, the lips, the heart," which obscures John of Damascus's use of parallelism: "venerate. . . with the eyes," "kiss with. . . lips," and "cleave to in the heart." This causes the statement to read as a list of organs (i.e. "the eyes, the lips, the heart") responsible for "kissing" the object, rather than as a diagrammatic representation of the process of veneration that the original text addresses. Furthermore, the limited

[7] Note that Nelson also appears to take "Shall I not embrace (περιπτύξομαι) with my eyes and lips that which is a wonder to the angels. . .?" from Anderson's translation since both break a single Greek sentence into two rhetorical questions, the latter portion of which is cited and to which we shall return later in this chapter. Nevertheless, Nelson rightly corrects Anderson by adding, "and lips," which was erroneously left out by Anderson in his translation. See John of Damascus, *Orationes de imaginibus tres*, I.47 (cf. II.43), trans. David Anderson, *On the Divine Images: Three Apologies against Those Who Attack the Divine Images* (Crestwood, NY: St. Vladimir's Seminary Press, 1980), 39. See also Nelson, "To Say and to See," 153; 165 n 56–57.

[8] John of Damascus, *Orationes de imaginibus tres*, II.10, trans. Anderson, *On the Divine Images*, 57–58.

understanding of προσκυνεῖν as prostration, rather than broadly encompassing the actions of veneration, further affects the understanding of these related kinesthetic and perceptual processes being discussed, which all are deployed in the process of veneration. Therefore, John of Damascus's characterization of the image's veneration is not as strange as it once was perceived, lacking perhaps an enchanting alterity, and instead attesting to an orderly process of both mental and kinesthetic approach toward the icon through the various stages of veneration.

John of Damascus is indeed sketching out a diagram of the icon's visuality here, but it is not one oriented around haptic sight, or even sight alone. Instead, it enumerates sight alongside other kinesthetic, haptic, and psychological steps in the practice of veneration, as the process of *proskynesis* entails. As the first iteration of the phrase in the passage reads:

> . . . all these we venerate (προσκυνοῦμεν) and embrace (περιπτυσσόμεθα) and kiss (καταφιλοῦμεν) with eyes and lips, and cleave to in our hearts,. . .

Here, John of Damascus composes an image recognizable to those familiar with contemporary Orthodox practices of greeting an icon upon an icon stand or screen, whereby the faithful approach the icon gazing at it, embrace it, and kiss it. Rather than a direct one-to-one parallelism between veneration and sight, kissing and touching, what John of Damascus is actually portraying is the approach toward an icon in the psychic disposition of the viewer.[9] This is both a somatic and cognitive practice as the reference to the "eyes and lips" makes clear. As both the phrases conclude though, this process is undertaken and results in cleaving to that image in one's heart – a reference that speaks to the resulting impression of the icon's image in the heart for future emulation. As we can see here as well, the paralleling of verbs suggests that καταφιλεῖν complements ἀσπάζεσθαι, thus playing with the literal kissing of the icon, and the spiritual, cognitive kissing with the heart. This emerges from Byzantine perceptual theory and is echoed by various sources ranging from the ninth to thirteenth century, including Photios's Homily 17, as has been discussed at length. As Michael Psellos explains in his commentary on Aristotle's *On Sense and Sensible Objects*, based on the popular commentary by Alexander of Aphrodisias, the heart is the center of perception and linked to the organs of perception through the conduits of the body, which pass on to the necessary parts the forms of the things seen, a matter to which we shall return in this chapter.

9 On the embrace of the image in the contemporaneous Jewish mystical tradition, see Rachel Neis, "Embracing Icons: The Face of Jacob on the Throne God," *Images* 1:1 (2007): 36–54.

Thus, to cleave or cling to the icon with one's heart alludes to its impression upon the soul for the emulation and remembrance of the image depicted. In this manner, the process that John of Damascus articulates is not strictly scientific or religious, but both. Yet it is also not merely a metaphor; instead it speaks to the proper psychosomatic entanglement through which image veneration operates.

It is possible to sketch out a coherent system within the defense of icons treatises of the Damascene regarding this interlacing of sight, touch, and psychic disposition. Elsewhere, for example, John of Damascus writes: "Many times I have seen those who long for someone, when they have seen his garment, greet it with their eyes and lips (ἀσπασαμένους τὸ ἱμάτιον ὀφθαλμοῖς τε καὶ χείλεσι), as if it were the one longed for himself."[10] Staged here as a broader process of greeting, the eyes and lips work in tandem to allude to this process of coming upon an object with an affective, psychosomatic response that deploys both sight and touch in its desire for the presence of that person who is inherently absent. Here this process of desiring the absent not only is in accord with his own image theory, but likewise places touching and seeing alongside a process of greeting (ἀσπασαμένους). Thus, it is not possible to read these two terms as speaking to a theory of vision alone, but rather to a holistic process of "greeting" an image in the mind and heart.

This separation stresses the importance of both touching and seeing in Byzantine ritual practices around icons, which is expressed in the careful language of Theodore of Stoudios. Theodore, like John of Damascus, often deploys the senses in unison in order to articulate the paired perception of the icon through sight and touch, given that each is particular to itself. As Theodore writes in one instance: "When He became flesh and entered into circumscription, the uncircumscribable one was seen, and the intangible and invisible one became subject to bodily sight and touch (θέᾳ καὶ ἁφῇ. . .σωματοειδεῖ)."[11] Here, touch and sight speak to the circumscribable nature of the Christ, who condescended to become visible *and* tangible. The language of sight and touch reappears throughout his treatises on icons, precisely to stress the perceptible physicality of the

[10] "Εἶδον πολλάκις ποθοῦντας, ἱμάτιον τοῦ ποθουμένου θεασαμένους, ὡς αὐτὸν τὸν ποθούμενον ἀσπασαμένους τὸ ἱμάτιον ὀφθαλμοῖς τε καὶ χείλεσι." John of Damascus, *Orationes de imaginibus tres*, III.10, ed. Kotter, 102; trans. Louth, 92.

[11] "ὡς ἐπειδὴ σὰρξ ἐγένετο καὶ εἴσω περιγραφῆς. ὤφθη ὁ ἀπερίγραπτος: καὶ θέᾳ καὶ ἁφῇ ὑποπίπτωκε σωματοειδεῖ ὁ ἀναφής τε καὶ ἀθέατος." Theodore of Stoudios, *Antirrhetici adversus iconomachos*, II.5, ed. PG 99:356A; trans. Catharine P. Roth, *On the Holy Icons* (Crestwood, NY: St. Vladimir's Seminary Press, 1981), 46.

incarnate Christ, justifying his circumscribability in colors.[12] Furthermore, the pairing also emerges regarding the treatment of ritual objects in the Old Testament to describe those that were regulated through prohibitions regarding their looking at (προσβλέπειν) and touching (προσάπτεσθαι).[13] Hence, this language speaks to a clear separation of touch and sight in Theodore's thinking, while nevertheless indicating that the two are necessarily deployed in unison for the perceptibility of Christ and the use of ritual objects alike.

Theodore also eloquently demonstrates to his reader that the dynamics of sight and tactility are not only relegated to physical sensation, but also to the mental acts of contemplation and intellection. As he writes:

> Οἷς μὲν τὰ καθόλου ὁρᾶται, νοῦς καὶ διάνοια· οἷς δὲ τὰ καθ' ἕκαστα, ὀφθαλμοὶ, οἱ τὰ αἰσθητὰ βλέποντες. Εἰ οὖν τὴν καθόλου ἡμῶν φύσιν ὁ Χριστὸς ἀνέλαβε, μὴ ἐν ἀτόμῳ δὲ θεωρουμένην, νῷ μόνῳ ἐστὶ θεωρητὸς καὶ διανοίᾳ ψηλαφητός. Ἀλλὰ μήν φησι πρὸς τὸν Θωμᾶν, Ὅτι ἑώρακάς με, πεπίστευκας· μακάριοι οἱ μὴ ἰδόντες, καὶ πιστεύσαντες. Καὶ αὖθις· Φέρε τὸν δάκτυλόν σου, ὧδε, καὶ ἴδε τὰς χεῖράς μου, καὶ φέρε τὴν χεῖρά σου, καὶ βάλε εἰς τὴν πλευράν μου· αἰσθητοῖς αἰσθητὰ παραβαλών. Αἰσθητὸς ἄρα ὁ Χριστός, ψηλαφητός, καὶ σωματικοῖς ὀφθαλμοῖς ὁρατός, καὶ διὰ τοῦτο περιγραπτός.
>
> Generalities are seen with the mind and thought; particular individuals are seen with the eyes, which look at perceptible things. If, therefore, Christ assumed our nature in general, not contemplated in an individual manner, He can be contemplated only by the mind and touched only by thought. But He says to Thomas, "Because you have seen me, you have believed; blessed are those who have not seen and yet believe." And He also says, "Put your finger here, and see my hands; and put out your hand, and place it in my side"; thus He associates perceptible things with perceptible things. So Christ is perceptible, tangible, and visible with bodily eyes; and therefore He is circumscribed.[14]

Here we see Theodore clarifying Christ's comprehensibility through perception versus mental contemplation (θεωρουμένην). Rhetorically, the passage demonstrates the language of tangibility and visibility operating both physically and mentally, alluding not only to a physical tangibility,

[12] For example, see Theodore of Stoudios, *Antirrhetici adversus iconomachos*, I.3, I.4, I.10, I.12, II.5, II.46, III.1, III.4, III.12, III.15, III.16, III.46, trans. Roth, 22, 23, 31, 33, 46, 72, 78, 79, 81, 83, 94.

[13] "Τοσούτου δεῖ τὸ καταπέτασμα, ἢ τὸ ἱλαστήριον, ἢ τὴν κιβωτὸν, ἢ τὰ Χερουβίμ, ἢ προσβλέπειν εἶναι παντός, ἢ προσάπτεσθαι... Ἀλλὰ γὰρ καὶ τὸ μὴ θεμιτόν, ἢ προσβλέπειν, ἢ προσάπτεσθαι τῶν προειρημένων ἁγίων, ἢ μόνον τοῦ ἱερέως, πολὺ τὸ ἀνῳκισμένον κατὰ τιμὴν ὑποφαίνει ὁ λόγος." See Theodore of Stoudios, *Antirrhetici adversus iconomachos*, II.37–38, ed. PG 99:377A–C; trans. Roth, 65–66.

[14] Theodore of Stoudios, *Antirrhetici adversus iconomachos*, III.16, ed. PG 99:397A; trans. Roth 83.

but also to that of the mind as it grasps and understands concepts (θεωρητὸς καὶ διανοίᾳ ψηλαφητός). Christ in the flesh was perceptible, tangible, and also visible with the bodily eyes (Αἰσθητὸς ἄρα ὁ Χριστός, ψηλαφητός, καὶ σωματικοῖς ὀφθαλμοῖς ὁρατός), but this is not to say that this tangibility is tied to his visibility, only that he is notably both visible and also tangible. In Photios's Homily 17, as we saw in Part II, where he discusses the operation of sight and hearing, he likewise uses a similar technical language of grasping and drawing near to characterize the cognitive acts of mental perception with words such as πρόληψις, κατάληψις, ἀντιλαμβάνειν, and εἵλκυσε that suggest forms of apprehension and comprehension.[15] This language of tactility in terms of cognition is seen consistently throughout texts on cognition and memory since the late antique period.[16] As such, we can deduce that in John of Damascus and Theodore of Stoudios, sight and touch are deployed in careful unison with one another to indicate acts of both visual and physical veneration, understanding that each sense contributes a sensation proper to itself.

Returning to John of Damascus, such matters are clarified with care in what is perhaps John's most nuanced reflection on the procedurality of veneration. In the third treatise, John writes:

> Καὶ ὥσπερ Ἰακὼβ δεξάμενος παρὰ τῶν ἀδελφῶν τοῦ Ἰωσὴφ τὸν χιτῶνα τὸν ποικίλον ᾑμαγμένον, ὅτε ἐπίπρασαν τὸν Ἰωσήφ, κατεφίλησε πάντως τὸν χιτῶνα μετὰ δακρύων καὶ τοῖς ὀφθαλμοῖς τοῖς ἰδίοις τοῦτον ἔθηκεν οὐ τὸ ἱμάτιον θρηνῶν, ἀλλὰ δι᾽ αὐτοῦ τὸν Ἰωσὴφ νομίζων καταφιλεῖν καὶ ἐν χερσὶν αὐτὸν κατέχειν, οὕτω καὶ χριστιανῶν παῖδες εἰκόνα <Χριστοῦ> ἢ ἀποστόλου ἢ μάρτυρος κατασπαζόμενοι τῇ σαρκὶ τῇ ψυχῇ νομίζομεν αὐτὸν ἐκεῖνον τὸν Χριστὸν ἢ τὸν μάρτυρα αὐτοῦ ἀσπάζεσθαι.
>
> And just as when his brothers sold Joseph, and Jacob received from them the blood-stained coat of many colors, he kissed the garment with tears, and set this to his own eyes, not mourning the garment, but through it thinking to kiss Joseph and hold him in his arms, so also the children of the Christians, when they physically embrace the icon [of Christ] or an apostle or a martyr, reckon that they honor Christ himself or the martyr spiritually.[17]

John eloquently recounts how the garment is "kissed" (κατεφίλησε) with tears and eyes, yet this observation is careful not to elide touch and sight: The garment is kissed "with tears" (μετὰ δακρύων), tears here in the plural

[15] Photios, *Homiliae*, 17, ed. Laourdas, 170–171.

[16] See, for example, Peter Lautner, "Mental Images," 220–50. For the medieval West, see Mary Carruthers, *The Craft of Thought: Meditation, Rhetoric, and the Making of Images, 400–1200* (Cambridge University Press, 1998).

[17] John of Damascus, *Orationes de imaginibus tres*, III.87, ed. Kotter, 179–80; trans. Louth, 132–133.

genitive going with μετὰ, whereas the act of seeing is depicted by the act of putting the garment up to the eyes in the indirect object dative (τοῖς ὀφθαλμοῖς τοῖς ἰδίοις τοῦτον ἔθηκεν). Yet, as John continues, the garment itself is not that which is receiving these actions, but rather the absent figure that is being represented in the mind so as to *think* – "through it" (δι' αὐτοῦ) – that one is kissing him (νομίζων καταφιλεῖν) and holding him in one's arms (ἐν χερσὶν αὐτὸν κατέχειν). This phrase is deployed precisely as a justification for the icon, which like the cloak, is not itself the object of devotion, but rather leads cognition toward the absent one. What this implies is that even if one were capable of kissing the icon with the eyes, one would not have any greater contact with the divine, given that the icon is merely the manifestation of something that is absent – a site of desire that only the fantasy of the mind can mediate. To believe anything more would be heretical. This point is driven home in the subsequent line where John strives to further clarify this by juxtaposing the physical embrace (κατασπαζόμενοι τῇ σαρκί) of an icon with the spiritual embrace that the worshiper reckons they are doing (τῇ ψυχῇ νομίζομεν. . . ἀσπάζεσθαι) – the emphatic prefix κατα- stressing here the physicality of the greeting, just as in καταφιλεῖν it transforms the abstract verb "to love" (φιλεῖν) into the concrete verb "to kiss." As such, the somatic embrace characterized through the term κατασπάζομαι is contrasted with the psychic embrace of ἀσπάζομαι, as we have seen in the first passage.

Compare this to a similar distinction within a secular context in Michael Psellos's *Chronographia*, where Psellos recounts Constantine IX Monomachos's perceptual engagement with Empress Zoe, given his love for his mistress Maria Skleraina. There Psellos captures the disjuncture between physical and perceptual embrace, writing:

> ἀλλ' ὅ γε αὐτοκράτωρ, οὐδ' ἐν αὐτῇ τῇ πρὸς τὴν βασιλείαν εἰσόδῳ, τῆς γυναικὸς ἐπιλέληστο· ἀλλὰ τοῖς μὲν αἰσθητοῖς ὄμμασι, τὴν βασιλίδα τεθέαται· τὴν δὲ ἐκείνης μορφήν, τοῖς ἀπὸ τῆς ψυχῆς συνήθρει καὶ συνελάμβανε· καὶ τὴν μέν, ἠγκάλιστο· τὴν δὲ ἐγκόλπιον εἶχεν ἐν τῇ ψυχῇ.
>
> But, the Emperor had not forgetten the woman in his ascent to royalty. But while with the physical eyes he beheld the Empress [Zoe], the form of that one [his mistress] he gazed and held; and when he had embraced [the Empress], he embraced in his soul the one in his heart.[18]

In this striking depiction, Psellos structures Constantine's lust and infidelity through the dissonances of his perceptual actions, physically (αἰσθητοῖς) embracing one and yet perceiving himself to be embracing another in his

[18] Michael Psellos, *Chronographia*, VI.52, ed. Reinsch, 128.

spirit (ἐν τῇ ψυχῇ). The statement relies heavily on parallelism to make the point regarding Constantine's duplicitous behavior, and yet rather than using an orderly echoing of verbs, Psellos uses a proliferation of terms in keeping not only with his personal style, but also so as to capture the various states and permutations of contemplation and embrace that occur in Constantine's soul. While he beholds (τεθέαται) and embraces (ἠγκάλιστο) the Empress with these common verbs, he emphatically looks upon and apprehends (συνήθρει καὶ συνελάμβανε) his mistress, holding fast to her (εἶχεν). Hence, Psellos here purposely suggests heightened states of contact and embrace occurring in Constantine's mind in order to emphasize Constantine's longing and desire. In the case of John of Damascus, the repetition of the similar verbs for the physical embrace (κατασπαζόμενοι τῇ σαρκί) and the spiritual embrace (τῇ ψυχῇ νομίζομεν... ἀσπάζεσθαι) stress the harmony of the two's operation, yet the perceptual process at work here is homologous to that of Constantine's own lustful thoughts.

The juxtaposition of these two homologous words with the same root should lead one to reconsider, then, the rendering of haptic-seeming terms such as to embrace (ἀσπάζεσθαι) or to enfold (περιπτύσσω) in the work of John of Damascus. The double operation of ἀσπάζεσθαι stresses that the word does not simply serve as analogy or metaphor, but rather has a technical weight to it that grapples and encompasses precisely the complex and nuanced mediation of the icon in terms of human cognition and the fantasy of the imagination. The need to embrace the icon physically and spiritually emerges precisely because the icon is a medium for the person represented, not a re-presentation of that person. In fact, John of Damascus makes this evident in his florilegia, where he recounts a passage from the *Vita* of John Chrysostom, where it is written that:

> Ἦν δὲ καὶ τὸ ἐκτύπωμα τοῦ αὐτοῦ ἀποστόλου Παύλου ἔχων ἐν εἰκόνι, ἔνθα ἀνεπαύετο ... Καὶ ἡνίκα διήρχετο τὰς ἐπιστολὰς αὐτοῦ, ἐνητένιζεν αὐτῇ καὶ ὡς ἐπὶ ζῶντος αὐτοῦ οὕτω προσεῖχεν αὐτῷ μακαρίζων αὐτόν, καὶ ὅλον αὐτοῦ τὸν λογισμὸν πρὸς αὐτὸν εἶχε φανταζόμενος καὶ διὰ τῆς θεωρίας αὐτῷ ὁμιλῶν.
>
> [John] had a depiction of the same apostle Paul in an icon, in a place where he used to rest... And when he had finished his epistles, he would gaze at it and attend to him as if he were alive and bless him, and bring the whole of his thoughts to him, imagining that he was speaking with him in his contemplation.[19]

[19] John of Damascus, *Orationes de imaginibus tres*, I.61 (cf. II.57, III.54), ed. Kotter, 161–162; trans. Louth, 55.

Through this citation, John of Damascus explains the very process that he has sketched out in the passages cited above, whereby the practices of gazing (ἐνητένιζεν), embracing (προσεῖχεν), and blessing (μακαρίζων) are choreographed as distinct acts leading to the cognitive emplacement of the person into the icon so as to imagine or fantasize (φανταζόμενος) that one is conversing with the person depicted there. This passage constructs the icon not as a site of presence or aesthetic fulfillment, but rather as a psycho-spiritual game of procedural and affective entanglement with the icon, just as with Joseph's garment. Not only does this passage help to clarify John of Damascus's perceptual theory, but as a fragment in the florilegia – originating from a seventh-century *Life of John Chrysostom* by George, Bishop of Alexandria, based on the earlier version by Palladius, a disciple of John Chrysostom – the excerpt also indicates that the process detailed by John of Damascus is not an idiosyncratic conception of icon veneration.

Thus, when we come across the other appearance in the text where such parallelism between sight and touch is constructed, we encounter the formula that should now be familiar:

> Οὐ γράψω καὶ λόγῳ καὶ χρώμασι τὸ τῶν μαρτύρων μαρτύριον καὶ ὀφθαλμοῖς καὶ χείλεσι περιπτύξομαι «τὸ θαυμαστὸν καὶ ἀγγέλοις καὶ πάσῃ τῇ κτίσει, ὀδυνηρὸν δὲ τῷ διαβόλῳ καὶ φοβερὸν δαίμοσιν,» ὡς αὐτὸς ὁ φωστὴρ τῆς ἐκκλησίας ἔφησεν;
>
> Shall I not depict in words and in colors the martyrdom of the martyrs and embrace with eyes and lips "what is wonderful to angels and the whole creation, painful to the devil and fearful to demons,"[20] as the same beacon of the Church said?[21]

This passage, namely, "embrace with eyes and lips" (ὀφθαλμοῖς καὶ χείλεσι περιπτύξομαι), is not to be read as a unification of sight and touch, but rather a procedural indication of the phenomenological practice of approaching the icon: first, contemplating it intently and taking grasp of it in the mind, and then kissing its material surface with the lips, while thinking in the mind and spirit to kiss what is being represented there in material form. Furthermore, here lips are not merely addressing the act of kissing, but also the act of speaking, going along with "in words and colors" to suggest another dimension of sensory deployment here. What such passages bring to bear on the conversation of sensual knowledge and

[20] Basil of Caesarea, *In Gordium martyrem*, ed. PG 31:501B.

[21] John of Damascus, *Orationes de imaginibus tres*, I.47 (cf. II.43), ed. Kotter, 152; trans. Louth, 46.

worship in Byzantium is therefore not the elision of the different senses, such as sight and touch, but rather the opposite: their heightened specificity as inexorably unique mediums of veneration that must be deployed in tandem alongside one another, without mixing or confusion.

Tactile Veneration

This misleadingly synesthetic style of writing, in which sight and touch would seem to be confused, evidences that an effort to describe the periphery of kinesthetic actions that complemented ocular worship is popular amidst pre-iconoclastic writers as well. A rhetorical language analogous to that of John of Damascus can be found in the fourth-century mystagogical homilies of Cyril of Jerusalem in which at first glance it would seem that Cyril as well ascribed to a haptic-iterant of sight. In Homily 5, Cyril writes on the Eucharistic sacrament and describes the proper mental and physical disposition of those approaching the rite of Communion. There he carefully describes the kinesthetic properties of the rite, such as placement of the hands, position of the arms, and bowing of the head. Yet he also describes the necessary sensual and psychic outlook of the communicant, who must elevate their senses – as the *sursum corda* in the pre-anaphoral rites suggests[22] – so as to not taste mere bread and wine, but the antitype of the body and blood of Christ. Enigmatically, he describes the way in which one should cup the hands when approaching the bread, respond "Amen" upon receiving it, and then, he instructs that one: "With care, sanctify the eyes with the touch of the sacred body, and partake, taking care not to lose any of it."[23] Cyril's statement in this shorthand description could easily appear to entail a haptic theory of sight, whereby the eyes are hallowed by the assumed visual touch (τῇ ἐπαφῇ) of the bread in their hands. After all, to modern readers it would seem awkward that one would press the bread upon the eyes, and the statement's lack of a clarifying preposition before τῇ ἐπαφῇ leaves ample room for confusion. Thus, with haptic extramission in mind, it would seem only natural to ascribe to Cyril this theory of vision.

[22] See Robert F. Taft, "The Dialogue before the Anaphora in the Byzantine Eucharistic Liturgy, II: The *Sursum Corda*," *Orientalia Christiana Periodica* 54 (1988): 47–77.

[23] "Μετ' ἀσφαλείας οὖν ἁγιάσας τοὺς ὀφθαλμοὺς τῇ ἐπαφῇ τοῦ ἁγίου σώματος μεταλάμβανε, προσέχων μὴ παραπολέσῃς τι ἐκ τούτου." Cyril of Jerusalem, *Mystagogiae*, 5.21, ed. Pierre Paris and Auguste Piédagnel, *Catéchèses Mystagogiques*, 2nd edition, Sources chrétiennes 126 bis (Paris: Éditions du Cerf, 1988), 170–172.

However, just as in the case of John of Damascus, further in the text this reading is clarified. In reality, the text evidences a detailed description of the gestures and practices of early Byzantine forms of veneration in regards to the Eucharistic gifts. In the following paragraph, dedicated to the communion of the wine from the chalice, Cyril commands: "While it is still moist upon the lips, touch it with the fingers, and sanctify the eyes, forehead, and other organs of sense."[24] Here, Cyril commands that one literally touch the eyes with the wine left upon the lips with one's hands, leading one to reread his earlier command following the communion of the bread as an indication that one should touch the bread to the organs of sight and so sanctify the eyes as well – an enigmatic statement whose indications seemed odd in isolation.

There is a pronounced concern with sensual experience within Cyril's catechetical homilies, but it is not one reliant upon any union or confusion of vision and touch. This anointing of the sensual organs serves to elevate the senses above the earthly realities, or as Cyril exhorts the congregants before Communion, to "not entrust judgment to your bodily palate, but to undoubting faith."[25] Thus it follows that Cyril uses the noun and participle forms of the verb ἐπαφίημι to describe a literal touching of the Eucharistic gifts with the sense organs. Cyril speaks to the clear differences between senses by urging the participant to touch each of the sense organs separately, understanding that if perception is to be elevated toward the celestial realities of the Divine Liturgy, all perception must be consecrated in this manner. Hence he asks that one touch the organs of sight (the eyes) and touch (the forehead) with the wine-moistened fingers, it having already touched the organ of taste in the mouth, and thus leaving only the nose and ears as the "other organs of sense." This act of touching entailed by ἐπαφίημι has nothing to do with perception itself, but rather with the ritual act of consecration through contact, as one would discuss regarding a relic. For if this were a matter of tasting, touching, seeing, hearing, or smelling the wine, then the mere act of having the wine upon the fingers would have satisfied touch and the forehead would not need to be consecrated with the bread in the hands. In chrismation, as Cyril recounts, the priest applies a special ointment or perfumed oil to the body "upon the forehead and upon the other organs of sense" (ἐπὶ μετώπου καὶ

[24] "Ἔτι δὲ τῆς νοτίδος ἐνούσης τοῖς χείλεσι χερσὶν ἐπαφώμενος καὶ ὀφθαλμοὺς καὶ μέτωπον καὶ τὰ λοιπὰ ἁγίαζε αἰσθητήρια." Cyril of Jerusalem, *Mystagogiae*, 5.22, ed. Paris and Piédagnel, 172.

[25] "Μὴ τῷ λάρυγγι τῷ σωματικῷ ἐπιτρέπητε τὸ κριτικόν, ἀλλὰ τῇ ἀνενδοιάστῳ πίστει." Cyril of Jerusalem, *Mystagogiae*, 5.20, ed. Paris and Piédagnel, 170.

τῶν ἄλλων σου χρίεται αἰσθητηρίων), i.e. the eyes, nostrils, mouth, ears, and other similar parts.[26] As Georgia Frank suggests, Cyril's ritual imperatives deploy this logic of chrismation so as to awaken the spiritual senses through their earthly counterparts as a rite of initiation.[27] In this manner, Cyril's homily alerts us to how such synesthetic descriptions of sensual experience are often deployed by these writers to convey the practical realities of worship through sensual practices.

In the ninth-century *Narratio* of the creation and arrival of the Mandylion of Edessa by Constantine VII Porphyrogennetos, we observe a similar application of this well-known contact-relic's use when it reaches King Abgar. There, as Glenn Peers has argued, Abgar's encounter with the Mandylion is "carefully constructed as well to mirror the Byzantine audience's own sensual desire for contact with that originary body."[28] There the narrative recounts how the relic cured Abgar, who had been afflicted by leprosy:

> λαβὼν τοίνυν ἀπὸ τοῦ ἀποστόλου τὸ τοιοῦτον ὁμοίωμα καὶ σεβασμίως αὐτὸ τῇ κεφαλῇ περιθεὶς καὶ τοῖς ὄμμασι καὶ τοῖς χείλεσι καὶ οὐδὲ τ' ἄλλα τῶν τοῦ σώματος μερῶν στερήσας τῆς τοιαύτης προσψαύσεως, ἔγνω παρευθὺ τὰ μέλη πάντα θαυμασίως ἀναρρωννύμενα καὶ τὴν εἰς τὸ κρεῖττον μεταβολὴν εἰσδεχόμενα.
>
> He received the likeness from the apostle and with great reverence put it round his head, on his eyes and on his lips, and did not omit any of the rest of his body. He knew immediately that his limbs had been miraculously healed, and changed for the better.[29]

Like the sequence of touch that occurs in Cyril's accounts, here we see this important relic-icon being utilized in the same manner, moving from the forehead, to the eyes, to the lips, and to the remaining parts of the body, as if following in the rites of chrismation so as to sanctify all avenues of perception, yet here for the purposes of curing these parts.

Consider, for example, the similar language and rhetoric used to describe early pilgrims' interactions with relics, which likewise betray what appears on the surface to be a confusion of touch and sight. Gregory of Nyssa, in his homily on St. Theodore, writes:

26 Cyril of Jerusalem, *Mystagogiae*, 3.3, ed. Paris and Piédagnel, 124.

27 Georgia Frank, "'Taste and See': The Eucharist and the Eyes of the Faith in the Fourth Century," *Church History* 70:4 (2001): 619–43, esp. 625.

28 Glenn Peers, *Sacred Shock: Framing Visual Experience in Byzantium* (University Park, PA: Pennsylvania State University Press, 2004), 127.

29 Constantine VII Porphyrogennetos, *Narratio de imagine Edessena*, ed. and trans. Mark Guscin, *The Image of Edessa* (Leiden: Brill, 2009), 28–29.

Καὶ τοῖς αἰσθητοῖς οὕτω φιλοτεχνήμασιν ἐνευπαθήσας τὴν ὄψιν, ἐπιθυμεῖ λοιπὸν καὶ αὐτῇ πλησιάσαι τῇ θήκῃ· ἁγιασμὸν καὶ εὐλογίαν τὴν ἐπαφὴν εἶναι πιστεύων. Εἰ δὲ καὶ κόνιν τις δοίη φέρειν τὴν ἐπικειμένην τῇ ἐπιφανείᾳ τῆς ἀναπαύσεως, δῶρον ὁ χοῦς λαμβάνεται, καὶ ὡς κειμήλιον ἡ γῆ θησαυρίζεται. Τὸ γὰρ αὐτοῦ τοῦ λειψάνου προσάψασθαι, εἴ ποτέ τις ἐπιτυχία τοιαύτη παράσχοι τὴν ἐξουσίαν, ὅπως ἐστὶ πολυπόθητον, καὶ εὐχῆς τῆς ἀνωτάτω τὸ δῶρον, ἴσασιν οἱ πεπειραμένοι, καὶ τῆς τοιαύτης ἐπιθυμίας ἐμφορηθέντες. Ὡς σῶμα γὰρ αὐτὸ ζῶν καὶ ἀνθοῦν οἱ βλέποντες κατασπάζονται, τοῖς ὀφθαλμοῖς, τῷ στόματι, ταῖς ἀκοαῖς, πάσαις προσάγοντες ταῖς αἰσθήσεσιν, εἶτα τὸ τῆς εὐλαβείας καὶ τὸ τοῦ πάθους ἐπιχέοντες δάκρυον.

Taking delight in the seeing of such works of art that can be observed, one longs for the rest, in particular to approach the tomb, trusting that touching it results in sanctification and blessing. . . For as if it is the same body, still alive and flourishing, those beholding it embrace it with the eyes, the mouth, the ears. And when they have approached it with all the senses, they pour tears out over it from piety and emotion.[30]

Gregory of Nyssa's homily aptly allows us to subvert the given understanding of these synesthetic constructions that rhetorically seem to suggest an elision between sight, touch, and the other senses. Here, the phrase "those beholding it embrace it with the eyes, the mouth, the ears" (οἱ βλέποντες κατασπάζονται τοῖς ὀφθαλμοῖς, τῷ στόματι, ταῖς ἀκοαῖς) features a strikingly similar syntax to John of Damascus's misleading "to venerate and kiss them with eyes and lips, and cleave to them in the heart" (προσκυνοῦμεν καὶ περιπτυσσόμεθα καὶ καταφιλοῦμεν καὶ ὀφθαλμοῖς καὶ χείλεσι καὶ καρδίᾳ ἀσπαζόμεθα). Gregory even deploys the similarly emphatic "embrace" (κατασπάζονται) as did John of Damascus to stress the "physical embrace" (κατασπαζόμενοι τῇ σαρκὶ) of the devotional object. Yet, if we were to read this sentence as indicating a haptic visuality, we would not only have to justify an elision between sight and touch to make this sentence work, but also a confusion between sight, touch, taste, and hearing.[31] Gregory's statement allows us little room for misinterpretation, given that the acting subjects are specifically described as "viewers" (οἱ βλέποντες), yet they are said to embrace the object with their eyes, mouth, and ears, or, more precisely, "sense of

[30] Gregory of Nyssa, *De sancto Theodoro*, ed. PG 46:740A; trans. Johan Leemans, "A Homily on Theodore the Recruit" in Leemans et al., *"Let Us Die that We May Live,"* 85.

[31] On the sensual practices around relics, see Patricia Cox Miller, "'The Little Blue Flower is Red': Relics and the Poetizing of the Body," *Journal of Early Christian Studies* 8:2 (2000): 213–236. See also Patricia Cox Miller, *The Corporeal Imagination: Signifying the Holy in Late Ancient Christianity* (Philadelphia: University of Pennsylvania Press, 2009).

hearing" (ταῖς ἀκοαῖς). The viewers shed tears only once they have approached the tomb with all the senses.

This may be corroborated furthermore by the fourth-century text of Egeria's pilgrimage to the Holy Land, which renders these practices once again for us, but this time in Latin, allowing for a useful comparison across linguistic boundaries. In recounting the practice of pilgrims venerating the relics of the True Cross on Good Friday at the Holy Sepulcher, Egeria writes:

> Ac sic ergo omnis populus transit unus et unus toti acclinantes se, primum de fronte, sic de oculis tangentes crucem et titulum, et sic osculantes crucem pertranseunt, manum autem nemo mittit ad tangendum.
>
> Thus all the people go past one by one. They stoop down, touch the holy Wood first with their forehead and then with their eyes, and then kiss it, but no one puts out their hand to touch it.[32]

Thus, again we can observe the indication of a ritual practice of veneration whereby viewers literally touch the True Cross with their forehead, eyes, and lips. This must be understood here to be a literal act of touching the relic to their sensory organs, a procedure that is once again in keeping with local and contemporary practices of chrismation, starting with the forehead, standing in for the sense of touch, and then moving on through the other bodily senses. Thus, in the context of Gregory and Cyril's homilies, we can come to understand that what we have at work in these late antique and early Byzantine writers is a careful detailing of the forms of ritual practice that surround the veneration of holy objects, ranging from icons to relics and even the Eucharist.

In the early seventh century, the poems of Sophronios, patriarch of Jerusalem from 634 to 638, attest to a similar and telling veneration of holy sites in the Holy Land. Notably, upon reaching the Church of the Nativity in Bethlehem, Sophronios exults:

> Λιπαρῇ πλακὶ προθήσω
> θεὸν ἣ δέδεκτο βρέφος,
> βλέφαρα στόμα κράνον τε
> κομίσαι χάρισμ' ἐκεῖθεν.

[32] Anonymous, *Itinerarium Egeriae*, 37.3, ed. P. Geyer and O. Cuntz, *Itineraria et alia geographica*, Corpus Christianorum, Series Latina 175 (Turnhout: Brepols, 1965), 81; trans. John Wilkinson, *Egeria's Travels to the Holy Land*, revised edition (Jerusalem: Ariel Publishing House, 1981), 137 (37.3).

The shining slab which received the infant God,
I will touch,
With my eyes, my mouth, my forehead,
To gain its blessing.[33]

This account manifests once more the process of a haptic, tactile veneration that elicits a direct physical contact between the viewer and the slab upon which the Christ child was placed. However, Sophronios's account is particularly enlightening not only because it attests to the endurance of such practices well into the seventh century, but also since the term translated as eyes is more precisely "eyelids" (βλέφαρα). Thus, it specifically clarifies that it is neither the eyeball itself nor the power of sight that grasps (δέδεκτο) the gleaming slab, but rather the closed eyelids of the pilgrims as they sanctify their bodies through these representative sites upon the body, including also their mouths (στόμα) and the center of the forehead (κράνον). Hence, it would be better to read those two lines as indicating that the pilgrims shall touch the slab "to the eyelids, mouth, and forehead," representing once again practices of veneration that seemingly sanctify representatives of the sensory organs through literal touch, rather than through the operation of the senses themselves. In other words, sight is sanctified by touching the eyes or the eyelids, rather than by merely gazing upon the object itself or by relying on any association with sight as a species of touch.

The oddity of some of these practices may lead to some unexpected readings in our modern eyes. Elsewhere, in the *Narratio* of the Mandylion, the narrative recounts the alleged practices surrounding the relic precisely during Lent in Edessa. Here, the author recounts that the relic was not always available for viewing, and was kept in a chest behind closed doors so that it would not come to be approached too often as if by habit and thus weaken the intensity of coming face to face with the image. However, the image was exhibited twice a week on Wednesday and Friday. As the narrative explains:

> ἐβλέπετο μὲν παρὰ παντὸς τοῦ συνεληλυθότος πλήθους καὶ ταῖς εὐχαῖς ἕκαστος ἐξιλεοῦντο τὴν ἐκείνης ἀκατάληπτον δύναμιν, οὐ μὴν δὲ ἠφίετό τινι προσεγγίσαι ἀλλ᾽ οὐδὲ χείλεσιν ἢ ὄμμασι τοῦ ἱεροῦ προσψαῦσαι μορφώματος, ὡς ἐντεῦθεν τοῦ θείου φόβου τὴν πίστιν αὔξοντος φοβερωτέραν καὶ φρικωδεστέραν τὴν πρὸς τὸ τιμώμενον τιμὴν ἀποδείκνυσθαι.

[33] Sophronios, *Anacreontica*, 19.45–48, ed. M. Gigante, *Anacreontica*, Testi per esercitazioni accademiche 10–12 (Rome: Gismondi, 1957), 119; trans. John Wilkinson, *Jerusalem Pilgrims Before the Crusades* (Warminster, England: Aris & Phillips, 1977), 92.

> The image was beheld by all the congregation, and everyone propitiated the image's limitless power in their prayers, although nobody was allowed to draw near and touch the holy form with their lips or eyes. Thus did fear of God increase their faith and make the people shudder with a greater fear for the revered object.[34]

This passage presents us with a conundrum, stating that none were allowed to touch the holy form with their lips or eyes (οὐδὲ χείλεσιν ἢ ὄμμασι τοῦ ἱεροῦ προσψαῦσαι μορφώματος). It presents the emphatically haptic and tactile προσψαῦσαι that indicates this prohibition of touching the object, combined with sight. Surely, it is feasible to read this as some inkling of a haptic visuality or at the very least as an idiomatic indication of one touching this with one's gaze. However, this prohibition does not make logical sense given that the passage has clearly explicated the image's concealment and revelation previously, and now addresses its display. Thus, despite its oddity to the modern reader, what we appear to have here is an indication that viewers were not allowed to touch the relic to their eyes or lips, only to gaze at it from afar, unable to draw close to it (οὐ μὴν δὲ ἠφίετό τινι προσεγγίσαι). This seems to evidence practices of tactile veneration, which like Egeria's observation on the veneration of the True Cross during Holy Week, involved pressing the object to the sense organs, particuarly to the eyes and lips, in keeping with the similar rites evidenced by Cyril of Jerusalem and Gregory of Nyssa. Hence, what these processes of touch come to index are the haptic practies of engaging with icons, relics, and other holy matter, which seem to have had a surprisingly uniform sequence of ritual unfolding.

These practices of relic and icon veneration appear to have consolidated around the practices of pilgrims visiting the Holy Land, which inaugurated what Robert Wilken has called a "new tactile piety."[35] This tactile piety indicated an emeshing of hearing, seeing, and touching as pilgrims traveled to the Holy Land and its environs, listening to the recitation of the Bible stories, seeing the important sights where these events took place, and touching the relics and holy sights with hands, lips, and, as we have seen, other parts of the body. Touch in this respect even appears to have various degrees and processes of ritual action. For example, in Jerome's letter to Marcella, in the name of Paula and Eustochium, he describes the tactile

[34] Constantine VII Porphyrogennetos, *Narratio de imagine Edessena*, ed. and modified trans. Guscin, 64–65.

[35] Robert L. Wilken, *The Land Called Holy: Palestine in Christian History & Thought* (New Haven: Yale University Press, 1994), esp. 115.

practices of veneration, writing: "Everywhere we venerate the tombs of the martyrs; we apply their holy ashes to our eyes; we even touch them, if we may, with our lips."[36] Note, that given what we have learned, the ash is applied to the eyes in an act that appears to consecrate the senses, whereas the desire to kiss, when allowed, appears to fufill the pilgrim's desire to intimately greet the sites. Thus, suggesting that even within touch there were different degrees, orders, and functions that contact upon skins enacted, just as we see regarding the various senses themselves. The same could be said about the degrees and operations of vision itself.

Georgia Frank has surveyed with great nuance these early practices of veneration and their relation to multi-sensory piety, alongside contemporaneous theories of perception.[37] Her study on the affinities between sight and touch reveal to us likewise the functional importance of the "eye of faith," or the operation of the mind's eye, in enlivening the sights and places visited. As Jerome recounts in his letter to Eustochium about Paula, upon seeing the cave of the Nativity in Bethlehem:

> me audiente iurabat cernere se fidei oculis infantem pannis inuolutum uagientem in praesepe, deum magos adorantes, stellam fulgentem desuper, matrem uirginem, nutricium sedulum, pastores in nocte uenientes...
>
> I heard her swear to me that she could perceive with the eyes of faith (*fidei oculis*) the infant Lord wrapped in swaddling clothes, crying in the manger, the wise men worshipping God, the star shining overhead, the virgin mother, the attentive foster-father, the shepherds coming by night...[38]

The "eyes of faith" articulates and contextualizes what read as almost hallucinatory, visionary experiences within these pilgrimage sites and sights. In the letter to Marcella, Jerome continually articulates how pilgrims witness the events that occurred at each site associated with the various embodied practices of veneration. Compare, for example, these two passages that demonstrate how sensory contact unfolds into these imaginary experiences: "As often as we enter it we see the Savior in His

[36] "martyrum ubique sepulchra ueneramur et sanctam fauillam oculis adponentes, si liceat, etiam ore contingimus." Jerome, *Epistula*, 46.8, ed. Isidorus Hilberg, *Corpus Scriptorum Ecclesiasticorum Latinorum*, vol. LIV (Vienna: F. Tempsky, 1910), 338; trans. W. H. Fremantle, "Paula and Eustochium to Marcella" in Philip Schaff and Henry Wace (eds.), *Nicene and Post-Nicene Fathers*, second series, vol. VI (New York: Christian Literature Company, 1893), 63.

[37] Georgia Frank, *The Memory of the Eyes: Pilgrims to Living Saints in Christian Late Antiquity* (Berkeley: University of California Press, 2000), esp. 102–133.

[38] Jerome, *Epistula*, 108.10, ed. Isidorus Hilberg, *Corpus Scriptorum Ecclesiasticorum Latinorum*, vol. LL (Vienna: F. Tempsky, 1912), 316; modified trans. W. H. Fremantle, "To Eustochium" in Philip Schaff and Henry Wace (eds.), *Nicene and Post-Nicene Fathers*, second series, vol. VI (New York: Christian Literature Company, 1893), 199.

grave clothes and if we linger we see again the angel sitting at His feet, and the napkin folded at His head";[39] or, "Then shall we touch with our lips (*lambere*) the wood of the cross, and rise in prayer and resolve upon the Mount of Olives with the ascending Lord? Shall we see Lazarus come forth bound with grave clothes, shall we look upon the waters of Jordan purified for the washing of the Lord?"[40] Triggered either by sight or by touch, both these actions stirred the imaginations of these embodied pilgrims. Of note in the first account is the disclaimer, "and if we linger for a while in that place" (*et paululum ibidem commorantes*), suggesting that these events are replayed before them if they dwell upon the site, presumably contemplating and imagining the events that occurred there, rather than hastily viewing the sites and moving along. As Frank aptly puts it, "whereas tourists see the markers of the biblical events, pilgrims 'linger' to see the event itself."[41]

Beyond the cooperation of sight and touch in pilgrimage, the ability to perceive the unfolding of the biblical narratives in the mind was paramount. This almost suggests that these indications of sight, such as "we see the Savior," "we see again the angel," "we shall see Lazarus," "we shall look upon the waters," more often refer to the visions of the imagination overlayed upon the vision of the sights themselves. Even the act of looking upon the waters of the Jordan is anachronic to the pilgrim here since they are the waters "purified for the washing of the Lord," rather than the waters that *were* purified by virtue of the washing of the Lord. Hence, sight in these instances almost seems to be haptic, so to speak, because it suggests the very act of grappling and grasping the sensory data of the sites, and perceiving them through the biblical narratives and images already engraved in the memory of the pilgrims. This is perhaps best captured by the words of the fourth-century bishop, Asterios of Amaseia, who when describing the pilgrims at the oak of Mamre in his *Encomium on Phocas*, eloquently muses on the various cognitive actions that go to work in this process:

39 "quod quotienscumque ingredimur, totiens iacere in sindone cernimus saluatorem et paululum ibidem commorantes rursum uidemus angelum sedere ad pedes eius et ad caput sudarium conuolutum." Jerome, *Epistula*, 46.5, ed. Hilberg, *Corpus Scriptorum*, 334; trans. Fremantle, *Nicene and Post-Nicene Fathers*, 62.

40 "crucis deinde lignum lambere et in olivueti monte cum ascendente domino uoto et animo subleuari? uidere exire Lazarum fasceis conligatum et fluenta Iordanis ad lauacrum domini puriora?" Jerome, *Epistula*, 46.13, ed. Hilberg, *Corpus Scriptorum*, 343; modified trans. Fremantle, *Nicene and Post-Nicene Fathers*, 65.

41 Frank, *The Memory of the Eyes*, 108.

Ὥσπερ δὲ οἱ πλησιάζοντες τῇ δρυῒ τῇ Μαμβρῇ ἢ τῷ σπηλαίῳ τῷ ὠνητῷ, ὅπερ Ἐφραῒμ ὁ Χετταῖος εἰς ταφὴν τῆς Σάρρας ἀπέδοτο, ἔνθα δὴ καὶ αὐτὸς σὺν τοῖς παισὶν ὁ πατριάρχης ἀπόκειται, εὐθέως μετὰ τῆς θέας τῶν τόπων ἐν τοῖς λογισμοῖς ἀνανεοῦνται τὴν φαντασίαν καὶ βλέπουσι τῇ διανοίᾳ τὸν πιστὸν πατριάρχην, τὴν ἀπαρχὴν τῆς εὐσεβείας, τὴν ἐν ἐκείνῳ περιτομήν· λογίζονται δὲ καὶ τοὺς τῆς ἐκείνου ῥίζης ἀποβλαστήσαντας, τὸν Ἰσαάκ, τὸν Ἰακώβ, καὶ μετὰ τῆς μνήμης τῶν ἀνδρῶν ὅλης τῆς ἐπ᾽ αὐτοὺς ἱστορίας γίνονται θεαταί· οὕτως κἀγὼ σήμερον τὸν τίμιον σηκὸν τοῦ τρισμακαρίου Φωκᾶ καταλαβὼν ἐκ τοῦ τόπου πληροῦμαι τῆς μνήμης πάντων ὁμοῦ τῶν ἐπ᾽ αὐτῷ λεγομένων διηγημάτων· βλέπω τὸν κηπουρὸν τὸ ἐπιτήδευμα, τὸν ἄπλαστον τὴν ψυχήν, τὸν φιλόξενον, τὸν τῆς παραλίας ἐξαίρετον, τὸν τῆς μεσογείου εὐεργέτην, τῶν ἁγίων τὸν ἅγιον καὶ τῶν δεδοξασμένων διὰ Χριστὸν ἐνδοξότερον.

People who approach the oak of Mamre or the cave bought for the burial of Sarah from Ephrem the Chattean, where the patriarch himself rests with his children, after their observation of the places renew the representation with their thought and see the faithful patriarch with their mind, the firstling of faith, his circumcision. And they consider also the offshoots of the same root, Isaac, Jacob, and so with the recollection of these men they become spectators of the complete history of the patriarchs. In the same way do I approach today the venerable sepulcher of the thrice-blessed Phocas, and by the spot itself am I filled with the recollection of all the stories about him. I see the gardener by occupation, the uneducated soul, the hospitable fellow, the jewel of this coastal region, the benefactor of the inner country, that most holy man, famous more than all those who are famous because of Christ.[42]

Asterios provides his listener with an insight into the process of pilgrimage whereby it is not merely an act of revelation, yet one of renewal that revives the images of the imagination (ἀνανεοῦνται τὴν φαντασίαν) in order to see the things in the mind (τῇ διανοίᾳ). This imaginative thinking is tied to an understanding of the narratives and stories associated with the sites, as suggested by the use of the term λογισμοῖς here to denote thinking given its connotation with language and reasoning. Much like the pilgrims to the Holy Land, for whom perceiving the sites leads precisely to a refreshing of the images generated through years of listening to the biblical stories, Asterios recounts how the people see (βλέπουσι) the patriarch himself with their minds (τῇ διανοίᾳ). And as he continues, with the active recollection (μετὰ τῆς μνήμης) of these holy figures, the pilgrims "become

42 Asterios of Amaseia, *Homiliae*, 9.2 (Encomium on Phokas), ed. Cornelis Datema, *Homilies I–XIV* (Leiden: Brill, 1970); modified trans. Boudewijn Dehandschutter, "A Homily on Phocas" in Leemans et al., *"Let Us Die that We May Live,"* 168–169.

spectators" (γίνονται θεαταί) in the process, seeing those deeds vividly unfolding before them.

These practices of perceiving the invisible, both the narratives of the past and the presence of the divine, are not limited to pilgrimage alone in contemporary thinking, but rather evidence a widespread perceptual practice whereby the psychic investment in veneration was one regulated by the mind's imaginative faculties. When speaking about the Eucharist, for example, John Chrysostom familiarly states:

> Ὅταν γὰρ ἴδῃς τὸν Κύριον τεθυμένον καὶ κείμενον, καὶ τὸν ἱερέα ἐφεστῶτα τῷ θύματι καὶ ἐπευχόμενον, καὶ πάντας ἐκείνῳ τῷ τιμίῳ φοινισσομένους αἵματι, ἆρα ἔτι μετὰ ἀνθρώπων εἶναι νομίζεις καὶ ἐπὶ τῆς γῆς ἑστάναι, ἀλλ' οὐκ εὐθέως ἐπὶ τοὺς οὐρανοὺς μετανίστασαι καὶ πᾶσαν σαρκικὴν διάνοιαν ἐκβάλλων, γυμνῇ τῇ ψυχῇ καὶ τῷ νῷ καθαρῷ περιβλέπεις τὰ ἐν οὐρανοῖς; Ὢ τοῦ θαύματος· ὢ τῆς τοῦ Θεοῦ φιλανθρωπίας. Ὁ μετὰ τοῦ Πατρὸς ἄνω καθήμενος κατὰ τὴν ὥραν ἐκείνην ταῖς ἁπάντων κατέχεται χερσὶ καὶ δίδωσιν αὑτὸν τοῖς βουλομένοις περιπτύξασθαι καὶ περιλαβεῖν, ποιοῦσι δὲ τοῦτο πάντες διὰ τῶν ὀφθαλμῶν τότε.
>
> When you see the Lord sacrificed, and laid upon the altar, and the priest standing and praying over the victim, and all the worshippers empurpled with that precious blood, can you then think that you are still among men, and standing upon the earth? Are you not, on the contrary, straightway translated to Heaven, and casting out every carnal thought from the soul, do you not with disembodied spirit and pure reason contemplate the things which are in Heaven? Oh! What a marvel! What love of God to man! The one seated above with the father, at that time, is held by all hands and gives himself to those desiring to clasp and embrace him, but this they all do through the eyes.[43]

This language of clasping (περιπτύξασθαι) and embracing (περιλαβεῖν) certainly resonates with the logic of John of Damascus, when he asks that we "embrace with eyes and lips" (ὀφθαλμοῖς καὶ χείλεσι περιπτύξομαι) the icon. This haptic language describing ritual behaviors appears in close proximity with noetic forms of ocular veneration, whereby the imagination is deployed so as to partake in the divine realities that are inaccessible to our bodily senses. There is an insightful discrepancy here since the text tells us emphatically that all hands may hold the Eucharistic body, but only the mind's eye can see the realities of that body, whose outward form in the

[43] John Chrysostom, *De sacerdotio*, 3.4, ed. A.-M. Malingrey, *Sur le Sacerdoce*, Sources chrétiennes 272 (Paris: Éditions du Cerf, 1980), 144; modified trans. W. R. W. Stephens, "Six Books on the Priesthood" in Philip Schaff (ed.), *Nicene and Post-Nicene Fathers*, first series, vol. IX (Buffalo: Christian Literature Publishing, 1889), 46–47.

amnos remains unchanged despite its consecration into the body of Christ.[44] Hence, as the hand embraces the physical body of Christ in the Eucharist, his visual form is perceptible to the eyes alone through speculation and imagination. Hence the language of grasping and embracing here is performing a dual action, referring to the fact that while one clasps and embraces the Christ in one's hands via the Eucharist, the action of clasping and embracing that sacrificed Christ, which they witnessed upon the altar, can only be done through this visual-cognitive contemplation. In these diverse sources, the eyes come to operate as a synecdoche for the sensory and perceptual process of veneration, which demands that the imagination be active and invested so as to complement the physical realities. When John Chrysostom writes that those holding the Eucharist seek to clasp and embrace the Christ, but may do so only "through the eyes" (διὰ τῶν ὀφθαλμῶν), here the use of ὀφθαλμός, the physical eye, is striking. Rather than a term such as sight (ὄψις) or seeing (θέα), the focus on the physical eye produces a generative friction and disjuncture that emphasizes that "through the eyes" represents a more complex process of perception, beyond what is available to the sensations of the eye itself.

As Georgia Frank comes to carefully conclude regarding these similar interweavings in the practices of Holy Land pilgrimages, "Sight and touch remained discrete senses, but in late antique Christian piety their functions converged to create the conditions for a biblical realism."[45] This "biblical realism" also helps to explain the double entendre in John Chrysostom's phrasing, whereby both the physical hands and the synecdoche of the eyes "clasp and embrace" (περιπτύξασθαι καὶ περιλαβεῖν), given that while the physical hands do just that, the workings of the mind likewise need to clasp and embrace "through" the eyes. Thus the language does double duty, suggesting that the words be read with their cognitive uses as well, understanding περιπτύξασθαι to suggest a physical and mental embrace as in the uses discussed above by John of Damascus and περιλαβεῖν to denote "comprehension" proper, a common use of the latter term.

In the case of John of Damascus, as Charles Barber has meticulously argued, "there was a trace of presence in the icon," and that hence it could indeed be treated as if it were that which it represented since Christ in redeeming matter had allowed for it to manifest the divine.[46] Yet,

[44] See Aden Kumler, "The Multiplication of the Species: Eucharistic Morphology in the Middle Ages," *RES: Anthropology and Aesthetics* 59/60 (2011): 179–91.

[45] Frank, *The Memory of the Eyes*, 133.

[46] Barber, "From Image into Art," 7.

nevertheless, the Damascene also makes explicitly clear that an icon "is a likeness depicting a prototype, but having some difference from it; the image is not the archetype in every way."[47] Therefore, it does not bring forth unmediated presence, but is different from the prototype. This crucial difference between the sign and the referent is not lost upon Byzantine thinkers, and figures such as Patriarch Nikephoros and Theodore of Stoudios would then work further to establish the icon properly as what Barber aptly terms a "directed absence."[48] While the Eucharist presents John Chrysostom with an undisputed model for the true presence of Christ, here perception is limited by its inability to perceive in it the form of Christ's body, and yet when John Chrysostom turns his attention to the icon in the narrative of his *Vita*, quoted by John of Damascus, he treats the image there "as if" he were living (ὡς ἐπὶ ζῶντος αὐτοῦ). And yet this treatment is characterized precisely by John bringing "the whole of his thoughts to him, imagining that he was speaking with him in his contemplation" (ὅλον αὐτοῦ τὸν λογισμὸν πρὸς αὐτὸν εἶχε φανταζόμενος καὶ διὰ τῆς θεωρίας αὐτῷ ὁμιλῶν). Even in these instances where the image seems to betray the disciplined, sober degrees of the icon's veneration as a "directed absence," the perceptual operation toward this perhaps more-present, pre-iconoclastic icon does not differ. Here, again, veneration is made possible by the same imaginative investment in the image, where thoughts are drawn to it, and the viewer imagines (φανταζόμενος) that they are speaking with the one depicted.

Even if we might debate to what extent this imagination operates, whether it be perceptual in the classical sense or more hallucinatory as in the pejorative early Christian responses to *phantasia*, it is clear that the icon's operation keeps within its contemporaneous perceptual bounds. Here, the icon does not exceed the limits of the imagination, and the author is careful to clarify this by appending the line with the disclaimer, "through his contemplation" (διὰ τῆς θεωρίας). This essentially reiterates the statement that the viewer imagines the dialogue with the icon through the act of contemplation (θεωρίας), just as one might perceive the embrace of the body of Christ in the Eucharist "through the eyes" (διὰ τῶν ὀφθαλμῶν) in John Chrysostom, which always denotes an invested visual and mental contemplation. Thus, while John of Damascus fruitfully

[47] "Εἰκὼν μὲν οὖν ἐστιν ὁμοίωμα χαρακτηρίζον τὸ πρωτότυπον μετὰ τοῦ καί τινα διαφορὰν ἔχειν πρὸς αὐτό· οὐ γὰρ κατὰ πάντα ἡ εἰκὼν ὁμοιοῦται πρὸς τὸ ἀρχέτυπον." John of Damascus, *Orationes de imaginibus tres*, I.9, ed. Kotter, 83–84; modified trans. Louth, 25.

[48] Barber, *Figure and Likeness*, 121.

elaborates for us the perceptual procedures of the imagination, beyond that his contribution to this conversation lies in his ability to eloquently build on perceptual language to also sketch out bodily, kinesthetic processes of worship that accompany these mental acts of cognition and perception. Even despite his nuanced belief in the icon's relative degree of presence, John of Damascus seems to fall short in his ability to violate and transgress these bounds of mere human perception and action. The icon might be treated then "as if" it were alive, but this "as if" function was relegated to the viewer's psychic disposition toward the icon in how they mentally invested themselves through fantasy in the image. The icon does not speak, it does not respond, it does not behave on its own – except in miraculous and extraordinary circumstances – for if that were the case the icon would be alive, it would not be *as if* it were alive.

Imagination and Veneration

In asserting the prowess of the imagination in Byzantine practices of veneration, there is a challenge given the objections that the imaginative functions of the imagination presented early Christian thinkers. Consider, for example, Evagrios's admonition in his text, *On Prayer*: "When you are praying, do not shape (σχηματίζῃς) within yourself an image of the Deity, and do not let your intellect be stamped (τυπωθῆναι) with the impress of any form; but approach the Immaterial in an immaterial manner, and then you will understand."[49] Evagrios's words deploy the Stoic language of impression to suggest a familiarity with the cognitive functions of perception and imagination through the use of the language of stamping the mind (τυπωθῆναί . . .τὸν νοῦν). However, he carefully disavows this operation as impeding the soul's proper confrontation with the divinity through prayer, for which one should aspire to be moved to tears by the experience but not seek to have a visual representation of said divinity. This objection to the imagination would be in keeping with Evagrios's own political project against idol worship and pagan practices of veneration. Indeed, Elizabeth Clark has described Evagrios as a "quintessential iconoclast, radicalizing and internalizing the historical anti-idolatry campaign," whose thinking resisted the operation of *phantasia* as a source for a

[49] "Μὴ σχηματίζῃς τὸ Θεῖον ἐν ἑαυτῷ προσευχόμενος, μηδὲ πρὸς μορφήν τινα συγχωρήσῃς τυπωθῆναί σου τὸν νοῦν· ἀλλ' ἄϋλος τῷ ἀΰλῳ πρόσιθι, καὶ συνίσεις." Evagrios, *De oratione*, 66 (67), ed. PG 79:1181; trans. G. E. H. Palmer, Philip Sherrard, and Kallistos Ware, *The Philokalia: The Complete Text*, vol. 1 (London: Faber and Faber, 1979), 63.

vision of the divine.[50] As Verity Platt has eloquently demonstrated, the imagination was paramount to the functions of epiphany in the Second Sophistic, and thus, writing in the late fourth century, Evagrios appears to have responded quite clearly to this intellectual context.[51]

Nevertheless, the operation of the imaginative faculty is difficult to assess given that while many writers support its operation in the veneration of icons, including the important voices of John of Damascus and Theodore of Stoudios, its troubled heritage in early Christianity's desire to distance itself from paganism has often led thinkers to avoid the language of *phantasia* despite clear indications of its operation. Even some modern scholars have attempted to distance the icon from the imagination, given that it not only moderates the degree of presence necessary in the icon, but also challenges what one might refer to as iconic exceptionalism, a belief that the icon is fundamentally different from other forms of representation by virtue of the exceptional manner in which it connects the viewer to the divine. Thereby, necessitating (or, rather, depending on) an articulate difference from pedestrian aspects of human thought and cognition that could feasibly apply to any other form of visual or literary art, pagan or otherwise. For example, Fr. Maximos (Nicholas) Constas has recently suggested that while the "imagination" appears thirty times in writings of Theodore of Stoudios and nearly one hundred times in Nikephoros, its appearance is invariably negative, reflecting on the imagination's "untrustworthy capacity."[52] There, Constas argues with no degree of uncertainty that: "Ordinary imagination based on sense perception, then, plays virtually no part in the Byzantine theory of the icon."[53] However, the operation of the imagination is clearly attested in quite articulate and eloquent ways beyond the use of *phantasia* or other cognate terms.[54] Notions of

[50] Elizabeth Clark, *The Origenist Controversy* (Princeton University Press, 1992), 84. See also Jean Gouillard, "Contemplation et imagerie sacrée dans le christianisme byzantin," *École pratique des hautes études, Section des sciences religieuses. Annuaire* 86 (1977–1978): 29–50.

[51] See Verity Platt, *Facing the Gods: Epiphany and Representation in Greco-Roman Art, Literature and Religion* (Cambridge University Press, 2011), esp. 293–332.

[52] Maximos Constas, "Beyond the Veil: Imagination and Spiritual Vision in Byzantium," unpublished paper, University of Chicago Divinity School Workshop (March 2014; revised April 2015), 5. Available at www.academia.edu/11995072/Beyond_the_Veil_Imagination_and_Spiritual_Vision_in_Byzantium_University_of_Chicago_Workshop_Paper_March_2014_.

[53] Constas, "Beyond the Veil," 6.

[54] In an earlier article, Constas himself argues quite eloquently for the necessity and function of the imagination in relation to the icon, writing: "The sacred image presupposes a sacred imagination responsive to the visual promptings of the icon which can assist in recollecting and focusing the power of human memory, imagination, and hope." See Nicholas Constas, "Icons and the Imagination," *Logos: A Journal of Catholic Thought and Culture* 1:1 (1997): 114–127.

theophanic contemplation, discussions of the mind's eye or the eye of faith often overlap precisely with the established technical operation of the imagination as evidenced by contemporary writers well versed in the cognitive theories of their time, from Nemesios to Photios, John of Damascus to Theodore of Stoudios.

The most eloquent articulation of the imagination in relation to icons thus comes from the latter, Theodore of Stoudios, who in his letter to a certain Naukratios, an iconoclast, expounds on the necessity of the icon through the parallel necessity of the imagination for cognition and perception. In that letter, Theodore writes:

> μάτην δὲ ἡμῖν ὅσον ἐπὶ τῇ τοιᾷδε προτάσει καὶ ἡ σταυροειδὴς εἰκών, μάτην καὶ ἡ λογχοειδής, μάτην καὶ ἡ σπογγοειδὴς (ἐπείπερ καὶ ταῦτα μιμήματα, κἂν οὐκ ἀνθρωπόμορφα), μάτην καὶ ὅσα ἄλλα αἰσθηταῖς εἰκόσι, Διονυσαϊκῶς εἰπεῖν, παραδέδοται ἡμῖν, δι᾽ ὧν ἐπὶ τὰς νοητὰς κατὰ τὸ δυνατόν, φησίν, ἀναγόμεθα θεωρίας. ἔπειτα καὶ ἡ μία τῶν πέντε τῆς ψυχῆς δυνάμεων ἡ φαντασία, φαντασία δὲ δόξειέν τις εἰκών· ἰνδάλματα γὰρ ἀμφότερα. οὐκ ἀνωφελὴς ἄρα ἡ εἰκών, τῇ φαντασίᾳ ἐοικυῖα. εἰ δὲ ἀνωφελὴς ἡ προὐργιαίτερα, πολλῷ οὖν μᾶλλον ἡ ὑφειμενεστέρα, μάτην συνυπάρχουσα τῇ φύσει· καὶ εἰ μάτην, καὶ αἱ σύστοιχοι αὐτῆς, αἴσθησις, δόξα, διάνοια, νοῦς. οὕτω φυσιολογῶν ὁ λόγος ἄνουν ὑπαγορεύει διὰ θεωρίας ἐπαγωγικωτέρας τὸν τῆς εἰκόνος ἤτοι φαντασίας ἐξουδενωτήν. ἐγὼ δὲ καὶ ἄλλως ἄγαμαι τὴν φαντασίαν· φασὶ γάρ τινες γυναῖκα κατὰ τὸν τῆς συλλήψεως καιρὸν Αἰθίοπα φαντασθεῖσαν Αἰθίοπα ἀποτεκεῖν. τοιοῦτον καὶ τὸ κατὰ τὸν πατριάρχην Ἰακὼβ ἐπὶ τῷ τῶν ῥάβδων λεπισμῷ, ἐξ ἧς φαντασίας ἐποικίλλοντο τὰ ἐν τοῖς θρέμμασι τικτόμενα. καὶ ὢ τοῦ θαύματος, ὅτι πραγματικῆς ἐνεργείας ἡ εἰκονοειδὴς φαντασία ὦπται ἀποτέλεσμα.

> Following such a premise, the representation of the cross would be also futile, not to mention the spear and the sponge (yes, they are also imitations, though not in human shape), and futile would be also all sorts of stimuli coming from sensible images, which, to put it in Dionysios' words, were given to us to facilitate the superior noetic contemplation, as far as possible. I should add that imagination is one of the five faculties of the soul, imagination and icons are moreover somehow alike: both are simulacra. Therefore, icons, inasmuch as they are similar to imagination, cannot be useless. Indeed if we declare useless the first term [i.e. the imagination], all the more so the second term [i.e. the icons] will be useless, being futile by its very nature. And if so, we must conclude that the other interrelated faculties of the soul, like sense perception, opinion, discursive thought, and intelligence, are also futile. Likewise, the doctrine of the physicians demonstrates through a more inductive reasoning that people who denigrate the imagination talk nonsense. Anyway, I greatly admire imagination: for they say that a woman, on conceiving her child, imagined a black man and gave birth to a black baby. A similar story is told also about Patriarch Jacob,

> when he pilled white strakes in the rods: thanks to the imagination of them, the animals brought forth spotted cattle. What a thing of wonder it is that the visual imagination provides for the fulfillment of deeds in actuality.[55]

Not only does Theodore's letter to Naukratios evidence his understanding of perception and cognition unfolding through the processes of sensation (αἴσθησις), opinion (δόξα), imagination (φαντασία), discursive thought (διάνοια), and intelligence (νοῦς), following Aristotle's *De anima* and John of Damascus in the *Expositio fidei*, but it demonstrates how that foundation grounded his theorization of the icon.[56] While Constas attempts to argue that this "exception proves the rule," it is difficult to decontextualize Theodore's statements from the extensive body of literature discussed in this chapter that carefully sees such imaginative practices at work in the articulation of veneration at holy sites, in the Eucharist, and before icons.

Theodore's reasoning here is that since the imagination is a faculty of the soul itself, and the imagination shares a simulacral likeness to the prototype, it is comparable to the icon's own relation to its prototype, and therefore the icon is as undeniable to veneration as the faculty of the imagination is to perception. However, the poignancy of Theodore's argument here is not only that it demonstrates the degrees of actuality possible through the imagination, but the mere fact that the imagination is offered up as a matter of fact. Were Theodore to perceive the imaginative faculty as a problematic or even an object of contention, he would have had to first argue for its validity in these circumstances before going on to share his thoughts with Naukratios. Thus, as argued previously, while the fantastical images of the imagination might have posed a threat to some writers, like Evagrios, it is clear that the imaginative faculty of the mind was generally accepted as a fact of human psychology within the works of Theodore of Stoudios and John of Damascus, who going back to Nemesios of Emesa likewise noted its operations in his defenses on icons and in his *Expositio fidei*.

In his letter to Naukratios, however, Theodore does not tell us exactly how it is that the imagination operates alongside the icon in this passage. That is to say, while he compares the imagination to the icon, he does not

55 Theodore of Stoudios, *Epistulae*, 380, ed. Georgios Fatouros, *Epistulae*, Corpus Fontium Historiae Byzantinae, Series Berolinensis 31, vol. 1 (Berlin: De Gruyter, 1992), 516–517; modified trans. Aglae Pizzone, "Theodore and the Black Man: Imagining (through) the Icon in Byzantium" in Andreas Speer and Philipp Steinkrüger (eds.), *Knotenpunkt Byzanz* (Berlin: De Gruyter, 2012), 47–70, at 53–54.

56 See Aristotle, *De anima*, 428a3–5, ed. and trans. Hett, 158–159; John of Damascus, *Expositio fidei*, II.22, ed. Kotter, 89.

explicitly recount how the imagination itself is deployed in the processes of veneration, though given his sources and framing of the material it is certainly the process that we have come to be familiar with here, and which one might presume Evagrios criticized in his allusion to the stamping of the mind (τυπωθῆναί. . .τὸν νοῦν). While I contend that the evidence is indeed ample regarding the imaginative contemplation of the faithful, one of the more eloquent depictions of this process is found in a mid-sixth-century poem by Agathias upon an icon of the Archangel Michael at Plate, preserved in the *Greek Anthology*.[57] There, Agathias writes:

> Ἄσκοπον ἀγγελίαρχον, ἀσώματον εἴδεϊ μορφῆς,
> ἆ μέγα τολμήεις, κηρὸς ἀπεπλάσατο. ἔμπης οὐκ ἀχάριστον,
> ἐπεὶ βροτὸς εἰκόνα λεύσσων
> θυμὸν ἀπιθύνει κρέσσονι φαντασίῃ·
> οὐκέτι δ᾽ ἀλλοπρόσαλλον ἔχει σέβας, ἀλλ᾽ ἐν ἑαυτῷ
> τὸν τύπον ἐγγράψας ὡς παρεόντα τρέμει·
> ὄμματα δ᾽ ὀτρύνουσι βαθὺν νόον· οἶδε δὲ τέχνη
> χρώμασι πορθμεῦσαι τὴν φρενὸς εἰκασίην.
>
> Invisible chief of the angels, incorporeal in the appearance of his form,
> Oh, how greatly daring was the wax that molded [him].
> Yet it is not without grace, for a mortal looking upon the icon
> Directs their passion to a higher imagination;
> They no longer hold a capricious reverence, but in themselves
> They inscribe the type, and tremble as if [the Archangel] were present.
> The eyes rouse the depths of the mind; And so art is able
> By colors to carry over the imagined likeness.[58]

The poem begins with a couple of negations, "invisible" (ἄσκοπον) and "incorporeal" (ἀσώματον), which speak to the mystery of the icon's ability to represent in material form that which is immaterial. Here, matter – that is, the wax itself of the encaustic icon – is praised for achieving what seems impossible, and which culminates in another negation, the double negative of "not without grace" (οὐκ ἀχάριστον) that solidifies the effective paradox of representation. This grace that emerges from the icon is dependent, however, not merely on that daring wax alone, but rather on the fact that the image causes the mortal gazing upon it to direct their passions to a "higher imagination" (κρέσσονι φαντασίῃ). Here, the imagination is elevated by virtue of the icon, as the author avoids any possible negative

[57] On this poem, see Aglae Pizzone, "Toward a Self-Determined and Emotional Gaze: Agathias and the Icon of the Archangel Michael" in Sergei Mariev and Wiebke-Marie Stock (eds.), *Aesthetics and Theurgy in Byzantium* (Berlin: De Gruyter, 2013), 75–103.

[58] Anonymous, *Anthologia Graeca*, 1.34, ed. and modified trans. Paton, 32–33.

understanding of the imagination, by using the comparative adjective that speaks to its loftier operation. This imagination then ends the viewer's wavering, capricious, or, quite literally, back-and-forth (ἀλλοπρόσαλλον) reverential approach (σέβας) to the image. Thus, the viewer impresses the image upon themselves (ἐν ἑαυτῷ τὸν τύπον ἐγγράψας), which causes them to shudder as if in the presence of the Archangel himself as the eyes rouse the mind.

Beginning from the impossibility of the icon, the reader of the inscription is led through its power and its cognitive effects, rooted in the powers and operations of perception. Here, as one might expect, the imagination results in the inscription of imprints into the soul, harking back to the language of the Stoic imagination. The presence of the Archangel is qualified by that similar structuring of "as if he were present" (ὡς παρεόντα), just as in John of Damascus we witness the vividness of icons manifesting themselves with the lines, "as if it were the one longed for himself" (ὡς αὐτὸν τὸν ποθούμενον), or "as if he were alive" (ὡς ἐπὶ ζῶντος αὐτοῦ). It is the impression of the types into oneself as a product of the imagination that makes one tremble *as if* the depicted were present, which validates for Agathias the fact that the "eyes rouse the depths of the mind" (ὄμματα δ' ὀτρύνουσι βαθὺν νόον). And thus the art and skill of the image (τέχνη) is able by virtue of colors to "carry over the imagined likeness" (πορθμεῦσαι τὴν φρενὸς εἰκασίην). Here, the likeness (εἰκασίην) is described as being a product of the psychic faculties, perhaps mental or imaginative (φρενός), though a precise translation of the term is difficult since it broadly denotes a passion or impulse formed in the body. Thus, the term aptly captures the manner in which the icon serves as a medium between the invisible and incorporeal Archangel and the human viewer, given that by virture of colors, the imperceptible is given a form that may be impressed upon the soul. Yet the poem does not place this power in the impudent wax that molded the figure, but rather in the imagination that rouses the depths of the mind through the stimuli of the eyes. It is for that reason, that is, it is because of the imaginative faculty, that art is able to transmit the signified.

In the eleventh century, Michael Psellos provides a similar defense of the icon, which stresses the action of the impression of visual form as the action that enacts the operation of the icon. In his oration on the "Usual" Miracle at the Blachernai, Psellos writes:

> εἰκονίζοντες δὲ καὶ ἐξομοιοῦντες ἐμφαινομένην τοῖς ὁμοιώμασιν ἔχομεν, καὶ μάλισθ' ὅταν μὴ χρώμασι τοσοῦτον τυπῶμεν, ἀλλὰ πόθοις ἀρρήτοις καὶ ταῖς πρὸς αὐτὴν δι' ἀρετῶν οἰκειώσεσι. καὶ τοῖς μὲν τῶν φωστήρων

πλάσμασιν οὐκ ἂν τις αὐτὰ ἴδοι τὰ παραδείγματα, εἰ μὴ ὅσον βραχύ τι γνοίη περὶ τῶν ἀρχετύπων ἀπὸ τῶν εἰκασμάτων· τὴν δέ γε παρθένον ὁμοῦ τε γράφομεν καὶ τὴν καρδίαν περὶ τὴν γραφὴν ἐντιθέμεθα καὶ τὰ μὲν ὑπ᾽ ὄμμασι φαινομένην ὁρῶμεν, ὁποῖον δὴ τὸ φαινόμενον, τὰ δὲ ἐν ψυχῇ πάσχοντες ἐντυπούμεθα· οὕτω τις πρὸς αὐτὴν ἡμῖν οἰκείωσις ἐμπέφυκεν ἄρρητος καὶ πρὸς τὸ ἡμεδαπὸν γένος ἐκείνῃ ἀπορρητότερα συμπάθεια. καί, ἵνα τι τολμήσας ἐρῶ, πρὸς μὲν τὰς καθάρσεις τῶν ἡμετέρων ψυχῶν καὶ αὖθις πρὸς τὰς ἐκ τῶν παθῶν κηλῖδας θεὸς ἐφέστηκε καὶ ἀφέστηκε, καὶ τρόπον ἐνόπτρου πρὸς τὰς ἐκεῖθεν αὐγὰς καθεστήκαμεν, στίλβοντες μὲν ταύτας δεχόμενοι, ἀμαυρούμενοι δὲ τούτων στερούμενοι· ἡ δὲ συμπαθὴς τῆς θεομήτορος φύσις καὶ τὸ φιλάνθρωπον ὑπερφυὲς ἔχουσα ἅπασιν ὁμοίως ἐμφαίνεται, ὅσοις τε ἡ ψυχὴ διαυγάζει καὶ ὅσοις ὁ νοῦς ἔτι τεθόλωται, καὶ ἴδοι ἂν τις αὐτὴν αὐτοπτήσας ὁπόσον ἰδεῖν δύναιτο, οὐ τῶν ὑψηλῶν μόνον καὶ μετεώρων τὴν ἀρετήν, ἀλλά τι καὶ ἀπερίεργον γύναιον κατόπιν τῇ εἰκόνι αὐτῆς ἐφεπόμενον καὶ οὐδ᾽ ἀκριβῶς τὸν ὕμνον αὐτῇ διαρθροῦν. οὐχὶ γὰρ τὸν ἄριστον χαρακτῆρα τῶν λέξεων, ἀλλὰ τὸν τῶν ψυχῶν ἐπαινεῖ καὶ ἀσπάζεται, οὐδὲ τὴν συνθήκην τῶν ὀνομάτων, ἀλλὰ τὴν ἐμμελῆ ἁρμονίαν τῶν τρόπων ἐξοικειοῦται.

When representing and likening her [the Theotokos], we have her manifested by means of these likenesses [i.e. works of art], but [she becomes even more apparent] when we figure her not so much by means of colors but through secret yearnings and through becoming close to her virtuous nature. One might not see the patterns themselves by means of the fabrications of their radiances, even though in some small measure one might know of the archetypes from the likenesses. But we depict the Virgin at the moment when we set [our] heart about the image: we see her partly visible to our eyes, as a phenomenon indeed, and partly feel her impressed into our soul. Thus, we possess a mysterious affection for her and her sympathy for us is even more mysterious . . . The sympathetic nature of the Mother of God, who possesses a supernatural philanthropy, appears the same way to all, both to those whose soul is radiant and to those whose mind is still confused. One might see her with one's own eyes in so far as it is possible to see, not only those with the highest and mightiest virtue but also a simple woman attending her icon from the rear [of the crowd] and incorrectly reciting her hymn. For she [the Theotokos] does not praise and embrace the best verbal portrayal, but the best living portrayal, nor does she gather to herself well-turned phrases, but rather the most apt disposition.[59]

Struggling with the fact that the "simple woman" (ἀπερίεργον γύναιον) venerating the icon in the rear of the crowd is unable to properly recite the

[59] Micheal Psellos, *Orationes hagiographicae*, 4.56–82, mod. ed. Elizabeth A. Fisher, *Orationes hagiographicae* (Leipzig: Teubner, 1994), 202–203; trans. Barber, *Contesting the Logic of Painting*, 88–89.

hymn to the Theotokos, Psellos is forced to reason that she is able to partake in the miracle by other means. Following Charles Barber's reading of the scene, the presumed disability of the figure serves as an important given for Psellos, since the visible, audible aspects of image veneration clearly are not the ones that enable the activity of the icon, since this woman may partake in it as well.[60] Hence, his apt resolution is that the Theotokos manifests herself not through the outward actions of veneration, but rather through its interior processes. This observation, when placed in the context of a model of veneration that is not purely or even wholly dependent on kinesthetic outward actions, however, is not innovative. Even in the context of relics, Gregory of Nazianzos once observed that "the bodies of the martyrs have the same powers as their holy souls, whether one touches them or just venerates them."[61] Gregory's words seem to enforce the sense that even within the context of relic worship, it was the non-haptic act of spiritual veneration that elicited benefits from the relics.

Psellos's argument hinges on the suggestion that "we depict (γράφομεν) the Virgin at the moment when we set [our] heart about the image (περὶ τὴν γραφὴν ἐντιθέμεθα): we see her partly visible to our eyes, as a phenomenon indeed, and partly feel her impressed into our soul (ἐν ψυχῇ πάσχοντες ἐντυπούμεθα)." The passage places the power of revelation not on the image alone, but on the psychic disposition of the viewer as they encounter the icon as a perceptible phenomenon. Yet the encounter with the Virgin occurs not so much because of the image itself, but because of the "ineffable longings" (πόθοις ἀρρήτοις) that the viewer brings to her image. By virtue of the process of perception, the viewer then comes to feel the Virgin's image impressed into their soul, which suggests that at that moment perception has been completed and the imprints have now been deposited in the memory. The focus on the heart (καρδίαν) here has a technical importance, given that in Psellos's summary on perception, he focuses on the heart as the central operant in the processing and transmission of sensory data. There, Psellos, building on Alexander of Aphrodisias's commentary on Aristotle's *De anima*, writes:

> Μίαν δὲ καὶ τῷ ἀριθμῷ τὴν αἰσθητικὴν ψυχὴν λέγει, καὶ ταύτην ἐν τῇ καρδίᾳ οἰκίζει, ἀφ' ἧς ἐπὶ τὸν ἐγκέφαλον ἡ τῆς αἰσθήσεως γίνεται διάδοσις. πόροι γὰρ ἐντεῦθεν ἐπὶ τὸν ἐγκέφαλον διατείνουσι τρεῖς, εἶτα ἀπὸ τοῦ

[60] Barber, *Contesting the Logic of Painting*, 88–91.

[61] "ὧν καὶ τὰ σώματα μόνον ἴσα δύνανται ταῖς ἁγίαις ψυχαῖς, ἢ ἐπαφώμενα, ἢ τιμώμενα." Gregory of Nazianzos, *Contra Julianum imperatorem* 1 (*Orat.* 4), ed. PG 35:592.

ἐγκεφάλου ὁ μὲν αὐτῶν ἐπὶ τὴν ὄψιν, ὁ δὲ ἐπὶ τὴν ἀκοήν, ὁ δὲ ἐπὶ τὴν ὄσφρησιν καθήκουσιν. οἱ δὲ τῆς ἀφῆς καὶ τῆς γεύσεως ἐπ' εὐθείας ἀπὸ τῆς καρδίας τέτανται, ἀλλ' οὐ διὰ τῆς ἐπὶ τὸν ἐγκέφαλον ὁδοῦ.

But also, [Aristotle] says that the soul of perception is one by number, and that it dwells in the heart, from which the distribution of perception begins [going] on to the brain. For three passages from there on to the brain conduct [perception], then from the brain one of each reaches on to sight, and on to hearing, and on to smell. The passages of touch and taste stretch on a direct route from the heart, but not on the road through the brain.[62]

Rather than understanding the heart as a representation of emotion, one can read the heart here to signify the seat of perception. Thereby, the act of "setting the heart about the image" (τὴν καρδίαν περὶ τὴν γραφὴν ἐντιθέμεθα) indicates the action of turning one's attention to the perception of the image with one's whole psychic attention, which involves the heart, the brain, and the various conduits that unite these together as one.

The dynamic unity of perception throughout its various parts therefore also encompasses the operations of the soul's faculties involved in perception, including the imagination. Later in his oration, Psellos alludes to the various faculties placed into motion by the perception of the divine form in the icon of the Virgin. As he writes:

περὶ δέ γε θεαγωγίας ψυχῶν τοσοῦτον ἐνταῦθα ῥητέον, ὡς ἐκ διαφόρων ἡμῶν συγκειμένων δυνάμεων, κρειττόνων καὶ χειρόνων, ὑψηλοτέρων καὶ ταπεινοτέρων, νοῦ λέγω καὶ διανοίας καὶ φαντασίας καὶ τῶν γε ἑτέρων, ὅταν μὲν ἡ ψυχὴ πᾶσα ἐπιπνέηται καὶ ὁ νοῦς πρῶτος τὴν ἔλλαμψιν ὑποδέξαιτο, ἐν ὀργάνου τάξει τἆλλα κινεῖται, ὑπεσταλμένα τὴν ἑαυτῶν ζωήν, ὅθεν καὶ ἀπαρακολουθήτως ἔχει τότε πρὸς ἑαυτὸν ὁ θεόπτης· εἰ δὲ ἡ διάνοια κινηθείη θεαγωγικῶς ἢ τὸ πνεῦμα τῆς φαντασίας, ἀναλλοιώτως ὁ νοῦς τούτων ὑπερκαθήμενος γνοίη ἂν τὰς θείας κινήσεις καὶ ἑρμηνεύσοι αὐτάς.

However, this much must be said at this point regarding the evocation of God by souls: since we are composed of faculties that differ from one another – superior and inferior, loftier and more humble (I am speaking of intellect, of discursive thought, of imagination, and of the others) – whenever the soul is inspired as a whole and the intellect is first to receive the illumination, the other [faculties] are set in motion as [the intellect's] instrument, since they are restricted in the life that is theirs; this is why [the intellect] as it sees God is then also unaware of itself. If in contrast the [faculty of] discursive thought or the spirit of the imagination has been set in motion by evoking God, the

[62] Michael Psellos, *Opuscula psychologica*, 8, ed. O'Meara, 16.10–15.

> intellect, which is placed above these, would recognize the divine movements and interpret them, without being subject to change.[63]

Psellos here acknowledges the role in which the faculties of the soul are motivated in the attempt to commune with the divine, with the soul being inspired as a whole (ἡ ψυχὴ πᾶσα ἐπιπνέηται). As would be expected, Psellos understands these faculties to be five in number, a fact that is attested in his poem on Cosmas Indicopleustes, where he writes, "For every soul is disposed by nature to have five powers: intellect, sensation, discursive thought, opinion, and imagination," which parallel the five senses.[64] Thus, in the account of the miracle, Psellos suggests that not only is the intellect moved by the approach toward the divine, but the other faculties are likewise set in motion; and when the process begins in other portions of the soul, such as in the "spirit of the imagination" (τὸ πνεῦμα τῆς φαντασίας), the other higher faculties, such as the intellect, are positioned to recognize and interpret those lower actions. As such, Psellos understands perception to be wholly unified, understanding that if one faculty of the soul is stimulated, all the others will take action as well in the process to contribute their own, unique properties to the whole. Here, in particular, Psellos seems to foreground intellection and imagination as key players in the recognition of the divine.

While perhaps somewhat underplayed, it is clear that Psellos is giving the imagination a great deal of power in his consideration of the Virgin's icon and its effects on the human soul. Not only may we understand the imprints upon the soul to be a by-product of perception mediated by the imagination, but similarly the focus on the heart within Psellos's own work suggests a unified attention here to the operations of perception as it makes its way through the various psychic faculties and the various bodily organs. Perhaps the most articulate statement on the role of the imagination comes early on in the text, soon before the excerpt above discussing the image's impression, when Psellos writes: "Those of us who express devotion to her in foreshadowings, images, icons, and through an 'unlike likeness,' so to

[63] Michael Psellos, *Orationes hagiographicae*, 4.706–716, ed. Fisher, 228; trans. Elizabeth A. Fisher, *Michael Psellos on Symeon the Metaphrast and on the Usual Miracle at Blachernae: Annotated Translations with Introductions* (Washington, DC: Center for Hellenic Studies, 2014), n.p. http://chs.harvard.edu/CHS/article/display/5488.

[64] "ψυχὴ γὰρ πᾶσα πέφυκε δυνάμεις ἔχειν πέντε, / νοῦν, αἴσθησιν, διάνοιαν, δόξαν καὶ φαντασίαν." Michael Psellos, *Poemata*, 54.141–142, ed. L. G. Westerink, *Poemata* (Stuttgart: Teubner, 1992), 333.

speak, we imagine (φανταζόμεθα) her inimitable nature."[65] In other words, those who approach a representation of the Virgin approach an "unlike likeness" (ἀνομοίου… ὁμοιότητος), which is formally similar to the Virgin, yet different to her in nature: such as her foreshadowings or shadows (σκιαῖς), images or simulacra (ἰνδάλμασι), or icon or likeness (εἰκόσι). Yet, in order to commune with the Virgin across the gap of these paradoxical 'unlike likenesses,' the viewer must imagine her nature, which cannot be imitated (τὴν ἀμίμητον φύσιν αὐτῆς). Only through the act of imagining (φανταζόμεθα) can the mortal come to perceive that which is inherently unrepresentable, yet which the icon makes manifest partially. As Psellos went on to say, we see her visibly before the eyes in the icon, in as much as she is a visible phenomenon (ὑπ' ὄμμασι φαινομένην ὁρῶμεν, ὁποῖον δὴ τὸ φαινόμενον), yet on the other hand we see her when we come to feel her imprints in our perception. The icon mediates, by virtue of its formal likeness, between the Virgin and our imagination of her. The latter is the only one that can properly grasp her "inimitable nature" (ἀμίμητον φύσιν), since by virtue of her unity to Christ, as Psellos states in the first section of his oration, the Virgin is "transcendent and incomprehensible" (ὑπερκειμένην καὶ ἀκατάληπτον). Thus, it is only through the imagination that the viewer may come to perceive the nature that lies beyond her form, just as the viewer embraces the Christ in the Eucharist in the mind, or imagines themselves to converse with the figure in the icon.

In this central operation of *phantasia*, one finds that Byzantine veneration, despite the complaints of many early Christian thinkers, shares more with Graeco-Roman pagan image veneration than many might have liked to acknowledge. In Philostratus the Younger's *Life of Apollonius of Tyana*, a book praised in Photios's *Bibliotheca* for its protagonist's philosophical way of life,[66] the protagonist argues that "viewers of the works of painters necessitate a mimetic faculty" (τοὺς ὁρῶντας τὰ της γραφικῆς ἔργα μιμητικῆς δεῖσθαι) in order to properly understand and conceive the things being depicted.[67] The argument here is that the viewers themselves must create, as Verity Platt has said, "a complex web of images and words in which it is the viewer/reader's power to form an abstract idea of the image

65 "ὁπόσοι δὲ ἐν σκιαῖς καὶ ἰνδάλμασι καὶ εἰκόσι ταύτῃ προσκείμεθα, καὶ διὰ τῆς ἀνομοίου, οὕτως εἰπεῖν, ὁμοιότητος τὴν ἀμίμητον φύσιν αὐτῆς φανταζόμεθα." Michael Psellos, *Orationes hagiographicae*, 4.46–49, ed. Fisher, *Orationes hagiographicae*, 202; trans. Fisher, *Michael Psellos on Symeon the Metaphrast*, n.p.

66 Photios, *Bibliotheca*, 44, ed. Henry, 1:28–30.

67 Philostratus, *Vita Apollonii*, II.22.5, ed. and trans. Christopher P. Jones, *Apollonius of Tyana*, 2 vols., Loeb Classical Library 16–17 (Cambridge, MA: Harvard University Press, 2005), 1:184–185.

in his mind" for the sake of being able to "view intelligently" (μὴ ἀνοήτως ὁρῶσιν) these things.[68] As is later developed, this notion of a mimetic faculty on the part of the viewer structures Philostratus's notion of *phantasia*, whereby it is an act of "viewing with imagination."[69] Ella Birmelin has explicated this conundrum between the roles of *mimesis* and *phantasia* in Philostratus, wherein *phantasia* is greater than *mimesis*, because while *mimesis* can only depict what it has seen, *phantasia* on the other hand can generate images of what it has not seen through what it has.[70] This turn in the conception of *phantasia* is in keeping with the transformations of the term during the late antique period and the Second Sophistic.[71]

The operation of this imaginative faculty thus was attested not only in the operation of viewing religious imagery, but also in the production of it. In the *Life of Apollonius of Tyana*, the role of *phantasia* in the making of art is perhaps best exemplified by Apollonius's statement that a great artist constructs images of the Gods through the work of the imagination and their understanding of the natural world, since the artist did not ascend into heaven to portray their likeness from a model.[72] Thus, Apollonius comments: "Imitation will create what it knows, but Imagination will also create what it does not know, conceiving it with reference to the real."[73] While such a sentiment, openly embracing the fantastical nature of imagination, might have been precisely what Christian authors sought to avoid, Basil of Caesarea in *De spiritu sanctu* similarly writes in the fourth century, regarding artists, that they "either design the product in their mind beforehand and then apply their creative vision, their imagination (φαντασίαν), to the work, or, looking at an already existing model (ἐκκείμενον παράδειγμα), they direct their activity in accordance with its likeness."[74] Though this does not explicitly address the facture of a religious

[68] Verity Platt, "Virtual Visions," 134. Cf. Philostratus, *Vita Apollonii*, II.22.4, ed. and modified trans. Jones, I:184–185.

[69] Verity Platt, "Virtual Visions," 134. See also Karl Thein, "Gods and Painters: Philostratus the Elder, Stoic Phantasia and the Strategy of Describing," *Ramus* 31:1/2 (2002): 136–145, esp. 142. Cf. Philostratus, *Vita Apollonii*, II.22.4, ed. and modified trans. Jones, II:154–159.

[70] Ella Birmelin, "Die kunsttheoretischen Gedanken in Philostrats Apollonios," *Philologus* 88 (1933): 149–180 and 392–414, esp. 396. See also Graeme Miles, "Reforming the Eyes: Interpreters and Interpretation in the *Vita Apolloni*" in Kristoffel Demoen and Danny Praet (eds.), *Theios Sophistès: Essays on Flavius Philostratus' Vita Apollonii* (Leiden: Brill, 2008), 129–160.

[71] Gerard Watson, *Phantasia in Classical Thought* (Galway University Press, 1988), 59–95.

[72] On some passing thoughts regarding this text in relation to Byzantine image theory, see Gilbert Dagron, "Holy Images and Likeness," *Dumbarton Oaks Papers* 45 (1991): 23–33.

[73] "μίμησις μὲν γὰρ δημιουργήσει, ὃ εἶδεν, φαντασία δὲ καὶ ὃ μὴ εἶδεν, ὑποθήσεται γὰρ αὐτὸ πρὸς τὴν ἀναφορὰν τοῦ ὄντος." Philostratus, *Vita Apollonii*, VI.9.2, ed. and trans. Jones, II:154–155.

[74] "Ἢ γὰρ προαναζωγραφήσας τῇ διανοίᾳ τὸ κατασκεύασμα, οὕτως εἰς ἔργον τὴν φαντασίαν ἤγαγεν· ἢ πρὸς ἤδη ἐκκείμενον παράδειγμα ἀποβλέπων, καθ' ὁμοίωσιν ἐκείνου τὴν ἐνέργειαν

image, it speaks to a certain similarity with Philostratus over time regarding the creative faculty of the artist, who works between the imagination, given models, and knowledge of the natural world.

Likewise, it should not go without mention that the same suspicions of the imagination that may have afflicted early Christian thinkers were likewise present in Philostratus. In his narrative, Pythagoreanism personified promises to Apollonius:

> καθαρῷ δὲ ὄντι σοι καὶ προγιγνώσκειν δώσω καὶ τοὺς ὀφθαλμοὺς οὕτω τι ἐμπλήσω ἀκτῖνος, ὡς διαγιγνώσκειν μὲν θεόν, γιγνώσκειν δὲ ἥρωα, σκιοειδῆ δ᾽ ἐλέγχειν φαντάσματα, ὅτε ψεύδοιντο εἴδη ἀνθρώπων.
>
> Being pure, you will receive from me the gift of foreknowledge, and I will so fill your eyes with radiance that you will recognize gods, know heroes, and unmask insubstantial imaginations when they disguise themselves in human form.[75]

These lines speak to the tension between the promise of a theophanic recognition of the gods against the deceptive musings of the imagination (φαντάσματα). The distinguishing factor here comes about through foreknowledge (προγιγνώσκειν), which resonates with the operations of preconception (προλῆψις), supposition (ὑπόληψις), or opinion (δόξα) in the process of perception that initially come to discern the sensible object, as discussed previously. This works precisely to unmask, refute, expose, or, rather, to cross-examine (ἐλέγχειν) the delusions of those "shadowy imaginations" (σκιοειδῆς… φαντάσματα) that distract the viewer from proper theophany. This promise seems to echo the very concerns regarding the imagination that early Christian writers contemplated, particularly when referring to the images of the imagination (φαντάσματα) themselves, rather than the faculty of the imagination (φανταστικόν) or its operations.

Therefore, returning to the question of veneration and imagination in Byzantium, it would seem that one could turn to Verity Platt's own observations on Philostratus to characterize the state of the imagination in the Iconoclastic and post-Iconoclastic periods:

> In a remarkable feat of philosophical syncretism, Apollonius thus unites the notion of *phantasia* as an unmediated encounter with the divine not only with Platonic concepts about abstract reasoning, but also with Stoic formulations of rational perception as *phantasia katalēptikē* (in which *phantasiai*

κατευθύνει." Basil of Caesarea, *De spiritu sanctu*, III.5.22–25, ed. B. Pruche, *Sur le Saint-Esprit*, 2nd edition, Sources chrétiennes 17 bis (Paris: Éditions du Cerf, 1968), 266; modified trans. Watson, *Phantasia in Classical Thought*, 86–87.

[75] Philostratus, *Vita Apollonii*, VI.11.6, ed. and trans. Jones, II:124–125.

> are subjected to a rigorous process of examination before being granted assent), and *phantasia logikē* (in which this process is explicitly related to structures of language). According to this model, the cognitive reliability required by sacred images is attained through active mental processes rather than the passive reception of epiphanies sent by the gods.[76]

A similar syncretism occurs in the retention of the language and logic of imprints to describe the manifestation of perceptual forms in the soul throughout the Middle Byzantine period following iconoclasm, combined with the understanding, dating well before iconoclasm, that the veneration of either holy sites, relics, or images involved a certain contemplation that deployed the operations of the imaginative faculty of the soul in order to perceive and witness the events having occurred there or the person to whom one directed their veneration.

While idol worship and a desire to distance Christianity from related pagan practices might have seemed particularly pressing to a figure like Evagrios, by the time of the Iconoclastic period and the theses posed by John of Damascus, the concern had shifted to an investigation as to what human cognition offers and how material objects may contribute in the process of veneration. Even the Second Iconoclastic Council of 815 would conclude that the production of icons was "unfit for veneration and useless" (ἄχρηστον τὴν τῶν εἰκόνων ποίησιν ὁρίζομεν), yet they explicitly go on to say: "We refrain, however, from calling them idols since there is a distinction between different kinds of evil."[77] Hence, even the Iconoclastic councils were less concerned with the possibility of idolatry as the veneration of false or deceptive images, than they were with their simple uselessness (ἄχρηστον) and the misguided attention given to them, which might detract from proper veneration.

Therefore, the need to distance oneself from pagan images or practices was less salient as a threat in iconoclastic discourse. After all, as the definition of the Seventh Ecumenical Council suggests, one should bring to icons incense and lights, "as was the pious custom of the ancients" (καθὼς καὶ τοῖς ἀρχαίοις εὐσεβῶς εἴθισται), which one may read as the customs of the early Christians and those before them. Furthermore, the evidence is overwhelming that the operative function of "viewing with imagination," as Platt refers to it, continues around the veneration of icons

[76] Platt, *Facing the Gods*, 326.

[77] "ἄχρηστον τὴν τῶν εἰκόνων ποίησιν ὁρίζομεν, εἴδωλα δὲ ταύτας εἰπεῖν φεισάμενοι ἔστι γὰρ καὶ κακοῦ πρὸς κακὸν ἡ διάκρισις." Anonymous, *Decretum concilii iconomachi*, Frag. 16, ed. Paul J. Alexander, "The Iconoclastic Council of St. Sophia (815) and its Defintion (*Horos*)," *Dumbarton Oaks Papers* 7 (1953): 35–66, esp. 60; trans. Mango, *Art of the Byzantine Empire*, 169.

as the critical manner in which the image is made animate, beyond mere material alone, and discerned by the mental faculties. In placing emphasis on the operation of either *mimesis* or *phantasia*, Philostratus and many others were stressing, as Graeme Miles writes, "the active role of the viewer/interpreter," which is best manifested through the cognitive investment that such mental faculties require.[78] As Andrew Louth has commented in regards to Theodore of Stoudios's defense of the imagination, such a cognitive faculty "would be rendered useless if it had no role in our apprehension of the divine, which it has in the process of looking and understanding images."[79] Here, Louth's words echo Platt's by acknowledging the role that the imagination plays in perception and comprehension, understanding that one must not passively glance or gaze at an icon, but rather contemplate it with the investment of the imagination. On this matter, even the iconoclasts seemed to agree since at the First Iconoclastic Council of Hiereia in 754, it was resolved that worshippers should, "reproduce in themselves [the saint's] virtues as actually living images, with the aid of what has been recorded about them in books, in order to be stimulated to zeal like theirs,"[80] which, as Jaś Elsner has observed, suggests that the "true icon is not a painted image but the virtues of the saints painted in oneself."[81] The imprinting of the icon in the soul, just as the deposition of perceived types into the memory, ensured the future emulation of these images as living icons.

78 Miles, "Reforming the Eyes," 154–155.

79 Andrew Louth, *Greek East and Latin West: The Church AD 681–1071* (Crestwood, NY: St. Vladimir's Seminary Press, 2007), 130.

80 "καὶ οὐχὶ δὴ μᾶλλον τὰς τούτων ἀρετὰς διὰ τῶν ἐν γραφαῖς περὶ αὐτῶν δηλουμένων οἷόν τινας ἐμψύχος εἰκόνας ἐν ἑαυτῷ ἀναζωγραφεῖ, καὶ πρὸς τὸν ὅμοιον αὐτοῖς ἐκ τούτου διεγείρεται ζῆλον." Anonymous, *Council of Hieria*, ed. J. D. Mansi, *Sacrorum conciliorum nova et amplissima collectio*, vol. XIII (Paris: H. Welter, 1902), 345C–D; modified trans. Milton Anastos, "The Ethical Theory of Images Formulated by the Iconoclasts in 754 and 815," *Dumbarton Oaks Papers* 8 (1954): 151–160, esp. 155.

81 Elsner, "Iconoclasm as Discourse," 380.

CHAPTER 11

Synesthesia and Remediation

Byzantine writers' predilection for synesthetic language is a matter that deserves further consideration for the effects it had on how both language and perception could be understood to occur across the boundaries of the individual senses. Synesthesia may be defined as "occurring when stimulation of one sensory modality automatically triggers a perception in a second modality, in the absence of any direct stimulation to this second modality."[1] More simply, the term denotes in contemporary cognitive science a condition whereby a person may associate one sensory input with another one simultaneously, as when a person sees a color upon hearing a sound. In art and rhetoric, this phenomenon is one that, while enjoying a thriving history in modern and contemporary art, stretches back well into antiquity and aptly brings to the foreground the intertwined theories of sensation, perception, and rhetorical representation.[2]

In the literary arts, metaphors, for example, operate as a source of what John E. Harrison and Simon Baron-Cohen have described as "pseudo-synesthesia."[3] This underlines the manner in which rhetoric often strives to transgress sensory boundaries, while nevertheless allowing one to distinguish such uses from the cognitive condition of synesthesia itself, which at times may or may not apply to a particular author. This nuance is important for clarifying the terms and approaches toward historical

[1] John E. Harrison and Simon Baron-Cohen, "Synaesthesia: An Introduction" in John E. Harrison and Simon Baron-Cohen (eds.), *Synaesthesia: Classic and Contemporary Readings* (Cambridge, MA: Blackwell, 1997), 3–16, at 3.

[2] On synesthesia and visual culture, see Vivian Sobchack, "What My Fingers Knew: The Cinesthetic Subject, or Vision in the Flesh," *Carnal Thoughts: Embodiment and Moving Image Culture* (Berkeley: University of California Press, 2004), 53–84; Cretien van Campen, *The Hidden Sense: Synesthesia in Art and Science* (Cambridge, MA: MIT Press, 2007); Richard Cytowic, *The Man Who Tasted Shapes* (Cambridge, MA: MIT Press, 1998); Kevin T. Dann, *Bright Colors Falsely Seen: Synaesthesia and the Search for Transcendental Knowledge* (New Haven: Yale University Press, 1999); Lawrence E. Marks, *The Unity of the Senses: Interrelations Among the Modalities* (New York: Academic Press, 1978).

[3] Harrison and Baron-Cohen, "Synaesthesia," 8–12.

material that seems to evidence a synesthetic logic in which a metaphor or passing description might transgress sensory bounds, like the idea of a sweet and soft voice or the grazing touch of the gaze. For our purposes here, this distinction alerts us to the ways in which culturally widespread metaphors, turn of phrases, and other cross-sensory associations can themselves lead to the construction of a synesthetic experience within a sociocultural group, which is nevertheless not foregrounded on the psychological phenomenon itself.

Yet the operation of such cross-sensory metaphors remains unexamined in terms of the operative functions they serve as models of and models for cognition and perception. As such, this chapter surveys the appearance of metaphors and constructions that deploy more than one sensory feature simultaneously in what one might call a synesthetic fashion. Then it turns its attention to consider the various ways in which this synesthetic language affects Byzantine systems of representation. In other words, how can we consider this proliferation of cross-sensory metaphor and language to be in dialogue with the goals and aims of the *enargeia* of *ekphrasis*, which precisely desires to transgress the sensory limits of specific mediums, as discussed in Chapter 9? My proposition here is that, on the one hand, confusions of haptic sight are in part a by-product of this language, and on the other hand, the goals of representation were contoured around the idea that the senses could dialogue with one another in perception.

The goal of this chapter is to wrap up the last of the three lines of investigation at the core of this book, which I have been bringing to a synthetic close in this final part. In Chapters 1 and 9, we studied the medium of sight and its strategies of mediation. In Chapters 2 and 10, we considered closely the problem of sight's tactility, and conversely sight's deployment alongside tactility in veneration. Finally, in this chapter we expand on the third's concerns with the commonalities of the various senses and the role of the common sense in collecting and collating diverse sensory data. Here, the goal is to nuance and expand the articulation of the common sense, and place this cognitive operation in dialogue with the kaleidoscopic variety of synesthetic language in the Byzantine world. Using the common sense to argue for a broader cross-sensory operation, we are able to witness how the desire to transgress sensory bounds is not only rooted in the theories of perception that we have already laid out, but also fundamental to the expected aims of artistic representation. This is to say, the ability for the common sense to stitch together various sensory inputs, and also use one input to recall another, is foundational to the way in which artistic representation in any one medium operates.

The sight of the icon not only cues up the visual appearance of a person's form, but also allows us to imagine their fleshed-out presence being offered up to the various senses. John of Damascus and others offered us instances in the previous chapter where one might imagine oneself to kiss the cloak of the one depicted or dialogue with them. This suggests that the visual stimuli of the icon, working hand-in-hand with the imagination, allow us to partake in the imprints of the other senses, which are stored in the memory and associated around particular things through the common sense. For example, when looking upon a pot of honey, either in real life or in representation, these operations allow us to feel viscerally that honey is sweet, sticky, amber, with the slight scents of its flowers. And, we could even imagine the sound of flowing viscous honey coming out of a jar, or, perhaps, even the flutter of the bees and the smells of the meadow from which it came. Thus, the act of looking with the imagination, of looking with *phantasia*, allows the common sense to remediate one sensory medium into another. In other words, *ekphrasis* can describe sights through words, just as visual images can call up the imprints of touch or aurality in the mind. In the last two chapters, we have been closing in on these operations, and here we are able to contextualize the phenomenon culturally both through rhetorical practices as well as through late antique and Byzantine thinking on human cognition via the common sense and the imagination.

The Common Sense

Synesthetic metaphors were first given sustained attention in an appendix to W. Bedell Stanford's *Greek Metaphor* (1936), entitled "On Synaesthetic or Intersensal Metaphor."[4] There, Stanford understood statements such as Cicero's "murmur of the sea" (*murmur maris*) or Demetrios's "silvery voice" (λευκὴ φωνή) as a "type of metaphor [that] depends upon the sensuous charm of composite images aroused by words *in the imagination*,"[5] the implication being that this language operates with the imagination to conjure up understandings that resonate across sensual boundaries. Recent studies have furthered these investigations in the classical world to understand the role and function of synesthesia in these sources, including, particularly, in a monograph by Christoph Catrein on

[4] W. Bedell Stanford, *Greek Metaphor: Studies in Theory and Practice* (Oxford University Press, 1936), 47–62.

[5] Stanford, *Greek Metaphor*, 47.

the Roman world and an edited volume by Shane Butler and Alex Purves primarily on Greek antiquity.[6] More recently, James Porter's work has likewise demonstrated the extent to which aesthetic experience in ancient Greece can never be limited to a single sense, but rather thrived on the transgression of sensory bounds and the affiliative relations made across the individual senses.[7] Even verbal *ekphrases* of paintings, such as the late antique *Imagines* of Philostratus the Elder, as Michael Squire notes, "do not appeal to sight and sound alone," but there are also frequent invocations of scent and smell, as well as of touch and taste.[8] To see with the imagination did not merely involve visualizing. A wide bouquet of sensory experiences was also cued that was not limited to the sonic experience of the text or the mental visuals it conjured up: the "mental image" or *phantasia* formed by the text was not only visual, but encompassed a variety of senses. In the Byzantine world, however, this phenomenon has been discussed almost exclusively in the context of synesthesia's most literal understanding from the Greek, detailing the manner in which sensation occurs "together" in multi-sensory experience, as in the studies of Bissera Pentcheva.[9] One notable exception is found in the work of Liz James, which seeks to unite the simultaneity of sensory experiences with their more nuanced cognitive and phenomenological interweaving.[10]

Nevertheless, similar synesthetic metaphors can be found in Byzantine rhetoric wherever any allusions to the senses might appear. For instance, in the Coptic story of Abba Pambo, the narrator recounts, "I, Pambo, the most unworthy, smelled the sweet fragrance of that brother [Abba Pamoun] a mile before I got to his home."[11] And, the *Souda* defines "Sweet" (Ἀδύ, the Doric form of ἡδύ), by quoting Theocritus, *Idylls* 1.1, "sweet is the whisper and that pine-tree, goatherd" (ἁδύ τι τὸ ψιθύρισμα καὶ ἁ πίτυς, αἰπόλε, τήνα).[12] Sweetness reappears in antique and

6 Christoph Catrein, *Vertauschte Sinne: Untersuchungen zur Synästhesie in der römischen Dichtung* (Munich: K. G. Saur, 2003); Shane Butler and Alex Purves (eds.), *Synaesthesia and the Ancient Senses* (Durham: Acumen Publishing, 2013).

7 See James I. Porter, *The Origins of Aesthetic Thought in Ancient Greece: Matter, Sensation, and Experience* (Cambridge University Press, 2010), esp. 45–47, 64–6.

8 Michael Squire, "Apparitions Apparent: Ekphrasis and the Parameters of Vision in the Elder Philostratus's *Imagines*," *Helios* 40:1–2 (2013): 97–140, esp. 112.

9 See Pentcheva, *The Sensual Icon*, esp. 227 n 5.

10 See James, "Senses and Sensibility," esp. 528.

11 "ⲁⲛⲟⲕ ⲡⲁⲙⲃⲱ ⲡⲉⲓ ⲉⲗⲁⲭⲉⲓⲥⲧⲟⲥ ⲁⲓϣⲱⲗⲙ̄ ⲙ̄ ⲡⲉⲥϯ ⲛⲟⲩϥⲉ ⲙ̄ ⲡⲥⲟⲛ ⲉⲧ ⲙ̄ⲙⲁⲩ ⲛ̄ⲁ ⲟⲩⲙⲏⲗⲓⲟⲛ ⲙ̄ⲡⲁ · ϯⲡⲱϩ ⲉ ⲣⲟϥ." Anonymous, *The Story of Abba Pambo*, ed. E. A. Wallis Budge, *Coptic Martyrdoms*, vol. 1 (London: British Museum, 1914), 130; trans. Tim Vivian, *Journeying into God: Seven Early Monastic Lives* (Minneapolis: Augsburg Fortress, 1996), 31.

12 Anonymous, *Souda*, "Ἀδύ," alpha 536, (cf. alpha 3220, tau 511, psi 98), ed. and trans. Jennifer Benedict, *Suda On Line*, www.stoa.org/sol-entries/alpha/536.

Byzantine sources across sensory bounds, often appearing as a descriptor for taste as well as smell, or even more rarely "sweet-glancing" (γλυκυδερκής), that is "sweet in sight" (ἡδὺς ἐν τῷ ὁρᾶν).[13] Here what we encounter is primarily an understanding that there are some common things that the senses can perceive. As the *Souda* recounts, following John Philoponos, these may include things like movement, rest, number, figure, magnitude, and so on, which various senses can perceive.[14] These "common sensibles" (κοινὰ αἰσθητά), as John Philoponos explains, "are called 'common' not because each one of them falls under all the senses but because each one falls under many several and not [just] under a single one, and at the same time some fall under all."[15] This means that there are aspects of sensation that by nature are able to cross sensory boundaries, being perceptible to one or more of the individual senses.

In relation to sweetness, for example, the connection is drawn between taste and smell by Aristotle, who articulates that:

> Ἔστι δ', ὥσπερ χυμὸς ὁ μὲν γλυκὺς ὁ δὲ πικρός, οὕτω καὶ ὀσμαί. ἀλλὰ τὰ μὲν ἔχουσι τὴν ἀνάλογον ὀσμὴν καὶ χυμόν (λέγω δὲ οἷον γλυκεῖαν ὀσμὴν καὶ γλυκὺν χυμόν), τὰ δὲ τοὐναντίον. ὁμοίως δὲ καὶ δριμεῖα καὶ αὐστηρὰ καὶ ὀξεῖα καὶ λιπαρά ἐστιν ὀσμή. ἀλλ' ὥσπερ εἴπομεν, διὰ τὸ μὴ σφόδρα διαδήλους εἶναι τὰς ὀσμὰς ὥσπερ τοὺς χυμούς, ἀπὸ τούτων εἴληφε τὰ ὀνόματα καθ' ὁμοιότητα τῶν πραγμάτων.
>
> Just as flavours are sweet and bitter, so are smells. In some things the smell corresponds to the taste, e.g., both smell and taste are sweet; in others they are contrasted. Smell, like flavour, may be pungent, rough, acid or oily. But, as we have said, smells being, unlike flavours, not easily differentiated, they have taken their names from flavours on the ground of a correspondence between them.[16]

Thus, we can understand that the smell of "sweet fragrance" evidences a synesthetic logic established by the ancient understandings of commonalities between the senses and what types of common sensibles they are able to perceive. In the late antique world, Susan Ashbrook Harvey has eloquently demonstrated how such sweet odors can come to be culturally and

[13] Anonymous, *Souda*, "Γλυκυδερκής," gamma 313, ed. and trans. Jennifer Benedict, *Suda On Line* www.stoa.org/sol-entries/gamma/313.

[14] *Souda*, alpha iota, 326, ed. Adler, and trans. Marcelo Boeri, *Suda On Line*, www.stoa.org/sol-entries/alphaiota/326.

[15] "κοινὰ δέ φησιν εἶναι ταῦτα πασῶν, οὐχ ὅτι ἓν ἕκαστον αὐτῶν πάσαις ταῖς αἰσθήσεσιν ὑποπίπτει, ἀλλ' ὅτι ἕκαστον πλείοσι καὶ οὐ μιᾷ, τινὰ δὲ καὶ πάσαις ἅμα." John Philoponos, *In Aristotelis libros de anima commentaria*, ed. Hayduck, 311.1; trans. William Charlton, *On Aristotle's On the Soul 2.1–6* (London: Duckworth, 2005), 120.

[16] Aristotle, *De anima*, 421a28–421b2, ed. and trans. Hett, 120–121.

religiously associated with salvific substances, like relics or the unwashed bodies of holy ascetics.[17] Within the olfactory imagination of the early Christian world, this could likewise mean that sweetness and aromas could be extended as an epistemology all on their own. That is to say, beyond a synesthetic confusion of smell and taste, rooted in cultural notions of sensation, these could also come to indicate the presence or knowledge of the holy. For instance, 2 Corinthians 2:14–16 describes the spread of "the aroma of the knowledge" (τὴν ὀσμὴν τῆς γνώσεως αὐτοῦ) of Christ and the "pleasing-aroma of Christ" (Χριστοῦ εὐωδία). Just as for Pambo the "sweet fragrance" of the ascetic indicated his presence from afar, so too we can see how such sensual perceptions could lead to a knowledge or understanding of the miraculous that lies beyond immediate presence.

These metaphors, if we should even call them that, begin to push the bounds of even the common sensibles, and speak to the power of this language's stirrings in the imagination to produce forms of knowing and perceiving that were not always concordant with the sense alone that provided them to the intellect. As W. Bedell Stanford speculatively muses in his work on Greek metaphor:

> Presumably if Aristotle had been pressed for a statement of where and how these transferences and interchanges from sense to sense happened, he would have referred to the vague processes of the φαντασία... or to the *sensus communis.*[18]

We can understand that the activity of these sensual transferences occurs within the operations of the imagination as it processes and perceives words and images, but Stanford here also draws attention to the importance of the so-called common sense (*sensus communis* or κοινὴ αἴσθησις), as discussed earlier.[19] Unlike common sensibles (κοινὰ αἰσθητά), which describe the sensible qualities of a perceptible that can be perceived by various senses (such as movement that can be discerned through seeing,

[17] Susan Ashbrook Harvey, *Scenting Salvation: Ancient Christianity and the Olfactory Imagination* (Berkeley: University of California Press, 2006), 1–56, 201–206.

[18] Stanford, *Greek Metaphor*, 50.

[19] See Jacques Brunschwig, "Les multiples chemins aristotéliciens de la sensation commune," *Revue de Métaphysique et de Morale* 96 (1991): 455–474; J. Brunschwig, "En quel sens le sens commun est-il commun?" in C. Viano and G. Romeyer Dherbey (eds.), *Corps et âme: sur le De Anima d'Aristote* (Paris: J. Vrin, 1996), 189–218; David W. Hamlyn, "Koine Aisthesis," *The Monist* 52 (1968): 195–209; Danielle Lories, "Des sensibles communs dans le *De anima* d'Aristote," *Revue philosophique de Louvain* 89 (1991): 401–420; Deborah Modrak, "Koinê Aisthêsis and the Discrimination of Sensible Differences in *de Anima* III 2," *Canadian Journal of Philosophy* 11 (1981): 405–423.

touch, or hearing),[20] the common sense is the syncretic process of the mind as it processes perceptual data and turns the perceptible object into a mental object encompassing all its features and sensations together as one.[21]

The root of the idea of the common sense lies in Aristotle's often sporadic remarks on the matter, which are subsequently developed in the thinking of figures like Theophrastus, Alexander of Aphrodisias, Plotinos, Plutarch of Athens, John Philoponos, Simplikios, and Priscian of Lydia. As discussed in an earlier chapter, such figures focused carefully on the simultaneity and/or distinction between the operations of the common sense and the *phantasia* in the work of Aristotle. One of the more eloquent descriptions of the phenomenon is found in Aristotle's *De anima*:

> τῶν δὲ κοινῶν ἤδη ἔχομεν αἴσθησιν κοινήν, οὐ κατὰ συμβεβηκός· οὐκ ἄρ᾽ ἐστὶν ἰδία· οὐδαμῶς γὰρ ἂν ᾐσθανόμεθα ἀλλ᾽ ἢ οὕτως ὥσπερ εἴρηται τὸν Κλέωνος υἱὸν ἡμᾶς ὁρᾶν. τὰ δ᾽ ἀλλήλων ἴδια κατὰ συμβεβηκὸς αἰσθάνονται αἱ αἰσθήσεις, οὐχ ᾗ αὐταί, ἀλλ᾽ ᾗ μία, ὅταν ἅμα γένηται ἡ αἴσθησις ἐπὶ τοῦ αὐτοῦ, οἷον χολὴν ὅτι πικρὰ καὶ ξανθή· οὐ γὰρ δὴ ἑτέρας γε τὸ εἰπεῖν ὅτι ἄμφω ἕν.
>
> But we have already a common faculty which apprehends common sensibles directly. Therefore there is no special sense for them. If there were, we should have no perception of them... The senses perceive each other's proper objects incidentally, not in their own identity, but acting together as one, when sensation occurs simultaneously in the case of the same object, as for instance of bile, that it is bitter and yellow; for it is not the part of any single sense to state that both objects are one.[22]

This excerpt clarifies the relation of the common sense to the other senses, explaining how and why the common sensibles are not simply detected by a unique sense, but rather stitched together into one identity from the information that is uniquely culled from the various individual senses. The commentators build on these, at times cryptic and limited remarks, in order to collate and compose a more robust and articulate description of the common sense that clarifies its relationship to the individual senses, the operations of the imagination, and its overall place within the context of perception.

20 See Stanislas Cantin, "La perception des sensibles communs au moyen du mouvement d'après Aristote," *Laval théologique et philosophique* 17 (1961): 9–21.

21 Gregorić, *Aristotle on the Common Sense*, esp. 1–16, 129–144. See also L.A. Kosman, "Perceiving That We Perceive: *On the Soul* III.2," *Philosophical Review* 84 (1975): 499–519.

22 "Aristotle, *De anima*, 425a27–425b3, ed. and trans. Hett, 144–145.

In the commentary on *De anima* III.3, Philoponos pays a great deal of attention to clarifying the difference between the imagination and the common sense. This is somewhat unique among the extant commentators, though as attested in the work of Philoponos and others, Plutarch of Athens also focused attention on this matter.[23] There Philoponos suggests clearly that while the actions of the imagination occur after sensation, the activity of the common sense coincides and overlaps with the act of sensation.[24] His reasoning is that since the imagination requires sensory data to occur, it must precede the reception of sensation's forms to take action, yet the common sense acts alongside sensation so as to produce the imprints of sense for the imagination to use. For example, the common sense is the faculty that allows one to recognize that honey is both yellow and sweet, even though sight cannot recognize sweetness, and taste cannot recognize yellow.[25] The imagination for Philoponos works through the activation of what he describes earlier in his commentary on *De anima* as these residual (ἐγκατα-λείμματα) imprints through recollection.[26] Using these imprints, the imagination may then work alongside the common sense so as to stitch together the various sensory inputs as one. As Philoponos explains:

> τὸ τὴν κοινὴν αἴσθησιν δεξαμένην ἄμφω τοὺς τύπους, τὸν μὲν ἐκ τῆς γεύσεως, τὸν δὲ ἐκ τῆς ὄψεως, εἶτα τῇ ὑστεραίᾳ τοῦ ἐκ τῆς ὄψεως τύπου ἀνακινηθέντος τῷ θεαθῆναι ἡμῖν μέλι, λοιπὸν ἀνακινεῖσθαι καὶ τὸν ἐκ τῆς γεύσεως τύπον, καὶ ὡς ἔστιν εἰπεῖν ἀναχλίζεσθαι θάτερον ὑπὸ θατέρου τύπον, καὶ ταύτῃ γινώσκειν ὅτι γλυκύ ἐστι, κἂν μὴ γευσώμεθα.
>
> The common sense received both imprints (τύπους), the one from taste and the one from sight, and the following day when the imprint from sight is stirred up, and the one imprint is, so to speak, *réchauffé* by the other, and thereby we recognize that it is sweet, even though we do not taste it.[27]

The common sense allows for the ability to recognize that these two sensory inputs are connected, but critically, as Philoponos notes, "the imagination stirs up the imprints" (ἡ φαντασία ἀνακινεῖ τοὺς ἐν αὐτῇ

[23] See Peter Lautner, "Plutarch of Athens on κοινὴ αἴσθησις and *Phantasia*," *Ancient Philosophy* 20 (2000): 425–446.

[24] John Philoponos, *In de anima*, 477.20–512.4, esp. 509.25–30, ed. M. Hayduck, *Ioannis Philoponi in Aristotelis de anima libros commentaria* (Berlin: Reimer, 1897), 477–512, esp. 509; trans. William Charlton, *On Aristotle's "On the Soul 3.1–8"* (Ithaca: Cornell University Press, 2000), 53–88, esp. 85.

[25] See John Philoponos, *In Aristotelis libros de anima commentaria*, ed. Hayduck, 455.15–34; trans. Charlton, 30–31.

[26] John Philoponos, *In Aristotelis libros de anima commentaria*, 158.19, ed. Hayduck, 158; trans. van der Eijk, 80. See Lautner, "Philoponean Accounts of *Phantasia*," 165, 168–170.

[27] John Philoponos, *In Aristotelis libros de anima commentaria*, 455.25–30, ed. Hayduck, 455; trans. Charlton, 31.

τύπους).[28] In other words, while the common sense recognizes yellow and sweet to be linked in honey, it is nevertheless the imagination that cues the subsequent imprints the next time that honey is encountered. Therefore, it is still the imagination that is tasked with the activation of the impressions collected by the common sense.

At stake in these debates regarding the operation of the common sense is a fundamental question regarding the unity of sensation itself, which for a thinker like Plotinos also translated into the necessary unity of the soul. In the *Enneads*, Plotinos provides us with an eloquent meditation on distinction and unification of the senses:

> εἴ τι μέλλει αἰσθάνεσθαί τινος, ἓν αὐτὸ δεῖ εἶναι καὶ τῷ αὐτῷ παντὸς ἀντιλαμβάνεσθαι, καὶ εἰ διὰ πολλῶν αἰσθητηρίων πλείω τὰ εἰσιόντα εἴη ἢ πολλαὶ περὶ ἓν ποιότητες, κἂν δι᾽ ἑνὸς ποικίλον οἷον πρόσωπον. οὐ γὰρ ἄλλο μὲν ῥινός, ἄλλο δὲ ὀφθαλμῶν, ἀλλὰ ταὐτὸν ὁμοῦ πάντων. καὶ εἰ τὸ μὲν δι᾽ ὀμμάτων, τὸ δὲ δι᾽ ἀκοῆς, ἕν τι δεῖ εἶναι, εἰς ὃ ἄμφω. ἢ πῶς ἂν εἴποι, ὅτι ἕτερα ταῦτα, μὴ εἰ τὸ αὐτὸ ὁμοῦ τῶν αἰσθημάτων ἐλθόντων; δεῖ τοίνυν τοῦτο ὥσπερ κέντρον εἶναι, γραμμὰς δὲ συμβαλλούσας ἐκ περιφερείας κύκλου τὰς πανταχόθεν αἰσθήσεις πρὸς τοῦτο περαίνειν, καὶ τοιοῦτον τὸ ἀντιλαμβανόμενον εἶναι, ἓν ὂν ὄντως. εἰ δὲ διεστὼς τοῦτο γένοιτο, καὶ οἷον γραμμῆς ἐπ᾽ ἄμφω τὰ πέρατα αἱ αἰσθήσεις προσβάλλοιεν, ἢ συνδραμεῖται εἰς ἓν καὶ τὸ αὐτὸ πάλιν, οἷον τὸ μέσον, ἢ ἄλλο, τὸ δὲ ἄλλο, ἑκάτερον ἑκατέρου αἴσθησιν ἕξει· ὥσπερ ἂν εἰ ἐγὼ μὲν ἄλλου, σὺ δὲ ἄλλου αἴσθοιο. καὶ εἰ ἓν εἴη τὸ αἴσθημα, οἷον πρόσωπον, ἢ εἰς ἓν συναιρεθήσεται – ὅπερ καὶ φαίνεται· συναιρεῖται γὰρ καὶ ἐν αὐταῖς ταῖς κόραις· ἢ πῶς ἂν τὰ μέγιστα διὰ ταύτης ὁρῷτο; ὥστε ἔτι μᾶλλον εἰς τὸ ἡγεμονοῦν ἰόντα οἷον ἀμερῆ νοήματα γίγνεσθαι – καὶ ἔσται ἀμερὲς τοῦτο· ἢ μεγέθει ὄντι τούτῳ συμμερίζοιτο ἄν, ὥστε ἄλλο ἄλλου μέρος καὶ μηδένα ἡμῶν ὅλου τοῦ αἰσθητοῦ τὴν ἀντίληψιν ἴσχειν. ἀλλὰ γὰρ ἕν ἐστι τὸ πᾶν· πῶς γὰρ ἂν καὶ διαιροῖτο;
>
> If anything is going to perceive anything, it must itself be one and perceive every object by one and the same means, both if a number of impressions are received through many sense-organs, or many qualities are perceived in one thing, or if through one sense-organ a complex thing, for example a face, is perceived. For there is not one perception of the nose and another of the eyes, but one and the same perception of all together. And if one perception comes through the eyes and another through hearing, there must be some one thing to which both come. Or how could one say that these sense-perceptions are different, if they did not all come together to one and the same [recipient]? This then must be like a centre, and the

[28] John Philoponos, *In Aristotelis libros de anima commentaria*, 455.23, ed. Hayduck, 455; trans. Charlton, 31.

> sense-perceptions from every quarter, lines coming together from the circumference of the circle, must reach it, and that which apprehends them must be of this kind, really one. But if this were extended, and the sense-perceptions arrived at something like the terminal points at both ends of a line, either they will run together again at one and the same point, like the middle of the line, or the two different terminal points will each have a perception of something different (as if I perceived one thing and you another). And if the object of perception was one, a face for instance, either it will be gathered together into a unity – which is what does obviously happen; for it is gathered together in the pupils of the eyes themselves: or how could the largest things be seen through the pupil of our eye? So still more when they reach the ruling principle they will become like partless thoughts – and this ruling principle will be partless; or if this is a size the sense-objects would be divided up along with it, so that each part would perceive a different part of the object and none of us would apprehend the perceptible thing as a whole. But the whole is one: for how could it be divided?[29]

Using metaphors, such as the face and the circle, Plotinos underlines the unity of perception by virtue of the way in which a person recognizes complex perceptibles in the world, which despite carrying various individual parts are still understood as a united whole. Yet, this applies not only to the parts of a whole in the context of sight, but also to the various senses, like sight and hearing: as Plotinos observes, "there must be some one thing to which both come." This is a crucial realization for Plotinos since it allows not simply for union, but for the discrimination and distinction between each of the separate senses. For, as he writes, "how could one say that these sense-perceptions are different, if they did not all come together to one and the same [recipient]?"

In other words, the common sense rather than muddling the senses, instead allows for the self-reflexive understanding that one is not only perceiving, but that one is perceiving through various and distinct conduits that all lead to a cognitive center, like the center of a circle, a metaphor to which the earlier commentators Alexander of Aphrodisias, John Philoponos, Themistios, and Simplikios would also resort in explicating the matter. Plotinos even tackles here a question of scale, understanding that the largest things can be seen through the pupils, presumably despite being over-sized for their scale, since even fragmentary views are united into one in the soul's perception. This operation of union occurs when sensations arrive at "the ruling principle" (ἡγεμονοῦν) where they will become

[29] Plotinos, *Enneades*, IV.7.6.4–29, ed. and trans. Armstrong, *Enneads*, IV:352–356.

"like partless thoughts" (οἷον ἀμερῆ νοήματα). Thus, the common sense provides a certain self-consciousness in regards to perception, which values the uniqueness of the senses, while nevertheless stitching together various sensory impressions (coming from the same and different organs of sensation) into a unitary whole in the mind.

In his sixth-century commentary on Theophrastus's "On Sense-Perception," Priscian of Lydia neatly summarizes the common sense. While lacking the poetic complexity of Plotinos, this later work demonstrates the continued elaboration and consolidation of the common sense, which in Plotinos is understood more broadly as the unity of sensation. As Priscian writes:

> ἡ δὲ κοινὴ αἴσθησις οὔτε ἡ αὐτὴ ταῖς κατὰ μέρος οὔτε πάντῃ ἑτέρα. κατὰ γὰρ τὴν πασῶν συναίρεσιν καὶ τὴν εἰς ἓν ἀμέριστον ἀποκορύφωσιν· διὸ καὶ ἑκάστῃ πως συναισθήσεται ὅτι αἰσθάνεται, οὐχ ᾗ μεμέρισται ἀλλ᾽ ᾗ συνῆπται τῇ μιᾷ.
>
> But the common sense is neither the same as the particular ones nor entirely different. For it is by way of the synthesis of all [the senses] and their concentration into an undivided one; hence in a way each [sense] is conscious that it perceives, not as having been divided off but as joined together in the one.[30]

Priscian's summary demonstrates the force of the "common sense" since its unitary function in synthesizing sensation leads to the production of consciousness, or at least what Pavel Gregorić cautiously refers to as the "rudiments of a theory of consciousness," whose very name comes from that function of unifying perception (συναισθήσεται).[31] The common sense came precisely with an awareness that one is perceiving, and it is this quality of self-awareness that contributes to its importance for the discussion of consciousness.[32] Unlike Bissera Pentcheva, who uses *synaisthesis* (συναίσθησις) merely to give form to an esthetic, poly-sensory experience, which deploys the various senses at once, the late antique and Byzantine commentators understood the term as denoting not merely sensation, but rather the cognitive models of sensible-things that perception produces through the imagination in order to allow for cognition, memorization, and, of course, perception itself. The views espoused by the commentators, specifically John Philoponos, Alexander of Aphrodisias,

[30] Priscian, *Metaphrasis in Theophrastum*, ed. Bywater, 22.1–22.5; trans. Huby, 31.

[31] See Gregorić, *Aristotle on the Common Sense*, 8.

[32] See C. H. Kahn, "Sensation and Consciousness in Aristotle's Psychology," *Archiv für Begriffsgeschichte* 48:1 (1966): 43–81.

and Priscian of Lydia become the standard for the definition of the common sense in Byzantium.

Throughout the early Christian and Byzantine period, reference to the common sense is made, noting its central operation in the perception of sensible objects. Notably, its necessity in perception is almost taken for granted, as in one fourth-century homily by Gregory of Nazianzos where he lists the "common faculty of the senses" (κοινὸν αἰσθήσεων δύναμις) alongside a series of universal truths experienced by all persons.[33] In the eleventh century, Michael Psellos foregrounds the faculty in one of his entries on the soul in his *De omnifaria doctrina*, describing the three "passions of the soul" (πάθος τῆς ψυχῆς): the highest being conception (νόησις), the middle one being discursive thought (διάνοια), and the lowest being opinion (δόξα), which he defines as the "cognitive faculty of sensation through the medium of the common sense" (τὸ τῶν αἰσθητῶν γνωστικὸν διὰ μέσης τῆς κοινῆς αἰσθήσεως).[34] Here, the common sense is precisely associated with the lower operation of processing sensory information, before its suitability for higher orders of intellectual reason and thought, a location associated by the commentators either with or preceding the imagination.

Michael Psellos's discourse featuring diverse and varied arrangements of philosophical arguments provides two paragraphs devoted to explicating the common sense before moving on to a discussion of the imagination. The text, which is a compendium of paraphrases from the work of John Philoponos, quickly distills Philoponos's reading of Aristotle as follows:

> Ὅτι ἔστι κοινὴ αἴσθησις, δείκνυσιν οὕτως Ἀριστοτέλης. ἐπεὶ οὐ <μόνον> ἴσμεν τὰ ὁμοειδῆ ὡς λευκὸν καὶ μέλαν, ἀλλὰ καὶ τὰ ἀνομοιοειδῆ ὡς λευκὸν καὶ γλυκὺ καὶ λευκὸν ἵππον, καὶ δύναται εἶναι <τῶν> δύο αἰσθήσεων οὔ, ἄλλη τις τῶν πέντε διακεκριμένη οὔ (οὐ γάρ εἰσιν ἓξ αἰσθήσεις), κοινὴ ἄρα. αὕτη δὲ σῶμα οὔ· οὐ γὰρ μεριστῶς ἐνεργεῖ, ἐν ταὐτῷ γὰρ κρίνει λευκὸν καὶ μέλαν. πάλιν πᾶν σῶμα ἁπτόμενον δρᾷ, αὕτη οὐ δρᾷ. ἔτι καὶ οὐκ ἐν χρόνῳ γινώσκει. ἄλλως καὶ ταὐτῷ οἶδε τὴν διαφορὰν γλυκέος λευκοῦ.
>
> Ἡ κοινὴ αἴσθησις μία μὲν τῷ ὑποκειμένῳ, πολλαὶ δὲ τῇ σχέσει κατὰ Ἀριστοτέλην· ὡς ἀπὸ κοινῆς γὰρ πηγῆς τῆς κοινῆς αἰσθήσεως προ{σ}έρχονται ἄλλαι αἰσθήσεις κατὰ μέρος, ὥσπερ τὸ κέντρον τοῦ κύκλου καθ' αὑτὸ μὲν ἕν, ὅτι δέ ἐστιν ἀρχὴ εὐθειῶν πρὸς τὸν κύκλον ἐκβαλλομένων πρὸς τὸν κύκλον πολλά. ἔτι ὑλικῶς οὐχ ἅμα τὰ ἐναντία ἀντιλαμβάνεται ἡ

[33] Gregory of Nazianzos, *De moderatione in disputando* (*Orat.* 32), ed. PG 36:200.

[34] "Τριττόν ἐστι τὸ πάθος τῆς ψυχῆς. καὶ τὸ μὲν ἀκρότατον αὐτῆς, τὸ τοῖς νοητοῖς ἡνωμένον, νόησις ὠνόμασται· τὸ δὲ ἔσχατον, τὸ τῶν αἰσθητῶν γνωστικὸν διὰ μέσης τῆς κοινῆς αἰσθήσεως, δόξα λέγεται· τὸ δὲ μέσον, τὸ ἀνιέναι καὶ κατιέναι πεφυκός, διάνοια." Michael Psellos, *De omnifaria doctrina*, 31, ed. Westerink, 30.

κοινὴ αἴσθησις, κριτικῶς δὲ ἅμα, ἐπεὶ καὶ αἱ ἐναντίαι δόξαι οὐ περὶ ἐναντίων, ἀλλ' εἰς τὸ αὐτὸ τὰ ἐναντία δοξάζουσαι, ὡς τὸ λευκὸν διακρίνει καὶ οὐ συγκρίνει.

Thus, there is the common sense, just as Aristotle demonstrates. Since we know not only that things of the same kind differ, just as white and black, but even that heterogeneous things do, just as a white sweet [thing], and a white horse, and it is not a power of the two senses, but it does belong to some other one of the five senses to discern (for there are not six senses), so it is common. But it is not a body. For it does not function dividedly, for in this it discerns white and black. Again all bodies act by contact, but it does not act this way. Thus it does not act in time. Instead, it knows in this the difference between sweet and white.

The common sense is one in subject, but many in arrangement according to Aristotle: that is, the various other senses issue from the common sense as if from a common [i.e. single] spring. Just as if it were the center of a circle which is one according to itself, so it is the beginning of the lines going toward the circle and also the manifold of lines [going] towards the circle. Thus, materially the common sense does not grasp the contraries together, but in discerning seizes both, since contrary opinions are not about contraries, but the contraries opine toward itself, just as it discerns white but does not combine.[35]

Psellos's account here is written in a stilted style, condensing a great deal of complicated and nuanced metaphors into a quick succession in order to depict the common sense for his reader, at times even notably altering Philoponos's points. The key point that Psellos wishes to stress is that the common sense is not a sixth sense in the material sense, but rather a faculty of the mind that discerns the various sensory inputs arriving from the senses. This results in an understanding of the common sense as singular and multiple at the same time, since it is a faculty into which all sensation flows, but also composed of all the unique senses. Psellos critically wishes to stress then that while the common sense discerns the senses together, it does not conflate them as one so as to maintain the nuance of their individual characteristics.

In late Byzantium, the common sense continues to appear in the works of Byzantine thinkers and commentators, such as in Nikephoros Gregoras's fourteenth-century commentary on Synesios's *On Dreams*. Synesios in his own work expounds the power of both the common sense and the imagination, stating first that: "And if to have the vision of a god with

[35] Michael Psellos, *Opuscula psychologica*, 13, ed. O'Meara, 64. Cf. John Philoponos, *In Aristotelis libros de anima commentaria*, ed. Hayduck, 477.3–481.30; trans. Charlton, 52–57.

one's eyes is a blessed thing, to apprehend him by the imagination is a loftier kind of vision. This is in fact the sense of senses, because the imaginative *pneuma* is the most general organ of senses and the first body of the soul."[36] This text underlines the privileged position of the imagination in accessing a vision of the divine, and also characterizes it as the "sense of senses," almost suggesting a kinship with the common sense, which he will soon thereafter address, stating:

> ἀκοὴ δὲ καὶ ὄψις οὐκ εἰσὶν αἰσθήσεις, ἀλλ᾽ αἰσθήσεως ὄργανα τῆς κοινῆς ὑπηρέτιδες, οἷον πυλωροὶ τοῦ ζῴου διαγγέλλουσαι τῇ δεσποίνῃ τὰ θύραθεν αἰσθητά, ὑφ᾽ ὧν θυροκοπεῖται τὰ ἔξωθεν αἰσθητήρια... διανέμει δὲ τὰς δυνάμεις ἄλλην κατ᾽ ἄλλο, καὶ προὔκυψαν ἐκ τοῦ ζῴου χωρὶς ἑκάστη, καὶ εἰσὶν οἷον εὐθεῖαί τινες ἐκ κέντρου ῥυεῖσαι καὶ εἰς τὸ κέντρον συννεύουσαι.
>
> Hearing and sight are not senses but organs of sense, servants of the common sense, doorkeepers of the living being, as it were, who report to their mistress the sense-impressions from outside, which knock on the door of the external organs of sense... And, it distributes its powers among various organs, and they emerge from the living being each one separately; they are like straight lines flowing from the center and returning to the center.[37]

As Anne Sheppard notes, Synesios engages and depicts the common sense almost like the Stoic "ruling part" (ἡγεμονικόν), similar to what we saw as well in Plotinos.[38] This suggests a possible overlap or adjacency between the operations of the imagination and the common sense.

In his comments on Synesios's description of the common sense through the circle metaphor, Nikephoros Gregoras writes at one point, addressing the common sense directly, "Indeed, [Synesios] says the imagination is the common sense and properly the divine perception, seeing that it unites in an unmediated fashion with the intellectual soul."[39]

[36] "εἰ δὲ τὸ αὐτοπτῆσαι θεὸν χρῆμα εὔδαιμον, τὸ διὰ φαντασίαν ἑλεῖν πρεσβυτέρας αὐτοψίας ἐστίν. αἴσθησις γὰρ αἰσθήσεων αὕτη, ὅτι τὸ φανταστικὸν πνεῦμα κοινότατόν ἐστιν αἰσθητήριον καὶ σῶμα πρῶτον ψυχῆς." Synesios, *De insomniis*, 4.135b, ed. and trans. Donald A. Russell and Heinz-Günther Nesselrath, *On Prophecy, Dreams and Human Imagination: Synesius,* De insomniis (Tübingen: Mohr Siebeck, 2014), 20–21.

[37] Synesios, *De insomniis*, 5.136a, ed. and modified trans. Russell and Nesselrath, 20–21.

[38] Anne Sheppard, "*Phantasia* in *De insomniis*" in Donald A. Russell and Heinz-Günther Nesselrath (eds.), *On Prophecy, Dreams and Human Imagination* (Tübingen: Mohr Siebeck, 2014), 97–110, at 108.

[39] "τὴν μέντοι φαντασίαν λέγει κοινὴν αἴσθησιν καὶ κυρίως αἴσθησιν καὶ θειοτέραν, ἅτε ἀμέσως ὁμιλοῦσαν τῇ νοερᾷ ψυχῇ." Nikephoros Gregoras, *Explicatio in librum Synesii De insomniis*, 136B (153.12), ed. P. Pietrosanti, *Explicatio in librum Synesii 'De insomniis'* (Bari: Levante Editori, 1999), 30.

His words here seem to allude precisely to Synesios's statements above, which praise the imagination for its ability to partake in a loftier perception of the divine compared to the earthly senses alone, precisely because, as Gregoras states, it allows for perception to unite with the more spiritual parts of the soul. Likewise, the focus on the circle metaphor in Gregoras's gloss once again demonstrates a certain proclivity among Byzantine writers to deploy this particular visual as a depiction of the common sense, focusing on its operation as a unitary and collating function within the soul and perception. The same metaphor, for example, appears in the twelfth commentary of Michael of Ephesus on the *Parva Naturalia* and the late thirteenth-/early fourteenth-century paraphrase of *De Anima* by Sophonias.[40]

Earlier in his commentary on Synesios, Gregoras explains the operation of the various faculties of the soul by articulating how they each mediate between certain particular qualities of said soul. As he writes: "Just as on the one hand thought (διάνοια) is the medium between the mind and the soul, on the other hand opinion (δόξα) is the medium between the soul and the imagination, and sensation is the medium between the imagination and the body."[41] Then it seems that the common sense operates on the higher end of the imagination, functioning almost at the level of opinion (δόξα) so as to unite in an unmediated fashion with the soul. This would not be unheard of since not only did earlier commentators at times suggest that the imagination was bifurcated between a lower and a higher function, but also because the operation of the common sense is similar to the opinion's role in producing initial judgments and recognitions of what is being perceived. While various commentators differed on the exact place of opinion, here it appears as a medium between the imagination and the loftier operations of the soul, though this hierarchy is not linearly structured since thought (διάνοια) is also given intermediary access to the soul from the mind. Gregoras from there goes on to describe this transmission upwards of information to the loftier soul as the difference between the lower and higher air, which in its highest portions bears an elemental fire that warms it, but in the lower regions tends toward a

[40] See Michael of Ephesus, *In parva naturalia commentaria*, 47.30–48.2, ed. P. Wendland, *In parva naturalia commentaria*, Commentaria in Aristotelem Graeca 22.1 (Berlin: Reimer, 1903), 47–48; Sophonias, *In Aristotelis libros de anima paraphrasis*, ed. M. Hayduck, *In libros Aristotelis de anima paraphrasis*, Commentaria in Aristotelem Graeca 23.1 (Berlin: Reimer, 1883), 114.24–28.

[41] "οἷον ὥσπερ ἐστὶ μεταξὺ μὲν τοῦ νοῦ καὶ τῆς ψυχῆς ἡ διάνοια, μεταξὺ δὲ ψυχῆς καὶ φαντασίας ἡ δόξα, μεταξὺ δὲ φαντασίας καὶ σώματος ἡ αἴσθησις." Nikephoros Gregoras, *Explicatio in librum Synesii De insomniis*, 141 C (163.06), ed. Pietrosanti, 56.

watery moisture.[42] Thus, Gregoras has to some extent collapsed the imagination with the common sense by placing them at that hegemonic center of perception through which the earthly senses are (after earlier processes of mediation) then transmitted in an unmediated fashion toward the higher operation of the soul, associated with divine perception, attempting to depict the chains of mediation and contact occurring throughout the process.

Theodore Metochites, a contemporary of Gregoras, would likewise come to share his thoughts on the common sense, following the practices and theories of his colleagues and predecessors. In the case of Metochites, his statements on the common sense come down to us particularly in his paraphrase of Aristotle's *De memoria*, composed of quotations as well as his own statements, in which he summarizes much of Aristotle's text. Particularly pertinent for our purposes here is the following passage, which explicates the relationship between the imagination and the common sense:

> τὸ φάντασμα δ', ὡς εἴρηται, τῆς κοινῆς αἰσθήσεώς ἐστι πάθος. ἐπὶ γὰρ τῇ κοινῇ αἰσθήσει ἕπεται, ἣν ἐν τοῖς Περὶ ψυχῆς διορίζεται, καθ' ἣν μεγέθους καὶ κινήσεως ἡ ψυχὴ αἰσθάνεται καὶ ταῦτα γνωρίζει, καθὼς δὴ καὶ τὸν χρόνον αὐτόν, ὅτι καὶ τὸν χρόνον αὐτὸν τῇ κοινῇ καὶ πρώτῃ αἰσθήσει καταλαμβάνει καὶ γνωρίζει ἡ ψυχὴ κατὰ τὸ φανταστικόν.
>
> The mental image, as [Aristotle] said, is the experience of the common sense. For it follows upon the common sense, which in the *De anima* he distinguished as that according to which the soul senses largeness and movement and knows these things, to the extent [it knows] time itself, since it comprehends time itself by the common and primary sense, and the soul knows [this] according to the faculty of the imagination.[43]

From there on, Metochites turns to memory, which draws together thought alongside the mental images produced for the preservation of knowledge. Metochites's passage helps contextualize Gregoras's own statements equating the common sense with the imagination, since the images of the imagination are staged here as the "experience" (πάθος) of the common sense. The nuances between these various authors speak to the

[42] "ὥσπερ καὶ ὁ ἀήρ, δύο ἔχων ἀκρότητας καὶ λήξεις, κατὰ μὲν τὴν ἄνω λῆξιν, τὴν τῷ στοιχειακῷ πυρὶ πλησιάζουσάν ἐστι θερμός· κατὰ δὲ τὴν κάτω καὶ πρὸς τῷ ὕδατι ὑγρός." Nikephoros Gregoras, *Explicatio in librum Synesii De insomniis*, 141 C (163.06), ed. Pietrosanti, 56.

[43] Theodore Metochites, *Paraphrasis in Aristotelis De memoria et reminiscentia*, ed. David Bloch, "Theodoros Metochites on Aristotle's *De Memoria*: An Edition," *Cahiers de l'Institut du Moyen-Âge Grec et Latin* 76 (2005): 3–30, at 15–16.

long-attested debate of the commentators and later thinkers on where exactly to place the operations of the imagination.

Since the common sense in these various texts has an intimately close relationship with sensation, collecting into itself its imprints and collating them, this means that the imagination could thus be seen to be overlapping this process, or there can be another function taking its place such as preconception (προλήψις), supposition (ὑπόληψις), or opinion (δόξα), as we have discussed before. Yet another possibility is that the imagination simply follows this process. Here we see the common model where the two seem to overlap or at least operate in such close proximity that they would seem to be as one. Thus the mental image (φάντασμα) arising from sensation is a product of the experiences of the common sense, which seems almost to stand in the place of the imagination in that it not only produces these images, but also stands in that ruling part of the soul to which all sensation is sent.

Yet, as Metochites makes clear here, the soul knows that it perceives time precisely through the operation of the imagination in relation to the common sense. This is why he refers to it not only as the common sense here, but also as the first sense-faculty (κοινῇ καὶ πρώτῃ αἰσθήσει), which Aristotle explains in *De memoria* as the basic necessity in thought and memory to understand continuity and time so as to be able to discern that one has perceived.[44] Thus the common sense takes primacy alongside the perception of time for Aristotle and Metochites since, as we have discussed, the common sense is notable as being conscious of the fact that it is perceiving, a matter which it shares with the so-called primary sense's perception of time. Thus, what Metochites's text offers us is a succinct expansion of the notion of the common sense, which not only speaks to its unitary features, but also places it alongside the operations of perception in direct dialogue with the imagination.

Finally, Nikephoros Choumnos, another contemporary of Metochites and his bitter rival,[45] gives us one of the more eloquent accounts of sensory unity and the common sense in his treatise *On the Nutritive and Sensitive Soul*, though he does not explicitly engage the term the "common sense" (κοινὴ αἴσθησις) itself. There he discusses the unity of the senses, as one would expect, between his explication of sense perception and the

[44] Aristotle, *De memoria et reminiscentia*, 449b24–450a11 and 452b7–453a5, ed. and trans. W. S. Hett, *Parva Naturalia*, Loeb Classical Library 288 (Cambridge, MA: Harvard University Press, 1936), 290–291 and 307–309.

[45] See Ihor Ševčenko, *Études sur la polémique entre Théodore Métochite et Nicéphore Choumnos* (Brussels: Byzanthion, 1962).

imagination.[46] Choumnos boasts that he has successfully unified Plato's theory of vision with that of Aristotle through a dialectic synthesis, as Börje Bydén has carefully elucidated, fusing Aristotle's notion of the transparent with Plato's extramissive rays.[47] This is founded upon a contemplation on mirrors and reflections, which he uses to summarize his findings on sight. From there, Choumnos moves on to consider the various interrelations between the senses, understanding how they parallel and relate to one another.[48] A primary focus here is the question of magnitude (μεγέθος) and movement (κίνησις), which feature prominently in Aristotle's conceptualization of the common sense. This closely follows Aristotle's *De anima*, stressing key points, such as the observation that all senses are based on discerning a series of binaries or contradictions, such as sweet and bitter in taste, white and black in sight, and so on.[49]

Through Plotinos, Michael Psellos, Nikephoros Gregoras, Theodore Metochites, and Nikephoros Choumnos, we can observe variations and subtle nuances regarding the understanding and emphasis of the common sense in Byzantine thought. The groundwork for the concept derives closely from Aristotle, yet the sources within the Aristotelian corpus vary widely. Furthermore, while John Philoponos appears to be the most popular commentator for the subject matter, as was usually the case, the late Byzantine thinkers evidence not only further sources, but also a return to the corpus of Aristotle as in the case of Choumnos and Metochites, which allowed for a greater flexibility in their conceptualizations even within the context of commentaries and paraphrases. It is upon this intellectual foundation that sensation and perception were understood to be processed and regulated by the faculties of the mind via a syncretic process. In this process, models of sensibles were construed in the mind by the imagination by virtue of the compiling of sensual information allowed for by the common sense. This common sense stitched together the data from the disparate sensory organs, retaining their uniqueness, yet also providing a sense as to how all these sensations came together to form their referents in the world, as well as how they at times overlapped to form common observations, like magnitude or motion.

[46] See Verpeaux, *Nicéphore Choumnos*, esp. 143–145.

[47] Börje Bydén, *Theodore Metochites'* Stoicheiosis Astronomike *and the Study of Natural Philosophy and Mathematics in Early Palaiologan Byzantium* (Gothenburg: Acta Universitatis Gothoburgensis, 2003), 206–207.

[48] See Nikephoros Choumnos, *Opera philosophica*, 6, ed. Konstantinos P. Chrestou, *Τὸ φιλοσοφικὸ ἔργο τοῦ Νικηφόρου Χούμνου* (Thessaloniki: Ekdoseis Kyromanos, 2002), 92–142, esp. 130–138.

[49] Cf. Aristotle, *De anima*, 422b24–28, ed. and trans. Hett, 128–129.

Synesthesia in Practice and Perception

In this act of composition, of stitching together the various sensibles through the ruling power of the common sense, we can understand the manner in which sensations could not only be collated into a holistic cognitive image of the perceived entity, but also how this action of stitching together might allow one to make associations between the senses in their construction of an object and the way it operated. For example, in differentiating Christian worship from pagan practices of sacrifice in his *Stromata*, Clement of Alexandria draws a parallel between the rising of incense and Christian prayer, saying:

> ἡ σύμπνοια δὲ ἐπὶ τῆς ἐκκλησίας λέγεται κυρίως. καὶ γάρ ἐστιν ἡ θυσία τῆς ἐκκλησίας λόγος ἀπὸ τῶν ἁγίων ψυχῶν ἀναθυμιώμενος, ἐκκαλυπτομένης ἅμα τῇ θυσίᾳ καὶ τῆς διανοίας ἁπάσης τῷ θεῷ.
>
> The word "conspiration" is that properly used for the Church. For the Church's sacrifice is indeed speech rising like incense, from holy souls, while every thought of the heart is laid open to God along with the sacrifice.[50]

Here Clement parallels the congregation's simultaneous exhalations in prayer with the smoke that rises from the pagan altars, yet in the former it is the prayers, i.e. the words (λόγος), that rise up as that smoke. Thus, what begins as a parallel between the pagan's bloody, smoky sacrifice and the Christian's bloodless "sacrifice of praise" (Hebrews 13:15), is amplified through a sensual contemplation on the materiality of these two distinct offerings: words that are a smoky breath, like the incense shed, are contrasted with the burnt offerings on the pagan altar. Hence that breath becomes an incense "compounded of many tongues and voices in the way of prayer," as he says later.[51] This allows an association, through a shared materiality, between the physicality of the spoken word as an exhalation and the smoke of incense.

Such examples demonstrate not only rich and concerted contemplation of the sensual materiality of the physical world, but also an understanding of sensible objects as sharing a variety of perceptible features beyond their

[50] Clement of Alexandria, *Stromata*, 7.6.32, ed. L. Früchtel, O. Stählin, and U. Treu, *Clemens Alexandrinus*, vol. 3, second edition (Berlin: Akademie Verlag, 1970), 24; trans. H. Chadwick and J. E. L. Oulton, *Alexandrian Christianity: Selected Translations of Clement and Origen* (Philadelphia: Westminster Press, 1954), 112.

[51] "τὸ ἐκ πολλῶν γλωσσῶν τε καὶ φωνῶν κατὰ τὴν εὐχὴν συγκείμενον." Clement of Alexandria, *Stromata*, 7.6.34, ed. L. Früchtel et al., 26; trans. Chadwick and Oulton, 114.

primary function. In the case of Clement, it is not the smell of incense that enables the metaphor, but its visual connection to smoke rising from the pagan altar, and that smoke's origin in burnt matter. In associating the smoke of incense to prayer, there is a sense that words are spoken by breath, that this breath rises and diffuses into the air, that it is warm to the touch. As such, we can consider these metaphors to be rhetorical analogues to the common sense's operation within the mind, birthed from the imagination's ability to construct such associations through its habitual perception of the world's multi-sensory objects. In his *Vita* of St. Onnophrios, Paphnutios provides us with a similar construction, when he recounts the saint's imperative to "proclaim my memory as fragrant incense to the brethren" (ⲡⲉϫⲁϥ ⲟⲛ ⲛⲁⲓ ϫⲉ ⲉⲕϣⲁⲛⲃⲱⲕ ⲉⲕⲏⲙⲉ ⲧⲁϣⲉⲟⲉⲓϣ ⲙ̄ ⲡⲁ ⲣ̄ ⲡⲙⲉⲉⲩⲉ · ⲉⲩⲥϯ ⲛⲟⲩϥⲉ ⲛ̄ ⲛⲉⲥⲛⲏⲩ).[52] Not only does this text resonate with contemporary practices of veneration, including the offering of incense before the holy, as Harvey has discussed, but it also plays with the understanding of speech's breath as somehow being akin to incense, as if the aroma of the knowledge of the saint commingled with the hotness and steam of warm breath upon the skin.

These fragrant, sweet odors are often directly associated precisely with knowledge beyond the inherent sensual overlap of smell and taste. For example, in *The Alexiad* Anna Komnene recounts a discussion with her mother about her intimidation at studying the works of Maximos the Confessor, yet her mother comforts her, saying:

> ἐπαινετὴν οἶδα τὴν δειλίαν ταύτην· καὶ οὐδ᾽ αὐτὴ ἀτρέμας ταῖς βίβλοις ταύταις πρόσειμι, ἀλλ᾽ ὅμως ἀποσπᾶσθαι τούτων οὐ δύναμαι. σὺ δέ μοι μικρὸν ἀνάμεινυν καὶ ταῖς ἄλλαις ἐγκύψασα βίβλοις πρότερον καὶ τῆς τούτων ἀπογεύσῃ ἡδύτητος.
>
> "Your reluctance is commendable, I'm sure," she replied, and "I myself do not approach such books without a tremble. Yet I cannot tear myself away from them. Wait a little and after a close look at other books, believe me, you will taste the sweetness of these."[53]

[52] Paphnutius, *Life of Onnophrius*, ed. E. A. Wallis Budge, "The Life of Apa Onnophrios the Anchorite," *Coptic Martyrdoms*, vol. 1 (London: British Museum, 1914), 205–224, 455–473. 216; trans. Tim Vivian, *Histories of the Monks of Upper Egypt and the Life of Onnophrius*, revised edition (Piscataway, NJ: Gorgias Press, 2009), 158.

[53] Anna Komnene, *Alexias*, v.9, ed. A. Kambylis and D. R. Reinsch, *Alexias*, Corpus Fontium Historiae Byzantinae, Series Berolinensis 40/1 (Berlin: De Gruyter, 2001), 166; trans. E. R. A. Sewter, *The Alexiad*, revised edition (New York: Penguin Classics, 2003), 150.

The book makes her shudder; it makes her taste sweetness. The sensual effects are made palpable for the reader, who understands the power of words, uttered or upon the page, to affect their readers, so much so indeed that they might "taste sweetness" (ἀπογεύσῃ ἡδύτητος). We might reason precisely that as the reader reads, their mouths masticating the words as they pronounce them quietly or aloud, it would seem to them that the delight enjoyed by the mind apprehending the discourse imagines this pleasure to be coming from the words themselves churning upon the tongue.

This association between speech itself and taste is also expressed in Michael Psellos's *Chronographia*, where Michael VI said of Psellos, "I taste the honey of your lips" (σου τοῦ τῶν χειλέων... ἀπογεύομαι μέλιτος), and regarding Constantine X Doukas's admiration for him, Psellos writes: "He respected me more than anyone else and 'drank his fill of waters at my fount': to him they were as nectar" (ἐσέβετό τε διαφερόντως τῶν ἄλλων· καὶ διέκειτο ἀναπιμπλάμενος ἐμοῦ ὥσπερ νέκταρος).[54] This imagery of words as fluid and flowing waters, nectar, honey, or even gold are popular tropes in the Byzantine imaginary. Not only does John Chrysostom's epithet (and that of his pagan precedent Dio Chrysostom before), the "golden-mouthed" (Χρυσόστομος), carry the suggestion of a synesthetic tie between sound and sight, as audible words hewn of shimmering gold, but Theodore of Stoudios, for example, described him as the one "from whose lips flow the truly golden words" (οὗ ἐκ τῶν χειλέων τὰ χρυσότατα ῥυΐσκεται ἔπη).[55] Hence, drawing an association not only between sight and sound, but also taste, Theodore understands words to be visual, sonic, and fluid. These associations were certainly ascribed to John Chrysostom's words and portraits of him depict the scroll of his homilies transforming into rushing streams of water which the faithful imbibe, as in his author portrait for his homilies on the Pauline epistles (Milan, Biblioteca Ambrosiana, A 172 sup., fol. 263v).[56] Yet, the same was true of other divinely inspired discourses, which might be said to "flow forth from the tongue of your initiates as from a fountain," and whose streams might be

[54] Michael Psellos, *Chronographia*, VII.16 and VII.25, ed. Reinsch, 214, 260–261; trans. Sewter, 284 and 342.

[55] Theodore of Stoudios, *Antirrhetici adversus iconomachos*, II.29, ed. PG 99:373B; trans. Roth, 62.

[56] On such imagery and the related "Fountain of Life" motif, see Tania Velmans, "L'iconographie de la 'Fontaine de Vie' dans la tradition Byzantine à la fin du Moyen Âge," in André Grabar and Jean Hubert (eds.), *Synthronon* (Paris: Librairie C. Klincksieck, 1968), 119–34; Paul Underwood, "The Fountain of Life in Manuscripts of the Gospel," *Dumbarton Oaks Papers* 5 (1950): 41–138.

ingested by the faithful as "a new immortal drink," as in the opening inscription to Michael the Monk's Gospel book in Istanbul.[57]

In other instances, moments of melodic inspiration were likewise recounted through the sense of taste, such as the understanding that Romanos the Melodist was inspired by the Virgin for his hymns on the birth of Christ by being fed a scroll with the hymns.[58] In a poetic epigram on the silver revetment of an icon of the Theotokos *Kyriotissa*, for example, the Virgin's inspiration is described as the "sweet drink of wisdom" (τῶν τῆς σοφίας γλυκερῶν κερασμάτων).[59] In the *Vita* of Theodore Sykeon, the author recounts how Theodore wished to imitate David in hymnography and attempted to learn the psalms by heart, but continually failed. In his desperation, he asks God for help through an icon in a chapel of the martyr Christopher near his village. Immediately, Theodore "felt a sweetness more pleasant than honey poured into his mouth" (ᾔσθετο γλυκύτητα ἡδύτερον μέλιτος ἐγχυθεῖσαν ἐν τῷ στόματι αὐτοῦ), recognized it as God's grace, "partook of the sweetness" (μεταλαβὼν τῆς γλυκύτητος), and thus was able to finish memorizing the psalms in a matter of days.[60] Paul Speck has argued that this was an interpolation added to the text during iconoclasm to stress the divine intervention resulting from the operation of his prayer to the icon.[61] While Speck's suggestion may be dubious and thus must be taken lightly, if true the taste of honey would seem to result not only from the pronunciation of his prayer, but from the stimulus of the icon, both visual and divine.

In her discussion of such occurrences, Pentcheva has insightfully drawn attention to the resonances between this type of sweet taste and the line "Taste and see that the Lord is good" (γεύσασθε καὶ ἴδετε ὅτι χρηστὸς ὁ

57 "Ἀλλ' ὢ βλύσας ἄβυσσον ἐνθέαν λόγων, Ψυχὴν ἐμὴν ἄκιμον ἐν καιρῷ δίκης, Ἔιης ποτίζων καινὸν ἄμβροτον πόμα, Ὁ νοῦς μαθητὰς εἶπας ἐκπίνειν τότε." Robert S. Nelson, "Michael the Monk and His Gospel Book" in *Actes du XVe Congrès International d'Études Byzantines*, vol. II (Athens: Association Internationale des Etudes Byzantines, 1981), 575–582, at 580.

58 As seen in a miniature from the 'Menologion' of Basil II (976–1025) depicting Romanos the Melodist receiving the inspiration for his hymn on the birth of Christ. In this image, the Theotokos literally inserts words into the composer's mouth by giving him a scroll to eat (Vatican City, Biblioteca Apostolica Vaticana, gr. 1613, fol. 78r). On this image, see Bissera Pentcheva, "Visual Textuality: The Logos as Pregnant Body and Building," *RES: Aesthetics and Anthropology* 45 (2004): 225–38, esp. 232.

59 Anonymous, *Anthologia Marciana*, 334, ed. Lambros, 181.

60 George of Sykeon, *Vita sancti Theodori Syceotae*, 13, ed. A.-J. Festugière, *Vie de Théodore de Sykeôn*, vol. I, Subsidia Hagiographica 48 (Brussels: Société des Bollandistes, 1970), 289; trans. Elizabeth Dawes, *Three Byzantine Saints: Contemporary Biographies of St. Daniel the Stylite, St. Theodore of Sykeon and St. John the Almsgiver* (Oxford: Blackwell, 1948), 95–96.

61 Paul Speck, "Wunderheilige und Bilder: Zur Frage des Beginns der Bilderverehrung," *Poikilia Byzantina* 11:3 (1991), 163–247, esp. 245–246.

κύριος) from Psalm 34, which intricately plays with the phonetic pronunciation in Medieval Greek of "good" (χρηστός) and "Christ" (Χριστός). Thus, in tasting and seeing, the congregation experience not only the goodness of God, but his revelation in Christ, who made himself perceptible to taste and sight through the Eucharist.[62] Yet this perceptibility in the Eucharist, as we saw in John Chrysostom's comments on the sacrifice earlier, was not immediately visible as such. As Aden Kumler has intricately explained, in the case of the medieval western world, for figures like Thomas Aquinas "the language of prayer exults in the failure of the phenomenal: being trumps seeming," given that while the Eucharist is ontologically changed, its own visual morphology and taste remained the same as bread.[63] This structures the problems and limits faced by perception in the medieval world, given that rather than believing in the mere efficacy of the representation of the divine through the sensual realm, writers in the medieval east and west strove to understand that physical perception must be accompanied by spiritual perception in order to perceive beyond the limits of the senses – even when the divine might be ontologically present, as in the Eucharist.

For Pentcheva, however, transgression between the senses in her definition of "synesthesis" is the action of sensing together the sensible world at hand.[64] Her theory acknowledges that Byzantine ritual and veneration engaged all the senses and that these senses "are engaged simultaneously."[65] Liz James suggests a similar operation, but rightly understands this as a product of a cognitive synesthesia, stressing that the "senses did not act in isolation. They interacted in creating a sensory effect," which she supports with the example of Theodore of Sykeon tasting honey upon hearing the Scriptures.[66] On the surface, this concept certainly resonates with Aristotle's common sense, but Aristotle and subsequent commentators' useful theorizations on the matter are wholly absent from Pentcheva's theorization, which is focused on the sensual experience of art. The crucial distinction between Pentcheva's "synesthesis" and Aristotle's common sense is that while Aristotle's interest lies in the ways in which perception and cognition make cross-sensory associations for individual objects, coming to know them over time as being composed of a variety of sensual experience that noetically comprises a person's knowledge of that object, for Pentcheva "synesthesis" is merely

[62] Pentcheva, *The Sensual Icon*, 43. [63] Kumler, "The Multiplication of the Species," 179.
[64] See Bissera Pentcheva, "The Performative Icon," *The Art Bulletin* 88:4 (2006): 631–655, esp. 650.
[65] Pentcheva, "The Performative Icon," 650. [66] James, "Senses and Sensibility," 528.

about simultaneous perception coming from the various types of sensory data that the image offers up. Hence, what we find in Pentcheva's work is an endemic confusion between the sensible world at hand and the perceptual understandings of that world as if sensation alone could communicate heavenly realities. What Pentcheva fails to acknowledge is the disjuncture between what is actually sensibly present and those heavenly realities and experiences that emerge through perception in the imagination via a form of culturally trained synesthesia.

In the sources discussed here, we can come to understand various permutations on the acquisition of knowledge that are accompanied by a gustatory sensation of sweetness, indicating either the miraculous intervention of divine inspiration, as in the case of Theodore Sykeon and Romanos the Melodist, or alluding to the synesthetic manifestation of goodliness in the mouth when encountering pleasant or rousing discourses, as in the case of Michael Psellos and Anna Komnene. Yet we cannot understand that these perceptions are indeed occurring through sensation alone. Surely, we might allow that through divine intervention Theodore Sykeon tasted honey, but likewise we cannot say that Michael VI tasted honey from Psellos's lips – and yet the two feature a similar cross-sensory depiction of knowledge acquisition. Thus, what I suggest here is that we must understand these cross-sensory experiences as being a by-product of the *habitus* of perception through involved acts of the imagination and the common sense. Through these culturally trained perceptual metaphors, classical and Byzantine subjects came to experience the world synesthetically (in its modern definition), so the acquisition of exceptionally good knowledge could be accompanied by the sweetness of taste or smell.

The association of taste and knowledge is pervasive in secular literature, as in the latter examples, but also as a manner of indicating knowledge acquisition, as when Michael Psellos states in the opening to a didactic poem on medicine that he wishes to "produce a small appetite" (μικρὰν τεκεῖν ὄρεξιν ἀνδράσι φίλοις) for medicine, or John Mauropous's similar claim to give in his works "a small taste of an abundant bouquet" (γεῦμα μικρὸν δαψιλοῦς ἀνθοσμίου), a comment that alludes not only to taste, but also to smell.[67] Yet there is no sweetness necessarily in these lessons, nor is there a

[67] Michael Psellos, *Poemata*, 9.533, ed. L. G. Westerink, *Poemata* (Stuttgart: Teubner, 1992), 208; John Mauropous, *Epigrammata*, 1.29, ed. P. de Lagarde, *Joannis Euchaitorum Metropolitae quae in codice Vaticano Graeco 676 supersunt*, Abhandlungen der Historisch-Philologischen Classe der Königlichen Gesellschaft der Wissenschaften zu Göttingen Bd. 28 (Amsterdam: Hakkert, 1882), 1. See Floris Bernard and Kristoffel Demoen, "Giving a Small Taste" in Floris Bernard and Kristoffel

fragrant aroma – the bouquet has merely been offered up to the senses, but whether it is full of a sweet taste or fragrance is not expressed, since this is something which only the perceiving, reading subject can judge and assess.

Not every lesson was accompanied with the flow of honey. The sweetness of Psellos's lips is an indicator of an exceptional state of enlightenment. In such heightened states of learning, it was understood that in perception we might cross-associate these sensory inputs through the action of the common sense as it vividly forged its cross-sensory associations in the mind while apprehending and grappling with the noetic image of the outside world. Were this just a common occurrence in perception, John Chrysostom's epithet would be nothing but pedantic, and the claims of Psellos's flowing nectar would be nothing but an expected commonality, not meriting explicit mention – nor worthy as standing in for those truly divine moments of divine inspiration where God's intervention manifested itself through the sweet taste of honey. Hence, this sweetness was precisely an event in the perceptual world, an event that occurred only in exceptional cases where discourse and sensory experience became so vivid and engaging as to transgress in the mind its usual bounds and limits. Hence, I likewise am suggesting that this was a product not of sensation, but of perception. That is to say, just as the "as if" function of *ekphrasis* seems to place things before the eyes, while always confronting its inexorable inability to do just that, the same is true of these moments where sensory bounds were seemingly transgressed.[68]

Remediation

As has been shown earlier, the classical, late antique, and Byzantine writings on the senses often strove with the utmost clarity to state that

Demoen (eds.), *Poetry and Its Contexts in Eleventh-Century Byzantium* (Burlington, VT: Ashgate, 2013), 3–15, esp. 3–4.

[68] In her 2012 volume, *Kissing Architecture*, Sylvia Lavin argues that the nature of contemporary architecture is one of an ever-proliferating plurality of mediums and media. In these instances, practices like installation art blur the boundaries of traditional mediums and their generative specificity. However, it is crucial to note that such conditions of plurality cannot occur if the mediums have indeed lost their uniqueness and specificities. Lavin's "kiss" from her title could not occur were it not for the intermediary space between particular media, which is what allows us to perceive an interval of difference and enables their erotic contact to occur. The same must be said for the Byzantine's predilection for synesthetic language, which derives its efficacy not from mixing and conflation, but rather through the purposeful transgression of the traditional bounds of sensory mediums. This wrought language of cross-sensory action always seeks not only to remediate information across the senses, but also to enact suspended, heightened states of sensory perception. See Sylvia Lavin, *Kissing Architecture* (Princeton University Press, 2012), 22–65.

while sensibles may be shared across the senses and while the common sense and the imagination might collate these in the mind into one, nevertheless each sensory input was particular and specific to itself and to the type of knowledge it could communicate. Hence, it would be unfeasible to read these various religious and secular metaphors as actually suggesting that the sight or sound of poetry or education could lead one to the sensorial smell or taste of sweetness. These were states of vividness and actualization that occurred within the realm of the imagination, and which thus we can understand to have a striking similarity with the rhetorical theories of *ekphrasis* and *enargeia*, which sought the same cross-sensory operation with the mind. Thus I propose that we come to understand these exceptional events of sensory transgression to be born precisely out of the contemporaneous Byzantine theories of rhetorical representation.

We might contrast these forms of sensory transgression with the rhetorical function of hyperbole, which even for us today operates as a violation of truth for the sake of producing a more heightened state of experience. The ninth-century rhetorician George Choiroboskos defined hyperbole in his treatise *On Poetic Tropes* as follows:

> Ὑπερβολή ἐστι φράσις ὑπερβαίνουσα τὴν ἀλήθειαν αὐξήσεως χάριν, ὡς ὅταν τις τὸν γοργῶς τρέχοντα εἴπῃ, ὅτι τρέχει, ὡς ὁ ἄνεμος· ἢ ὑπερβολή ἐστι λόγος ὑπεραίρων τὴν ἀλήθειαν ἐμφάσεως ἢ ὁμοιώσεως ἕνεκα.
>
> Hyperbole is an expression that goes beyond the truth for the sake of amplification, as when one says of someone running vigorously that 'he runs like the wind.' Or hyperbole is a statement that exaggerates the truth, for the sake of emphasis or comparison.[69]

In language, hyperbole takes on precisely the function of overstepping (ὑπερβαίνουσα and ὑπεραίρων) the usual bounds of sensibles and truth. In his case study on the so-called flight of Symeon, for example, Henry Maguire has analyzed how Byzantine writers and artists composed images of Symeon running to receive Christ in the Presentation in the Temple, where Symeon is depicted as flying or dancing in his rush to arrive at the Christ.[70] The presence of hyperbole in the basic rhetorical training of Byzantine subjects, and its definition as a transgression for emphasis or comparison cautions us to consider with care its own potential applications to instances of perception and presence that might seem to indicate a Byzantine sensual or perceptual reality. As we can see from the example

[69] George Choiroboskos, *Περὶ τρόπων ποιητικῶν*, 19, ed. L. Spengel, *Rhetores Graeci*, vol. III (Leipzig: Teubner, 1856), 252–253; trans. Maguire, *Art and Eloquence*, 84.

[70] Maguire, *Art and Eloquence*, 84–90.

offered up by Choiroboskos, hyperbole often creates its emphatic effect through a form of synesthesia, where running becomes flying: from the medium of ground to the medium of air, from seeing running with the physical eyes, but seemingly perceiving flying in the mind's eye. Hence, the pervasiveness of synesthetic language necessitates that we reconsider instances where sight might operate like touch, or hearing might operate like taste. This is not to say that such synesthetic language must be disregarded as a pedantic oratorical flourish, but rather that its effect must be understood as having a specific and particular function within its linguistic medium – as language and using its own tactics.

Just as a swift runner might seem to fly, the lascivious gaze of a lustful adulterer might also seem to stroke the flesh of the desired one. Take for instance John Chrysostom's words in his homily on Matthew 5:27–28 regarding adultery, where he warns: "Even if you have not touched them with your hand (ἥψω τῇ χειρί), you have caressed them with your eyes (ἐψηλάφησας τοῖς ὀφθαλμοῖς)."[71] Here, this unambiguous collapse of sight and touch serves the purpose of emphasizing that acts of adultery are not relegated to physical contact alone, but also to the mental energies exhausted upon coveting a lover through one's lusting and desirous gaze, as when Michael Psellos details Constantine IX Monomachos's embrace of Zoe while imagining himself adulterously grasping his mistress in the *Chronographia*, discussed earlier. Haptic sight here is taking on a very precise and unique function in the context of this homily, which, as hyperbole is defined, "exaggerates the truth, for the sake of emphasis or comparison," here for the sake of both emphasis and comparison. As well-crafted rhetoric, John is able to create a hyperbolic description of sight that captures the moral implications that he is trying to emphasize regarding the gaze by riffing on the language used to describe the operation of mental faculties in conjunction with visual perception, which grasp and apprehend sensible objects and bring them into the confines of the mind. It would be hasty though to consider this as evidence for haptic sight within this context.

As John says elsewhere, "You carry out the sinful deed, if not through intercourse (διὰ τῆς μίξεως), then through the eyes."[72] Certainly, we would not be inclined to read this as a confusion of sexual intercourse and sight,

[71] "Εἰ γὰρ καὶ μὴ ἥψω τῇ χειρί, ἀλλ' ἐψηλάφησας τοῖς ὀφθαλμοῖς." John Chrysostom, *In Matthaeum*, ed. PG 57:257.

[72] "καὶ ἀπήρτισταί σοι τὰ τῆς ἁμαρτίας, εἰ καὶ μὴ διὰ τῆς μίξεως, ἀλλὰ διὰ τῆς ὄψεως." John Chrysostom, *Quod regulares feminae viris cohabitare non debeant*, 1.67–69, ed. Jean Dumortier, *Les cohabitations suspectes* (Paris: Les Belles Lettres, 1955), 97–98.

as if indicating that the physiology of sight could encompass some form of sexual intercourse. In such statements, John is precisely comparing sight and touch as a clever rhetorical play for emphasis: One the one hand, John is utilizing the established haptic language of cognition to metaphorically parallel perception itself to the process of coveting and grasping another person outside of wedlock. On the other hand, he is likewise suggesting that the energies of lust and desire – even if relegated only to a gaze – are nevertheless charged with the sin of adultery itself, even if a hand is never laid upon the object of desire.[73]

Compare this to Gregory of Nazianzos's language in his second theological oration, where he writes:

> πολλὰ περὶ ὄψεως ἀρρήτως κοινωνούσης τοῖς ὁρατοῖς, καὶ μόνῳ τῷ βούλεσθαι καὶ ὁμοῦ κινουμένης, καὶ ταὐτὸν τῷ νοὶ πασχούσης· μετὰ γὰρ τοῦ ἴσου τάχους ἐκεῖνός τε μίγνυται τοῖς νοουμένοις καὶ αὕτη τοῖς ὁρωμένοις. πολλὰ περὶ τῶν ἄλλων αἰσθήσεων, αἳ παραδοχαί τινές εἰσι τῶν ἔξωθεν, λόγῳ μὴ θεωρούμεναι.
>
> There are many facts about sight and its mysterious communion with objects. It is set going along with, and only by, the will. Sight is in the same case with mind, for it joins its objects with just the same speed as does the mind its thoughts. There are many facts about the rest of the senses, which act as receivers of external impressions unseen by the eye of reason.[74]

For Gregory, this passage comes within the context of a discussion of the various senses and seeks to marvel precisely at the mysterious workings of sight's ability to commune with the object seen, which sets it apart from the workings and operation of other forms of perception, such as sound. This "mysterious communion" (ἀρρήτως κοινωνούσης) is stated precisely as a mixing or union between the thing seen and the eyes, and between the thing seen and the intellect. Here we might even understand how John Chrysostom's comparison of sight and intercourse (μίξεως) can be actually understood as an allusion to the perceptual communion and grasp of the mind, but also to the sensual mixing of the visual rays in sight as they join (μίγνυται) themselves and bring the object to the mind. Hence, even in an instance that demonstrates the metaphorical force of such statements, we can see authors playing with possible language related to sensation and perception. This is not to say that we should understand John as advocating that sexual intercourse might occur through sight, but rather that he

[73] On the sexual gaze in John Chrysostom, see Blake Leyerle, "John Chrysostom on the Gaze," *Journal of Early Christian Studies* 1:2 (1993): 159–74.

[74] Gregory of Nazianzos, *Orat.* 28.22 (*De theologia*), ed. Barbel, 106; trans. Wickham, 54.

constructs the moral thesis of his writings through language that alludes to the actual functions and operations of perception. Thus we must understand such phrases within their broader context as doing important psycho-spiritual work rooted on the rhetorical training in hyperbole and inter-sensual metaphors.

Such hyperbolic statements and metaphors appear in relation to sight most prominently around the problem of the "evil eye" in popular thought with Patristic writers.[75] That is, the belief that the eyes of envious and malicious people shoot out rays, like arrows, that pierce and corrode the bodies of those they carry ill-will toward. As Thomas Rakoczy has traced, coming originally from ancient Egypt and Mesopotamia, the evil eye articulated itself in the ancient Greek world as a concern for the harmful gaze coming from the gods, who might endeavor to harm humans through their desiring and envious gaze.[76] This was believed to be a very palpable threat, even to non-human beings, such as horses, who being prized and precious property could also be subjected to the detrimental effects of the evil eye, as attested, for example, by inscriptions upon amulets warding off the evil eye's effects.[77] In his Homily 11, "On Envy," Basil of Caesarea explains the functions of the evil eye while seeking to disprove its fallacious logic and placing blame on the envy that is said to generate that harmful gaze, rather than any magical belief in the powers of the optical rays.[78] There, Basil writes:

> Ἤδη δὲ καὶ τῶν ἰοβόλων αὐτῶν ὀλεθριώτεροι εἶναι οἱ νοσοῦντες τὴν βασκανίαν ὑπονενόηνται· εἴπερ τὰ μὲν διὰ πληγῆς ἐνίησι τὸν ἰὸν, καὶ κατὰ μικρὸν σηπεδόσι τὸ δηχθὲν ἐπινέμεται· τοὺς δὲ φθονερούς τινες οἴονται καὶ δι᾽ ὀφθαλμῶν μόνων τὴν βλάβην ἐπιβάλλειν· ὥστε τὰ εὐεκτικὰ σώματα, καὶ ἐκ τῆς κατὰ τὴν ἡλικίαν ἀκμῆς εἰς τὴν ἄκραν ὥραν ὑπερανθήσαντα, τήκεσθαι παρ᾽ αὐτῶν καταβασκαινόμενα, καὶ ὅλον ἀθρόως συναναιρεῖσθαι τὸν ὄγκον, οἷον ῥεύματός τινος ὀλεθρίου ἐκ τῶν φθονερῶν

[75] On the evil eye in Patristic thought, see Matthew W. Dickie, "The Fathers of the Church and the Evil Eye" in Henry Maguire (ed.), *Byzantine Magic* (Washington, DC: Dumbarton Oaks, 1995), 9–34. On the origins and history of the evil eye, see Thomas Rakoczy, *Böser Blick: Macht des Auges und Neid der Götter: Eine Untersuchung zur Kraft des Blickes in der griechischen Literatur* (Tübingen: Gunter Narr, 1996). See also Edward Gifford, *Evil Eye: Studies in the Folklore of Vision* (New York: Macmillan, 1958); Walton Brooks McDaniel, "The Pupula Duplex and Other Tokens of an 'Evil Eye' in the Light of Ophthalmology," *Classical Philology* 13 (1918): 335–46; Matthew Dickie, "Heliodorus and Plutarch on the Evil Eye," *Classical Philology* 86:1 (1991): 17–29.

[76] See Rakoczy, *Böser Blick*, esp. 63–72, 247–270.

[77] See Anne McCabe, *A Byzantine Encyclopedia of Horse Medicine: The Sources, Compilations, and Transmission of the Hippiatrica* (Oxford University Press, 2007), 151–152.

[78] See Vasiliki Limberis, "The Eyes Infected by Evil: Basil of Caesarea's Homily, 'On Envy'," *Harvard Theological Review* 84:2 (1991): 163–184.

ὀφθαλμῶν ἀπορρέοντος, καὶ λυμαινομένου καὶ διαφθείροντος. Ἐγὼ δὲ τοῦτον μὲν τὸν λόγον ἀποπέμπομαι, ὡς δημώδη καὶ τῇ γυναικωνίτιδι παρεισαχθέντα ὑπὸ γραϊδίων· ἐκεῖνο δέ φημι, ὅτι οἱ μισόκαλοι δαίμονες, ἐπειδὰν οἰκείας ἑαυτοῖς εὕρωσι προαιρέσεις, παντοίως αὐταῖς πρὸς τὸ ἴδιον ἀποκέχρηνται βούλημα· ὥστε καὶ τοῖς ὀφθαλμοῖς τῶν βασκάνων εἰς ὑπηρεσίαν χρῆσθαι τοῦ ἰδίου θελήματος.

Even so, however, persons who suffer from this malady of envy (βασκανίαν) are supposed to be even more dangerous than poisonous animals, since these inject their venom by piercing their victim; then, gradually, putrefaction spreads over the infected area, but some people believe that envious people inflict harm merely through their eyes, so that healthy persons in the full flower and vigor of their prime are made to pine away under their spell, suddenly losing all their plumpness, which dwindles and wastes away by some flowing emission of destruction from the eyes of those being envious. I reject this explanation as pedestrian, something introduced by old women into the women's quarters. But this I do say: the devils, who are enemies of all that is good, use for their own ends such free acts as they find congenial to their wishes. In this way, they make even the eyes of envious persons serviceable to their own purposes.[79]

While he does not place this popular belief in dialogue with contemporary theories of vision, in this case the rays, despite being emitted by the eyes, are described as being sent out toward those people being envied through the use of the term "emission" (ἀπορρέοντος). Given Basil's exhortation against envy, the eyes of the envious are depicted as foreign and external to the reader, who might fear or be suspicious of the evil eye's harms upon their body. This positions the eyes of the envious as if they themselves were objects in the world, distanced and exterior from the righteous viewer who becomes victim to it.

Basil's words should be considered alongside those of John Chrysostom's on the same phenomenon as espoused in his homily on Galatians 3:1. John's reasoning is that the logic of the evil eye would be unfeasible, given that "the eye cannot be evil, since it is merely a limb" (ὀφθαλμὸς γὰρ οὐκ ἂν εἴη πονηρός, αὐτὸ τὸ μέλος). He goes on to say that the eye simply sees, thus evil seeing is simply caused from within by the corrupt knowledge that muddles and contaminates perception, given that the eye merely provides a passive organ through which contemplation can stream into the soul.[80] As Matthew Dickie concludes regarding this passage, John

[79] Basil of Caesarea, *Homilia de invidia* (Homily 11), ed. PG 31:380; modified trans. Wagner, 469–470.

[80] "Ὅταν δὲ φθόνον ἀκούσῃς ἐνταῦθα, καὶ ἐν τῷ Εὐαγγελίῳ ὀφθαλμὸν πονηρὸν τὸ αὐτὸ δηλοῦντα, μὴ τοῦτο νομίσῃς, ὅτι ἡ τῶν ὀφθαλμῶν βολὴ τοὺς ὁρῶντας βλάπτειν πέφυκεν· ὀφθαλμὸς γὰρ οὐκ

Chrysostom's theory of sight is resolutely inactive, where the eye simply serves as a receptor for the sensory data that streams into it.[81] Thus, it would be unlikely that John Chrysostom would be arguing elsewhere for a haptic iterant of sight, as his statements on the sexual gaze might superficially suggest. He evidently sees the eye as a mere receptor in vision, and thus seems to subscribe to some variant of an intromissive theory of sight that here plays a resolutely passive role in the influx of sensory data. The agency and action of the evil eye is thus relegated to the overwhelming passions of the people in their envy and in their belief that such intense desire might lead to physical harm through the desiring eye alone.

Like John Chrysostom, Basil similarly attacks the theory of the evil eye as being a crude popular lore, passed on through baseless gossip, "something introduced by old women into the women's quarters." While he merely explains the beliefs of others in his homily, characterizing the notion of the evil eye as the belief that "some deadly efflux emanates from the eyes" (ῥεύματός τινος ὀλεθρίου ἐκ τῶν φθονερῶν ὀφθαλμῶν ἀπορρέοντος), Basil does not offer us much in his homily to decipher his thesis on visual perception. Yet, like John, he immediately discredits this theory as being a logical fallacy. Instead, it is the work of demons stirred on by human envy that takes possession of the eyes for their own purposes; rather than simply discrediting the evil eye as being even a physiological possibility, Basil goes on to suggest that it is only this demonic outside power which could be seen to use eyes in such a manner. Granted, Basil does not concede that the evil eye is real, but rather that just like any other erring aspect of humanity, industrious demons will put the flesh to use as their own devices. Thus he cautions his listeners not to fear such idle gossip about vision, but rather to fear the underlying psychic groundings that would result in the conditions for the evil eye as those are the true threat to the human soul.

What the evil eye offers us then is a demonstration of how sight in such a heightened state is articulated as transgressing and overstepping its established bounds so as to argue against the power and danger of its

ἂν εἴη πονηρὸς, αὐτὸ τὸ μέλος· ἀλλ᾽ ἐνταῦθα ὁ Χριστὸς οὕτω τὸν φθόνον λέγει. Ὀφθαλμῶν μὲν γὰρ τὸ ἁπλῶς ὁρᾷν μόνον, τὸ δὲ πονηρῶς ὁρᾷν τῆς ἔνδον διεστραμμένης γίνεται γνώμης. Ἐπειδὴ γὰρ διὰ τῆς αἰσθήσεως ταύτης εἰς τὴν ψυχὴν ἡμῶν εἰσρεῖ τῶν ὁρωμένων ἡ θεωρία, καὶ ὡς ἐπὶ πολὺ ἐν πλούτῳ μάλιστα ὁ φθόνος τίκτεται, ὁ δὲ πλοῦτος ἀπὸ ὀφθαλμῶν ὁρᾶται, καὶ αἱ δυναστεῖαι καὶ αἱ δορυφορίαι· διὰ τοῦτο πονηρὸν ὀφθαλμὸν ἐκάλεσεν, οὐ τὸν ὁρῶντα, ἀλλὰ τὸν μετὰ βασκανίας ὁρῶντα ἀπὸ τῆς κατὰ ψυχὴν πονηρίας." John Chrysostom, *In epistulam ad Galatas commentarius*, ed. PG 61:648.

81 Dickie, "The Fathers of the Church," 22–23.

underlying psychic causes. We have observed similar conditions where sight seems or at least seeks to overstep its bounds in instances inciting a heightened emotional state. This is exemplified in cases of love, where it would seem as if the gaze or visage of one's beloved shoots out arrows or rays that forcefully impinge upon the viewing lover. However, as Douglas Cairns has noted, there is an asymmetry between the evil eye and *eros*, given that while the evil eye can "straightforwardly transmit what those individuals are feeling and achieve the desired effect on their target," love on the other hand cannot transmit its emotions.[82] Authors, like Achilles Tatius, certainly suggested that the gaze of lovers could enact forms of coupling and intercourse from afar, but this was far less palpable than the evident harm that the myth of the evil eye extolled. Likewise, Helen Lovatt has demonstrated how haptic iterations of sight emerge in ancient epic within the context of warriors and enemies, whose gaze not only appears to be haptic, but has the power to penetrate and peer into the very flesh of its object of sight.[83] These epic heroes are often characterized as having blazing, shining, glowing, fiery eyes that manifest their active prowess and allow them to smite their enemies. Through such instances, one is confronted with conditions in which sight appears to operate beyond its prescribed function, as if being capable not only of touch, but of a whole host of actions that enact the will of its subjects.

In such instances where we see the purposeful transgression of sensory boundaries, they function on the level of hyperbole and metaphor that operate as if they indeed have effects upon the world beyond a turn of phrase, particularly as evidenced in the widespread cultural phenomenon of the evil eye's pseudo-synesthesia. Such instances build on described or depicted human activities within language to suggest the violation of neat sensory boundaries. In such instances, both in literature and culture, running might become flying, and seeing might become touching – either sexual or destructive, as with the adulterous or envious gazes. However, the operation which hyperbole seeks to *depict in* words through a form of metaphor, the function of rhetoric overall seeks to *achieve through* words. For instance, as Stratis Papaioannou has shown, Eustathios of Thessaloniki in his twelfth-century commentaries on Homeric poetry repeatedly underscores rhetorical vividness as being achieved

[82] Douglas L. Cairns, "Looks of Love and Loathing: Cultural Models of Vision and Emotion in Ancient Greek Culture," *Mètis* n.s. 9 (2011): 37–50, esp. 48.

[83] Helen Lovatt, *The Epic Gaze: Vision, Gender and Narratives in Ancient Epic* (Cambridge University Press, 2013), esp. 310–346. See also Steven H. Lonsdale, "If Looks Could Kill: παπταίνω and the Interpenetration of Imagery and Narrative in Homer," *The Classical Journal* 84:4 (1989): 325–333.

through metaphor.[84] This is in keeping with Aristotle's own rhetorical theory in which metaphor functioned in a similar way to *ekphrasis*, whereby "things are set before the eyes by words that signify actuality" (πρὸ ὀμμάτων ταῦτα ποιεῖν, ὅσα ἐνεργοῦντα σημαίνει). These words strikingly echo the language used by Aphthonios and fellow Byzantine rhetoricians to describe *ekphrasis*'s ability to bring images "under the eyes vividly" (ὑπ' ὄψιν ἄγων ἐναργῶς).[85] Thus, what might be culled from these various intersecting concepts is that metaphor, hyperbole, *ekphrasis*, and other related rhetorical theories are grounded upon the ability of or desire for literature to overstep the bounds of its medium, to transgress its own sensual limitations through the perceptual actions of the imagination.

The critical and common operator in rhetoric's ability to transgress and remediate medial and sensory bounds is the generation of "vividness" (ἐνάργεια). While variously translated, *enargeia* comes to define a condition whereby the things depicted seem to be as if they were in actuality.[86] In other words, *enargeia* summarizes and manifests the "as if" function of *ekphrasis* in its ability to convey the discourse as if it were unfolding before the eyes of the readers or listeners. Thus, the condition of *enargeia* is fundamentally one of remediation, a term articulated by Jay David Bolter and Richard Grusin to describe "the representation of one medium in another," which its authors articulate precisely by turning to *ekphrasis*'s ability to produce literary descriptions of the visual world through words.[87] Remediation could thus be understood to be at work precisely when the imagination grasps and apprehends the sensible objects in the outside world, drawing them into the seat of consciousness where it may remediate them into noetic images composed of various sensory inputs via the common sense. Aristotle, for instance, clarifies that "the soul is like the hand" (ὥστε ἡ ψυχὴ ὥσπερ ἡ χείρ ἐστιν), which apprehends sensible objects in the world and draws them into the mind via its powers of remediation: from external material object to internal immaterial image.[88] For example, a "stone does not exist in the soul, but only the form of the

[84] Stratis Papaioannou, "Byzantine *Enargeia*," *Byzantinoslavica* 69:3 (2011): 48–60, esp. 55–57.

[85] Aristotle, *Rhetorica*, 1411b24–25, ed. and trans. Freese, 404–405; Aphthonios, *Progymnasmata*, 216.20–24, ed. Rabe, 36–37.

[86] See Ruth Webb, "Sight and Insight: Theorizing Vision, Emotion and Imagination in Ancient Rhetoric" in Michael Squire (ed.), *Sight and the Ancient Senses* (London: Routledge, 2016), 205–219.

[87] Jay David Bolter and Richard Grusin, *Remediation: Understanding New Media* (Cambridge, MA: MIT Press, 1999), 45.

[88] Aristotle, *De anima*, 432a1–2, ed. and modified trans. Hett, 180–181.

stone" (οὐ γὰρ ὁ λίθος ἐν τῇ ψυχῇ, ἀλλὰ τὸ εἶδος) lies therein.[89] Hence the act and action of perception is fundamentally one of remediation, not only in the collations of sensory data inherent in the common sense, but also in the act of transitioning from mere sensation to perception itself.

Yet critically and perhaps most significant here is the point that remediation is birthed from a double logic between immediacy and hypermediacy: the double-bind of a culture's desire "both to multiply its media and to erase all traces of mediation: ideally, it wants to erase its media in the very act of multiplying them."[90] The function of *enargeia* seeks to do just that. In attempting to multiply the forms by which *ekphrasis* communicates, from mere words to images as well, the rhetorical aim is to erase all traces of the mediation inherent in this proliferation so as to erase the traces of the communicated events being a mediated representation, rather than something unfolding actually (ἐναργῶς) before one's eyes. In a sense, the goals of Pentcheva's theory of "synesthesis" are not unlike this cruelly optimistic rhetorical theory in which the proliferation of sensory data – often communicated through nested levels of remediation, such as the comprehension that sight might also be a form of touch or a form of taste – aims at enacting a state of presence and immediacy for the icon by virtue of the proliferation of mixed media. As such, this "double logic," as Bolter and Grusin refer to it, manifests itself precisely as a double-bind, whereby hypermediation and immediacy are deployed simultaneously for the sake of exceeding the ordinary bounds of representation. Yet, as I have shown here, the Byzantine subject critically and actively subverted this double-bind and their complicity with its bad faith and cruel optimism by precisely articulating that it was in the end this interval that representation mediates. That is to say, it is the disjuncture between the remediation of various sensory forms with a desire for presence and immediacy that representation attempts, all the while understanding its impossibility. In acknowledging the "as if" function of art and rhetoric, Byzantium understood the limits of its own representation, understanding precisely that the medium of art was a condition of mediating betwixt the desire for manifestation and its inherent impossibility.

In these terms, remediation operates as the common overarching drive behind Byzantine representational theories, both verbal and visual, manifested through the operative effect of *enargeia.* The *Souda* articulates this term precisely through a synesthetic metaphor, defining it as:

[89] Aristotle, *De anima*, 431b30–432a1, ed. and trans. Hett, 180–181.
[90] Bolter and Grusin, *Remediation*, 5.

> Ἐνάργεια: ἡ τῶν λόγων λευκότης καὶ φανότης. Ἐνέργεια δὲ ἡ ἐν λόγοις, ἢ ἡ ἀθρόα προσβολή.
>
> *Enargeia*: The whiteness and brightness of words. But actualization [*energeia*] is the activity in words; or their collective impact.[91]

In order to convey the graphic operation of *enargeia* in creating a condition of vividness or graphic quality, the author has resorted to explaining this through the visual language of whiteness (λευκότης) and brightness (φανότης). Thus, in explaining the function of *enargeia* to place images before the eyes vividly, the author has defined it not by exegetically explaining what it does, but rather by representing the very synesthetic process whereby the spoken word becomes intense, gleaming light.

The *Souda*'s entry alludes to *ekphrasis*'s ability to transgress sensory bounds and generate images from words through the imagination, but as the entry explains, vividness (ἐνάργεια) is also the "activity" or "energy" (ἐνέργεια) of words or, rather, "actualization in words." These two terms, ἐνάργεια and ἐνέργεια, have long been entwined and confused in Greek scholia, which may possibly have originated as a scribal error or a confusion between the two similar words.[92] In his twelfth-century commentary on Homer's *Iliad*, Eustathios of Thessaloniki generatively deploys this confusion to state that the ancient critics had noted that Homer had used every possible source to articulate the "activity (ἐνέργειαν) of the battle or the graphic quality (ἐνάργειαν) of the narrative."[93] Hence, as René Nünlist notes, the former ends up depending on the latter, wherein "activity" (ἐνέργεια) is a quality of the specific actions, but that "visual vividness" (ἐνάργεια) a product of the narrative itself. Hence the *Souda*'s entry alludes to this division by depicting *enargeia* as the graphic quality of

[91] Anonymous, *Souda*, epsilon 1126, "Ἐνάργεια," ed. and trans. Boeri, *Souda On Line*, www.stoa.org/sol-entries/epsilon/1126.

[92] Nina Otto for one has dismissed the possibility that the two terms meant the same thing, or that the early usage of ἐνέργεια in Aristotle was simply a misappropriation of ἐνάργεια, while Alessandra Manieri has suggested a greater overlap and carefully elucidated the parallels between these two terms and their rhetorical as well as philosophical operations. See Nünlist, *The Ancient Critic at Work*, 197; Meijering, *Literary and Rhetorical*, 36–37. See also Otto, *Enargeia*, 71–76; Manieri, *L'immagine poetica*, esp. 97–104; Lucia Calboli Montefusco, "Ἐνάργεια et ἐνέργεια: l'évidence d'une démonstration qui signifie les choses en acte" in Mireille Armisen-Marchetti (ed.), *Demonstrare: voir et faire voir: forme de la démonstration à Rome* (Toulouse: Presses Universitaires du Mirail, 2005), 43–58.

[93] "Ἰστέον δὲ ὅτι παρασημειοῦνται οἱ παλαιοὶ ἐνταῦθα, ὡς πανταχόθεν ὁ ποιητὴς ἐκίνησε τὴν τῆς μάχης ἐνέργειαν ἢ τὴν τοῦ λόγου ἐνάργειαν." Eustathios of Thessaloniki, 419.28–29, *Commentarii ad Homeri Iliadem*, ed. M. van der Valk, *Commentarii ad Homeri Iliadem pertinentes*, vol. III (Leiden: Brill, 1979), 419.

words, through precisely visual language, whereas *energeia* is the actuality and action occurring in the words and their collective impact.

The use of visual language, often associated with light, to describe the effects of language is common throughout the Byzantine period and speaks precisely to that graphic quality of narrative vividness. In his homily on baptism, Gregory of Nazianzos uses similar language to convey the illuminations (φωτισμῶν) offered by Scripture. There he evocatively says:

> Πάλιν ὑπομνήσω σε τῶν φωτισμῶν, καὶ πολλάκις ἐκ τῶν θείων λογίων τούτους ἀναλεγόμενος. Αὐτός τε γὰρ ἡδίων ἔσομαι τῇ τούτων μνήμῃ, (τί γὰρ φωτὸς ἥδύτερον, τοῖς φωτὸς γευσαμένοις;) καὶ σὲ περιαστράψω τοῖς λεγομένοις.
>
> Again I will remind you of illuminations and will often gather examples of them from the divine Scriptures. For I myself will take pleasure in the memory of these things – for what is more pleasant than light to those who have tasted light? – and I will surround you with lightning flashes by my words.[94]

Here we see that the pleasure of Scripture comes from its form of vividness through illumination, understood metaphorically as enlightenment and in ekphrastic terms as the ability to manifest this enlightenment through the perception of light. Yet the light that emanates from words can here be tasted as well – and later in the homily Gregory even asks that we take delight in these "divine words sweeter than honey" (τοῖς γλυκυτέροις μέλιτος λόγοις).[95] Here, the gustatory lighting of his words wholly surrounds (περιαστράψω) the viewer/taster/listener/reader just like the sound of his words during the sermon surrounds the audience.

This collation of sight with mental processes is common in the Church Fathers and continues well beyond their works as a way of depicting the importance of visual thinking in cognition, which is often described with verbs and nouns indicating a sense of grasping to convey comprehension as an apprehension. Consider Maximos the Confessor's statement in his *Ambigua*, where he writes:

> Οὔτε γὰρ δίχα φωτὸς τῶν αἰσθητῶν ὀφθαλμὸς ἀντιλαμβάνεσθαι δύναται, οὔτε νοῦς χωρὶς γνώσεως Θεοῦ θεωρίαν δέξασθαι πνευματικήν. Ἐκεῖ τε γὰρ τῇ ὄψει τὸ φῶς τῶν ὁρατῶν τὴν ἀντίληψιν δίδωσι, καὶ ἐνταῦθα τῷ νῷ τὴν γνῶσιν τῶν νοητῶν ἡ τοῦ Θεοῦ ἐπιστήμη χαρίζεται.
>
> For as the eye cannot, without light, grasp sensible things, neither can the mind, apart from the knowledge of God, receive spiritual contemplation.

[94] Gregory of Nazianzos, *Orat.* 40.36 (*In sanctum baptisma*), ed. PG 36:411; trans. Nonna Verna Harrison, *Festal Orations* (Yonkers, NY: St. Vladimir's Seminary Press, 2008), 131.

[95] Gregory of Nazianzos, *Orat.* 40.38 (*In sanctum baptisma*), ed. PG 36:413.

> For there light gives to sight the perception of visible things, and here the vision of God grants to the mind the knowledge of things intelligible.[96]

This example draws one's attention to the interactions between perceptual processes of apprehension (ἀντιλαμβάνεσθαι) and the mind's own parallel and synchronous processes, captured through the perception of light. Such comparisons continue throughout Byzantine literature and homiletics, as we can see for example in Gregory Referendarios's homily on the arrival of the Mandylion of Edessa in Constantinople in 944. There he writes:

> Just as it is impossible, when the sun has set, for the pupils of the eyes to be lit up if they are not struck by the sun's rays, so our inner spiritual eye of the soul (ἔνδον... νοητὸν ὀφθαλμὸν τῆς ψυχῆς) cannot see what is right without the brightness of the divine scriptures.[97]

Here, Gregory Referendarios is careful to not conflate visual seeing with spiritual seeing, making the common metaphor, while nevertheless endeavoring to be clear and careful in the extent of his homily's vividness on the conflation of sight and scriptural enlightenment – clarifying that he refers here to the "inner noetic eye of the soul" (ἔνδον... νοητὸν ὀφθαλμὸν τῆς ψυχῆς), not the actual physical one.

Reading across Gregory of Nazianzos, Maximos the Confessor, and Gregory Referendarios we can again see this process whereby the physiology of sensation and the psychology of perception are paralleled with the knowledge communicated therein, and this process comes to be a metaphor for knowledge itself. In other words, here, the perception achieved through sight is understood as occurring with light, and thus given the understanding that discourse's goal is to place things before the eye, the metaphor of this light that enables vision is used to metonymically capture the process of achieving knowledge itself. Yet Gregory of Nazianzos pushes this beyond the basic comparison, so as to stress that the intensity of Scripture is such that we do not merely experience the simple light of bodily sensation, but rather the thunderous, cross-sensory light of ecstatic perception – a process that we can resolutely understand as occurring not on the level of sensation, but during perception in the mind.

96 Maximos the Confessor, *Ambigua ad Joannem*, 10, ed. PG 91:1156B; trans. Andrew Louth, *Maximus the Confessor* (London: Routledge, 1996), 126.

97 "Ὡς γὰρ ἡλίου ὑπὸ γῆν ὄντος, ἀδῦνατον τὰς κόρας τῶν ὀμμάτων ταῖς ἡλιακαῖς ἀκτῖσι μὴ βαλλομένας φωτίζεσθαι, οὕτως οὐδε τὸν ἔνδον ἡμῖν νοητὸν ὀφθαλμὸν τῆς ψυχῆς δύνασθαι τὸ συμφέρον ὁρᾶν ἄνευ τῆς τῶν θείων φωταυγείας γραφῶν." Gregory Referendarios, *Sermon*, ed. and trans. Guscin, 84–85.

We can witness here the complexity, nuance, agility, and precision with which these writers are purposely mixing and transgressing sensual bounds, which reflect the necessary acts of remediation inherent within discourse itself. In this sense then, these authors are not suggesting the belief in the transgression of such senses, but rather deploying language's performative overstepping of the bounds of representation. Yet what is most interesting across these examples is how such inter-sensual metaphors gain momentum and develop layers of intricacy. These authors are clearly building upon common metaphors and always trying to push those even further beyond their limits: hence, we go from a comparison between knowledge and sight to a comparison between knowledge and light to a comparison between knowledge and lightning. This tactic is one that ensures the vividness of *enargeia*, and one not limited to rhetoric alone but also to the image.

If rhetoric was to place images before the eyes, the image was to speak eloquently from the wall. This meant that sight might also speak: for example, John Chrysostom elsewhere writes, "you have spoken them more clearly with your eyes,"[98] and John of Damascus explains at length:

> Τί τούτων τηλαυγέστερον πρὸς ἀπόδειξιν, ὅτι βίβλοι τοῖς ἀγραμμάτοις εἰσὶν αἱ εἰκόνες καὶ τῆς τῶν ἁγίων τιμῆς ἀσίγητοι κήρυκες ἐν ἀήχῳ φωνῇ τοὺς ὁρῶντας διδάσκουσαι καὶ τὴν ὅρασιν ἁγιάζουσαι;
>
> What could demonstrate more clearly than these passages that images are books for the illiterate and silent heralds of the honor of the saints, teaching those who see with a soundless voice and sanctifying the sight?[99]

In both these instances, we see conflations of sight and sound which, just like the conflation of sexual intercourse with sight, we would not hasten to take as an indicator that the Byzantines saw haptically, sexually, or sonically. Yet, here, John of Damascus is self-reflexively weaving sound and sight as a justification for the image as a remediation of the Gospel text. There is a telling parallelism between "images are… silent heralds" (αἱ εἰκόνες… ἀσίγητοι κήρυκες) and "those who see with soundless voice" (ἐν ἀήχῳ φωνῇ τοὺς ὁρῶντας), which structures a conflation of sight and hearing, image and sound.

Beyond perception, however, this was an admirable feature of the image as well. As Theodore Prodromos writes of the art of the painter Eulalios (Εὐλάλιος), whose name meant eloquent (εὔλαλος):

[98] "ἐκάλεσας φωνῆς λαμπρότερον διὰ τῶν ὀφθαλμῶν." John Chrysostom, *Quod regulares feminae viris cohabitare non debeant*, 1.63–64, ed. Dumortier, 97.

[99] John of Damascus, *Orationes de imaginibus tres*, I.47 (cf. II.43), ed. Kotter, 151; trans. Louth, 46.

τοσοῦτος Εὐλάλιος τὴν τέχνην, ὡς ἱστορεῖν πως τὰς γραφὰς φερωνύμως, λαλοῦντά τε χρώματα συγκεραννύειν. Οὐ τῆς γραφικῆς ἐστὶ τὸ χρῆμα, ξένε, ἀλλ' ἡ περιλάλητος ἀνθρώποις κόρη τὴν Εὐλαλικὴν ἰθύνασα γραφίδα, εὔλαλον οὕτω καὶ τὸ χρῶμα δεικνύει.

Such is the art of Eulalios that he can make paintings worthy of his name, and mix colors that can talk. Yet, stranger, this has not come about through the art of painting, but the maiden who is talked about among men has guided the brush of Eulalios, and made his color so eloquent.[100]

This notion that art might be able to speak is pervasive throughout texts on *ekphrasis*, homilies, and image theory, understanding that images operate as a remediation of words. For instance, Gregory of Nyssa described the depiction of the martyr's deeds in his homily on St. Theodore "as if it were a book that uttered speech" (ὡς ἐν βιβλίῳ τινὶ γλωττοφόρῳ), going on to state that "painting, even if it is silent, is capable of speaking from the wall" (οἶδε γὰρ καὶ γραφὴ σιωπῶσα ἐν τοίχῳ λαλεῖν).[101] It is the ability of painting to seem as if it could speak that attested to its vividness, a concept often expressed by the term "inspirited" (ἔμψυχος), or related terms that spoke to its perceived animate and life-like nature.[102] Thus, if the technique of *ekphrasis*' goal is to bring images below the eyes, it is the goal of images to bring voices to the ear.

In praising the life-like nature of the Theotokos in the apse of Hagia Sophia in his Homily 17, Photios precisely expresses this complicated sentiment through the double-negative: "You might say she is not incapable of speaking," if you pressed her with a question.[103] For, as Photios states, "to such an extent have the lips been made flesh by the colors, that they appear merely to be pressed together."[104] While the iconoclasts

[100] Theodore Prodromos, *Poems*, "Εἰς τὸν αὐτὸν," ed. E. Miller, "Poésies inédites de Théodore Prodrome," *Annuaire de l'Association pour l'encouragement des études grecques* 17 (1883): 16–64 at 33; trans. Cyril Mango, *Art of the Byzantine Empire*, 231.

[101] Gregory of Nyssa, *De sancto Theodoro*, ed. PG 46:740A; trans. Mango, *Art of the Byzantine Empire*, 36–37.

[102] On the concept of inspirited painting, see Hans Belting, *Likeness and Presence: A History of the Image Before the Era of Art*, trans. Edmund Jephcott (University of Chicago Press, 1994), 261–296; Robin Cormack, "Living Painting" in Elizabeth Jeffreys (ed.), *Rhetoric in Byzantium* (Burlington, VT: Ashgate, 2003), 235–253; Charles Barber, "Living Painting, or the Limits of Pointing? Glancing at Icons with Michael Psellos" in C. Barber and D. Jenkins (eds.), *Reading Michael Psellos* (Leiden: Brill, 2006), 117–130; Stratis Papaioannou, "Animated Statues: Aesthetics and Movement" in *Reading Michael Psellos*, 95–116; Pentcheva, *The Sensual Icon*, 191–198; Glenn Peers, "Real Living Painting: Quasi-Objects and Dividuation in the Byzantine World," *Religion and the Arts* 16 (2012): 433–460.

[103] "εἴποις ἂν αὐτὴν (μηδ' εἴ τις ἐπερωτῴη, πῶς δὲ παρθενεύεις καὶ τέτοκας;) μηδ' ἂν τὸ φθέγξασθαι παραιτήσασθαι." Photios, *Homiliae*, 17, ed. Laourdas, 167.

[104] "Οὕτω διεσαρκώθη τὰ χείλη τοῖς χρώμασιν. <ὥστε> καὶ συνεπτύχθαι μόνον." Photios, *Homiliae*, 17, ed. Laourdas, 167.

understood the depiction of Christ and saints in silent form as something to be anathematized given the prohibition of idols by Scripture, the iconophiles understood this silence to be an ontological requisite of the icon so as to produce an image that stressed that while it was the same in signification, it was nevertheless different in essence, and hence did not take upon itself any idolatrous animacy. Yet, for Photios, while we understand that the Virgin cannot speak, the image nevertheless exists in a state where you "might say" (εἴποις) that she could – or, literally, not that she cannot. This double negative then critically helps to work against remediation's resultant double-bind, whereby immediacy is contrasted against hypermediacy, where presence and mediation are naively said to be able to function as one.

Yet this process of remediation is inherent in the foundations of Byzantine image theory, which saw images as resulting from words, words which in turn resulted from observation. As Theodore of Stoudios argues:

> Ὄψις ἀκοῆς προτέρα καὶ τῇ θέσει τοῦ τόπου, καὶ τῇ αἰσθητικῇ ἀντιλήψει· Πρότερον γάρ τις ἰδών τι, ἀκοῇ τὸ ὁραθὲν διεπόρθμευσεν. Οἷον φέρε εἰπεῖν, εἶδεν Ἡσαΐας τὸν Κύριον καθήμενον ἐπὶ θρόνου δόξης, καὶ τοῦτον ὑπὸ τῶν ἑξαπτερύγων Σεραφὶμ κυκλούμενον καὶ αἰνούμενον. Καὶ πάλιν Ἰεζεκιὴλ τὸ ὄχημα τοῦ Θεοῦ, τὰ Χερουβὶμ ἀντενίζει. Ἀλλὰ γὰρ καὶ οἱ θεσπέσιοι μαθηταί, πρότερον ἰδόντες τὸν Κύριον, δευτερεύοντα τὸν λόγον διεχάραξαν. Εἰ τοίνυν ταῦτα τοῦτον ἔχει τὸν τρόπον, καί, ὅσον ἂν ἀναποδίσῃς, εὕροις τὴν λογογραφίαν εἰς ἱστορίαν καταλήγουσαν, ἀνάγκη πᾶσα ἀναιρουμένης τῆς ἱστορίας Χριστοῦ, προαναιρεῖσθαι τὴν λογογραφίαν αὐτοῦ. Καὶ χαραττομένης τῆς δευτέρας προχαράττεσθαι τὴν προτέραν.
>
> Sight precedes hearing both in the location of its organs and in the perception by the senses. For one first sees something and then transmits the sight to the sense of hearing. For example, Isaiah saw the Lord sitting on the throne of glory, surrounded and praised by the six-winged seraphim. Likewise Ezekiel gazed upon the chariot of God, the cherubim. Moreover, even the divine-voiced disciples first saw the Lord and later wrote out the message. If, therefore, this is how it is, and however far you go back, you would find the written word originating in observation, then it is undoubtedly necessary that if the sight of Christ is removed, the written word about Him must be removed first; and if the second is sketched out, the first must be sketched beforehand.[105]

Here, Theodore begins by sketching out the foundations of perception, clear to elucidate the telling distinction between physical sensation and

[105] Theodore of Stoudios, *Antirrhetici adversus iconomachos*, III.2, ed. PG 99:392A; trans. Roth, 78.

psychological perception, its bodily location (τῇ θέσει τοῦ τόπου) versus its perceptual apprehension (τῇ αἰσθητικῇ ἀντιλήψει). Then, considering the origin of Scripture in the observation of the wondrous deeds, Theodore argues that to deny the image is to deny the word first, given that the image is a remediation of the word, which is in itself an inscriptive remediation of the deeds and actions of God as observed and then written down by humanity. Thus we can appreciate that for Theodore, then, the image pulsates between image and text, sight and hearing, because its ontological foundation lies precisely in representation's ability to perpetually mediate and remediate realities that are not present to the senses – whether this be past events or the heavenly divinity.

The roots of Byzantine representation, which as I have shown here share a common history and theoretical framework with rhetoric, thus enabled a manifestation of the divine that could be perpetually remediated not only in the literal depiction of deeds in words and images vividly, but also in all forms of sensual experience in the world. As when John of Damascus writes:

> Ὁρῶμεν γὰρ εἰκόνας ἐν τοῖς κτίσμασι μηνυούσας ἡμῖν ἀμυδρῶς τὰς θείας ἐμφάσεις· ὡς ὅτε λέγομεν τὴν ἁγίαν τριάδα, τὴν ὑπεράρχιον, εἰκονίζεσθαι δι' ἡλίου καὶ φωτὸς καὶ ἀκτῖνος· ἢ πηγῆς ἀναβλυζούσης καὶ πηγαζομένου νάματος καὶ προχοῆς· ἢ νοῦ καὶ λόγου καὶ πνεύματος τοῦ καθ' ἡμᾶς· ἢ ῥόδου φυτοῦ καὶ ἄνθους καὶ εὐωδίας.
>
> For we see images in created things intimating to us dimly reflections of the divine; as when we say that there is an image of the holy Trinity, which is beyond any beginning, in the sun, its light and its ray, or in a fountain welling up and the stream flowing out and the flood, or in our intellect and reason and spirit, or a rose, its flower and its fragrance.[106]

For John, as with earlier thinkers such as Plotinos and Pseudo-Dionysios the Areopagite, the world is filled with images of the divine throughout the realms of the different senses, hence the Holy Trinity may be perceived in the sun's rays by sight, the sonic and tactile flood of a gushing stream, the noetic images in the mind that constitute cognition, the fragrance of a flower, and so on. The image of the divine can be perpetually remediated across sensory bounds, given that this perpetual act of transgression is what is inherently at stake in the theory of representation.

The act of perception is a process of remediating sense perception into an image adequate to the functions and needs of the mind, where the

[106] John of Damascus, *Orationes de imaginibus tres*, 1.11, ed. Kotter, 85; trans. Louth, 26.

common sense collates various sensory inputs to create an image that the imagination can think with and that memory can impress upon its tablets. Classical and Byzantine rhetorical theories responded to these needs by producing a theory of representation whose goal was precisely to deploy the imagination so as to vividly make manifest the narratives recounted, using the images stored within the mind for the sake of rendering a visual narrative that unfolded before the mind's eye with the hopes that engaged listeners, as Michael Psellos once put it, might "think that what was being said they saw as if it were happening" (ᾤοντο γὰρ τὰ λεγόμενα ὁρᾶν ὡς γινόμενα).[107] Inter-sensory metaphors and other similar rhetorical forms of remediation in the texts surveyed here, however, suggest that beyond mere vision of the stories told, other sensory perceptions could be conjured up by the imagination and that other sensory inputs beyond hearing and reading could also convey narratives and images. For image and discourse are themselves constantly caught in chains of remediation between the observable (and, at times, divinely ineffable) world that gave root to their representations, and to their various permutations and remediations afterwards. Thus, the theory of the icon's representation comes to evidence a basic epistemic commonality with the system of perception laid out by classical rhetorical theories of representation, which grappled with the power of vividness and actuality, while always acknowledging that full, unmediated presence could never be achieved, only its momentary illusion in the imagination. In allowing perception to constantly remediate sensation across sensory and medial bounds, discourse and the icon alike strove for an illusion of immediacy through a form of hypermediacy. Yet Byzantine thinkers avoided the potential double-bind and cruel optimism of this logic by always acknowledging the "as if" function of remediation, where all representation seems to operate as a form of metaphor, rather than as a sincere claim to presence or immediacy. Thus Byzantine representation is inherently rooted on a form of failure of representation: a praise for the illusions of the imagination, which always made vividly manifest that which was fundamentally not there.

[107] Michael Psellos, *De Euripide et Georgio Piside judicium*, ed. Dyck, 40–50. On this passage, see Papaioannou, *Michael Psellos*, 116–117.

Conclusion: Tempted to Touch

How and why the historiographic predilection for haptic sight developed in art historical writing (roughly, since the mid 1990s to the present) is a problem that has both a historical and historiographic answer. Building on the influential arguments for the image in the "era before art," promoted by Hans Belting, as one bereft of mediation in exchange for presence, haptic sight strove to instill in the Byzantine viewer a striking alterity to our own modern understanding of optics and the image alike.[1] The conceit here is that neither the Byzantine history of vision nor its modern historiography can be considered in isolation for they are one and the same, since the articulation of the Byzantines' visuality has always been construed in accordance with the ontology ascribed to their images and the assumed agency of their viewers.

When Robert Nelson favored haptic extramission as the predominant visuality of the Byzantine world, he was not merely favoring a *possibility* afforded by the archive, but also favoring a visuality appropriate to contemporary ontologies of the image at the time of his essay's publication in 2000.[2] In an attempt to distance art historical discourse from patterns of artistic intention and focus on reception theory, Nelson's visuality of the Byzantine icon enabled a tacit suppression, collapse, and erasure of the medium in keeping with current historiographic trends: including, but not limited to the rise of visual studies in its various cultural and physiological permutations and the broader moves from epistemology in the 1980s and

[1] Belting, *Likeness and Presence*.

[2] See, for example, the focus on touching given in an essay by James Hall and others in a volume devoted to the question of presence, which emerged from a conference of the Association of Art Historians held at the University of Edinburgh in 2000: James Hall, "Desire and Disgust: Touching Artworks from 1500 to 1800" in Rupert Shepherd and Robert Maniura (eds.), *Presence: The Inherence of the Prototype within Images and Other Objects* (Burlington, VT: Ashgate, 2006), 145–160.

toward phenomenology in the early 1990s. The production and enforcement of a haptic theory of extramission, however, also enabled a broader philosophical synthesis between the utility of the cult image and the gaze of the viewer, uniting into one force both the era of art's spectatorship and the cult image's numinous presence.

In 1999, for example, the historian of modern art, Rosalind Krauss published her landmark *"A Voyage on the North Sea": Art in the Age of the Post-Medium Condition*, a text that attempted to rehabilitate the faltering notion of medium-specificity in art history. In this volume, Krauss worked against the mid-century critic Clement Greenberg's idea that modern art thrived on tactics of medium-specificity and medium-reflexivity, which stressed the flatness of the painted-picture plane and the opticality alone of the visual arts.[3] Krauss moves beyond the Greenbergian reduction of medium to the mere physical, material support of the work of art (e.g. paint on canvas) in order to consider the concept of a medium as a condition of possibility: that is to say, the system of rules, guidelines, and training that enables the artistic production of form to become perceptible as such. As Krauss suggests, since medieval guilds the division of the arts enabled specific mediums to emerge through the specialization of skills rather than a material support itself, even if at times such skill-sets coincided or clustered around similar materials.

Yet Krauss expands this definition further to consider how the articulation of medium has been expanded to encompass a variety of different structures and concepts that have enabled individual artists to pursue and produce their own artistic projects. For example, Krauss convincingly argues that for Ed Ruscha the medium is the automobile. In his *Every Building on the Sunset Strip* (1966), the automobile operates as the generative mechanism, the condition of possibility that enables the work of art to emerge. As such, Ruscha's medium is not the camera or the artist's book, which would be traditionally identified as the material mediums of this project, but rather the underlying conceptual system that makes it all possible: the car which drove him down the Sunset Strip and delineated the project's unfolding. Thus, by reformulating the medium as an epistemological field for the generation of art, Krauss urged art historians to consider what other strategies and systems of artistic production and reception could be said to enable the articulation of the artwork. This coincided in medieval art history, however, with an inverse trajectory as an

[3] See Caroline Jones, *Eyesight Alone: Clement Greenberg's Modernism and the Bureaucratization of the Senses* (University of Chicago Press, 2006).

opposition to the "era of art" via Hans Belting took hold in the early to mid 1990s. In this context, the icon came to mirror desires for the startling alterity of the premodern image as the bearer of cultish, unmediated presence. While Krauss and Belting follow different trajectories, both understood that what was at stake in the artwork at the end of the millennium was the issue of its medium and said medium's mediation. Therefore, one can argue that the privileging of extramission in the medieval field at large emerged primarily from a historiographic desire to see the premodern, Byzantine image as a site of full and unmediated presence in the wake of Belting's influential argument, where the unmediated contact with the image's signified – or, rather in this context, the referent itself – was reified into the medieval image's visuality.

The pairing of tactility and sight is precisely a condition of perpetual perceptual temptation. This alludes to the ritual processes of body and mind that are used to bridge that mediating, intervallic gap between representation and that which is represented: that is, between signifier, signified, and referent. Thus, the apparent tactility of Byzantine visuality emerges paradoxically from that *necessary* inability to touch with sight, which is met in the faithful with the insuppressible desire to do just that. We are continuously tempted by all those tactile qualities that objects offer up to our eyes, but which sight itself simultaneously denies us from partaking in. To suggest then that sight is a species of touch is thus to say that sight could perhaps supplant touch, or make it wholly unnecessary. As Liz James said, with the interchange and conflation of touch and sight: "Touching an icon becomes a form of seeing and vice versa."[4] Though this is certainly not the point of Liz James's statement, nor of Nelson's argument, it is worthwhile considering an important implication here: when seeing becomes a form of touching, this might suggest that Byzantines were or, at least, *could be* satisfied with eyesight alone since it already encompassed touch. In this case, the Byzantines would have had to approach images just like modern viewers in a museum for which the gaze alone must satisfy all the requirements that a painted panel can offer as "Art," touching being explicitly prohibited except in extraordinary circumstances that precisely seek to break this prohibition. This, however, was certainly not the case in Byzantium where objects demanded to be touched, precisely through their tantalizingly sensual effects, a process which undoubtedly often began with sight tempting one to touch.

[4] James, "Senses and Sensibility," 528.

In 1989, Robert Nelson himself passionately argued against an unmediated visuality for the Byzantine world in his article, "The Discourse of Icons, Then and Now." Observing the taxidermy of the Dumbarton Oaks Psalter and New Testament in its museum case, bereft of the opportunity to be touched, its pages turned, its words recited, and its images kissed, Nelson eloquently wrote:

> The lucite, the special lighting, the museum guards standing by and even Dumbarton Oaks' highly polished floors further signify aesthetic importance, as well as high monetary value in our culture. Second, this is a book in name only, for it cannot be read. The viewer cannot take it in his or her hands and turn the pages, and moreover most of its viewers do not know Greek. . . The book has become an aesthetic object to be valued in and for itself, not as a means of transmitting divine messages. In sum, the Byzantine manuscript has much in common with a butterfly on display in a natural history museum. Both have had their wings forced open and pinned down. Both have been labeled, which is to say classified, according to the intellectual structures of their keepers. And finally both are far from their original context. The butterfly is dead, of course, but so is the manuscript. . .[5]

In this passage, we can observe the epistemological foundations that over the next decade led to the development of the haptic gaze of Nelson's Byzantine viewer. We may see this gaze as a strategy of emancipation for the Byzantine objects subjected to the harsh and confining rays of the modern museum, which no matter how much one might be tempted to touch, needing that object so much, one can do nothing more than gaze upon its surface. It is in this frustration with the modernist gaze that I argue the haptic theory of vision for Byzantine art emerges given that it seeks to allow for sight to partake in that touch, which was so intrinsic and necessary to the Byzantine user's rites of veneration. By saying that sight was haptic, Nelson subverted the lithe gaze that we as museum goers are limited to enjoying, suggesting that since Byzantines believed that they could touch with their gaze, we too could allow ourselves to suspend our disbelief and imagine that we too could grasp these objects and kiss them with our adoring eyes. At least this way, these objects would not be so lonely in such a radically different world, where they are bereft of their own power and action.

In considering the "sensual affinities between sight and touch" within late antique practices of pilgrimage and veneration, Georgia Frank

[5] Robert S. Nelson, "The Discourse of Icons, Then and Now," *Art History* 12:2 (1989): 144–157, esp. 145.

eloquently depicts for us the evocative power that haptic sight has in bridging the gap between worshipers and all those objects that were on display to the senses:

> When pilgrims reach across the crowded bodies at the holy places to touch a sacred stone, column, or fragment of the True Cross, that gesture did not subvert or even invert the sensorium. Nor did that outstretched hand mime a dissatisfaction with vision's power. Instead, the pilgrims extended a hand that embodied the visual ray of optics. To see and touch, then, were not exclusive activities but rather convergent senses. That sight and touch were so polarized in philosophical and theological hierarchies of the senses did not prevent pilgrims from using both senses in concert at the holy places. Sight was not replaced by touch; it had always been a form of touch.[6]

While I differ from Frank's passing suggestion that sight "had always been a form of touch," she makes a beautifully nuanced point regarding the paradoxical division and unison of sight and touch. Rightly, Frank acknowledges that philosophy and theology distanced the two senses from one another, not only in their own unique characteristics, but also in their placement along the sensory hierarchy. But most importantly, Frank's statement here is poignant because it reminds us that by virtue of their operation in unison, particularly in veneration, one could say that they were one and the same in terms of their simultaneity. Sight and touch, as we have seen, worked together intimately and often simultaneously. These "convergent senses" therefore worked with and alongside one another for the full experience of sensible objects. Like the story of Doubting Thomas, while word of mouth leads one to become an eye witness, the eye witness desires to confirm what they have seen through touch.

Nevertheless, I would argue that the lucite barrier of the Dumbarton Oaks Museum case, separating the viewer and the object, is perhaps the most accurate metaphor for Byzantine visuality. This is not because one might boldly reach out to the object with one's visual rays and graze its surface, but because this impenetrable transparent medium enables sight in its illumination, yet also makes us painfully conscious of the fact that sight alone cannot partake in touch. Just as in the Byzantine world, then, the viewer desires to touch the supple pages of the holy object, wishing to break the bounds between themselves and the object, the earthly and the divine, so as to reach out and partake in an unmediated experience with that object or person which they so admire. Yet they are always confronted by the fact that sight is not touch; that something is lacking in this visual

[6] Frank, *Memory of the Eyes*, 132.

experience, which just contributes to fueling the insatiable desire to grasp that which is divine and out of reach with the grasping of the hands and of the mind as well. In museums, our lucite barriers are a manifestation of our temptation to touch, and seek to curtail just that by presenting us a mediated view of our art objects on display so that we do not succumb to tactility. This desire, sparked by the failure to present the work of art in an unmediated fashion before all our senses, precisely represents the theoretical and theological construction of the icon's epistemology for us as modern viewers. The lucite case serves for us as a metaphorical analogue to the imperceptibility of the divine always inherent in the icon, even when touching, kissing, and seeing were all possible at once.

This concern that sight inevitably leads one toward a temptation to touch is not foreign to the Byzantine viewer. In John Chrysostom's Homily 16 on Genesis 2:25, John recounts the taunting of the serpent, who speaks to humanity's temptation to touch given their sight of what lies at hand. As John writes:

> τὶ τὸ ὄφελος τῆς ἐν τῷ παραδείσῳ διαγωγῆς, ὅταν μὴ ἐξῇ τῶν ἐν αὐτῷ ἀπολαύειν, ἀλλὰ ταύτῃ μάλιστα πλείονα τὴν ὀδύνην ἔχειν, ὅταν θέα μὲν ᾖ, ἀπόλαυσις δὲ ἡ ἀπὸ τῆς μεταλήψεως μὴ γίνηται;
>
> What is the advantage of life in the garden when you are not free to enjoy the things in it, but are even worse off in incurring the more intense pain of having sight of things but missing out on the enjoyment that comes from partaking of them?[7]

While we might continue to compile a florilegium of further conflations of sensory data and negations of the haptic quality of sight, John Chrysostom's devilish words alert us to the bristling tension and desire that emerges in light of sight. There is a reason our museums are replete with placards and guards chastising those who might graze the object's surface with anything more than their optical rays. This makes sight the perfect sensory medium for contemplating not only acts of salvation through veneration, but also the damnation that comes from the unbridled lust to exceed sight and actually touch with the hands as well. The serpent's sentiments, depicted here by John, tacitly yet eloquently argue for the epistemic separation of touch and sight. For if sight were touch the expulsion from Paradise would have never happened. Eve's painful desire would have been satisfied by haptic eyesight alone. Yet, likewise, this statement also alerts us to the inevitable desire that sight stirs up in us,

[7] John Chrysostom, *In Genesim*, 16, ed. PG 53:127C.

precisely urging us to touch, whether in the Byzantine church or the modern museum. Sight tempts our gaze with so many of those tactile qualities that an object possesses, all those sensibles common to both sight and touch, but which touch alone can confirm. Since vision can only see tactile factures, but cannot also feel them, it taunts us, presenting us with all that is present to our eyes, but inaccessible to our touch: just as the icon taunts us with all that is partially perceptible to our eyes, but inaccessible to our earthly hands. The sight of the icon might compel us to touch the holy figure's face and flesh, but in doing so we are confronted with the lifeless skin of painted or sculpted matter, only stoking the flames of desire for that person. It is only in our longing and desire, which is given tentative gratification in the imagination and always left unsatisfied in the flesh.

Not unrelated is the process of the common senses: just as the sight of fire stirs up the touch of its heat and the threat of its pain, and impels us precisely not to touch it. Just as the sight of honey can lead us to experience the sensory imprints of its taste, smell, touch, and sound, those imprints can likewise make us long for honey's sweetness to come into our mouth. In knowing that one sense in the world might inevitably lead to another, one knows honey and fire not only by their sight, but also by virtue of all these other sensory characteristics: a fire that is not hot proves itself to not be fire, but only something that looks like fire. And, as John Philoponos noted, "a person who has not tasted Colophonian resin, but only seen it, may say it is honey because it resembles it in color."[8] This means that we might be in for a shock to discover that the pot of what we believed to be honey is resin when we taste it. In the same way, what one might think to be a Christian icon might actually prove to be a pagan idol.

This tension was vividly captured when Gregory Melissenos, then the emperor's confessor, landed in Venice in 1438 and made his way to Ferrara as part of a Constantinopolitan delegation. When the Patriarch of Constantinople, Joseph II, asked the Pope for the use of a church in order to celebrate the Easter liturgy, Melissenos spoke out in criticism, saying:

> Ἐγὼ ὅτε εἰς ναὸν εἰσέλθω Λατίνων, οὐ προσκυνῶ τινα τῶν ἐκεῖσε ἁγίων, ἐπεὶ οὐδὲ γνωρίζω τινά· τὸν Χριστὸν ἴσως μόνον γνωρίζω, ἀλλ᾽ οὐδ᾽ ἐκεῖνον προσκυνῶ· διότι οὐκ οἶδα πῶς ἐπιγράφεται, ἀλλὰ ποιῶ τὸν σταυρόν μου καὶ προσκυνῶ. Τὸν σταυρὸν οὖν, ὃν αὐτὸς ποιῶ, προσκυνῶ καὶ οὐχ ἕτερόν τι τῶν ἐκεῖσε θεωρουμένων μοι.

[8] "ἀμέλει τοι ἐντεῦθεν καὶ ἀπάτη γίνεται, καὶ τῆς Κολοφωνίας ὁ μὴ γευσάμενος, ἀλλὰ μόνον αὐτὴν θεασάμενος εἴποι μέλι εἶναι διὰ τὸ προσεοικέναι τῷ χρώματι." Philoponos, *In Aristotelis libros de anima commentaria*, ed. Hayduck, 455.30–32; trans. Charlton, 31.

> When I enter a Latin church, I do not revere any of the images of saints that are there, because I do not recognize any of them. At the most, I may recognize Christ, but I do not revere Him either, since I do not know in what terms he is inscribed. So I make the sign of the cross and I revere this sign that I have made myself and not anything that I see here.[9]

This protest, as Robert Nelson has shown, demonstrates the extent to which a Latin church could be deemed wholly unsuitable for a Greek worshiper, who saw (in part) the lack of inscriptions stating the names of the depicted holy figures: This was seen as being in opposition to the tenets of the Acts of the Seventh Ecumenical Council; and in the Latin church, the *Libri Carolini* had explicitly rejected the decisive role ascribed to such inscriptions in the East.[10] To not know "how it was inscribed" (πῶς ἐπιγράφεται), speaks to certain hesitation around images that may be different to what they outwardly would seem. With no sensory recourse to confirmation, Melissenos merely performs the sign of the cross. Here his touch does not confirm the outward appearance of the object of sight, but merely forges its own image in the sign of the cross, which he then tells us he goes on to revere.

In order to assert the reality of an object or its adherence to our imagined expectations, we confirm that reality through our other senses before assenting to the visual presentation of its form alone. Melissenos's comments are a poignant reminder of one's reticence to assent, and the challenge that art poses. This is a challenge with a longstanding history in ancient thinking about the arts, such as in the competition between Parrhasius and Zeuxis recounted in Pliny the Elder's *Natural History*, wherein Zeuxis exhibited an image of "some grapes so true to nature that birds flew up to the wall of the stage" (*uvas pictas tanto successu, ut in scaenam aves advolarent*).[11] This was defeated by Parrhasius's painting of a curtain, which fooled Zeuxis, and extolled Parrhasius's skill given that whereas Zeuxis had merely deceived the birds, Parrhasius had deceived a painter.[12] Similar tropes occur throughout ancient *ekphrases*, such as in

[9] Sylvester Syropoulos, *Historiae*, 4.46, ed. V. Laurent, *Les «Mémoires» du Grand Ecclésiarque de l'Église de Constantinople Sylvestre Syropoulos sur le concile de Florence (1438–1439)* (Paris: Centre National de la Recherche Scientifique, 1971), 250; trans. Cyril Mango, *Art of the Byzantine Empire, 312–1453* (Englewood Cliffs, NJ: Prentice Hall, 1972), 254.

[10] Robert S. Nelson, "Image and Inscription: Pleas for Salvation in Spaces of Devotion" in Liz James (ed.), *Art and Text in Byzantine Culture* (Cambridge University Press, 2007), 100–119, esp. 101–103.

[11] Pliny the Elder, *Historia naturalis*, 35.36.65, ed. and trans. H. Rackham, *Natural History*, vol. x, Loeb Classical Library 394 (Cambridge, MA: Harvard University Press, 1952), ix:308–309.

[12] Pliny the Elder, *Historia naturalis*, 35.36.66, ed. and trans. Rackham, ix:310–311.

Philostratus the Elder's *Imagines*, where it is uncertain "whether a real bee has been deceived by the painted flowers or whether we are to be deceived into thinking that a painted bee is real, I do not know" (οἷς καὶ μέλιττα ἐφιζάνει τις, οὐκ οἶδα εἴτ' ἐξαπατηθεῖσα ὑπὸ τῆς γραφῆς, εἴτε ἡμᾶς ἐξηπατῆσθαι χρὴ εἶναι αὐτήν).[13] These events generate instances of viewing where viewers and animals are left uncertain as to whether the art is artifice or reality.

In the Stoic's struggle with the assent to cognitive presentations of *phantasia*, Diogenes Laertius offers the story about the philosopher Sphaerus, who is deceived into thinking that wax pomegranates are the real thing. In this account, we can appreciate the handling of this uncertainty by the philosopher's response. As the story goes:

> βουλόμενος ὁ βασιλεὺς ἐλέγξαι αὐτόν, κηρίνας ῥόας ἐκέλευσε παρατεθῆναι τοῦ δὲ Σφαίρου ἀπατηθέντος ἀνεβόησεν ὁ βασιλεὺς ψευδεῖ συγκατατεθεῖσθαι αὐτὸν φαντασίᾳ. πρὸς ὃν ὁ Σφαῖρος εὐστόχως ἀπεκρίνατο, εἰπὼν οὕτως συγκατατεθεῖσθαι, οὐχ ὅτι ῥόαι εἰσίν, ἀλλ' ὅτι εὔλογόν ἐστι ῥόας αὐτὰς εἶναι διαφέρειν δὲ τὴν καταληπτικὴν φαντασίαν τοῦ εὐλόγου.
>
> The king, wishing to refute him, ordered some waxen pomegranates to be put on the table. Sphaerus was taken in and the king cried out, "You have given your assent to a presentation which is false." But Sphaerus was ready with a neat answer. "I assented not to the proposition that they are pomegranates, but to another, that there are good grounds for thinking them to be pomegranates. Certainty of presentation and reasonable probability are two totally different things."[14]

The king ridicules the philosopher for having given assent (συγκατατεθεῖσθαι) to a "false presentation" (ψευδεῖ... φαντασίᾳ), but the philosopher responds that he did not assent to their veracity as pomegranates, but to their verisimilitude that presented good grounds to believe them to be pomegranates. This, as Karl Thein has elucidated, demonstrates a difference in Stoic thought between absolute certainty and what is plausible or probable.[15]

The challenge of such false presentations is aptly addressed in Nemesios of Emesa's *De natura hominis* when he struggles precisely with this similar challenge of false objects, intended to deceive. There, Nemesios writes:

> εἰ γὰρ οὐχ ὑπὸ μόνου τοῦ χρώματος καὶ σχήματος εἰδοποιεῖται τὸ μῆλον, ἀλλὰ καὶ τῆς ὀσμῆς καὶ τῆς κατὰ τὴν γεῦσιν ποιότητος, οὐχὶ καὶ τούτων ἡ

[13] Philostratus the Elder, *Imagines*, 1.23.2, ed. and trans. Arthur Fairbanks, *Imagines*, Loeb Classical Library 256 (Cambridge, MA: Harvard University Press, 1931), 88–91.

[14] Diogenes Laertius, *Vitae philosophorum*, 7.6.177, ed. and trans. Hicks, 284–285.

[15] Thein, "Gods and Painters," 136–145.

ὄψις ἀντιλαμβανομένη οἶδεν ὅτι μῆλόν ἐστιν, ἀλλ' ἡ ψυχὴ τὴν μνήμην ἀποσῴζει τὴν ἐκ τῆς ὀσφρήσεώς τε καὶ γεύσεως, ἅμα τε τῷ θεάσασθαι συνεπινοεῖ καὶ ταῦτα τῷ σχήματι καὶ τῷ χρώματι. ὅταν οὖν κήρινον μῆλον νομίσωμεν ἀληθινὸν εἶναι μῆλον, οὐχ ἡ ὄψις ἐστὶν ἡ ἐξαπατωμένη, ἀλλ' ἡ διάνοια. ἡ γὰρ ὄψις ἐν τοῖς ἰδίοις αἰσθητοῖς οὐ διεσφάλη· τό τε γὰρ χρῶμα καὶ τὸ σχῆμα συνεῖδεν.

For if an apple is not identified by colour and shape alone, but also by its smell and its characteristic taste, sight knows that it is an apple not by grasping these as well, but [because] the soul calls up the memory gained from smelling and tasting and, at the time of observation, attends to these along with shape and colour. So when we believe that an apple made of wax is a real apple, it is not sight that is deceived but thought. For sight was not mistaken about its specific objects of sense; for it recognised both the colour and shape.[16]

As when he argued, just before this, against the idea that sight can sense the heat of fire, Nemesios here explains how it is not sight that has been deceived into believing that the waxen apple is false, for it adheres to the outward visual appearance of an apple. Instead, it is thought (διάνοια) that has erred in hastily believing that the wax apple is real without confirming this fact through the other senses, given that an apple is not color and shape alone. In Nemesios's model, certainly the viewer has assented to the false cognitive presentation of the wax fruit, yet Nemesios's theory nevertheless thrives on the importance of plausibility and probability in sense perception. Upon seeing the skin of the apple, sight has accurately portrayed its color and form, but since an apple is not just this appearance, it must be submitted to the further prodding of the senses. Unless one is deceived into believing that the wax is an apple, like that the resin is honey.

For Nemesios, the wax apple demonstrates the necessity for the other senses to always be put to use in tandem with the perception of painting or any other art form whose aim is to deceive the viewer. As Nemesios writes:

ποτὲ μὲν οὖν ἡ ὄψις τῆς ἐπιμαρτυρίας χρῄζει τῆς τῶν ἄλλων αἰσθήσεων, ὅταν ᾖ τὸ φαινόμενον ἐπιτετεχνημένον πρὸς ἀπάτην, ὡς ἐπὶ τῆς γραφῆς· γραφικῆς γὰρ ἔργον ἀπατῆσαι τὴν ὄψιν ἐξοχαῖς οὐκ οὔσαις καὶ κοιλώμασιν, εἰ τὸ πρᾶγμα τοιαύτην ἔχει τὴν φύσιν. ἔνθα χρεία πρὸς διάγνωσιν μάλιστα μὲν τῆς ἁφῆς, ἔσθ' ὅτε δὲ καὶ τῆς γεύσεως ἢ ὀσφρήσεως, ὡς ἐπὶ τοῦ κηρίνου μήλου. ποτὲ δὲ καθ' ἑαυτὴν ἐναργῶς παρίστησι τὰ φαινόμενα, ὅταν μὴ πόρρωθεν ὁρᾷ.

[16] Nemesios, *De natura hominis*, 7, ed. Morani, 61; trans. Sharples and van der Eijk, 108.

> So sometimes sight needs additional evidence from the other senses, when what is viewed is crafted in order to deceive, as is the case with pictures: for what painting does is to deceive sight with non-existent projections and hollows, if that is the nature of the thing. Hence for discernment there is need of grasping, especially by touch, but sometimes also by taste and smell, as in the case of a waxen apple. But sometimes sight by itself vividly presents things seen, when it sees them from not far off.[17]

Certainly, as Nemesios goes on to tell us from here, sight can err on its own under certain circumstances: as when the object is far away, moving quickly, or when one must look through something that obscures sight, such as mist, water, or smoke. Yet the responsibility to discern falls ultimately upon the mind, which should not be distracted, as that too can be a cause for thought to misinterpret or miss the imprints of sight. Overall, sight communicates "graphically vividly the things appearing before it" (ἐναργῶς παρίστησι τὰ φαινόμενα). Here, the language of "vividness" (ἐναργῶς) is deployed by Nemesios to speak to that form of perception that is unimpaired, unhindered, and self-evident. *Enargeia* stands in for that impression that is passed on to perception "so graphically vivid" that it leads the mind to assent. This use speaks to the power of sight to – usually – lead us flawlessly into unerring assent. This is the visual vividness that Photios suggested was inherently present in vision and for which reason visual perception could readily transmit the forms on to the memory without having to first "soberly judge" them before assent. As Nemesios tells us, this is often the case with sight, but at times – either viewing conditions or the deceits of art – necessitates that we seek out further confirmation: namely through the other senses, particularly touch and taste.

While sight might be exceptional at many things, and readily passes vivid visual images to cognition for assent, just like the other senses it has its limitations in discerning the truth of things on its own. Often, touch and taste (which is usually deemed a species of touch) are the primary senses deployed to confirm the imprints of sight. It is this logic that seems to have motivated Pliny the Elder to state: "Among the senses, that of touch in man ranks before all the other species, and taste next; but in the remaining senses he is surpassed."[18] While he does not state that touch and taste are superior to sight or hearing, he stresses that compared to other

[17] Nemesios, *De natura hominis*, 7, ed. Morani, 62; trans. Sharples and van der Eijk, 108–109.

[18] "Ex sensibus ante cetera homini tactus, dein gustatus; reliquis superatur a multis." Pliny the Elder, *Historia naturalis*, 10.88.191, ed. and trans. Rackham, III:412–413.

animals humanity's keen sense of touch and taste surpasses all, possibly because of its keen ability to help discern and confirm the other senses. And, as Pliny goes on to say, when it comes to the other senses, man is easily surpassed: the eagles have clearer sight, vultures better smell, and moles clearer hearing.

For Nemesios, the deceit of the apple's sight requires taste and smell to assuredly confirm its nature, just as the image of honey must be confirmed to be false only by tasting the resin, given that touch itself might lead one to a similar conclusion to that from sight. Just as in the story of Doubting Thomas, it is touch that confirms the visual perception of the risen Christ as not simply some phantasmal deception of sight. Rather than expressing a doubt for the risen Christ, Thomas expresses a doubt for the sincerity of his perception. These matters are stressed in Nicholas Mesarites's *ekphrasis* of the scene in the Church of Holy Apostles, when Peter attempts to sway Thomas regarding Christ's resurrection, stating that "that which was seen was not an illusion of the sight, no phantom of the midday brightness" (ὡς οὐκ ἦν τὸ ὁραθὲν ὄψεως περιπλάνημα, οὐ φάσμα νυκτερινόν, οὐ δαιμόνιόν τι μεσημβρινόν).[19] And, upon the moment that Thomas penetrates the side-wound, Mesarites having suggested that Thomas has already realized that Christ was who he said he was, he writes: "his eyes opened to their widest, free from all rheum and flux and murky accretion" (ἠνεῳγμένους ἐς τὸ παντελὲς ἔχων τοὺς ὀφθαλμούς, καθαροὺς λήμης πάσης καὶ ἐπιρροίας καὶ λιγνυώδους ἐπισυμβάματος).[20] In the most famous instance of touch in Christianity, Mesarites highlights Thomas's eyes, going to great pains to describe Thomas's unobstructed vision as he touches Christ. This touch, therefore, comes not in response to a doubt of faith; Thomas appears to already have been convinced of his identity, but it is a confirmation of perception: optimal viewing conditions, conjoined with touch.

This is why delineating a separation between sight and touch is so crucial for the Byzantine world, because it is through the power of touch that one can come to confirm the imprints of sight. If touch were already a part of sight, or sight already possessed the qualities of touch, touch would be unable to confirm the imprints of sight. Thus, rather than thinking that sight is not touch because touch is lesser, we should endeavor to think that sight cannot be touch because touch (and by extension taste) is uniquely positioned to confirm or deny sight's impressions. This does not mean that

[19] Nicholas Mesarites, *Descriptio ecclesiae SS. Apostolorum*, 33.3, ed. and trans. Downey, 886, 912.
[20] Nicholas Mesarites, *Descriptio ecclesiae SS. Apostolorum*, 34.5, ed. and trans. Downey, 888, 913.

touch is superior to sight, but simply that the keenness of touch often makes it the first sense one turns to in order to confirm one's sight. It is this temptation to touch, before assenting to any dubious impressions from sight alone, that generates and sustains the longing for the figure depicted in the icon or the manuscript that lies within a museum case. In both instances, one is actively denied the ability to understand how this foreign object of our gaze alone feels, to understand its material makeup and confirm its textures and contours beyond the deceits of the shine, shimmer, and glare of the icon's gold or the case's glass.[21]

Therefore, the experience of gazing upon that Byzantine manuscript in a lucite case is not that different to gazing upon the icon of Christ in a Byzantine church. Both show us only a shadow of the living, animate body that inherently lies outside human perception, discernible in its divine or historical realities only through the noetic acts of the imagination that seek to emplace us into the space of its history or divinity. Both objects demand, as it is put in the *Vita* of John Chrysostom cited by John of Damascus, that we "gaze (ἐνητένιζεν) at it and attend to him (προσεῖχεν) as if he were alive and bless him (μακαρίζων), and bring the whole of [our] thoughts to him, imagining that he was speaking with him through his contemplation." Otherwise, the butterfly remains dead, splayed and pinned down to the specificity of its optical medium as nothing more than paint on flat canvas for the good modernist to admire.

It was in that same article, "The Discourse of Icons, Then and Now," that Robert Nelson discussed John of Damascus's imperative that we, as cited there in 1989, "embrace [images] with the eyes, the lips, the heart; bow before them; love them…"[22] As we can see here, there is a striking difference in Nelson's thinking about the operation of this sentence in 1989 from that in 2000, which would then read: "Kiss (ἀσπάζεσθι) [*sic*] them with the eyes, the lips, the heart."[23] The former article features a wholly different translation that seizes and grasps the perceptual unfolding for which I have argued. In 1989, Nelson seems to render the complicated ἀσπάζεσθαι in two ways, simultaneously: not only suggesting a "cleaving" to in the heart as "love," but also the act of greeting the icon as reflected in "bow before them." While a looser translation than his 2000 reading, here Nelson sought to foreground the complex practice of

[21] On the connection between the icon's gold ground and theories of perception, see Roland Betancourt, "The Icon's Gold: A Medium of Light, Air, and Space," *West 86th: A Journal of Decorative Arts, Design History, and Material Culture* 23:2 (2016): 252–280.

[22] Nelson, "Discourse of Icons," 148.

[23] Nelson, "To Say and to See," 153.

veneration that John of Damascus was describing, and which he himself was attempting to elucidate in relation to the museum's lucite case. Thus, concluding his analysis of John of Damascus's sentence, Nelson wrote in 1989: "In essence the image, the icon, is a mediator, a way for the believer to comprehend God and his teachings and a medium through which God and the believer interact."[24] This is not the Robert Nelson of 2000 with his emphatic focus on the haptic visuality of the Byzantine viewer, whose agentic sight allowed for the immediate contact with the icon's subject, collapsing sight and touch into one. Instead, what we see in the 1989 article is a scholar attempting to understand the icon as a sign, whose relation to its interpreter emerges not only discursively, but also from the inexorable absence present in the icon, which seeks to represent, rather than re-present, that which is inherently not there.

In 1989, Nelson is taking on textbook semiotics to construct an icon that is by nature a communicative medium, whereas by 2000, he is taking on a matured discourse on reception theory to comprehend how the icon itself exceeds the bounds of modern signification and semiotics. Yet, in both instances, we find a scholar attempting to come to terms precisely with how the contemporary discourses on medium and mediation in art history and critical theory could contribute to a nuanced understanding of Byzantine art and its relation to past and present publics. Byzantine visuality thus has always been as much a product of its historiographic and historical times.

* * *

Throughout this volume, I have considered the various facets of touch's operation in, alongside, and against sight. In the first two parts of this volume, I emphasized the uniqueness of each of the senses, their overlapping perceptibles and commonalities, and went on to consider their ability to construct mental models of the world through the faculty of the imagination. Likewise, I understood how the conflation of sight with touch emerged by virtue of the metaphors used to describe the operation of sight's contact with its objects and the language used to describe apprehension and comprehension in the unfolding process of perception. Then, in Part III, we turned our attention to considering the implications that these findings have for our conceptualization of veneration and its sensorial, perceptual, and performative practices, for our comprehension of

[24] Nelson, "Discourse of Icons," 149.

how medium and mediation were conceived in art and literature given contemporaneous articulations of the senses and understanding, and for our understanding of how cognition and cultural practices help one to stitch together the various abilities of each of the senses, constructing an understanding of our world and encouraging synesthetic metaphors that stress the vividness and actuality of our perceptions. These findings led us to reimagine how it is that we think about mediums, both artistic and sensory ones, and the crucial role that the cognitive aspects of perception, beyond bodily sensation alone, played in the conceptualization of images in the Byzantine world. As the field shifts back to various object-oriented ontologies and concerns regarding the materiality of artistic media, it is necessary to reconsider the discourses that brought us here, and also to expand the sources and conceptual frameworks at our disposal so as to generate new and innovative trajectories for the history of the Byzantine image and its viewer. In this process, we must make clear strides to avoid the endemic confusion of signifiers and signifieds with their referents, which haptic sight has in part encouraged.

To summarize the findings and conclusions here, on the one hand, we can say that there are three primary causes of the haptic language associated with Byzantine visuality to date in the historical record: First, the notion of haptic sight emerges from a conflation of terms that often seek to describe in shorthand either the processes of veneration that involved visual and tactile actions deployed together and in sequence, or the perceptual processes of the mind as it comprehends sensual data through a language of cognition that deploys verbs associated with touch and grasping. Second, we likewise see the diagramming of ritual acts that invoked the literal touching of visible objects to the five senses as a manner of awakening the senses to the spiritual realities at hand. In its shorthand, these practices often deploy non-specific sensual language, where a seer might likewise be described as touching an object to their eyes as well as other senses, hence resulting in understandable confusions. Third, there appears to be a broader trend of a synesthetic language pervasive in Byzantine rhetoric that seeks to transgress usual sensual boundaries for emphasis and to express complex, abstract matters. While this language is grounded in notions of sensual perception, it also is clear that it is by no means limited to conflations of sight and touch, but rather appears as remediations between all sensory inputs. Thus, these varying usages demand that we qualify the extent to which we see these conflations as being significant to how sensation occurs, and instead articulate them as purposeful transgressions for effect and emphasis. Such transgressions of the traditional sensory boundaries derive their efficacy not by claiming a perceptual synesthesia, but precisely because they do overstep familiar boundaries and suggest modes or states of perception that exceed the logic of the individual sense in a given circumstance.

On the other hand, three main points are to be made regarding the historiographic causes and effects of haptic visuality's predominance in Byzantine art history: First, the theory of haptic sight allowed the Byzantine art historian to collapse the mediating contours of a medium in favor of the unmediated presence that was stereotypically and too hastily ascribed to the icon. Second, haptic sight allowed the scholar to combine the utility of the cult image with the aesthetic appreciation of the image through eyesight alone, which a post-Renaissance theory of art favored. Third, the theory of haptic sight also allowed a physiological reification of the agency of the viewer, which is displaced by the reintroduction of the concept of a mediating middle – whether this be Aristotle's *transparent*, Plato's *united rays*, or Galen's *optical pneuma*. For the historical study of Byzantine art, these observations require a change in the language regarding visual devotion, allowing the processes of touch and sight to exist as mutually fruitful, yet distinct – understanding that a psychosomatic engagement with the icon required not only visual practices, but also other functions that deployed the other senses uniquely within their own medium specificities. As such, it is necessary not simply to see past metaphor, but also to not take metaphors literally, rather to understand that metaphors carry their own logics and serve as potent tools in analogical thinking. This analogical and synesthetic thinking not only structures corollary behaviors, such as contemplating, embracing, and kissing an icon, but also speaks to ways of articulating abstract notions. In this vein, the commonality of touch to all the senses, when read closely, clarifies itself not as an argument for haptic sight, but rather as a way of manifesting the ineffable form of contact that founds the aesthetic experience of perceptible features – against the cognitive and spiritual aesthetic experience of imperceptible forms, noetic and divine.

Bibliography

Primary Sources

Achilles Tatius, *Leucippe et Clitophon.*

Ed. E. Vilborg, *Leucippe and Clitophon* (Stockholm: Almqvist & Wiksell, 1955).

Trans. John J. Winkler, "Leucippe and Clitophon" in Bryan P. Reardon (ed.), *Collected Ancient Greek Novels* (Berkeley: University of California Press, 1989), 170–284.

Ed. and trans. S. Gaselee, *Leucippe and Clitophon*, Loeb Classical Library 45 (Cambridge, MA: Harvard University Press, 1969).

Aetios of Amida, *Iatricorum liber*, or *Tetrabiblion.*

Ed. Alexander Olivieri, *Libri medicinales*, 2 vols., Corpus Medicorum Graecorum 8.1–2 (Berlin: Akademie Verlag, 1935–1950).

Partial trans. Julius Hirschberg, *The Ophthalmology of Aëtius of Amida*, trans. Richey L. Waugh, Jr. (Ostend, Belgium: J. P. Wayenborgh, 2000).

Partial trans. Eric Gowling, "*Libri Medicinales* Book 1: A Translation with Commentary," unpublished PhD thesis, University of Glasgow (2015).

Partial trans. James V. Ricci, *The Gynaecology and Obstetrics of the VIth Century, A.D.* (Philadelphia: The Blakiston Company, 1950).

Aetios of Antioch (see also Pseudo-Plutarch), *De placitis reliquiae (Stobaei excerpta)* or *(Theodoreti et Nemesii excerpta).*

Ed. H. Diels, *Doxographi Graeci* (Berlin: Reimer, 1879).

Alcinous, *Didaskalikos.*

Ed. Pierre Louis, *Épitomé* (Paris: Les Belles Lettres, 1945).

Alexander of Aphrodisias, *De anima.*

Ed. I. Bruns, *Praeter commentaria scripta minora*, Commentaria in Aristotelem Graeca suppl. 2.1 (Berlin: Reimer, 1887), 1–100.

Trans. Athanasios P. Fotinis, *The* De anima *of Alexander of Aphrodisias* (Washington, DC: University Press of America, 1979).

Alexander of Aphrodisias, *De anima libri mantissa*, or *De anima liber alter.*

Ed. Robert W. Sharples, *De anima libri mantissa*, Peripatoi 21 (Berlin: De Gruyter, 2008).

Alexander of Aphrodisias, *In Aristotelis meteorologicorum libros commentaria.*

Ed. Michael Hayduck, *In Aristotelis meteorologicorum libros commentaria*, Commentaria in Aristotelem Graeca 3.2 (Berlin: Reimer, 1899).

Partial trans. Eric Lewis, *On Aristotle Meteorology 4* (London: Duckworth, 1996).
Alexander of Aphrodisias, *In librum de sensu commentarium.*
Ed. P. Wendland, *In librum de sensu commentarium*, Commentaria in Aristotelem Graeca 3.1 (Berlin: Reimer, 1901).
Trans. Alan Towey, *On Aristotle On Sense Perception* (London: Duckworth, 2000).
Anonymous, *Anthologia Graeca.*
Ed. and trans. W. R. Paton, *Greek Anthology*, 4 vols., Loeb Classical Library 67, 68, 84, 86 (Cambridge, MA: Harvard University Press, 1917–2014).
Anonymous, *Anthologia Marciana.*
Ed. Spyridon P. Lambros, "Ὁ μαρκιανὸς κῶδιξ 524," *Νέος Ελληνομνήμων*, 8:2 (1911): 123–192.
Anonymous, *Blachernae synod.*
Ed. PG 127:980D–984A.
Anonymous, *Decretum concilii iconomachi.*
Ed. Paul J. Alexander, "The Iconoclastic Council of St. Sophia (815) and Its Definition (*Horos*)," *Dumbarton Oaks Papers*, 7 (1953): 35–66.
Anonymous, *Doctrina patrum.*
Ed. F. Diekamp, *Doctrina patrum de incarnatione verbi* (Munich: Aschendorff, 1907).
Anonymous, *Itinerarium Egeriae.*
Ed. P. Geyer and O. Cuntz, *Itineraria et alia geographica*, Corpus Christianorum, Series Latina 175 (Turnhout: Brepols, 1965), 37–90.
Trans. John Wilkinson, *Egeria's Travels to the Holy Land*, revised edition (Jerusalem: Ariel Publishing House, 1981).
Anonymous, *Lexica Segueriana*, or *Synagoge.*
Ed. L. Bachmann, *Anecdota Graeca*, vol. 1 (Leipzig: Hinrichs, 1828).
Anonymous, *Pantokrator Typikon.*
Ed. Paul Gautier, "Le Typikon du Christ Sauveur Pantocrator," *Revue des Études Byzantines*, 32 (1974): 1–145.
Trans. Robert Jordan, "Pantokrator: Typikon of Emperor John II Komnenos for the Monastery of Christ *Pantokrator* in Constantinople" in John Thomas and Angela Constantinides Hero (eds.), *Byzantine Monastic Foundation Documents*, vol. II (Washington, DC: Dumbarton Oaks, 2000), 725–781.
Anonymous, *Sacrorum conciliorum.*
Ed. J. D. Mansi, *Sacrorum conciliorum nova et amplissima collectio*, 53 vols. (Paris: H. Welter, 1901–1927).
Anonymous, *Scholia in Iliadem.*
Ed. H. Erbse, *Scholia Graeca in Homeri Iliadem (scholia vetera)*, 5 vols. (Berlin: De Gruyter, 1969–1977).
Anonymous, *Souda*, or *Suda.*
Ed. A. Adler, *Suidae Lexicon*, 4 vols., Lexicographi Graeci 1.1–1.4 (Leipzig: Teubner, 1928–1935).

Ed. and trans. David Whitehead et al., *Suda On Line: Byzantine Lexicography*, www.stoa.org/sol (2000–2017).
Anonymous, *Story of Abba Pambo.*
Ed. and trans. E. A. Wallis Budge, "The Life of Apa Cyrus," *Coptic Martyrdoms*, vol. 1 (London: British Museum, 1914), 128–136, 381–389.
Trans. Tim Vivian, "The Story of Abba Pambo" in Tim Vivian (ed.), *Journeying into God: Seven Early Monastic Lives* (Minneapolis: Augsburg Fortress, 1996), 30–36.
Anonymous, *Vita Leonis.*
Ed. Alexander G. Alexakis and trans. Susan Wessel, *The Greek Life of St. Leo Bishop of Catania* (Brussels: Société des Bollandistes, 2011).
Anonymous, *Vita sanctae Athanasiae.*
Ed. L. Carras, "The Life of St. Athanasia of Aegina" in Ann Moffatt (ed.), *Maistor: Classical, Byzantine, and Renaissance Studies for Robert Browning* (Canberra: Australian Association for Byzantine Studies, 1984), 212–224.
Trans. Lee Francis Sherry, "Life of St. Athanasia of Aegina" in Alice-Mary Talbot (ed.), *Holy Women of Byzantium: Ten Saints' Lives in English Translation* (Washington, DC: Dumbarton Oaks, 1996), 137–158.
Anthemios of Tralles, *Περὶ παραδόξων μηχανημάτων.*
Ed. and trans. G. L. Huxley, *A Study in Later Greek Geometry*, Greek, Roman, and Byzantine Studies Monographs 1 (Cambridge, MA: Harvard University Press, 1959).
Aphthonios, *Progymnasmata.*
Ed. Hugo Rabe, *Progymnasmata*, Rhetores Graeci 10 (Leipzig: Teubner, 1926).
Trans. George A. Kennedy, *Progymnasmata: Greek Textbooks of Prose Composition and Rhetoric* (Atlanta: Society of Biblical Literature, 2003), 89–128.
Aristotle, *De anima.*
Ed. and trans. W. S. Hett, *On the Soul*, Loeb Classical Library 288 (Cambridge, MA: Harvard University Press, 1957).
Aristotle, *De memoria et reminiscentia.*
Ed. and trans. W. S. Hett, *Parva Naturalia*, Loeb Classical Library 288 (Cambridge, MA: Harvard University Press, 1957), 287–313.
Aristotle, *De sensu et sensibilibus.*
Ed. and trans. W. S. Hett, *Parva Naturalia*, Loeb Classical Library 288 (Cambridge, MA: Harvard University Press, 1957), 214–283.
Aristotle, *De somno et vigilia.*
Ed. and trans. W. S. Hett, *Parva Naturalia*, Loeb Classical Library 288 (Cambridge, MA: Harvard University Press, 1957), 318–345.
Aristotle, *Ethica Nicomachea.*
Ed. and trans. H. Rackham, *Nicomachean Ethics*, Loeb Classical Library 73 (Cambridge, MA: Harvard University Press, 1926).
Aristotle, *Poetica.*
Ed. Rudolf Kassel, *De arte poetica liber*, corrected edition (Oxford: Clarendon Press, 1966).

Ed. and trans. Stephen Halliwell, *Poetics*, Loeb Classical Library 199 (Cambridge, MA: Harvard University Press, 1995).

Aristotle, *Rhetorica*.

Ed. and trans. J. H. Freese, *Art of Rhetoric*, Loeb Classical Library 193 (Cambridge, MA: Harvard University Press, 1926).

Asterios of Amaseia, *Descriptio picturae sanctae Euphemiae*, or *Homilia* 11.

Ed. François Halkin, *Euphémie de Chalcédoine. Légendes Byzantines*, Subsidia Hagiographica 41 (Brussels: Société des Bollandistes, 1965), 4–8.

Trans. Boudewijn Dehandschutter, "Ecphrasis on the Holy Martyr Euphemia" in J. Leemans et al. (eds.), *"Let Us Die that We May Live": Greek Homilies on Christian Martyrs from Asia Minor, Palestine and Syria* (London: Routledge, 2003), 173–176.

Asterios of Amaseia, *Encomium on Phokas*, or Homily 9.

Ed. Cornelis Datema, *Homilies I-XIV* (Leiden: Brill, 1970), 114–127.

Trans. Boudewijn Dehandschutter, "A Homily on Phocas" in J. Leemans et al. (eds.), *"Let Us Die that We May Live": Greek Homilies on Christian Martyrs from Asia Minor, Palestine and Syria* (London: Routledge, 2003), 167–173.

Athanasios of Alexandria, *Vita Antonii*.

Ed. G. J. M. Bartelink, *Vie d'Antoine*, Sources chrétiennes 400 (Paris: Éditions du Cerf, 2004).

Athenaeus, *Deipnosophistae*.

Ed. and trans. S. Douglas Olson, *The Learned Banqueters*, 8 vols., Loeb Classical Library 204, 208, 224, 235, 274, 327, 345, 519 (Cambridge, MA: Harvard University Press, 2007–2011).

Augustine, *De trinitate*.

Ed. W. J. Mountain, *De trinitate libri XV*, Corpus Christianorum, Series Latina 50–50A (Turnhout: Brepols, 1968).

Partial ed. Johann Kreuzer, *De trinitate (Bücher VIII-XI, XIV-XV, Anhang: Buch V)* (Hamburg: Felix Meiner Verlag, 2001).

Trans. Stephen McKenna, *On the Trinity: Books 8–15*, ed. Gareth B. Matthews (Cambridge University Press, 2002).

Trans. Stephen McKenna, *The Trinity*, The Fathers of the Church 45 (Washington, DC: Catholic University of America Press, 1970).

Basil of Caesarea, *De spiritu sanctu*.

Ed. Benoît Pruche, *Sur le Saint-Esprit*, 2nd edition, Sources chrétiennes 17 bis (Paris: Éditions du Cerf, 1968).

Basil of Caesarea, *Homilia de invidia*, or Homily 11

Ed. PG 31:372–385.

Trans. M. Monica Wagner, "Concerning Envy" in M. Monica Wagner (ed.), *Ascetical Works*, reprint (Washington, DC: Catholic University of America Press, 1962), 463–474.

Basil of Caesarea, *Homiliae in hexameron*.

Ed. S. Giet, *Homélies sur l'hexaéméron*, 2nd edition, Sources chrétiennes 26 bis (Paris: Éditions du Cerf, 1968).

Trans. Bloomfield Jackson, "The Hexaemeron" in *Nicene and Post-Nicene Fathers*, second series, vol. VIII, ed. Philip Schaff and Henry Wace (New York: Christian Literature Company, 1895), 51–107.
Basil of Caesarea, *Homiliae in hexameron* 10–11, or *Homilies on the Origin of Man* 1–2.
Ed. Alexis Smets and Michel van Esbroeck, *Sur l'origine de l'homme (Hom. X et XI de l'Hexaéméron)*, Sources chrétiennes 160 (Paris: Éditions du Cerf, 1970).
Trans. Nonna Verna Harrison, "On the Origin of Humanity, Discourse 1" and "On the Origin of Humanity, Discourse 2" in Nonna Verna Harrison (ed.), *On the Human Condition* (Yonkers, NY: St. Vladimir's Seminary Press, 2005), 31–64.
Basil of Caesarea, *In Gordium martyrem*, or Homily 18.
Ed. PG 31:489–508.
Trans. Pauline Allen, "A Homily on the Martyr Gordius" in J. Leemans et al. (eds.), *"Let Us Die that We May Live": Greek Homilies on Christian Martyrs from Asia Minor, Palestine and Syria* (London: Routledge, 2003), 56–67.
Blemmydes, Nikephoros, *Epitome physica*.
Ed. PG 142:1004–1320.
Bryennios, Manuel, *Harmonica*.
Ed. and trans. G. H. Jonker, *The Harmonics of Manuel Bryennius* (Groningen: Wolters-Noordhoff, 1970).
Choiroboskos, George, *Περὶ τρόπων ποιητικῶν*.
Ed. Leonhard von Spengel, *Rhetores Graeci*, vol. III (Leipzig: Teubner, 1856).
Choumnos, Nikephoros, *Opera philosophica*.
Ed. Konstantinos P. Chrestou, *Τὸ φιλοσοφικὸ ἔργο τοῦ Νικηφόρου Χούμνου* (Thessaloniki: Ekdoseis Kyromanos, 2002).
Chrysostom, John, *De Sacerdotio*.
Ed. A.-M. Malingrey, *Sur le Sacerdoce*, Sources chrétiennes 272 (Paris: Éditions du Cerf, 1980).
Trans. W. R. W. Stephens, "Six Books on the Priesthood" in Philip Schaff (ed.), *Nicene and Post-Nicene Fathers*, first series, vol. IX (Buffalo: Christian Literature Publishing, 1889), 33–83.
Trans. Graham Neville, *On the Priesthood* (Crestwood, NY: St. Vladimir's Seminary Press, 1996).
Chrysostom, John, *In epistulam ad Galatas commentarius*.
Ed. PG 61:611–682.
Chrysostom, John, *In Genesim (homiliae 1–67)*.
Ed. PG 53:21–385 and PG 54:385–580.
Chrysostom, John, *In Matthaeum (homiliae 1–90)*.
Ed. PG 57:13–472 and PG 58:471–794.
Chrysostom, John, *Quod regulares feminae viris cohabitare non debeant*.
Ed. Jean Dumortier, *Les Cohabitations Suspectes* (Paris: Les Belles Lettres, 1955).
Cicero, *Academica*.

Ed. and trans. H. Rackham, *Academics*, Loeb Classical Library 268 (Cambridge, MA: Harvard University Press, 1933).

Clement of Alexandria, *Stromata*.

Ed. L. Früchtel, O. Stählin, and U. Treu, *Clemens Alexandrinus*, vol. III, 2nd edition (Berlin: Akademie Verlag, 1970).

Trans. H. Chadwick and J. E. L. Oulton, *Alexandrian Christianity: Selected Translations of Clement and Origen* (Philadelphia: Westminster Press, 1954).

Constantine VII Porphyrogennetos, *Narratio de imagine Edessena*.

Ed. and trans. Mark Guscin, "The *Narratio de Imagine Edessena*" in Mark Guscin (ed.), *The Image of Edessa* (Leiden: Brill, 2009), 7–69.

Constantine VII Porphyrogennetos, *Treatise on Imperial Military Expeditions*.

Ed. and trans. John Haldon, *Three Treatises on Imperial Military Expeditions* (Vienna: Verlag der Österreichischen Akademie der Wissenschaften, 1990).

Cyril of Jerusalem, *Mystagogiae*.

Ed. Pierre Paris and Auguste Piédagnel, *Catéchèses Mystagogiques*, 2nd edition, Sources chrétiennes 126 bis (Paris: Éditions du Cerf, 1988).

Democritus, *Fragmenta* and *Testimonia*.

Ed. H. Diels and W. Kranz, *Die Fragmente der Vorsokratiker*, vol. II, 6th edition (Berlin: Weidmann, 1952).

Dio, Cassius, *Historiae Romanae*.

Ed. U. P. Boissevain, *Historiarum Romanarum quae supersunt*, 3 vols. (Berlin: Weidmann, 1895–1901).

Diocles, *On Burning Mirrors*.

Ed. and trans. G. J. Toomer, *On Burning Mirrors: The Arabic Translation of the Lost Greek Original* (Berlin: Springer, 1976).

Doukas Laskaris, Theodore II, *Epistula*.

Ed. N. Festa, *Epistulae CCXVII* (Florence: Istituto di Studi Superiori Pratici e di Perfezionamento, 1898).

Empedocles, *Fragmenta* and *Testimonia*.

Ed. H. Diels and W. Kranz, *Die Fragmente der Vorsokratiker*, vol. I, 6th edition (Berlin: Weidmann, 1951).

Empiricus, Sextus, *Adversus Mathematicos*.

Ed. and trans. R. G. Bury, *Against the Logicians*, Loeb Classical Library 291 (Cambridge, MA: Harvard University Press, 1935).

Ed. and trans. R. G. Bury, *Against Professors*, Loeb Classical Library 382 (Cambridge, MA: Harvard University Press, 1949).

Empiricus, Sextus, *Pyrrhoniae hypotyposes*.

Ed. and trans. R. G. Bury, *Outlines of Pyrrhonism*, Loeb Classical Library 273 (Cambridge, MA: Harvard University Press, 1933).

Epiphanios of Salamis, *Panarion* (= *Adversus haereses*).

Ed. K. Holl, *Epiphanius*, vol. I, Die griechischen christlichen Schriftsteller 25 (Leipzig: Hinrichs, 1915).

Euclid, *Catoptrica*.

Ed. J. L. Heiberg, *Opera omnia*, vol. VII (Leipzig: Teubner, 1895), 286–342.

Euclid, *Optica*.

Ed. J. L. Heiberg, *Opera omnia*, vol. VII (Leipzig: Teubner, 1895), 2–120.
Euclid, *Opticorum recensio Theonis*.
Ed. J. L. Heiberg, *Opera omnia*, vol. VII (Leipzig: Teubner, 1895), 144–246.
Eusebios, *De ecclesiastica theologia*.
Ed. G. C. Hansen and E. Klostermann, *Werke*, vol. IV (Berlin: Akademie Verlag, 1972), 61–182.
Eustathios of Thessaloniki, *Commentarii ad Homeri Iliadem*.
Ed. M. van der Valk, *Commentarii ad Homeri Iliadem pertinentes*, 4 vols. (Leiden: Brill, 1971–1987).
Eustratios of Nicaea, *Dialogus de sacris imaginibus*, or *Oratione* 5.
Ed. Andronikos Demetrakopoulos, *Ἐκκλησιαστικὴ Βιβλιοθήκη*, vol. I (Leipzig: Othonos Vigandou, 1866), 127–151.
Eustratios of Nicaea, *In Aristotelis ethica Nicomachea vi commentaria*.
Ed. G. Heylbut, *Eustratii et Michaelis et anonyma in ethica Nicomachea commentaria*, Commentaria in Aristotelem Graeca 20 (Berlin: Reimer, 1892).
Eustratios of Nicaea, *Syllogistic Demonstrations*, or *Expositio de imaginibus*.
Ed. Andronikos Demetrakopoulos, *Ἐκκλησιαστικὴ Βιβλιοθήκη*, vol. I (Leipzig: Othonos Vigandou, 1866), 151–160.
Euthymios Zigabenos, *Panoplia Dogmatica*.
Ed. PG 130:20–1360.
Evagrios, *De malignis cogitationibus*.
Ed. Paul Géhin, Claire Guillaumont, and Antoine Guillaumont, *Sur les pensées*, Sources chrétiennes 438 (Paris: Éditions du Cerf, 1998).
Evagrios, *De oratione*.
Ed. Simon Tugwell, *De oratione* (Oxford: Faculty of Theology, 1981).
Cf. ed. PG 79: 1165–1200.
Trans. John Eudes Bamberger, *Chapters on Prayer* (Kalamazoo, MI: Cistercian Publications, 1972).
Evagrios, *Life of St. Pankratios of Taormina*.
Ed. C. J. Stallman, "The *Life* of St. Pancratius of Taormina," unpublished DPhil dissertation, 2 vols., Oxford University (1986).
Evagrios, *Practicus*.
Ed. Antoine Guillaumont and Claire Guillaumont, *Traité pratique, ou, le moine*, vol. II, Sources chrétiennes 171 (Paris: Éditions du Cerf, 1971).
Galen, *De placitis Hippocratis et Platonis*.
Ed. and trans. Phillip De Lacy, *On the Doctrines of Hippocrates and Plato*, Corpus Medicorum Graecorum 4.1.2 (Berlin: Akademie Verlag, 1980).
Galen, *De usu partium*.
Ed. G. Helmreich, *De usu partium*, 2 vols. (Leipzig: Teubner, 1907–1909).
Trans. Margaret T. May, *Galen on the Usefulness of the Parts of the Body*, 2 vols. (Ithaca, NY: Cornell University Press, 1968).
Galen, *In Hippocratis de natura hominis librum commentarii iii*.
Ed. J. Mewaldt, *In Hippocratis de natura hominis commentarii tria*, Corpus Medicorum Graecorum 5.9.1 (Leipzig: Teubner, 1914).

George of Sykeon, *Vita sancti Theodori Syceotae*.
Ed. A.-J. Festugière, *Vie de Théodore de Sykeôn*, vol. I, Subsidia Hagiographica 48 (Brussels: Société des Bollandistes, 1970).
Trans. Elizabeth Dawes, "St. Theodore of Sykeon," *Three Byzantine Saints: Contemporary Biographies of St. Daniel the Stylite, St. Theodore of Sykeon and St. John the Almsgiver* (Oxford: Blackwell, 1948), 88–192.
George the Synkellos, *Ecloga chronographica*.
Ed. A. A. Mosshammer, *Ecloga chronographica* (Leipzig: Teubner, 1984).
Gorgias of Leonti, *Encomium in Helenam*, or Fragment 11.
Ed. H. Diels and W. Kranz, *Die Fragmente der Vorsokratiker*, vol. II, 6th edition (Berlin: Weidmann, 1952), 288–294.
Trans. John Dillon and Tania Gergel, "Encomium of Helen" in John Dillon and Tania Gergel (eds.), *The Greek Sophists* (London: Penguin Books, 2003), 76–84.
Gregoras, Nikephoros, *Explicatio in librum Synesii De insomniis*.
Ed. P. Pietrosanti, *Explicatio in librum Synesii 'De insomniis'* (Bari: Levante Editori, 1999).
Gregoras, Nikephoros, *Historia Romana*.
Ed. L. Schopen, *Historiae Byzantinae*, vols. I–II (Bonn: Weber, 1829–1830).
Ed. I. Bekker, *Historiae Byzantinae*, vol. III (Bonn: Weber, 1855).
Gregoras, Nikephoros, *Solutiones questionum*.
Ed. P. L. M. Leone, "Nicephori Gregorae 'Antilogia' et 'Solutiones questionum'," *Byzantion*, 40 (1970): 488–513.
Gregory of Nazianzos, *Carmina de se ipso*, I.
Ed. PG 37: 970–1017.
Ed. and trans. Denis M. Meehan, *Three Poems: Concerning His Own Affairs, Concerning Himself and the Bishops, and Concerning His Own Life* (Washington, DC: Catholic University of America Press, 1987).
Gregory of Nazianzos, *Contra Julianum imperatorem* I, or *Orat.* 4.
Ed. PG 35:532–664.
Gregory of Nazianzos, *De moderatione in disputando*, or *Orat.* 32.
Ed. PG 36:173–212.
Gregory of Nazianzos, *De pace* I, or *Orat.* 6.
Ed. PG 35:721–752.
Gregory of Nazianzos, *De theologia*, or *Orat.* 28.
Ed. J. Barbel, *Die Fünf Theologischen Reden* (Düsseldorf: Patmos-Verlag, 1963), 62–126.
Trans. Lionel Wickham, *On God and Christ: The Five Theological Orations* (Yonkers, NY: St. Vladimir's Seminary Press, 2002), 37–67.
Gregory of Nazianzos, *In sanctum baptisma*, or *Orat.* 40.
Ed. PG 36:360–425.
Trans. Nonna Verna Harrison, *Festal Orations* (Yonkers, NY: St. Vladimir's Seminary Press, 2008), 99–142.
Gregory of Nyssa, *Contra Eunomium*.
Ed. W. Jaeger (ed.), *Opera*, vols. I.I and II.2 (Leiden: Brill, 1960).

Trans. Lenka Karfíková, Scot Douglass, and Johannes Zachhuber, *Contra Eunomium II* (Leiden: Brill, 2007).
Trans. Stuart G. Hall, Johan Leemans, and Matthieu Cassin, *Contra Eunomium III* (Leiden: Brill, 2010).
Gregory of Nyssa, *De opificio hominis.*
Ed. PG 44:124–256.
Trans. H. A. Wilson, "On the Making of Man" in Philip Schaff and Henry Wace (eds.), *Nicene and Post-Nicene Fathers*, second series, vol. v (New York: Charles Scribner's Sons, 1902), 387–427.
Gregory of Nyssa, *De sancto Theodoro.*
Ed. PG 46:736–748.
Cf. ed. G. Heil, J. Cavarnos, O. Lendle, and F. Mann, *Sermones*, vol. II (Leiden: Brill, 1990), 61–71.
Trans. Johan Leemans, "A Homily on Theodore the Recruit" in J. Leemans et al. (eds.), *"Let Us Die that We May Live": Greek Homilies on Christian Martyrs from Asia Minor, Palestine and Syria* (London: Routledge, 2003), 82–91.
Gregory of Nyssa, *In Canticum canticorum.*
Ed. H. Langerbeck, *Opera*, vol. VI (Leiden: Brill, 1960).
Trans. Casimir McCambley, *Commentary on the Song of Songs* (Brookline, MA: Hellenic College Press, 1987).
Trans. Richard A. Norris, *Homilies on the Song of Songs* (Atlanta: Society of Biblical Literature, 2012).
Gregory Referendarios, *Sermon.*
Ed. and trans. Mark Guscin, "The Sermon of Gregory Referendarius" in Mark Guscin (ed.), *The Image of Edessa* (Leiden: Brill, 2009), 71–88.
Hermogenes, *Progymnasmata.*
Ed. H. Rabe, *Opera* (Leipzig: Teubner, 1913).
Trans. George A. Kennedy, *Progymnasmata: Greek Textbooks of Prose Composition and Rhetoric* (Atlanta: Society of Biblical Literature, 2003), 73–88.
Hippocrates, *De visu.*
Ed. and trans. Elizabeth M. Crank, *Two Hippocratic Treatises:* On Sight *and* On Anatomy (Leiden: Brill, 2006), 38–47.
Hippolytus, *Refutatio omnium haeresium*, or *Philosophumena.*
Ed. M. Marcovich, *Refutatio omnium haeresium*, Patristische Texte und Studien 25 (Berlin: De Gruyter, 1986).
Irenaeus, *Adversus Haereses*, 1.
Ed. W. W. Harvey, *Libri Quinque Adversus Haereses*, vol. I (Cambridge University Press, 1857).
Jerome, *Epistula*, 46.
Ed. Isidorus Hilberg, *Corpus Scriptorum Ecclesiasticorum Latinorum*, vol. LIV (Vienna: F. Tempsky, 1910), 329–344.
Trans. W. H. Fremantle, "Paula and Eustochium to Marcella" in Philip Schaff and Henry Wace (eds.), *Nicene and Post-Nicene Fathers*, second series, vol. VI (New York: Christian Literature Company, 1893), 60–65.

Jerome, *Epistula*, 108.
Ed. Isidorus Hilberg, *Corpus Scriptorum Ecclesiasticorum Latinorum*, vol. lv (Vienna: F. Tempsky, 1912), 306–351.
Trans. W. H. Fremantle, "To Eustochium" in Philip Schaff and Henry Wace (eds.), *Nicene and Post-Nicene Fathers*, second series, vol. VI (New York: Christian Literature Company, 1893), 195–212.
John of Damascus, *Expositio fidei*.
Ed. P. B. Kotter, *Die Schriften des Johannes von Damaskos*, vol. II, Patristische Texte und Studien 12 (Berlin: De Gruyter, 1973).
Trans. S. D. F. Salmond, "An Exact Exposition of the Orthodox Faith" in Philip Schaff and Henry Wace (eds.), *Nicene and Post-Nicene Fathers*, second series, vol. IX (New York: Charles Scribner's Sons, 1902), 1–101.
John of Damascus, *Orationes de imaginibus tres*.
Ed. P. B. Kotter, *Die Schriften des Johannes von Damaskos*, vol. III, Patristische Texte und Studien 17 (Berlin: De Gruyter, 1975).
Trans. Andrew Louth, *Three Treatises on the Divine Images* (Crestwood, NY: St. Vladimir's Seminary Press, 2003).
Cf. trans. David Anderson, *On the Divine Images: Three Apologies Against Those Who Attack the Divine Images* (Crestwood, NY: St. Vladimir's Seminary Press, 1980).
John VII Grammatikos, *Antirrheticus*.
Ed. Jean Gouillard, "Fragments inédits d'un antirrhétique de Jean le Grammairien," *Revue des Études Byzantines* 24 (1966): 171–181.
John of Sardis, *Commentarium in Aphthonii progymnasmata*.
Ed. H. Rabe, *Commentarium in Aphthonii progymnasmata*, Rhetores Graeci 15 (Leipzig: Teubner, 1928).
Trans. George A. Kennedy, *Progymnasmata: Greek Textbooks of Prose Composition and Rhetoric* (Atlanta: Society of Biblical Literature, 2003), 173–228.
Josephus, Flavius, *De bello Judaico*.
Ed. B. Niese, *Opera*, vol. VI (Berlin: Weidmann, 1895).
Ed. and trans. H. St. J. Thackeray, *The Jewish War*, 3 vols., Loeb Classical Library 203, 487, 210 (Cambridge, MA: Harvard University Press, 1927–1928).
Kabasilas, Nicholas, *Explicatio divinae liturgiae*.
Ed. R. Bornert, J. Gouillard, P. Périchon, and S. Salaville, *Explication de la Divine Liturgie*, Sources chrétiennes 4 bis (Paris: Éditions du Cerf, 1967).
Trans. J. M. Hussey and P. A. McNulty, *Commentary on the Divine Liturgy* (Crestwood, NY: St. Vladimir's Seminary Press, 1960).
Komnene, Anna, *Alexias*.
Ed. A. Kambylis and D. R. Reinsch, *Alexias*, Corpus Fontium Historiae Byzantinae, Series Berolinensis 40/1 (Berlin: De Gruyter, 2001).
Trans. E. R. A. Sewter, *The Alexiad*, revised edition (New York: Penguin Classics, 2003).
Komnenos, Isaac, *De malorum subsistentia*.
Ed. James John Rizzo, *Περὶ τῆς τῶν κακῶν ὑποστάσεως (De malorum subsistentia)*, Beiträge zur Klassischen Philologie 42 (Meisenheim am Glan: Hain, 1971).

Laertius, Diogenes, *Vitae philosophorum.*
Ed. and trans. R. D. Hicks, *Lives of Eminent Philosophers*, 2 vols., Loeb Classical Library 184–185 (Cambridge, MA: Harvard University Press, 1925).
Leo of Chalcedon, *Epistle to Alexios I Komnenos.*
Ed. Alexander Lauriotes, *Ekklesiastike Aletheias* 20 (1900): 403AB
Leo the Physician, *De natura hominis synopsis.*
Ed. and trans. Robert Renehan, *Epitome on the Nature of Man*, Corpus Medicorum Graecorum 10.4 (Berlin: Akademie Verlag, 1969).
Longinus, *De sublimitate.*
Ed. and trans. W. H. Fyfe, *On the Sublime*, revised Donald Russell, Loeb Classical Library 199 (Cambridge, MA: Harvard University Press, 1995).
Lucian, *Imagines.*
Ed. and trans. A. M. Harmon, *Lucian*, vol. IV, Loeb Classical Library 162 (Cambridge, MA: Harvard University Press, 1925).
Mauropous, John, *Epigrammata.*
Ed. P. de Lagarde, *Joannis Euchaitorum Metropolitae Quae in Codice Vaticano Graeco 676 Supersunt*, Abhandlungen der Historisch-Philologischen Classe der Königlichen Gesellschaft der Wissenschaften zu Göttingen Bd. 28, reprint (Amsterdam: Hakkert, 1979).
Maximos the Confessor, *Ambigua ad Joannem.*
Ed. PG 91:1062–1417.
Partial trans. Andrew Louth, *Maximus the Confessor* (London: Routledge, 1996).
Ed. and trans. Nicholas Constas, *On Difficulties in the Church Fathers: The Ambigua*, 2 vols., Dumbarton Oaks Medieval Library 28–29 (Cambridge, MA: Harvard University Press, 2014).
Maximos the Confessor, *Capita de caritate.*
Ed. A. Ceresa-Gastaldo, *Capitoli sulla carita* (Rome: Editrice Studium, 1963).
Maximos the Confessor, *Mystagogia.*
Ed. Raffaele Cantarella, *La Mistagogia ed altri scritti* (Florence: Testi Cristiani, 1931).
Trans. Dom Julian Stead, *The Church, the Liturgy, and the Soul of Man: The Mystagogia of St. Maximus the Confessor* (Still River, MA: St. Bede's Publications, 1982).
Meletios the Monk, *De natura hominis.*
Ed. J. A. Cramer, *Anecdota Graeca E Codd. manuscriptis bibliothecarum Oxoniensium*, vol. III, reprint (Amsterdam: Hakkert, 1963).
Mesarites, Nicholas, *Descriptio ecclesiae SS. Apostolorum.*
Ed. and trans. Glanville Downey, "Nikolaos Mesarites: Description of the Church of Holy Apostles," *Transactions of the American Philosophical Society* n.s. 47:6 (1957): 855–924.
Metochites, Theodore, *Γνωμικαί σημειώσεις.*
Ed. G. Müller and T. Kiessling, *Miscellanea philosophica et historica* (Leipzig: Teubner, 1821).

Ed. and trans. K. Hult, *Theodore Metochites on Ancient Authors and Philosophy (Semeioseis gnomikai 1–26 et 71)* (Gothenburg: Acta Universitatis Gothoburgensis, 2002).

Metochites, Theodore, *Paraphrasis in Aristotelis De memoria et reminiscentia.*

Ed. David Bloch, "Theodoros Metochites on Aristotle's *De Memoria*: An Edition," *Cahiers de l'Institut du Moyen-Âge Grec et Latin* 76 (2005): 3–30.

Metochites, Theodore, *Poems.*

Ed. Ioannis Polemis, *Carmina*, Corpus Christianorum, Series Graeca 83 (Turnhout: Brepols, 2015).

Ed. M. Treu, *Dichtungen des Gross-Logotheten Theodoros Metochites* (Potsdam: Krämer, 1895).

Metochites, Theodore, *Στοιχείωσις ἀστρονομική.*

Partial ed. and trans. Börje Bydén, *Theodore Metochites'* Stoicheiosis Astronomike *and the Study of Natural Philosophy and Mathematics in Early Palaiologan Byzantium* (Gothenburg: Acta Universitatis Gothoburgensis, 2003).

Michael the Deacon, *Ekphrasis of Hagia Sophia.*

Ed. and trans. Cyril Mango and John Parker, "A Twelfth-Century Description of St. Sophia," *Dumbarton Oaks Papers* 14 (1960): 233–245.

Michael of Ephesus, *In ethica Nicomachea IX-X commentaria.*

Ed. G. Heylbut, *In ethica Nicomachea commentaria*, Commentaria in Aristotelem Graeca 20 (Berlin: Reimer, 1892).

Michael of Ephesus, *In libros de partibus animalium commentaria.*

Ed. M. Hayduck, *In libros de partibus animalium, de animalium motione, de animalium incessu commentaria*, Commentaria in Aristotelem Graeca 22.2 (Berlin: Reimer, 1904).

Michael of Ephesus, *In parva naturalia commentaria.*

Ed. P. Wendland, *In parva naturalia commentaria*, Commentaria in Aristotelem Graeca 22.1 (Berlin: Reimer, 1903).

Nemesios, *De natura hominis.*

Ed. M. Morani, *De natura hominis* (Leipzig: Teubner, 1987).

Trans. R. W. Sharples and P. J. van der Eijk, *On the Nature of Man* (Liverpool University Press, 2008).

Nicholas of Myra (= Nicholas Rhetor), *Progymnasmata.*

Ed. J. Felten, *Progymnasmata*, Rhetores Graeci 11 (Leipzig: Teubner, 1913).

Trans. George A. Kennedy, *Progymnasmata: Greek Textbooks of Prose Composition and Rhetoric* (Atlanta: Society of Biblical Literature, 2003), 129–172.

Nicholas and Theodore of Andida, *Προθεωρία Κεφαλαιώδης.*

Ed. PG 140:418-468.

Nikephoros I, *Antirrheticus.*

Ed. PG 100:205–534.

Nikephoros I, *Refutatio et eversio definitionis synodalis anni 815.*

Ed. J. Featherstone, *Refutatio et eversio definitionis synodalis anni 815*, Corpus Christianorum, Series Graeca 33 (Turnhout: Brepols, 1997).

Niketas David Paphlagon, *Vita patriarchae Ignatii.*
Ed. and trans. Andrew Smithies, *The Life of Patriarch Ignatius* (Washington, DC: Dumbarton Oaks, 2013).
Oribasios, *Collectiones medicae.*
Ed. J. Raeder, *Collectionum Medicarum Reliquiae*, vol. IV, Corpus Medicorum Graecorum 6.2.2 (Leipzig: Teubner, 1933).
Origen, *De Principiis.*
Ed. H. Görgemanns and H. Karpp, *Vier Bücher von den Prinzipien* (Darmstadt: Wissenschaftliche Buchgesellschaft, 1976).
Trans. G. W. Butterworth, *On First Principles* (New York: Harper & Row, 1966).
Pachymeres, George, *Quadrivium.*
Ed. P. Tannery, *Quadrivium de Georges Pachymère*, Studi e Testi 94 (Vatican City: Biblioteca Apostolica Vaticana, 1940).
Paphnutius, *Life of Onnophrius.*
Ed. E. A. Wallis Budge, "The Life of Apa Onnophrios the Anchorite," *Coptic Martyrdoms*, vol. I (London: British Museum, 1914), 205–224, 455–473.
Trans. Tim Vivian, *Histories of the Monks of Upper Egypt and the Life of Onnophrius*, revised edition (Piscataway, NJ: Gorgias Press, 2009), 143–166.
Pardos, Gregory, *Commentarium in Hermogenis librum περὶ μεθόδου δεινότητος*
Ed. C. Walz, *Rhetores Graeci*, vol. VII.2 (Stuttgart: Cotta, 1834).
Paul of Aegina, *Epitomae medicae libri septem.*
Ed. J. L. Heiberg, *Paulus Aegineta*, 2 vols., Corpus Medicorum Graecorum 9.1–9.2 (Leipzig: Teubner, 1921–1924).
Trans. Francis Adam, *The Seven Books of Paul Ægineta*, 3 vols. (London: Sydenham Society, 1844–1847).
Philo of Alexandria, *De Abrahamo.*
Ed. and trans. F. H. Colson, *Philo*, vol. VI, Loeb Classical Library 289 (Cambridge, MA: Harvard University Press, 1935).
Philo of Alexandria, *Legum allegoriarum libri I-III.*
Ed. and trans. F. H. Colson and G. H. Whitaker, *Philo*, vol. I, Loeb Classical Library 226 (Cambridge, MA: Harvard University Press, 1929).
Philoponos, John, *In Aristotelis libros de anima commentaria.*
Ed. M. Hayduck, *In Aristotelis de anima libros commentaria*, Commentaria in Aristotelem Graeca 15 (Berlin: Reimer, 1897).
Trans. Philip J. van der Eijk, *On Aristotle's On the Soul 1.1–2* (Ithaca, NY: Cornell University Press, 2005).
Trans. Philip J. van der Eijk, *On Aristotle's On the Soul 1.3–5* (Ithaca, NY: Cornell University Press, 2006).
Trans. William Charlton, *On Aristotle's On the Soul 2.1–6* (London: Duckworth, 2005).
Trans. William Charlton, *On Aristotle's On the Soul 2.7–12* (London: Duckworth, 2005).
Trans. William Charlton, *On Aristotle's On the Soul 3.1–8* (Ithaca, NY: Cornell University Press, 2000).

Trans. William Charlton, *On Aristotle's On the Soul 3.9–13* (Ithaca, NY: Cornell University Press, 2000).

Philostratus the Elder, *Imagines*.

Ed. and trans. Arthur Fairbanks, *Imagines*, Loeb Classical Library 256 (Cambridge, MA: Harvard University Press, 1931).

Philostratus, Flavius, *Vita Apollonii*.

Ed. and trans. Christopher P. Jones, *Apollonius of Tyana*, 2 vols., Loeb Classical Library 16–17 (Cambridge, MA: Harvard University Press, 2005).

Photios, *Bibliotheca*.

Ed. R. Henry, *Bibliothèque*, 8 vols. (Paris: Les Belles Lettres, 1959–1977).

Photios, *Epistula et amphilochia*.

Ed. B. Laourdas and L. G. Westerink, *Epistulae et Amphilochia*, vols. I–VI.2 (Leipzig: Teubner, 1983–1988).

Partial trans. Despina Stratoudaki White, *Patriarch Photios of Constantinople: His Life, Scholarly Contributions, and Correspondence Together with a Translation of Fifty-two of His Letters* (Brookline, MA: Holy Cross Orthodox Press, 1981).

Photios, *Homiliae*.

Ed. B. Laourdas, *ΟΜΙΛΙΑΙ* (Thessaloniki: Hetaireia Makedonikōn Spoudōn, 1959).

Trans. Cyril Mango, *The Homilies of Photius Patriarch of Constantinople* (Cambridge, MA: Harvard University Press, 1958).

Photios, *Lexicon*.

Ed. C. Theodoridis, *Lexicon*, 3 vols. (Berlin: De Gruyter, 1982–2013).

Plato, *Phaedo*.

Ed. and trans. Harold North Fowler, *Phaedrus*, Loeb Classical Library 36 (Cambridge, MA: Harvard University Press, 1914).

Plato, *Respublica*.

Ed. Christopher Emlyn-Jones and William Preddy, *Republic*, 2 vols., Loeb Classical Library 237 and 276 (Cambridge, MA: Harvard University Press, 2013).

Plato, *Timaeus*.

Ed. and trans. R. G. Bury, *Timaeus*, Loeb Classical Library 234 (Cambridge, MA: Harvard University Press, 1929).

Pliny the Elder, *Historia naturalis*.

Ed. and trans. H. Rackham, *Natural History*, 10 vols., Loeb Classical Library 330, 352, 353, 370, 371, 392, 393, 418, 394, 419 (Cambridge, MA: Harvard University Press, 1938–1963).

Plotinos, *Enneades*.

Ed. and trans. A. H. Armstrong, *Enneads*, 7 vols., Loeb Classical Library 440–445, 468 (Cambridge, MA: Harvard University Press, 1966–1988).

Ed. P. Henry and H.-R. Schwyzer, *Opera*, vol. I (Leiden: Brill, 1951).

Trans. Stephen Mackenna and B. S. Page, *The Six Enneads* (London: Faber, 1966).

Porphyry, *Εἰς τὰ ἁρμονικὰ Πτολεμαίου ὑπόμνημα*.

Ed. and trans. Andrew Barker, *Porphyry's Commentary on Ptolemy's Harmonics: A Greek Text and Annotated Translation* (Cambridge University Press, 2015).

Cf. ed. I. Düring, *Kommentar zur Harmonielehre des Ptolemaios* (Gothenburg: Elanders, 1932).

Porphyry, *Vita Plotini.*

Ed. and trans. A. H. Armstrong, *Porphyry on Plotinus*, Loeb Classical Library 440 (Cambridge, MA: Harvard University Press, 1969).

Priscian, *Metaphrasis in Theophrastum.*

Ed. I. Bywater, *Prisciani Lydi Quae Extant*, Commentaria in Aristotelem Graeca suppl. 1.2 (Berlin: Reimer, 1886).

Trans. Pamela Huby, *On Theophrastus On Sense-Perception* (Ithaca, NY: Cornell University Press, 1997).

Prodromos, Theodore, *Poems.*

Ed. E. Miller, "Poésies inédites de Théodore Prodrome," *Annuaire de l'Association Pour l'encouragement des Études Grecques* 17 (1883): 18–64.

Proklos, *In primum Euclidis elementorum librum commentarii.*

Ed. G. Friedlein, *In primum Euclidis elementorum librum commentarii* (Leipzig: Teubner, 1873).

Psellos, Michael, *Chronographia.*

Ed. Diether Roderich Reinsch, *Chronographia*, 2 vols. (Berlin: De Gruyter, 2014).

Trans. E. R. A. Sewter, *Fourteen Byzantine Rulers* (New York: Penguin Books, 1966).

Psellos, Michael, *De Euripide et Georgio Piside judicium.*

Ed. and trans. Andrew R. Dyck, "Euripedes and George of Pisidia" in A. R. Dyck (ed.), *The Essays on Euripides and George of Pisidia and on Heliodorus and Achilles Tatius* (Vienna: Österreichische Akademie der Wissenschaften, 1986), 39–74.

Psellos, Michael, *De Heliodoro et Achille Tatio judicium.*

Ed. and trans. Andrew R. Dyck, "Heliodorus and Achilles Tatius" in A. R. Dyck (ed.), *The Essays on Euripides and George of Pisidia and on Heliodorus and Achilles Tatius* (Vienna: Österreichische Akademie der Wissenschaften, 1986), 89–118.

Psellos, Michael, *De omnifaria doctrina.*

Ed. L. G. Westerink, *De omnifaria doctrina* (Nijmegen: Centrale Drukkerij, 1948).

Psellos, Michael, *Epistula.*

Ed. K. N. Sathas, *Μεσαιωνική Βιβλιοθήκη*, vol. v (Venice: Typois tou Chronou 1876).

Psellos, Michael, *Epistula ad Michaelem Cerularium.*

Ed. U. Criscuolo, *Epistola a Michele Cerulario*, Hellenica et byzantina neapolitana 15, reprint (Naples: Bibliopolis, 1990).

Trans. Anthony Kaldellis and Ioannis Polemis, "Letter to the Patriarch Kyr Michael Keroullarios" in A. Kaldellis and I. Polemis (eds.), *Psellos and the Patriarchs: Letters and Funeral Orations for Keroullarios, Leichoudes, and Xiphilinos* (Notre Dame: University of Notre Dame, 2015), 37–48.

Psellos, Michael, *Opuscula logica, physica, allegorica, alia.*

Ed. J. M. Duffy, *Philosophica minora*, vol. 1 (Leipzig: Teubner, 1992).

Psellos, Michael, *Opuscula psychologica, theologica, daemonologica.*
Ed. D. J. O'Meara, *Philosophica minora*, vol. II (Leipzig: Teubner, 1989).
Psellos, Michael, *Orationes hagiographicae.*
Ed. Elizabeth A. Fisher, *Orationes hagiographicae* (Leipzig: Teubner, 1994).
Partial trans. Elizabeth A. Fisher, *Michael Psellos on Symeon the Metaphrast and on the Usual Miracle at Blachernae: Annotated Translations with Introductions* (Washington, DC: Center for Hellenic Studies, 2014).
Psellos, Michael, *Oratoria minora.*
Ed. A. R. Littlewood, *Oratoria minora* (Leipzig: Teubner, 1985).
Psellos, Michael, *Poemata.*
Ed. L. G. Westerink, *Poemata* (Stuttgart: Teubner, 1992).
Psellos, Michael, *Theologica.*
Ed. P. Gautier, *Theologica*, vol. I (Leipzig: Teubner, 1989).
Ed. J. M. Duffy and L. G. Westerink, *Theologica*, vol. II (Munich: De Gruyter, 2002).
Pseudo-Alexander of Tralles, *De oculis libri tres.*
Ed. T. Puschmann, *Nachträge zu Alexander Trallianus*, reprint (Amsterdam: Hakkert, 1963).
Pseudo-Dionysios the Areopagite, *De divinis nominibus.*
Ed. B. R. Suchla, *De divinis nominibus*, Patristische Texte und Studien 33 (Berlin: De Gruyter, 1990).
Trans. Colm Luibheid, "The Divine Names" in Colm Luibheid (ed.), *The Complete Works* (New York: Paulist Press, 1987), 47–131.
Pseudo-Plutarch (= Aetios of Antioch), *De placita philosophorum.*
Ed. Jürgen Mau, *Plutarchi moralia*, vol. V.2.1 (Leipzig: Teubner, 1971).
Trans. William W. Goodwin, *Plutarch's Morals* (Boston: Colonial Press, 1905).
Ptolemy, Claudius, *Apotelesmatica*, or *Tetrabiblos.*
Ed. E. Boer and F. Boll, *Opera quae exstant omnia*, vol. III.1 (Leipzig, 1940).
Ptolemy, Claudius, *Optics.*
Ed. A. Lejeune, *L'Optique de Claude Ptolémée dans la version latine d'après l'arabe de l'émir Eugène de Sicile* (Leiden: Brill, 1989).
Trans. A. Mark Smith, *Ptolemy's Theory of Visual Perception: An English Translation of the Optics* (Philadelphia: American Philosophical Society, 1996).
ibn Riḍwān, ʿAli, *Kifāyat at-ṭabīb fīmā ṣaḥḥa ladaya min al-tajārīb.*
Ed. and French trans. Jacques Grand'Henry, *Le livre de la méthode du médecin de 'Ali B. Ridwan (998–1067)*, 2 vols. (Louvain-la-Neuve: Université catholique de Louvain, Institut orientaliste, 1979).
Scholastikos, John (= John Scythopolis), *Prologus et scholia in Dionysii Areopagitae librum De divinis nominibus.*
Ed. B. R. Suchla, *Prologus et scholia in Dionysii Areopagitae librum "De divinis nominibus" cum additamentis interpretum aliorum*, Corpus Dionysiacum 4.1, Patristische Texte und Studien 62 (Berlin: De Gruyter, 2011).
Seth, Symeon, *Conspectus rerum naturalium.*
Ed. A. Delatte, *Anecdota Atheniensia et alia*, vol. II (Liège: H. Vaillant-Carmanne, 1939).

Seth, Symeon, *Oratio antirrhetica ad Galenum.*
Ed. and trans. Petros Bouras-Vallianatos and Sophia Xenophontos, "Galen's Reception in Byzantium: Symeon Seth and His Refutation of Galenic Theories on Human Physiology," *Greek, Roman, and Byzantine Studies* 55 (2015): 431–469.
Sikeliotes, John, *Commentarium in Hermogenis librum περὶ ἰδεῶν.*
Ed. C. Walz, *Rhetores Graeci*, vol. VI (Stuttgart: Cotta, 1834).
Simplikios, *In Aristotelis libros de anima commentaria.*
Ed. M. Hayduck, *In libros Aristotelis de anima commentaria*, Commentaria in Aristotelem Graeca 11 (Berlin: Reimer, 1882).
Trans. J. O. Urmson and P. Lautner, *On Aristotle On the Soul 1.1–2.4* (London: Duckworth, 1995).
Trans. Carlos Steel, *On Aristotle On the Soul 2.5–12* (London: Duckworth, 1997).
Trans. H. J. Blumenthal, *On Aristotle On the Soul 3.1–5* (London: Duckworth, 2000).
Trans. C. Steel and A. Ritups, *On Aristotle On the Soul 3.6–13* (London: Bristol Classical Press, 2013).
Sophonias, *In Aristotelis libros de anima paraphrasis.*
Ed. M. Hayduck, *In libros Aristotelis de anima paraphrasis*, Commentaria in Aristotelem Graeca 23.1 (Berlin: Reimer, 1883).
Partial trans. W. Charlton and F. Bossier, *On Aristotle's On the Intellect* (London: Duckworth, 1991).
Sophronios, *Anacreontica.*
Ed. M. Gigante, *Anacreontica*, Testi per esercitazioni accademiche 10–12 (Rome: Gismondi, 1957).
Trans. John Wilkinson, *Jerusalem Pilgrims Before the Crusades* (Warminster: Aris & Phillips, 1977), 90–92.
Stobaios, John, *Anthologium.*
Ed. O. Hense and C. Wachsmuth, *Anthologium*, 5 vols. (Berlin: Weidmann, 1884–1912).
Synesios, *De insomniis.*
Ed. and trans. Donald A. Russell and Heinz-Günther Nesselrath, *On Prophecy, Dreams and Human Imagination: Synesius,* De insomniis (Tübingen: Mohr Siebeck, 2014).
Cf. ed. N. Terzaghi, *Opuscula* (Rome: Polygraphica, 1944).
Syropoulos, Sylvester, *Historiae.*
Ed. V. Laurent, *Les «Mémoires» du Grand Ecclésiarque de l'Église de Constantinople Sylvestre Syropoulos sur le concile de Florence (1438–1439)* (Paris: Centre National de la Recherche Scientifique, 1971).
Themistios, *In Aristotelis libros de anima paraphrasis.*
Ed. R. Heinze, *In libros Aristotelis de anima paraphrasis*, Commentaria in Aristotelem Graeca 5.3 (Berlin: Reimer, 1899).
Trans. Robert B. Todd, *On Aristotle On the Soul* (London: Duckworth, 1996).

Theodore of Stoudios, *Antirrhetici adversus iconomachos.*
Ed. PG 99:327–436.
Trans. Catharine P. Roth, *On the Holy Icons* (Crestwood, NY: St. Vladimir's Seminary Press, 1981).
Cf. trans. Thomas Cattoi, *Writings on Iconoclasm*, Ancient Christian Writers (New York: Newman Press, 2015).

Theodore of Stoudios, *Epistulae.*
Ed. Georgios Fatouros, *Epistulae*, 2 vols., Corpus Fontium Historiae Byzantinae, Series Berolinensis 31 (Berlin: De Gruyter, 1992).

Theodoret of Cyrrhus, *Interpretatio in Psalmos.*
Ed. PG 80:857–1997.

Theon, Aelius, *Progymnasmata.*
Ed. L. Spengel, *Rhetores Graeci*, vol. II (Leipzig: Teubner, 1854).
Trans. George A. Kennedy, *Progymnasmata: Greek Textbooks of Prose Composition and Rhetoric* (Atlanta: Society of Biblical Literature, 2003), 1–72.

Theon of Alexandria, *Opticorum recensio Theonis.*
Ed. J. L. Heiberg, *Euclidis opera omnia*, vol. VII (Leipzig: Teubner, 1895), 144–246.

Theophrastus, *De sensu et sensibilibus*, or Fragment 1.
Ed. H. Diels, *Doxographi Graeci* (Berlin: Reimer, 1879).
Ed. and trans. G. M. Stratton, *Theophrastus and the Greek Physiological Psychology before Aristotle* (New York: Macmillan, 1927).

Theophrastus, *Fragmenta.*
Ed. F. Wimmer, *Opera, quae supersunt, omnia* (Paris: Didot, 1866).

Tornikios, George, *Orationes.*
Ed. J. Darrouzès, *Lettres et discours* (Paris: Centre National de la Recherche Scientifique, 1970).

Secondary Sources

Agapitos, P., "Teachers, Pupils, and Imperial Power in Eleventh Century Byzantium" in Yun Lee Too and Niall Livingstone (eds.), *Pedagogy and Power: Rhetorics of Classical Learning* (Cambridge University Press, 1998), 170–191.

Aghiorgoussis, Maximos, "Applications of the Theme 'Eikon Theou' (Image of God) according to Saint Basil the Great," *Greek Orthodox Theological Review* 21:3 (1976): 265–288.

Akbari, Suzanne Conklin, *Seeing through the Veil: Optical Theory and Medieval Allegory* (University of Toronto Press, 2004).

Alden Smith, R., *The Primacy of Vision in Virgil's Aeneid* (Austin: University of Texas Press, 2005).

Alexander, Paul J., *The Patriarch Nicephorus of Constantinople: Ecclesiastical Policy and Image Worship in the Byzantine Empire* (Oxford University Press, 1958).

Amand de Mendieta, Emmanuel, "Les deux homélies sur la création de l'homme que les manuscrits attribuent à Basile de Césarée ou à Grégoire de Nysse: le problème de leur rédaction" in *Zetesis: Album amicorum door vrienden en collega's aangeboden aan Prof. Dr. E. de Strycker ter gelegenheid van zijn 65e verjaardag* (Antwerp: De Nederlandsche Boekhandel, 1973), 695–716.

"Les neuf homélies de Basile de Césarée sur l'hexaéméron," *Byzantion* 48 (1978): 337–368.

Anastos, Milton, "The Ethical Theory of Images Formulated by the Iconoclasts in 754 and 815," *Dumbarton Oaks Papers* 8 (1954): 151–160.

Andriopoulos, D. Z., "Alcmaeon's Epistemological Framework," *Philosophical Inquiry* 38 (2014): 42–60.

Annas, Julia, "Truth and Knowledge" in Malcolm Schofield, Myles Burnyeat, and Jonathan Barnes (eds.), *Doubt and Dogmatism: Studies in Hellenistic Epistemology* (Oxford University Press, 1980), 84–104.

Anton, John P., "The Aristotelianism of Photius's Philosophical Theology" in Lawrence P. Schrenk (ed.), *Aristotle in Late Antiquity* (Washington, DC: Catholic University of America Press, 1994), 158–183.

Arabatzis, George, "Michael of Ephesus on the Empirical Man, the Scientist and the Educated Man (*In Ethica Nicomachea* x and *In Partibus Animalium* 1)" in Charles Barber and David Jenkins (eds.), *Medieval Greek Commentaries on the Nicomachean Ethics* (Leiden: Brill, 2009), 163–183.

"Michael of Ephesus and the Philosophy of Living Things (*In De Partibus Animalium*, 22.25–23.9)" in Börje Bydén and Katerina Ierodiakonou (eds.), *The Many Faces of Byzantine Philosophy* (Athens: Norwegian Institute at Athens, 2012), 51–78.

Παιδεία και Επιστήμη στον Μιχαήλ Εφέσιο. Εις περι ζώων μορίων Α 1,3–2,10 (Athens: Akadēmia Athēnōn, 2006).

"*Phantasia* et intellect chez Syméon Seth: sur les sources du *Conspectus rerum naturalium*" in M. C. Pacheco and J. F. Meirinhos (eds.), *Intellect et imagination dans la philosophie médiévale*, vol. 1 (Turnhout: Brepols, 2006), 483–493.

Armisen, Mireille, "La notion d'imagination chez les anciens: I – Les philosophes," *Pallas* 26 (1979): 11–51.

"La notion d'imagination chez les anciens: II – La rhétorique," *Pallas* 27 (1980): 3–37.

Babut, D., "Sur la notion d'imitation dans les doctrines esthétiques de la Grèce classique," *Revue des Études Grecques* 98 (1985): 72–92.

Baldes, R. W., "Democritus on the Nature of Perception of 'Black' and 'White'," *Phronesis* 23 (1978): 87–100.

"Democritus on Visual Perception: Two Theories or One?," *Phronesis* 20 (1975): 93–105.

Baldwin, Barry, "Beyond the House Call: Doctors in Early Byzantine History and Politics," *Dumbarton Oaks Papers* 38 (1984): 15–19.

"The Career of Oribasios," *Acta Classica* 18 (1975): 85–97.

Baltussen, H., "The Purpose of Theophrastus's "de Sensibus" Reconsidered," *Apeiron* 31:2 (1998): 167–199.

Theophrastus Against the Presocratics and Plato: Peripatetic Dialectic in the De sensibus (Leiden: Brill, 2000).

Barber, Charles, *Contesting the Logic of Painting: Art and Understanding in Eleventh-Century Byzantium* (Leiden: Brill, 2007).

"Eustratios of Nicaea on the Separation of Art and Theology" in Charles Barber and David Jenkins (eds.), *Medieval Greek Commentaries on the Nicomachean Ethics* (Leiden: Brill, 2009), 131–143.

Figure and Likeness: On the Limits of Representation in Byzantine Iconoclasm (Princeton University Press, 2002).

"From Image into Art: Art after Byzantine Iconoclasm," *Gesta* 34:1 (1995): 5–10.

"From Transformation to Desire: Art and Worship after Byzantine Iconoclasm," *The Art Bulletin* 75:1 (1993): 7–16.

"Living Painting, or the Limits of Pointing? Glancing at Icons with Michael Psellos" in C. Barber and D. Jenkins (eds.), *Reading Michael Psellos* (Leiden: Brill, 2006), 117–130.

Barber, Charles, and Stratis Papaioannou, *Michael Psellos on Literature and Art: A Byzantine Perspective on Aesthetics* (South Bend, IN: Notre Dame University Press, 2017).

Barion, C., "La rôle de la vue dans les arts de la mémoire latins" in L. Villard (ed.), *Études sur la vision dans l'antiquité classique* (Rouen: Publications des Universités de Rouen et du Havre, 2005), 199–214.

Bartsch, Shadi, *The Mirror of the Self: Sexuality, Self-Knowledge, and the Gaze in the Early Roman Empire* (University of Chicago Press, 2006).

"'Wait a Moment, *Phantasia*': Ekphrastic Interference in Seneca and Epictetus," *Classical Philology* 102:1 (2007): 83–95.

Beare, John I., *Greek Theories of Elementary Cognition: From Alcmaeon to Aristotle* (Oxford: Clarendon Press, 1906).

Bellosta, H., "Burning Instruments from Diocles to Ibn Sahl," *Arabic Sciences and Philosophy* 12 (2002): 285–303.

Belting, Hans, *Likeness and Presence: A History of the Image before the Era of Art*, trans. Edmund Jephcott (University of Chicago Press, 1994).

Benakis, L. G., "Commentaries and Commentators on the Works of Aristotle (except the Logical Ones) in Byzantium" in Burkhard Mojsisch and Olaf Pluta (eds.), *Historia philosophiae medii aevi* (Amsterdam: B. R. Grüner, 1991), 45–54.

Beneker, Jeffrey and Craig A. Gibson, *The Rhetorical Exercises of Nikephoros Basilakes: Progymnasmata from Twelfth-Century Byzantium*, Dumbarton Oaks Medieval Library 43 (Cambridge, MA: Harvard University Press, 2016).

Bernard, Floris, *Writing and Reading Byzantine Secular Poetry, 1025–1081* (Oxford University Press, 2014).

Bernard, Floris, and Kristoffel Demoen (eds.), *Poetry and Its Contexts in Eleventh-Century Byzantium* (Burlington, VT: Ashgate, 2013).

Betancourt, Roland, "The Icon's Gold: A Medium of Light, Air, and Space," *West 86th: A Journal of Decorative Arts, Design History, and Material Culture* 23:2 (2016): 252–280.

"A Liturgical Commentary in Verse: Introduction and Translation," *Orientalia Christiana Periodica* 81:2 (2015): 433–472.

"Medieval Art after Duchamp: Hans Belting's *Likeness and Presence* at 25," *Gesta* 55:1 (2016): 5–17.

"Stop Crossing Mediums: Grace Helbig and the YouTuber's Medium," *Journal of Popular Culture* 49:1 (2016): 196–223.

"Tempted to Touch: Tactility, Ritual, and Mediation in Byzantine Visuality," *Speculum* 91:3 (2016): 660–689.

"Why Sight Is Not Touch: Reconsidering the Tactility of Vision in Byzantium," *Dumbarton Oaks Papers* 70 (2016): 1–23.

Betts, E., "Towards a Multisensory Experience of Movement in the City of Rome" in R. Laurence and D. J. Newsome (eds.), *Rome, Ostia and Pompeii: Movement and Space* (Oxford University Press, 2011), 118–132.

Biernoff, Suzannah, *Sight and Embodiment in the Middle Ages* (New York: Palgrave Macmillan, 2002).

Blumenthal, H. J., *Aristotle and Neoplatonism in Late Antiquity: Interpretations of the De Anima* (London: Duckworth, 1996).

"John Philoponos and Stephanus of Alexandria: Two Neoplatonic Christian Commentators on Aristotle?" in D. J. O'Meara (ed.), *Neoplatonism and Christian Thought* (Albany: SUNY Press, 1982), 54–63.

"Neoplatonic Interpretations of Aristotle on 'Phantasia,'" *Review of Metaphysics* 31:2 (1977): 242–257.

"Photius on Themistius (Cod. 74): Did Themistius Write Commentaries on Aristotle?," *Hermes* 107 (1979): 168–182.

"Plotinus' Adaptation of Aristotle's Psychology: Sensation, Imagination, and Memory" in R. Baine Harris (ed.), *The Significance of Neoplatonism* (Norfolk, VA: International Society for Neoplatonic Studies, 1976), 41–58.

Plotinus' Psychology: His Doctrines of the Embodied Soul (The Hague: Martinus Nijhoff, 1971).

"Plutarch's Exposition of the *De Anima* and the Psychology of Proclus" in Bent Dalsgaard Larsen (ed.), *De Jamblique à Proclus* (Geneva: Fondation Hardt, 1975), 123–147.

Blundell, Sue, Douglas Cairns, Elizabeth Craik, and Nancy Sorkin Rabinowitz, "Introduction: Vision and Viewing in Ancient Greece," *Helios* 40:1–2 (2013): 3–37.

Blyth, P. H., "Apollodorus of Damascus and the *Poliorcetica*," *Greek, Roman, and Byzantine Studies* 33 (1992): 127–158.

Bolter, Jay David and Richard Grusin, *Remediation: Understanding New Media* (Cambridge, MA: MIT Press 1999).

Bouras-Vallianatos, Petros, "Clinical Experience in Late Antiquity: Alexander of Tralles and the Therapy of Epilepsy," *Medical History* 58:3 (2014): 337–353.

Bouras-Vallianatos, Petros and Sophia Xenophontos, "Galen's Reception in Byzantium: Symeon Seth and His Refutation of Galenic Theories on Human Physiology," *Greek, Roman, and Byzantine Studies* 55 (2015): 431–469.

Bradley, M. (ed.), *Smell and the Ancient Senses* (London: Routledge, 2015).

Brisson, L., "Plato's Theory of Sense Perception in the *Timaeus*: How It Works and What It Means," *Proceedings of the Boston Area Colloquium in Ancient Philosophy* 12 (1998): 147–174.

Brittain, Charles, *Philo of Larissa: The Last of the Academic Skeptics* (Oxford University Press, 2001).

Broadie, S., "Aristotle's Perceptual Realism," *The Southern Journal of Philosophy* 31 (1993): 137–159.

Browning, Robert, "Enlightenment and Repression in Byzantium in the Eleventh and Twelfth Centuries," *Past & Present* 69 (1975): 3–23.

"Further Reflections on Literacy in Byzantium" in S. Reinert et al. (eds.), *To Hellenikon: Studies in Honor of Speros Vryonis, Jr.* (New Rochelle, NY: Aristide D. Caratzas, 1993), 69–84.

"Literacy in the Byzantine World," *Byzantine and Modern Greek Studies* 4 (1978): 39–54.

"The Patriarchal School at Constantinople in the Twelfth Century," *Byzantion* 32 (1962): 167–202.

"The Patriarchal School at Constantinople in the Twelfth Century," *Byzantion* 33 (1963): 11–40.

"An Unpublished Funeral Oration on Anna Comnena" in Richard Sorabji (ed.), *Aristotle Transformed: The Ancient Commentators and Their Influence* (Ithaca, NY: Cornell University Press, 1990), 393–406.

Brubaker, Leslie, "Perception and Conception: Art, Theory, and Culture in Ninth-Century Byzantium," *Word & Image* 5:1 (1989): 19–32.

Brunschwig, Jacques, "En quel sens le sens commun est-il commun?" in C. Viano and G. Romeyer Dherbey (eds.), *Corps et âme: sur le* De anima *d'Aristote* (Paris: J. Vrin, 1996), 189–218.

"Les multiples chemins aristotéliciens de la sensation commune," *Revue de Métaphysique et de Morale* 96 (1991): 455–474.

Burkert, W., "Air-Imprints or Eidola: Democritus' Aetiology of Vision," *Illinois Classical Studies* 2 (1977): 97–109.

Butler, Shane and Alex Purves (eds.), *Synaesthesia and the Ancient Senses* (Durham: Acumen Publishing, 2013).

Bydén, Börje, *Theodore Metochites'* Stoicheiosis Astronomike *and the Study of Natural Philosophy and Mathematics in Early Palaiologan Byzantium* (Gothenburg: Acta Universitatis Gothoburgensis, 2003).

Cacouros, Michel, "La philosophie et les sciences du *trivium* et du *quadrivium* à Byzance de 1204 à 1453 entre tradition et innovation: les textes et l'enseignement, le cas de l'école du Prodrome (Pétra)" in Michel Cacouros and M.-H. Congourdeau (eds.), *Philosophie et sciences à Byzance de 1204 à 1453: les textes, les doctrines et leur transmission* (Leuven: Peeters, 2006), 1–51.

Cairns, Douglas L., "Bullish Looks and Sidelong Glances: Social Interaction and the Eyes in Ancient Greek Culture" in Douglas L. Cairns (ed.), *Body Language in the Greek and Roman Worlds* (Swansea: Classical Press of Wales, 2005), 123–155.

"Looks of Love and Loathing: Cultural Models of Vision and Emotion in Ancient Greek Culture," *Mètis* n.s. 9 (2011): 37–50.

Cantin, Stanislas, "La perception des sensibles communs au moyen du mouvement d'après Aristote," *Laval théologique et philosophique* 17 (1961): 9–21.

Carr, Annemarie Weyl, "Leo of Chalcedon and the Icons" in Doula Mouriki et al. (eds.), *Byzantine East, Latin West: Art-Historical Studies in Honor of Kurt Weitzmann* (Department of Art and Archaeology, Princeton University, 1995), 579–584.

"Reflections on the Medium of the Miraculous" in Sharon E. J. Gerstel (ed.), *Viewing Greece: Cultural and Political Agency in the Medieval and Early Modern Mediterranean* (Turnhout: Brepols, 2016), 33–52.

Carruthers, Mary, *The Book of Memory* (Cambridge University Press, 1990).

The Craft of Thought: Meditation, Rhetoric, and the Making of Images, 400–1200 (Cambridge University Press, 1998).

Caseau, Béatrice, "Christian Bodies: The Senses and Early Byzantine Christianity" in Liz James (ed.), *Desire and Denial in Byzantium* (Burlington, VT: Ashgate, 1999), 101–109.

Castelli, Elizabeth, "Asterius of Amasea: Ekphrasis on the Holy Martyr Euphemia" in Richard Valantasis (ed.), *Religions of Late Antiquity in Practice* (Princeton University Press, 2000), 464–468.

Catrein, Christoph, *Vertauschte Sinne: Untersuchungen zur Synästhesie in der römischen Dichtung* (Munich: K. G. Saur, 2003).

Charles, A., "L'imagination, miroir de l'âme selon Proclos" in M. P. M. Schuhl and M. P. Hadot (eds.), *Le Néoplatonisme* (Paris: Éditions du Centre National de la Recherche Scientifique, 1971), 241–248.

Chase, Michael, "Porphyry on the Cognitive Process," *Ancient Philosophy* 30 (2010): 383–405.

Cherniss, Harold, "Galen and Posidonius' Theory of Vision," *American Journal of Philology* 54 (1933): 154–161.

Chiaradonna, Riccardo, "Platonismo e teoria della conoscenza stoica tra II e III secolo D.C." in M. Bonazzi and C. Helmig (eds.), *Platonic Stoicism – Stoic Platonism: A Dialogue between Platonism and Stoicism in Antiquity* (Leuven University Press, 2007), 209–241.

"Platonist Approaches to Aristotle: From Antiochus of Ascalon to Eudorus of Alexandria (and Beyond)" in Malcolm Schofield (ed.), *Aristotle, Plato and Pythagoreanism in the First Century BC* (Cambridge University Press, 2013), 28–52.

Chinn, C. M., "Before Your Very Eyes: Pliny *Epistulae* 5.6 and the Ancient Theory of Ekphrasis," *Classical Philology* 102 (2007): 265–280.

Clark, Elizabeth, *The Origenist Controversy* (Princeton University Press, 1992).

Clark, Gordon H., "Plotinus' Theory of Sensation," *Philosophical Review* 51:4 (1942): 357–382.

Coakley, S., "The Resurrection and the 'Spiritual Senses'" in S. Coakley (ed.), *Powers and Submissions: Spirituality, Philosophy, and Gender* (Oxford: Blackwell, 2002), 130–152.

Colish, Marcia, *The Stoic Tradition from Antiquity to the Early Middle Ages*, 2 vols. (Leiden: Brill, 1985).

Conley, Thomas M., "Aristotle's *Rhetoric* in Byzantium," *Rhetorica: A Journal of the History of Rhetoric* 8:1 (1990): 29–44.

Constantinides, C. N., *Higher Education in Byzantium in the Thirteenth and Early Fourteenth Centuries* (Nicosia: Cyprus Research Center, 1982).

Constas, Maximos (Nicholas), "Beyond the Veil: Imagination and Spiritual Vision in Byzantium," unpublished paper, University of Chicago Divinity School Workshop (March 2014, revised April 2015).

"Icons and the Imagination," *Logos: A Journal of Catholic Thought and Culture* 1:1 (1997): 114–127.

Cormack, Robin, "Living Painting" in Elizabeth Jeffreys (ed.), *Rhetoric in Byzantium: Papers from the Thirty-fifth Spring Symposium of Byzantine Studies, Exeter College, University of Oxford, March 2001* (Burlington, VT: Ashgate, 2003), 235–253.

Corrigan, Kevin, *Evagrius and Gregory: Mind, Soul and Body in the 4th Century* (Burlington, VT: Ashgate, 2009).

Cribiore, Rafaella, "The Grammarian's Choice: The Popularity of Euripides' *Phoenissae* in Hellenistic and Roman Education" in Yun Lee Too and Niall Livingston (eds.), *Pedagogy and Power: Rhetorics of Classical Learning* (Cambridge University Press, 1998), 241–259.

Gymnastics of the Mind: Greek Education in Hellenistic and Roman Egypt (Princeton University Press, 2001).

Curd, P., "A New Empedocles? Implications of the Strasburg Fragments for Presocratic Philosophy," *Proceedings of the Boston Area Colloquium in Ancient Philosophy* 17 (2001): 27–50.

Cytowic, Richard, *The Man Who Tasted Shapes* (Cambridge, MA: MIT Press, 1998).

Dagron, Gilbert, "Holy Images and Likeness," *Dumbarton Oaks Papers* 45 (1991): 23–33.

Dain, Alphonse, "La transmission des textes littéraires classiques de Photius à Constantin Porphyrogénète," *Dumbarton Oaks Papers* 8 (1954): 31–47.

Dann, Kevin T., *Bright Colors Falsely Seen: Synaesthesia and the Search for Transcendental Knowledge* (New Haven: Yale University Press, 1999).

Darrigol, Olivier, *A History of Optics: From Greek Antiquity to the Nineteenth Century* (Oxford University Press, 2012).

Daryaee, Touraj, "Sight, Semen, and the Brain: Ancient Persian Notions of Physiology in Old and Middle Iranian Texts," *Journal of Indo-European Studies* 30:1/2 (2002): 103–128.

Daskas, Beatrice, "Nota sulla theotókos descritta da Fozio, Hom. XVII 2 (p. 167.14–17 Laourdas)," *ACME: Annali della Facoltà di Lettere e Filosofia dell'Università degli Studi di Milano* 64: 2 (2011): 339–351.

De Groot, Jean Christensen, *Aristotle and Philoponus on Light* (New York: Garland, 1991).

"Philoponus on *De Anima* II.5, *Physics* III.3, and the Propagation of Light," *Phronesis* 28:2 (1983): 177–196.

Delling, Gerhard, "The 'One Who Sees God' in Philo" in Frederick E. Greenspahn, Earle Hilgert, and Burton L. Mack (eds.), *Nourished with Peace: Studies in Hellenistic Judaism in Memory of Samuel Sandmel* (Chico, CA: Scholars Press, 1984), 27–41.

Dickie, Matthew W., "The Fathers of the Church and the Evil Eye" in Henry Maguire (ed.), *Byzantine Magic* (Washington, DC: Dumbarton Oaks, 1995), 9–34.

"Heliodorus and Plutarch on the Evil Eye," *Classical Philology* 86:1 (1991): 17–29.

Domański, Bolesław, *Die Psychologie des Nemesios* (Munich: Aschendorff, 1900).

Duffy, John, "Byzantine Medicine in the Sixth and Seventh Centuries: Aspects of Teaching and Practice," *Dumbarton Oaks Papers* 38 (1984): 21–27.

"Hellenic Philosophy in Byzantium and the Lonely Mission of Michael Psellos" in Katerina Ierodiakonou (ed.), *Byzantine Philosophy and Its Ancient Sources* (Oxford University Press, 2004), 139–156.

Dunbabin, K., and M. W. Dickie, "*Invidia rumpantur pectora*: The Iconography of *Phthonos/Invidia* in Graeco-Roman Art," *Jahrbuch für Antike und Christentum* 26 (1983): 7–37.

Dundes, A. (ed.), *The Evil Eye: A Folklore Casebook* (Madison: University of Wisconsin Press, 1981).

Düring, Ingemar, *Ptolemaios und Porphyrios über die Musik,* Göteborgs Högskolas Årsskrift 40 (Gothenburg: Elanders Boktryckeri Aktiebolag, 1934).

Dyson, Henry, *Prolepsis and Ennoia in the Early Stoa* (Berlin: De Gruyter, 2009).

Edwards, M., "A Portrait of Plotinus," *Classical Quarterly* 43 (1993): 480–490.

Elsner, Jaś, *Art and the Roman Viewer: The Transformation of Art from the Pagan World to Christianity* (Cambridge University Press, 1995).

"Iconoclasm as Discourse: From Antiquity to Byzantium," *The Art Bulletin* 94:3 (2012): 368–394.

"Introduction: The Genres of Ekphrasis," *Ramus* 31 (2002): 1–18.

"Naturalism and the Erotics of the Gaze" in N. B. Kampen (ed.), *Sexuality in Ancient Art* (Cambridge University Press, 1996), 247–261.

Roman Eyes: Visuality and Subjectivity in Art and Text (Cambridge University Press, 2007).

Elsner, Jaś, and Michel Meyer (eds.), *Art and Rhetoric in Roman Culture* (Cambridge University Press, 2014), 1–32.

Emilsson, Eyjólfur Kjalar, *Plotinus on Sense-Perception: A Philosophical Study* (Cambridge University Press, 1988).

Everson, Stephen, *Aristotle on Perception* (Oxford University Press, 1997).

Eyl, J., "Why Thekla Does Not See Paul: Visual Perception and the Displacement of *erōs* in the *Acts of Paul and Thekla*" in M. P. Futre Pinheiro, J. Perkins, and R. Pervo (eds.), *The Ancient Novel and Early Christian and Jewish Narrative: Fictional Intersections* (Groningen: Barkhuis Publishing, 2012), 3–19.

Fattori, M., and M. Bianchi (eds.), *Phantasia-imaginatio: Atti del V. colloquio internazionale (Centro di studio per il lessico intellettuale europeo), Roma 9–11 gennaio 1986* (Rome: Edizioni dell'Ateneo, 1988).

Featherstone, Jeffrey Michael, "Metochites's Poems and the Chora" in Holger A. Klein, Robert G. Ousterhout, and Brigitte Pitarakis (eds.), *The Kariye Camii Reconsidered* (Istanbul: Istanbul Research Institute, 2011), 213–237.

Frank, Georgia, *The Memory of the Eyes: Pilgrims to Living Saints in Christian Late Antiquity* (Berkeley: University of California Press, 2000).

"'Taste and See': The Eucharist and the Eyes of the Faith in the Fourth Century," *Church History* 70:4 (2001): 619–643.

Frankopan, Peter, "The Literary, Cultural and Political Context for the Twelfth-Century Commentary on the *Nicomachean Ethics*" in Charles Barber and David Jenkins (eds.), *Medieval Greek Commentaries on the Nicomachean Ethics* (Leiden: Brill, 2009), 45–62.

Frede, Dorothea, "The Cognitive Role of Phantasia in Aristotle" in M. Nussbaum and A. O. Rorty (eds.), *Essays on Aristotle's* De Anima (Oxford University Press, 1992), 279–295.

Frede, Michael, "Stoics and Skeptics on Clear and Distinct Impressions" in Myles Burnyeat (ed.), *The Skeptical Tradition* (Berkeley: University of California Press, 1983), 65–93.

Freeman, Elizabeth, *Time Binds: Queer Temporalities, Queer Histories* (Durham, NC: Duke University Press, 2010).

Frontisi-Ducroux, Françoise, "Eros, Desire and the Gaze" in Nathalie B. Kampen (ed.), *Sexuality in Ancient Art* (Cambridge University Press, 1996), 81–100.

Garzya, Antonio, et al., *Medici Bizantini* (Turin: Unione Tipografico – Editrice Torinese, 2006).

Gaul, Niels, "All the Emperor's Men (and His Nephews): *Paideia* and Networking Strategies at the Court of Andronikos II Palaiologos, 1290–1320," *Dumbarton Oaks Papers* 70 (2016): 245–270.

Thomas Magistros und die spätbyzantinische Sophistik: Studien zum Humanismus urbaner Eliten der frühen Palaiologenzeit (Wiesbaden: Harrassowitz, 2011).

Gibson, Craig A., *Libanius's Progymnasmata: Model Exercises in Greek Prose Composition and Rhetoric* (Atlanta: Society of Biblical Literature, 2008).

Gifford, Edward, *Evil Eye: Studies in the Folklore of Vision* (New York: Macmillan, 1958).

Gigineishvili, L. and Van Riel, G., "Ioane Petritsi: A Witness to Proclus' Works in the School of Psellus" in Alain-Philippe Segonds and Carlos Steel (eds.), *Proclus et la théologie platonicienne*, (Leuven: Leuven University Press, 2000), 571–587.

Glavinas, Apostolos, *Ἡ ἐπὶ Ἀλεξίου Κομνηνοῦ (1081–1118) περὶ ἱερῶν σκευῶν, κειμηλίων καὶ Ἁγίων εἰκόνων ἔρις (1081–1095)* (Thessaloniki: Kentron Vyzantinōn Ereunōn, 1972).

Goldhill, Simon, "The Naïve and Knowing Eye: Ecphrasis and the Culture of Viewing in the Hellenistic World" in S. Goldhill and R. Osborne (eds.), *Art and Text in Ancient Greek Culture* (Cambridge University Press, 1994), 197–223.

"Refracting Classical Vision: Changing Cultures of Viewing" in T. Brennan and M. Jay (eds.), *Vision in Context: Historical and Contemporary Perspectives on Sight* (New York: Routledge, 1996), 17–28.

"Review of Ruth Webb, *Ekphrasis, Imagination and Persuasion in Ancient Rhetorical Theory and Practice*," *Bryn Mawr Classical Review* 2009.10.03 (2009), http://bmcr.brynmawr.edu/2009/2009–10-03.html.

"The Seductions of the Gaze: Socrates and his Girlfriends" in P. Cartledge, P. Millett, and S. von Reden (eds.), *Kosmos: Essays in Order, Conflict and Community in Classical Athens* (Cambridge University Press, 1998), 105–124.

"What Is Ekphrasis for?," *Classical Philology* 102:1 (2007): 1–19.

(ed.), *Being Greek under Rome: Cultural Identity, the Second Sophistic, and the Development of Empire* (Cambridge University Press, 2001).

Gouillard, Jean, "Contemplation et imagerie sacrée dans le christianisme byzantin," *École Pratique des Hautes Etudes, Section des Sciences Religieuses. Annuaire* 86 (1977–1978): 29–50.

Grant, Mark, *Dieting for an Emperor: A Translation of Books 1 and 4 or Oribasius' Medical Compilations with an Introduction and Commentary* (Leiden: Brill, 1997).

Gregorić, Pavel, *Aristotle on the Common Sense* (Oxford University Press, 2007).

"Aristotle's 'Common Sense' in the Doxographic Tradition" in J. Allen, E. K. Emilsson, W.-R. Mann, and B. Morison (eds.), *Oxford Studies in Ancient Philosophy*, vol. XL (Oxford University Press, 2011), 111–131.

Greppin, J. A. C., "The Galenic Corpus in Classical Armenia: A Preliminary Report," *Society for Ancient Medicine* 9 (1982): 11–13.

Guilland, Rodolphe, *Essai sur Nicéphore Grégoras, l'homme et l'oeuvre* (Paris: P. Geuthner, 1926).

Guillory, John, "Genesis of the Media Concept," *Critical Inquiry* 36 (2010): 321–362.

Gutas, Dimitri, "Avicenna's Marginal Glosses on *De anima* and the Greek Commentatorial Tradition," *Bulletin of the Institute of Classical Studies* 47 (2004): 77–88.

Greek Thought, Arabic Culture (London: Routledge, 1998).

Hahm, David E., "Early Hellenistic Theories of Vision and the Perception of Color" in P. K. Machamer and R. G. Turnbull (eds.), *Studies in Perception: Interrelations in the History of Philosophy and Science* (Columbus: Ohio State University Press, 1978), 60–95.

Hall, James, "Desire and Disgust: Touching Artworks from 1500 to 1800" in Rupert Shepherd and Robert Maniura (eds.), *Presence: The Inherence of the*

Prototype within Images and Other Objects (Burlington, VT: Ashgate, 2006), 145–160.

Halperin, David J., "Plato and Erotic Reciprocity," *Classical Antiquity* 5 (1986): 60–80.

Halton, Thomas, "The Five Senses in Nemesius, *De Natura Hominis* and Theodoret, *De Providentia*," *Studia Patristica* 20 (1989): 94–101.

Hamlyn, David W., "Koine Aisthesis," *The Monist* 52 (1968): 195–209.

Hamm, D., "Sight to the Blind: Vision as Metaphor in Luke," *Biblica* 67:4 (1986): 457–477.

Harrison, John E., and Simon Baron-Cohen (eds.), *Synaesthesia: Classic and Contemporary Readings* (Cambridge, MA: Blackwell, 1997).

Harvey, Susan Ashbrook, *Scenting Salvation: Ancient Christianity and the Olfactory Imagination* (Berkeley: University of California Press, 2006).

Heiberg, Johan L., *Geschichte der Mathematik und Naturwissenschaften im Altertum*, Handbuch der Altertumswissenschaft 5.1.2 (Munich: C. H. Beck, 1926), 75.

Heller-Roazen, Daniel, *The Inner Touch: Archaeology of a Sensation* (Brooklyn, NY: Zone Books, 2007).

Henry, Patrick, "What Was the Iconoclastic Controversy about?," *Church History* 45:1 (1976): 16–31.

Hillsdale, Cecily, *Byzantine Art and Diplomacy in an Age of Decline* (Cambridge University Press, 2014).

Holman, S. R., "On Phoenix and Eunuchs: Sources for Meletius the Monk's Anatomy of Gender," *Journal of Early Christian Studies* 16:1 (2008): 79–101.

Huby, P. M. "The Transmission of Aristotle's Writings and the Places Where Copies of His Works Existed," *Classica et Mediaevalia* 30 (1969): 241–257.

Hunger, Herbert, *Die hochsprachliche profane Literatur der Byzantiner*, 2 vols. (Munich: Beck, 1978).

Schreiben und Lesen in Byzanz: Die byzantinische Buchkultur (Munich: C. H. Beck, 1989).

Ierodiakonou, Katerina, "Aristotle on Colours" in D. Sfendoni-Mentzou, J. Hattiangadi, and D. M. Johnson (eds.), *Aristotle and Contemporary Science*, vol. II (New York: P. Lang, 2000), 211–225.

"Basic and Mixed Colours in Empedocles and in Plato" in Marcello Carastro (ed.), *L'antiquité en couleurs: catégories, pratiques, représentations* (Grenoble: J. Millon, 2009), 119–130.

"The Byzantine Commentator's Task: Transmitting, Transforming, or Transcending Aristotle's Text" in Andreas Speer and Philipp Steinkrüger (eds.), *Knotenpunkt Byzanz: Wissensformen und kulturelle Wechselbeziehungen* (Berlin: De Gruyter, 2012), 199–209.

"Empedocles and the Ancient Painters" in Liza Cleland and Karen Stears, *Colour in the Ancient Mediterranean World*, British Archaeological Report Series 1267 (Oxford: Hedges, 2004), 91–94.

"Empedocles on Colour and Colour Vision," *Oxford Studies in Ancient Philosophy* 29 (2005): 1–37.

"Galen's Criticism of the Aristotelian Theory of Colour Vision" in Carlo Natali and Stephano Maso (eds.), *Antiaristotelismo* (Amsterdam: Hakkert, 1999), 123–141.
"On Galen's Theory of Vision," *Bulletin of the Institute of Classical Studies. Supplement* 114 (2014): 197–211.
"The Greek Concept of *Sympatheia* and its Byzantine Appropriation in Michael Psellos" in Paul Magdalino and Maria Mavroudi (eds.), *The Occult Sciences in Byzantium* (Geneva: La Pomme d'Or, 2006), 97–117.
"Hellenistic Philosophers on the Phenomenon of Changing Colours" in B. Holmes and K.-D. Fischer (eds.), *The Frontiers of Ancient Science: Essays in Honor of Heinrich von Staden* (Berlin: De Gruyter, 2015), 227–250.
"The Notion of *Enargeia* in Hellenistic Philosophy" in Ben Morison and Katerina Ierodiakonou (eds.), *Episteme, etc.: Essays in Honour of Jonathan Barnes* (Oxford University Press, 2011), 60–73.
"Plato's Theory of Colours in the *Timaeus*," *Rhizai* 2:2 (2005): 125–140.
"The Western Influence on Late Byzantine Aristotelian Commentaries" in Martin Hinterberger and Chris Schabel (eds.), *Greeks, Latins, and Intellectual History 1204–1500* (Leuven: Peeters, 2011), 373–384.
Ioppolo, Anna Maria, "Presentation and Assent: A Physical and Cognitive Problem in Early Stoicism," *Classical Quarterly* 40:2 (1990): 433–449.
James, Liz, "Art and Lies: Text, Image, and Imagination in the Medieval World" in Antony Eastmond and Liz James (eds.), *Icon and Word: The Power of Images in Byzantium* (Burlington, VT: Ashgate, 2003), 59–71.
Light and Colour in Byzantine Art (Oxford University Press, 1996).
"Senses and Sensibility in Byzantium," *Art History* 27:4 (2004): 523–537.
Jeffreys, Michael, "Literacy" in Robin Cormack, John F. Haldon, and Elizabeth Jeffreys (eds.), *The Oxford Handbook of Byzantine Studies* (Oxford University Press, 2008), 796–802.
Ježek, Václav, "Education as a Unifying and 'Uplifting' Force in Byzantium," *Byzantinoslavica* 65 (2007): 167–200.
Johansen, Thomas K., *Aristotle on the Sense-Organs* (Cambridge University Press, 1997).
Jones, Caroline, *Eyesight Alone: Clement Greenberg's Modernism and the Bureaucratization of the Senses* (University of Chicago Press, 2006).
Jung, Jacqueline, "The Tactile and the Visionary: Notes on the Place of Sculpture in the Medieval Religious Imagination" in C. Hourihane (ed.), *Looking Beyond: Visions, Dreams, and Insights in Medieval Art and History* (Princeton: Index of Christian Art, 2010), 203–240.
Kahn, C. H., "Sensation and Consciousness in Aristotle's Psychology," *Archiv für Begriffsgeschichte* 48:1 (1966): 43–81.
Kalogeras, Nikolaos M., "Byzantine Childhood Education and Its Social Role from the Sixth Century until the End of Iconoclasm," unpublished PhD dissertation, University of Chicago (2000).
Kambylis, Athanasios, "Zu Leos Schrift: Σύνοψις εἰς τὴν φύσιν τῶν ἀνθρώπων," *Hermes* 101:4 (1973): 483–495.

Kapriev, Georgi, "Die gottgebührenden Phantasien und die Gottesschau bei Photios" in *Intellect et Imagination dans la Philosophie Médiévale*, vol. I (Turnhout: Brepols, 2006), 473–481.

Kaster, R. A., "Notes on 'Primary' and 'Secondary' Schools in Late Antiquity," *Transactions of the American Philological Association* 113 (1983): 323–346.

Katsaros, V., "Προδρομικοί «θεσμοί» για την οργάνωση της ανώτερης εκπαίδευσης της εποχής των Κομνηνών από την προκομνήνεια περίοδο" in V. Vlyssidou (ed.), *The Empire in Crisis? Byzantium in the 11th Century, 1025–1081* (Athens: Hellenic National Research Foundation, 2003), 452–455.

Katsiampoura, G., "The *Quadrivium* of 1008 and Pachymeres' *Syntagma*: Comparing Two Byzantine *Quadrivia*" in Lucio Del Corso and Oronzo Pecere (eds.), *Libri di scuola e pratiche didattiche: dall'antichità al rinascimento*, vol. II (Cassino: Università di Cassino, 2010), 409–424.

Kazhdan, Alexander, "The Image of the Medical Doctor in Byzantine Literature of the Tenth to Twelfth Centuries," *Dumbarton Oaks Papers* 38 (1984): 43–51.

Keenan, Mary E., "St. Gregory of Nazianzus and Early Byzantine Medicine," *Bulletin of the History of Medicine* 9 (1941): 8–30.

"St. Gregory of Nyssa and the Medical Profession," *Bulletin of the History of Medicine* 15 (1944): 150–161.

Kemp, Simon and Garth J. O. Fletcher, "The Medieval Theory of the Inner Senses," *The American Journal of Psychology* 106:4 (1993): 559–576.

Kennedy, George A., *Greek Rhetoric under Christian Emperors* (Princeton University Press, 1983).

A New History of Classical Rhetoric (Princeton University Press, 1994).

Progymnasmata: Greek Textbooks of Prose Composition and Rhetoric (Atlanta: Society of Biblical Literature, 2003).

Kessler, Herbert L., *Seeing Medieval Art* (Orchard Park, NY: Broadview Press, 2004).

Knorr, W. R., "The Geometry of Burning-Mirrors in Antiquity," *Isis* 74 (1983): 53–73.

Koch Piettre, Renée, "Les dieux crèvent les yeux: l'*enargeia* dans la représentation du divin" in D. Mulliez (ed.), *Ateliers 21: la transmission de l'image dans l'antiquité* (Lille: Cahiers de la Maison de la Recherche, Université Charles-de-Gaulle, Lille 3, 1999), 11–21.

Koortbojian, Michael, "Mimesis or Phantasia? Two Representational Modes in Roman Commemorative Art," *Classical Antiquity* 24:2 (2005): 285–306.

Kosman, L. A., "Perceiving that We Perceive: *On the Soul* III.2," *Philosophical Review* 84 (1975): 499–519.

Kumler, Aden, "The Multiplication of the Species: Eucharistic Morphology in the Middle Ages," *RES: Anthropology and Aesthetics* 59/60 (2011): 179–191.

Kyriakis, M. J., "Student Life in Eleventh Century Constantinople," *Byzantina* 7 (1975): 375–388.

Lacoste, E., "Les Poliorcétiques d'Apollodore de Damas," *Revue des Études Grecques* 3 (1890): 230–281.

Ladner, G. B., "The Concept of the Image in the Greek Fathers and the Byzantine Iconoclastic Controversy," *Dumbarton Oaks Papers* 7 (1953): 1–34.
Lakey, Christopher, "The Materiality of Light in Medieval Italian Painting," *English Language Notes* 53:2 (2015): 119–136.
Lammert, F., "Zu den Poliorketikern Apollodoros und Athenaios und zur Poliorketik des Vitruvius," *Rheinisches Museum für Philologie* 87 (1938): 304–332.
Lampe, G. W. H., *A Patristic Greek Lexicon* (Oxford University Press, 1961).
Langslow, D. R., *The Latin Alexander Trallianus: The Text and Transmission of a Late Latin Medical Book* (London: Society for the Promotion of Roman Studies, 2006).
Lascaratos, J. and M. Tsirou, "Ophthalmological Ideas of the Byzantine Author Meletius," *Documenta Ophthalmologica* 74 (1990): 31–35.
Lauritzen, Frederick, "Psellos and Plotinos," *Byzantinische Zeitschrift* 107:2 (2014): 711–724.
Lautner, Peter, "The Distinction between Φαντασία and Δόξα in Proclus' *In Timaeum*," *Classical Quarterly* 52:1 (2002): 257–269.
"Mental Images in Porphyry's Commentary on Ptolemy's *Harmonics*," *Apeiron* 48:2 (2015): 220–250.
"Perception and Self-Knowledge: Interpreting Fr. 264 Smith," *Bulletin of the Institute of Classical Studies* 50 (2007): 77–90.
"Philoponean Accounts of *Phantasia*," *Acta Antiqua Academiae Scientiarum Hungaricae* 34 (1993): 159–170.
"Philoponus, *In de Anima III*: Quest for an Author," *Classical Quarterly* 42 (1992): 510–522.
"Plutarch of Athens on κοινὴ αἴσθησις and *Phantasia*," *Ancient Philosophy* 20 (2000): 425–446.
Lavin, Sylvia, *Kissing Architecture* (Princeton University Press, 2012).
Lebeck, Anne, "The Central Myth of Plato's *Phaedrus*," *Greek, Roman, and Byzantine Studies* 13 (1972): 267–290.
Lemerle, Paul, *Cinq études sur le XIe siècle byzantin* (Paris: Presses Universitaires de France, 1977).
Le premier humanisme byzantin: notes et remarques sur enseignement et culture à Byzance des origines au Xe siècle (Paris: Presses Universitaires de France, 1971).
Leyerle, Blake, "John Chrysostom on the Gaze," *Journal of Early Christian Studies* 1:2 (1993): 159–174.
Limberis, Vasiliki, "The Eyes Infected by Evil: Basil of Caesarea's Homily, 'On Envy'," *The Harvard Theological Review* 84:2 (1991): 163–184.
Lindberg, David C., *Theories of Vision from Al-Kindi to Kepler*, revised edition (University of Chicago Press, 1976; revised 1996).
Linnér, S., "Psellus' *Chronographia* and the *Alexias*: Some Textual Parallels," *Byzantinische Zeitschrift* 76 (1983): 1–9.

Lloyd, A. C., "The Aristotelianism of Eustratius of Nicaea" in J. Wiesner (ed.), *Aristoteles: Werk and Wirkung*, vol. II (New York: De Gruyter, 1987), 341–351.

Long, Anthony A., "Aisthesis, Prolepsis and Linguistic Theory in Epicurus," *Bulletin of the Institute of Classical Studies* 18:1 (1971): 114–133.

"Stoic Psychology and the Elucidation of Language" in Elizabeth Asmis and Giovanni Manetti (eds.), *Knowledge Through Signs: Ancient Semiotic Theories and Practices* (Brussels: Brepols, 1996), 938–950.

"Theophrastus' *De Sensibus* on Plato" in K. A. Algra, P. W. van der Horst, and D. T. Runia (eds.), *Polyhistor: Studies in the History and Historiography of Ancient Philosophy Presented to Jaap Mansfeld on his Sixtieth Birthday* (Leiden: Brill, 1996), 345–362.

"Thinking and Sense-Perception in Empedocles: Mysticism or Materialism," *Classical Quarterly* 16 (1966): 256–276.

Long, Anthony A. and David Sedley, *The Hellenistic Philosophers*, 2 vols. (Cambridge University Press, 1987).

Lonsdale, Steven H., "If Looks Could Kill: παπταίνω and the Interpenetration of Imagery and Narrative in Homer," *The Classical Journal* 84:4 (1989): 325–333.

Lo Presti, R., "Between Distinction and Separation: Rethinking the Centrality of the Brain in Alcmaeon's Theory of Sense-Perception and Cognition," *Antike Naturwissenschaft und ihre Rezeption* 19 (2009): 9–30.

Lorenz, K., "The Ear of the Beholder: Spectator Figures and Narrative Structure in Pompeian Wall-Painting," *Art History* 30 (2007): 665–682.

Lories, Danielle, "Des sensibles communs dans le *De anima* d'Aristote," *Revue Philosophique de Louvain* 89 (1991): 401–420.

Louth, Andrew, *Greek East and Latin West: The Church AD 681–1071* (Crestwood, NY: St. Vladimir's Seminary Press, 2007).

Lovatt, Helen, *The Epic Gaze: Vision, Gender and Narrative in Ancient Epic* (Cambridge University Press, 2013).

MacCoull, Leslie, "Notes on Philoponus' Theory of Vision," *Byzantion* 67:2 (1997): 558–562.

MacIsaac, D. Gregory, "*Phantasia* between Soul and Body in Proclus' Euclid Commentary," *Dionysius* 19 (2001): 125–136.

Magdalino, Paul, "The Byzantine Reception of Classical Astrology" in C. Holmes and J. Waring (eds.), *Literacy, Education and Manuscript Transmission in Byzantium and Beyond* (Leiden: Brill, 2002), 33–58.

"The Road to Baghdad in the Thought-World of Ninth-Century Byzantium" in Leslie Brubaker (ed.), *Byzantium in the Ninth Century: Dead or Alive? Papers from the Thirtieth Spring Symposium of Byzantine Studies, Birmingham, March 1996* (Burlington, VT: Ashgate, 1998), 195–213.

Maguire, Henry, *Art and Eloquence in Byzantium* (Princeton University Press, 1981).

Mango, Cyril, *The Homilies of Photius Patriarch of Constantinople* (Cambridge, MA: Harvard University Press, 1958).

(ed. and trans.), *Art of the Byzantine Empire, 312–1453: Sources and Documents* (Englewood Cliffs, NJ: Prentice Hall, 1972).
Manieri, Alessandra, "Colori, suoni e profumi nelle *Imagines*: principi dell'estetica filostratea," *Quaderni Urbinati di Cultura Classica* 63 (1999): 111–121.
L'immagine poetica nella teoria degli antichi: phantasia ed enargeia (Pisa: Istituti Editoriali e Poligrafici Internazionali, 1998).
Männlein-Robert, Irmgard, "Zum Bild des Phidias in der Antike: Konzepte zur Kreativität des bildenden Künstlers" in Thomas Dewender and Thomas Welt (eds.), *Imagination – Fiktion – Kreation: Das kulturschaffende Vermögen der Phantasie* (Munich: K. G. Saur, 2003), 45–67.
Manolova, Divna, "Discourses of Science and Philosophy in the Letters of Nikephoros Gregoras," unpublished PhD dissertation, Central European University (2014).
"Sophonias the Philosopher: A Preface of an Aristotelian Commentary: Structure, Intention, and Audience," unpublished MA thesis, Central European University (2008).
Mariev, Sergei, "Echi delle teorie ottiche antiche nelle omelie di Fozio," *Bulgaria Mediaevalis* 2 (2011): 71–80.
Marinis, Vasileios, "Piety, Barbarism, and the Senses in Byzantium," in Sally M. Promey (ed.), *Sensational Religion: Sensory Cultures in Material Practice* (New Haven: Yale University Press, 2014), 321–340.
Markopoulos, Athanasios, "De la structure de l'école byzantine: le maître, les livres et le processus éducatif" in B. Mondrain (ed.), *Lire et écrire à Byzance* (Paris: Association des Amis du Centre d'Histoire et Civilisation de Byzance, 2006), 85–86.
"Education" in Robin Cormack, John F. Haldon, and Elizabeth Jeffreys (eds.), *The Oxford Handbook of Byzantine Studies* (Oxford University Press, 2008), 785–795.
"Teachers and Textbooks in Byzantium: Ninth to Eleventh Centuries" in Sita Steckel, Niels Gaul, and Michael Grünbart (eds.), *Networks of Learning: Perspectives on Scholars in Byzantine East and Latin West, c. 1000–1200* (Berlin: Lit Verlag, 2014), 3–15.
Marks, Lawrence E., *The Unity of the Senses: Interrelations Among the Modalities* (New York: Academic Press, 1978).
Martin, Edward James, *A History of the Iconoclastic Controversy* (London: Society for Promoting Christian Knowledge, 1930).
Martin-Hisard, B., "Georgian Hagiography" in Stephanos Efthymiadis (ed.), *The Ashgate Research Companion to Byzantine Hagiography*, vol. 1 (Burlington, VT: Ashgate, 2011), 285–298.
McCabe, Anne, *A Byzantine Encyclopedia of Horse Medicine: The Sources, Compilations, and Transmission of the Hippiatrica* (Oxford University Press, 2007).
McCarty, W., "The Shape of the Mirror: Metaphorical Catoptrics in Classical Literature," *Arethusa* 22 (1989): 161–196.

McDaniel, Walton Brooks, "The Pupula Duplex and Other Tokens of an 'Evil Eye' in the Light of Ophthalmology," *Classical Philology* 13 (1918): 335–346.

Meijering, Roos, *Literary and Rhetorical Theories in Greek Scholia* (Groningen: Egbert Forsten, 1987).

Mergiali, Sophia, *L'enseignement et les lettrés pendant l'époque des Paléologues (1261–1453)* (Athens: Société des Amis du Peuple, Centre d'Études Byzantines, 1996).

Miles, Graeme, "Living as a Sphinx: Composite Being and Monstrous Interpreter in the 'Middle Life' of Michael Psellos" in Danijela Kambaskovic (ed.), *Conjunctions of Mind, Soul and Body from Plato to the Enlightenment* (New York: Springer, 2014), 11–24.

"Psellos and His Traditions" in Sergei Mariev (ed.), *Byzantine Perspectives on Neoplatonism* (Berlin: De Gruyter, 2017), 79–102.

"Reforming the Eyes: Interpreters and Interpretation in the Vita Apollonii" in Kristoffel Demoen and Danny Praet (eds.), *Theios Sophistès: Essays on Flavius Philostratus'* Vita Apollonii (Leiden: Brill, 2008), 129–160.

Miles, Margaret R., *Augustine on the Body* (Missoula, MT: Scholars Press, 1979).

"Vision: The Eye of the Body and the Eye of the Mind in Saint Augustine's De trinitate and Confessions," *Journal of Religion* 63:2 (1983): 125–142.

Miller, Patricia Cox, *The Corporeal Imagination: Signifying the Holy in Late Ancient Christianity* (Philadelphia: University of Pennsylvania Press, 2009).

"'The Little Blue Flower Is Red': Relics and the Poetizing of the Body," *Journal of Early Christian Studies* 8:2 (2000): 213–236.

Miller, Timothy S., *The Birth of the Hospital in the Byzantine Empire*, 2nd edition (Baltimore: Johns Hopkins University Press, 1997).

Modrak, Deborah, "Koinê Aisthêsis and the Discrimination of Sensible Differences in de Anima III 2," *Canadian Journal of Philosophy* 11 (1981): 405–423.

Moffatt, Ann, "Early Byzantine School Curricula and a Liberal Education" in *Byzance et les Slaves* (Paris: Association des Amis des Etudes Archéologiques des Mondes Byzantino-Slaves et du Christianisme Oriental, 1979), 275–288.

"Schooling in the Iconoclast Centuries" in Anthony Bryer and Judith Herrin (eds.), *Iconoclasm: Papers Given at the Ninth Spring Symposium of Byzantine Studies, University of Birmingham, March 1975* (Birmingham: Centre for Byzantine Studies, University of Birmingham, 1977), 85–92.

Montefusco, Lucia Calboli, "'Ενάργεια et ἐνέργεια: l'évidence d'une démonstration qui signifie les choses en acte" in Mireille Armisen-Marchetti (ed.), *Demonstrare: voir et faire voir: forme de la démonstration à Rome* (Toulouse: Presses Universitaires du Mirail, 2005), 43–58.

Morales, Helen, *Vision and Narrative in Achilles Tatius' Leucippe and Cleitophon* (Cambridge University Press, 2004).

Morgan, Teresa, *Literate Education in the Hellenistic and Roman Worlds* (Cambridge University Press, 1998).

Muradyan, G., "The Rhetorical Exercises (*Progymnasmata*) in the Old Armenian 'Book of Chreia' (Girk'Pitoyic'): Translation or Original Composition?," *Revue des Études Arméniennes* 27 (1998–2000): 399–415.

Neis, Rachel, "Embracing Icons: The Face of Jacob on the Throne of God," *Images* 1:1 (2007): 36–54.

The Sense of Sight in Rabbinic Culture: Jewish Ways of Seeing in Late Antiquity (Cambridge University Press, 2013).

Nelson, Robert S., "The Discourse of Icons, Then and Now," *Art History* 12:2 (June 1989): 144–157.

"Image and Inscription: Pleas for Salvation in Spaces of Devotion" in Liz James (ed.), *Art and Text in Byzantine Culture* (Cambridge University Press, 2007), 100–119.

"Michael the Monk and His Gospel Book" in *Actes du XVe Congrès International d'Études Byzantines*, vol. II (Athens: Association Internationale des Etudes Byzantines, 1981), 575–582.

"Review of Charles Barber, *Figure and Likeness: On the Limits of Representation in Byzantine Iconoclasm*," *The Art Bulletin* 85:4 (2003): 797–800.

"To Say and to See: Ekphrasis and Vision in Byzantium" in R. S. Nelson (ed.), *Visuality before and beyond the Renaissance: Seeing as Others Saw* (Cambridge University Press, 2000), 143–168.

Newhauser, Richard G. (ed.), *A Cultural History of the Senses in the Middle Ages* (London: Bloomsbury Academic, 2016).

Niehoff, Maren, *Philo on Jewish Identity and Culture* (Tübingen: Mohr Siebeck, 2001).

Nightingale, A. W., "The Philosopher at the Festival: Plato's Transformation of Traditional *theôria*" in Jaś Elsner and Ian Rutherford (eds.), *Pilgrimage in Graeco-Roman and Early Christian Antiquity: Seeing the Gods* (Oxford University Press, 2005), 151–180.

"On Wandering and Wondering: *Theôria* in Greek Philosophy and Culture," *Arion* 9:2 (2001): 23–58.

Nousia, Fevronia, *Byzantine Textbooks of the Palaeologan Period*, Studi e Testi, 505 (Vatican City: Biblioteca Apostolica Vaticana, 2016).

Nünlist, René, *The Ancient Critic at Work: Terms and Concepts of Literary Criticism in Greek Scholia* (Cambridge University Press, 2009).

Nutton, Vivian, "From Galen to Alexander, Aspects of Medicine and Medical Practice in Late Antiquity," *Dumbarton Oaks Papers* 38 (1984): 1–14.

"Galen in Byzantium" in Michael Grünbart et al. (eds.), *Material Culture and Well-Being in Byzantium* (Vienna: Österreichische Akademie der Wissenschaften, 2007), 171–176.

O'Brien, D., "The Effects of Simile: Empedocles' Theories of Seeing and Breathing," *Journal of Hellenic Studies* 90 (1970): 140–179.

"Théories atomistes de la vision: Démocrite et le problème de la fourmi céleste" in L. Benakis (ed.), *Proceedings of the First International Congress on Democritus* (Xanthi: Bouloukos, 1984), 27–57.

Ogle, B., "The Lover's Blindness," *American Journal of Philology* 41 (1920): 240–252.

O'Meara, Dominic J., "Aspects du travail philosophique de Michel Psellus" in C. F. Collatz et al (eds.), *Dissertatiunculae criticae: Festschrift für Günther Christian Hansen* (Würzburg: Königshausen und Neumann, 1998), 431–439.

Pythagoras Revived: Mathematics and Philosophy in Late Antiquity (Oxford University Press, 1989).

Onians, John, "Abstraction and Imagination in Late Antiquity," *Art History* 3:1 (1980): 1–24.

Otto, Nina, *Enargeia: Untersuchung zur Charakteristik alexandrinischer Dichtung* (Stuttgart: Franz Steiner Verlag, 2009).

Owens, J., "Aristotle on Common Sensibles and Incidental Perception," *Phoenix* 36 (1982): 215–236.

Palazzo, Éric, *L'invention chrétienne des cinq sens dans la liturgie et l'art au Moyen Âge* (Paris: Éditions du Cerf, 2010).

(ed.), *Les cinq sens au Moyen Âge* (Paris: Éditions du Cerf, 2016).

Papaioannou, Stratis, "Animated Statues: Aesthetics and Movement" in C. Barber and D. Jenkins (eds.), *Reading Michael Psellos* (Leiden: Brill, 2006), 95–116.

"Anna Komnene's Will" in D. Sullivan, E. Fisher, and S. Papaioannou (eds.), *Byzantine Religious Culture: Studies in Honor of Alice-Mary Talbot* (Leiden: Brill, 2011), 99–121.

"Byzantine *Enargeia* and Theories of Representation," *Byzantinoslavica* 69:3 (2011): 48–60.

"Der Glasort des Textes: Selbstheit und Ontotypologie im byzantinischen Briefschreiben (10. und 11. Jh.)" in W. Hörandner, J. Koder, and M. Stassinopoulou (eds.), *Wiener Byzantinistik und Neogräzistik: Beiträge zum Symposion Vierzig Jahre Institut für Byzantinistik und Neogräzistik der Universität Wien. Im Gedenken an Herbert Hunger (Wien, 4.-7. Dezember 2002)* (Vienna: Österreichische Akademie der Wissenschaften, 2004), 324 336.

Michael Psellos: Rhetoric and Authorship in Byzantium (Cambridge University Press, 2013).

Parisinou, E. *The Light of the Gods: The Role of Light in Archaic and Classical Greek Cult* (London: Duckworth, 2000).

Park, D., *The Fire within the Eye: A Historical Essay on the Nature and Meaning of Light* (Princeton University Press, 1999).

Payton, James R., "John of Damascus on Human Cognition: An Element in His Apologetic for Icons," *Church History* 65:2 (1996): 173–183.

Peers, Glenn, "Imagination and Angelic Epiphany," *Byzantine and Modern Greek Studies* 21 (1997): 113–131.

"Real Living Painting: Quasi-Objects and Dividuation in the Byzantine World," *Religion and the Arts* 16 (2012): 433–460.

Sacred Shock: Framing Visual Experience in Byzantium (University Park: Pennsylvania State University Press, 2004).

Pentcheva, Bissera, "Painting or Relief: The Ideal Icon in Iconophile Writing in Byzantium," *Zograf* 31 (2006–2007): 7–14.

"The Performative Icon," *The Art Bulletin* 88:4 (2006): 631–655.

The Sensual Icon: Space, Ritual, and the Senses in Byzantium (University Park: Pennsylvania State University Press, 2010).

"Visual Textuality: The Logos as Pregnant Body and Building," *RES: Aesthetics and Anthropology* 45 (2004): 225–238.

Perkams, Matthias, "Priscian of Lydia, Commentator on the *De anima* in the Tradition of Iamblichus," *Mnemosyne*, 4th series, 58:4 (2005): 510–530.

Petridou, G., "'Blessed Is He Who Has Seen': The Power of Ritual Viewing and Ritual Framing in Eleusis," *Helios* 40:1–2 (2013): 309–341.

Pizzone, Aglae, "Late Antique Φαντασία and the Greek Fathers: A Survey" in David Hernández de la Fuente (ed.), *New Perspectives on Late Antiquity* (Newcastle upon Tyne: Cambridge Scholars Publishing, 2014), 416–432.

"Theodore and the Black Man: Imagining (through) the Icon in Byzantium" in Andreas Speer and Philipp Steinkrüger (eds.), *Knotenpunkt Byzanz* (Berlin: De Gruyter, 2012), 47–70.

"Toward a Self-Determined and Emotional Gaze: Agathias and the Icon of the Archangel Michael" in Sergei Mariev and Wiebke-Marie Stock (eds.), *Aesthetics and Theurgy in Byzantium* (Berlin: De Gruyter, 2013), 75–103.

"When Homer Met *Phantasia*: Fiction, Epic Poetry, and Entertainment Literature in Byzantium," *British Academy Review* 19 (2012): 42–45.

Platt, Verity, *Facing the Gods: Epiphany and Representation in Greco-Roman Art, Literature and Religion* (Cambridge University Press, 2011).

"Making an Impression: Replication and the Ontology of the Graeco-Roman Seal Stone," *Art History* 29:2 (2006): 233–257.

"Virtual Visions: *Phantasia* and the Perception of the Divine in *The Life of Apollonius of Tyana*" in Ewen Bowie and Jaś Elsner (eds.), *Philostratus* (Cambridge University Press, 2009), 131–154.

Platt, Verity and Michael Squire (eds.), "The Art of Art History in Greco-Roman Antiquity," *Arethusa* 43:2 (2010): 133–369.

Plepelits, Karl, *Achilleus Tatios: Leukippe und Kleitophon* (Stuttgart: A. Hiersemann, 1980).

"Achilles Tatius" in Garth L. Schmeling (ed.), *The Novel in the Ancient World* (Leiden: Brill, 1996), 387–416.

Plett, H., *Enargeia in Classical Antiquity and the Early Modern Age: The Aesthetics of Experience* (Leiden: Brill, 2010).

Pollitt, J. J., "*Peri chromaton*: What Ancient Greek Painters Thought about Colors" in M. A. Tiverios and D. S. Tsifakis (eds.), *Color in Ancient Greece: The Role of Color in Ancient Greek Art and Architecture* (Thessaloniki: Aristotle University, 2002), 1–8.

Pormann, Peter E., *The Oriental Tradition of Paul of Aegina's Pragmateia* (Leiden: Brill, 2004).

Porter, James I., *The Origins of Aesthetic Thought in Ancient Greece: Matter, Sensation, and Experience* (Cambridge University Press, 2010).

Porter, Stanley, *Handbook of Classical Rhetoric in the Hellenistic Age, 330 B.C. – AD. 400* (Leiden: Brill, 1997).

Praechter, K., "Michael von Ephesus und Psellos," *Byzantinische Zeitschrift* 31 (1931): 1–12.

Prier, R. A., *Thauma idesthai: The Phenomenology of Sight and Appearance in Archaic Greek* (Tallahassee: Florida State University, 1989).

Rakoczy, Thomas, *Böser Blick: Macht des Auges und Neid der Götter: Eine Untersuchung zur Kraft des Blickes in der griechischen Literatur* (Tübingen: Gunter Narr, 1996).

Rapp, Christof, "Intentionalität und *Phantasia* bei Aristoteles" in Dominik Perler (ed.), *Ancient and Medieval Theories of Intentionality* (Leiden: Brill, 2001), 63–96.

Rashed, Roshdi, "A Pioneer in Anaclastics: Ibn Sahl on Burning Mirrors and Lenses," *Isis* 81.3 (1990): 464–491.

Reed, Baron, "The Stoics' Account of Cognitive Impression," *Oxford Studies in Ancient Philosophy* 23 (2002): 147–180.

Renehan, Robert, "Meletius' Chapter on the Eyes: An Unidentified Source," *Dumbarton Oaks Papers* 38 (1984): 159–168.

"On the Text of Leo Medicus: A Study in Textual Criticism," *Rheinisches Museum für Philologie* n.s. 113:1 (1970): 79–88.

Rinella, Michael A., *Pharmakon: Plato, Drug Culture, and Identity in Ancient Athens* (Lanham, MD: Lexington Books, 2010).

Rispoli, G. M., "φαντασία ed ἐνάργεια negli scolî all'*Iliade*," *Vichiana* 13 (1984): 311–339.

Robinson, T. M., *Plato's Psychology* (University of Toronto Press, 1970).

Roilos, Panagiotis, "Ancient Greek Rhetorical Theory and Byzantine Discursive Politics: John Sikeliotes on Hermogenes," in Teresa Shawcross and Ida Toth (eds.), *Reading in the Byzantine Empire and Beyond* (Cambridge University Press, 2018).

"*Phantasia* and the Ethics of Fictionality in Byzantium: A Cognitive Anthropological Perspective" in Panagiotis Roilos (ed.), *Medieval Greek Storytelling: Fictionality and Narrative in Byzantium* (Wiesbaden: Harrassowitz Verlag, 2014), 9–30.

"'Unshapely Bodies and Beautifying Embellishments': The Ancient Epics in Byzantium, Allegorical Hermeneutics, and the Case of Ioannes Diakonos Galenos," *Jahrbuch der Österreichischen Byzantinistik* 64 (2014): 231 246.

Rosenmeyer, T. G., "*Phantasia* und Einbildungskraft: Zur Vorgeschichte eines Leitbegriffes der europäischen Ästhetik," *Poetica* 18 (1986): 197–248.

Roueché, M., "Byzantine Philosophical Texts of the Seventh Century," *Jahrbuch der Österreichischen Byzantinistik* 23 (1974): 61–76.

Rousseau, Philip, "Human Nature and Its Material Setting in Basil of Caesarea's Sermons on Creation," *Heythrop Journal* 49 (2008): 222–239.

Rudolph, K., "Democritus' Ophthalmology," *Classical Quarterly* 62 (2012): 496–501.

"Democritus' Perspectival Theory of Vision," *Journal of Hellenic Studies* 131 (2011): 67–83.

Runia, David T., *Philo and the Church Fathers: A Collection of Papers* (Leiden: Brill, 1995).

"The Placita Ascribed to Doctors in Aëtius' Doxography on Physics" in J. Mansfeld and D. T. Runia (eds.), *Aëtiana: The Method and Intellectual Context of a Doxographer*, vol. III (Leiden: Brill, 2010), 515–575.

Rutherford, Ian, "*Theoria* and *Darsan*: Pilgrimage and Vision in Greece and India," *Classical Quarterly* 50 (2000): 133–146.

Sahas, Daniel J., *Icon and Logos: Sources in Eighth-Century Iconoclasm* (University of Toronto Press, 1986).

Saint-Pierre, Bernard, "La physique de la vision dans l'Antiquité: contribution à l'établissement des sources anciennes de l'optique médiévale," unpublished PhD dissertation, University of Montreal (1972).

Sambursky, Samuel, *Physics of the Stoics* (Princeton University Press, 1959).

Sandbach, Francis H., "Phantasia Katalēptikē" in A. A. Long (ed.), *The Problems in Stoicism* (London: Athlone, 1971), 9–21.

Savage-Smith, Emilie, "Galen's Lost Ophthalmology and the *Summaria Alexandrinorum*," *Bulletin of the Institute of Classical Studies. Supplement* 77 (2002): 121–138.

"Hellenistic and Byzantine Ophthalmology: Trachoma and Sequelae," *Dumbarton Oaks Papers* 38 (1984): 169–186.

Scarborough, John, "The Life and Times of Alexander of Tralles," *Expedition* 39:2 (1997): 51–60.

"Roman Medicine to Galen," *Aufstieg und Niedergang der Römischen Welt* 37:1 (1993): 3–48.

"Symposium on Byzantine Medicine: Introduction," *Dumbarton Oaks Papers* 38 (1984): ix-xvi.

"Theodora, Aetius of Amida, and Procopius: Some Possible Connections," *Greek, Roman, and Byzantine Studies* 53 (2013): 742–762.

Schacht, Joseph and Max Meyerhof, *The Medico-Philosophical Controversy between Ibn Butlan of Baghdad and Ibn Ridwan of Cairo* (Cairo: Egyptian University, 1937).

Schibille, Nadine, *Hagia Sophia and the Byzantine Aesthetic Experience* (Burlington, VT: Ashgate, 2014).

Schmitt, Charles B., "Aristotle as a Cuttlefish: The Origin and Development of a Renaissance Image," *Studies in the Renaissance* 12 (1965): 60–72.

Schofield, Malcolm, "Aristotle on the Imagination" in M. C. Nussbaum and A. O. Rorty (eds.), *Essays on Aristotle's* De Anima (Oxford University Press, 1995), 249–278.

Scholz, B., "*Sub oculos subiecto*: Quintilian on Ekphrasis and Enargeia" in V. Robillard and E. Jongeneel (eds.), *Pictures into Words: Theoretical and Descriptive Approaches to Ekphrasis* (Amsterdam: VU University Press, 1998), 73–99.

Sedgwick, Eve Kosofsky, *Touching Feeling: Affect, Pedagogy, Performativity* (Durham, NC: Duke University Press, 2003).

Sedley, David, "Empedocles' Theory of Vision and Theophrastus' *De sensibus*" in William W. Fortenbaugh and Dimitri Gutas (eds.), *Theophrastus: His*

Psychological, Doxographical, and Scientific Writings (New Brunswick, NJ: Transaction Publishers, 1992), 20–31.

"Zeno's Definition of *Phantasia Kataleptike*" in Theodore Scaltsas and Andrew S. Mason (eds.), *The Philosophy of Zeno: Zeno of Citium and His Legacy* (Larnaca: Municipality of Larnaca, 2002), 135–154.

Ševčenko, Ihor, "The Decline of Byzantium as Seen through the Eyes of Its Intellectuals," *Dumbarton Oaks Papers* 15 (1961): 169–186.

Études sur la polémique entre Théodore Métochite et Nicéphore Choumnos (Brussels: Byzanthion, 1962).

"Theodore Metochites, the Chora, and the Intellectual Trends of His Time" in Paul A. Underwood (ed.), *The Kariye Djami*, vol. IV (Princeton University Press, 1975), 17–92.

Sheppard, Anne, "*Phantasia* in *De Insomniis*" in Donald A. Russell and Heinz-Günther Nesselrath (eds.), *On Prophecy, Dreams and Human Imagination* (Tübingen: Mohr Siebeck, 2014), 97–110.

The Poetics of Phantasia: Imagination in Ancient Aesthetics (London: Bloomsbury, 2014).

Siegel, Rudolph E., *Galen on Sense Perception: His Doctrines, Observations, and Experiments on Vision, Hearing, Smell, Taste, Touch, and Pain, and Their Historical Sources* (Basel: Karger, 1970).

Silverman, A., "Color and Color-Perception in Aristotle's *De Anima*," *Ancient Philosophy* 9 (1989): 271–292.

"Plato on *Phantasia*," *Classical Antiquity* 10 (1991): 123–147.

Simon, Gérard, *Le regard, l'être et l'apparence dans l'optique de l'antiquité* (Paris: Éditions du Seuil, 1988).

Simpson, James, *Under the Hammer: Iconoclasm in the Anglo-American Tradition* (Oxford University Press, 2011).

Smith, A. Mark, *From Sight to Light: The Passage from Ancient to Modern Optics* (University of Chicago Press, 2014).

Ptolemy and the Foundations of Ancient Mathematical Optics: A Source Based Guided Study (Philadelphia: American Philosophical Society, 1999).

Ptolemy's Theory of Visual Perception (Philadelphia: American Philosophical Society, 1996).

Smith, M. M., *Sensing the Past: Seeing, Hearing, Smelling, Tasting, and Touching in History* (Berkeley: University of California Press, 2007).

Sobchack, Vivian, "What My Fingers Knew: The Cinesthetic Subject, or Vision in the Flesh" in *Carnal Thoughts: Embodiment and Moving Image Culture* (Berkeley: University of California Press, 2004), 53–84.

Sorabji, Richard, "Aristotle, Mathematics and Colour," *Classical Quarterly* 22 (1972): 293–308.

"Aristotle on Demarcating the Five Senses" in J. Barnes, M. Schofield, and R. Sorabji (eds.), *Articles on Aristotle*, vol. IV (London: Duckworth, 1979), 76–92.

"Intentionality and Physiological Processes: Aristotle's Theory of Sense-Perception" in M. Nussbaum and R. Rorty (eds.), *Essays on Aristotle's De Anima* (Oxford: Clarendon Press, 1992), 195–226.

"Universals Transformed: The First Thousand Years after Plato" in P. F. Strawson and Arindam Chakrabarti (eds.), *Universals, Concepts and Qualities* (Burlington, VT: Ashgate, 2006), 105–126.
Speyer, W., "Asterios von Amaseia," *Reallexikon für Antike und Christentum, Supplement* 4 (1986), 629–639.
"Die Euphemia-Reden des Asterios von Amaseia: Eine Missionsschrift für gebildete Heiden," *Jahrbuch für Antike und Christentum* 14 (1971), 39–47.
"Wunderheilige und Bilder: Zur Frage des Beginns der Bilderverehrung," *Poikilia Byzantina* 11:3 (1991): 163–247.
Spingou, Foteini, "Byzantine Collections and Anthologies of Poetry" in Wolfram Hoerandner, Andreas Rhoby, and Nikolaos Zagklas (eds.), *Brill Companion on Byzantine Poetry* (Leiden: Brill) (Forthcoming).
"Words and Artworks in the Twelfth Century and Beyond: The Thirteenth-Century Manuscript Marcianus gr. 524 and the Twelfth-Century Dedicatory Epigrams on Works of Art," unpublished DPhil dissertation, Oxford University (2012).
Squire, Michael, "Apparitions Apparent: Ekphrasis and the Parameters of Vision in the Elder Philostratus's *Imagines*," *Helios* 40:1–2 (2013): 97–140.
"Ecphrasis: Visual and Verbal Interactions in Ancient Greek and Latin Literature," *Oxford Handbooks Online* (2015), www.oxfordhandbooks.com.
The Iliad in a Nutshell: Visualizing Epic on the Tabulae Iliacae (Oxford University Press, 2011).
"Introductory Reflections: Making Sense of Ancient Sight" in Michael Squire (ed.), *Sight and the Ancient Senses* (London: Routledge, 2016), 1–35.
"Making Myron's Cow Moo?: Ecphrastic Epigram and the Poetics of Simulation," *American Journal of Philology* 131:4 (2010): 589–634.
Squire, Michael, and Jonas Grethlein, "'Counterfeit in Character but Persuasive in Appearance': Reviewing the *Ainigma* of the *Tabula Cebetis*," *Classical Philology* 109:4 (2014): 285–324.
Stanford, W. Bedell, *Greek Metaphor: Studies in Theory and Practice* (Oxford University Press, 1936).
Stephanou, Pelopidas, "La doctrine de Léon de Chalcédoine et de ses adversaires sur les images," *Orientalia Christiana Periodica* 12 (1946): 177–199.
Stewart, Columba, "Imageless Prayer and the Theological Vision of Evagrius Ponticus," *Journal of Early Christian Studies* 9:2 (2001): 173–204.
Sullivan, Denis F., *Siegecraft: Two Tenth-Century Instructional Manuals by "Heron of Byzantium"* (Washington, DC: Dumbarton Oaks, 2000).
Sullo, Nicole Paxton, "The Art of Memory in Byzantium during the Later Middle Ages," unpublished PhD dissertation, Yale University (in progress).
Taft, Robert F., "The Dialogue before the Anaphora in the Byzantine Eucharistic Liturgy, II: The *Sursum Corda*," *Orientalia Christiana Periodica* 54 (1988): 47–77.
Talbot, Alice-Mary, "Pilgrimage to Healing Shrines: The Evidence of Miracle Accounts," *Dumbarton Oaks Papers* 56 (2002): 153–173.

Tanner, Jeremy, *The Invention of Art History in Ancient Greece: Religion, Society and Artistic Rationalisation* (Cambridge University Press, 2006).
Tarrant, Harold, *Thrasyllan Platonism* (Ithaca, NY: Cornell University Press, 1993).
Temkin, Owsei, "Byzantine Medicine: Tradition and Empiricism," *Dumbarton Oaks Papers* 16 (1962): 95–115.
"On Galen's Pneumatology," *Gesnerus* 8 (1951): 180–189.
Hippocrates in a World of Pagans and Christians (Baltimore: Johns Hopkins University Press, 1993).
Thein, Karl, "Gods and Painters: Philostratus the Elder, Stoic Phantasia, and the Strategy of Describing," *Ramus* 31:1/2 (2002): 136–145.
Theobald, Werner, *Hypolepsis: Mythische Spuren bei Aristoteles* (Sankt Augustin, Germany: Academia Verlag, 1999).
Thomas, John P., *Private Religious Foundations in the Byzantine Empire* (Washington, DC: Dumbarton Oaks, 1987).
Tihon, Anne, "L'astronomie byzantine (du ve au xve siècle)," *Byzantion* 51 (1981): 603–624.
"Numeracy and Science" in Robin Cormack, John F. Haldon, and Elizabeth Jeffreys (eds.), *The Oxford Handbook of Byzantine Studies* (Oxford University Press, 2008), 803–819.
Todd, Robert B., "Galenic Medical Ideas in the Greek Aristotelian Commentators," *Symbolae Osloenses* 52:1 (1977): 117–134.
"Lexicographical Notes on Alexander of Aphrodisias' Philosophical Terminology," *Glotta* 52:3/4 (1974): 207–215.
"The Stoic Common Notions: A Re-examination and Reinterpretation," *Symbolae Osloenses* 48:1 (1973): 47–75.
"ΣΥΝΕΝΤΑΣΙΣ and the Stoic Theory of Perception," *Grazer Beiträge* 2 (1974): 251–261.
Toomer, G. J., *Diocles on Burning Mirrors* (New York: Springer, 1976).
Treadgold, Warren T., *The Nature of the Bibliotheca of Photios* (Washington, DC: Dumbarton Oaks, 1980).
Trentin, L., "Exploring Visual Impairment in Ancient Rome" in C. Laes, C. F. Goodey, and M. L. Rose (eds.), *Disabilities in Roman Antiquity: Disparate Bodies A Capite ad Calcem* (Leiden: Brill, 2013), 89–114.
Tuominen, Miira, "On Activity and Passivity in Perception: Aristotle, Philoponus, and Pseudo-Simplicius" in J. F. Silva and M. Yrjönsuuri (eds.), *Active Perception in the History of Philosophy: From Plato to Modern Philosophy* (London: Springer, 2014), 55–78.
Underwood, Paul, "The Fountain of Life in Manuscripts of the Gospel," *Dumbarton Oaks Papers* 5 (1950): 41–138.
van Campen, Cretien, *The Hidden Sense: Synesthesia in Art and Science* (Cambridge, MA: MIT Press, 2007).
van Hoorn, W., *As Images Unwind: Ancient and Modern Theories of Visual Perception* (Amsterdam: University Press Amsterdam, 1972).
Vasiliu, Anca, *Du Diaphane: image, milieu, lumière dans la pensée antique et médiévale* (Paris: J. Vrin, 1997).

Velmans, Tania, "L'iconographie de la 'Fontaine de Vie' dans la tradition Byzantine à la fin du Moyen Âge" in André Grabar and Jean Hubert (eds.), *Synthronon: art et archéologie de la fin de l'Antiquité et du Moyen âge* (Paris: Librairie C. Klincksieck, 1968), 119–134.

Verbeke, Gérard, *L'évolution de la doctrine du pneuma: du stoicisme à S. Augustin* (Paris: Desclée de Brouwer, 1945).

Jean Philopon: commentaire sur le De anima *d'Aristote*, trans. Guillaume de Moerbeke (Paris: Éditions Béatrice-Nauwelaerts, 1966).

"Levels of Human Thinking in Philoponus" in C. Laga, J. A. Munitiz, and L. Van Rompay (eds.), *After Chalcedon: Studies in Theology and Church History Offered to Professor Albert Van Roey for his Seventieth Birthday* (Leuven: Peeters, 1985), 451–470.

Verpeaux, J., *Nicéphore Choumnos: homme d'état et humaniste byzantin* (Paris: Editions A. et J. Picard, 1959).

Volk, Robert, *Der medizinische Inhalt der Schriften des Michael Psellos* (Munich: Institut für Byzantinistik und Neugriechische Philologie der Universität, 1990).

von Fritz, K., "Democritus' Theory of Vision" in E. A. Underwood (ed.), *Science, Medicine, and History: Essays on the Evolution of Scientific Thought and Medical Practice Written in Honour of Charles Singer*, vol. 1 (Oxford University Press, 1953), 83–99.

Walker, Andrew, "*Erōs* and the Eye in the *Love-Letters* of Philostratus," *The Cambridge Classical Journal* 38 (1993): 132–148.

Walker, Jeffrey, "*Pathos* and *Katharsis* in 'Aristotelian' Rhetoric: Some Implications" in Alan Gross and Arthur Walzer (eds.), *Reading Aristotle's Rhetoric* (Carbondale: Southern Illinois University Press, 2000), 74–92.

Watson, Gerard, "The Concept of 'Phantasia' from the Late Hellenistic Period to Early Neoplatonism," *Aufstieg und Niedergang der Römischen Welt* 36:7 (1994): 4766–4810.

Phantasia in Classical Thought (Galway University Press, 1998).

Webb, Ruth, "Accomplishing the Picture: Ekphrasis, Mimesis, and Martyrdom in Asterios of Amaseia" in Liz James (ed.), *Art and Text in Byzantine Culture* (Cambridge University Press, 2007), 13–32.

"The Aesthetics of Sacred Space: Narrative, Metaphor, and Motion in *Ekphraseis* of Church Buildings," *Dumbarton Oaks Papers* 53 (1999): 59–74.

"*Ekphrasis* Ancient and Modern: The Invention of a Genre," *Word & Image* 15:1 (1999): 7–18.

Ekphrasis, Imagination, and Persuasion in Ancient Rhetorical Theory and Practice (Burlington, VT: Ashgate, 2009).

"Imagination and the Arousal of Emotions in Greco-Roman Rhetoric" in Susanna Morton Braund and Christopher Gill (eds.), *The Passions in Roman Thought and Literature* (Cambridge University Press, 1997), 112–127.

"Sight and Insight: Theorizing Vision, Emotion and Imagination in Ancient Rhetoric" in Michael Squire (ed.), *Sight and the Ancient Senses* (London: Routledge, 2016), 205–219.

Webster, C., "Euclid's *Optics* and Geometrical Astronomy," *Aperion* 47 (2014): 526–551.

Wickert, Jacob, "Die Panoplia dogmatica des Euthymios Zigabenos: Untersuchung ihrer Anlage und ihrer Quellen, ihres Inhaltes und ihrer Bedeutung," *Oriens Christianus* 8 (1908): 278–388.

Wilken, Robert L., *The Land Called Holy: Palestine in Christian History & Thought* (New Haven: Yale University Press, 1994).

Wilson, Nigel G., "Aspects of the Transmission of Galen" in Guglielmo Cavallo (ed.), *Le strade del testo* (Bari: Adriatica, 1987), 47–64.

Scholars of Byzantium (Baltimore: Johns Hopkins University Press, 1983).

Wolska-Conus, Wanda, "Les écoles de Psellos et de Xiphilin sous Constantin IX Monomaque," *Travaux et mémoires* 6 (1976): 223–243.

"Stephanus d'Athènes et Stephanus d'Alexandrie: essai d'identification et de biographie," *Revue des Études Byzantines* 47 (1989): 5–89.

Young, Frances, "Typology" in Stanley E. Porter, Paul Joyce, and David E. Orton (eds.), *Crossing the Boundaries: Essays in Biblical Interpretation in Honour of Michael D. Goulder* (Leiden: Brill, 1994), 29–48.

Zagdoun, Mary-Anne, *La philosophie stoïcienne de l'art* (Paris: Centre National de la Recherche Scientifique, 2000).

Zahlfleisch, J., "Die Polemik Alexanders von Aphrodisias gegen die verschiedenen Theorien des Sehens," *Archiv für Geschichte der Philosophie* 8 (1895): 373–386.

Zanker, Graham, "*Enargeia* in the Ancient Criticism of Poetry," *Rheinisches Museum für Philologie* 124 (1981): 297–311.

Zipser, Barbara, "Deleted Text in a Manuscript: Galen *On the Eye* and the *Marc. gr. 276*," *Galenos: Rivista di Filologia dei Testi Medici Antichi* 3 (2009): 107–112.

"Pseudo-Alexander Trallianus. *De oculis*: Einleitung, Text, Übersetzung und Kommentar," unpublished PhD dissertation, Universität Heidelberg (2003).

"Überlegungen zum Text des Σύνοψις ἰατρικῆς des Leo Medicus," *Studia Humanitatis ac Litterarum Trifolio Heidelbergensi Dedicata* (Frankfurt am Main: Peter Lang 2004), 393–399.

"Zu Aufbau und Quellen der Σύνοψις ἰατρικῆς des Leo Medicus" in Thorsten Fögen (ed.), *Ancient Technical Texts* (Berlin: Walter de Gruyter, 2005), 107–115.

Index

The family name emerged only in middle Byzantium. Late antique and early Byzantine persons are normally listed by first name; later on, by family name.